SYMMETRICAL COMPONENTS

SYMMETRICAL COMPONENTS

AS APPLIED TO THE ANALYSIS OF UNBALANCED ELECTRICAL CIRCUITS

BY

C. F. WAGNER

Manager,
Central Station Engineering, Westinghouse Electric Corporation

AND

R. D. EVANS

Manager, Transmission Engineering,
Westinghouse Electric and Manufacturing Company; Member A.I.E.E

WITH AN INTRODUCTION BY

C. L. FORTESCUE

USED AS A TEXT IN THE DESIGN COURSE OF THE WESTINGHOUSE GRADUATE SCHOOL

ROBERT E. KRIEGER PUBLISHING COMPANY
MALABAR, FLORIDA
1982

Original Edition 1933
Reprint Edition 1982

Printed and Published by
ROBERT E. KRIEGER PUBLISHING COMPANY, INC.
KRIEGER DRIVE
MALABAR, FLORIDA 32950

Printed in the United States of America

Library of Congress Cataloging in Publication Data

Wagner, C. F. (Charles Frederick), 1895-
Symmetrical components, as applied to the analysis of unbalanced electrical circuits.

"Used as a text in the design course of the Westinghouse Graduate School."
Reprint. Originally published: 1st ed. New York ; London: McGraw-Hill, 1933.
Bibiliography: p.
Includes index.
1. Electric circuits. 2. Electric currents, Alternating--Polyphase. I. Evans, R. D. (Robert David), 1892-1947. II. Title.
[TK3226.W25 1982] 621.319'2 82-14051
ISBN 0-89874-556-X

PREFACE

The solution of unbalanced electrical circuits has been found to be practicable only by the method of symmetrical components. It is a very powerful analytical tool and is based on sound theory. The method has been applied to advantage in the solution of practically all phases of power system engineering, particularly in the investigation of conditions resulting from unsymmetrical transient disturbances.

The value of the method was recognized immediately upon its publication in 1918. There is now considerable literature on the subject of the solution of unbalanced polyphase circuits by this method. In spite of this there is no text which gathers together the fundamental theory on which the method is based and discusses the several principal applications. In the hope of meeting this need the present volume has been prepared.

The method of symmetrical components was discovered by Dr. C. L. Fortescue while investigating problems of single-phase railway systems. The early applications of the method were made principally by Dr. Fortescue and his associates, Messrs. Gilman, Peters, Chubb, and Dr. Slepian of the Westinghouse Electric and Manufacturing Company. During the past several years the authors have devoted their efforts principally to application problems of power systems. In this work they have found it necessary to extend the method particularly in connection with the determination of fault currents under unbalanced conditions and of stability under unsymmetrical faults.

This book is intended to meet the needs of the practicing engineer and the engineering student. It presupposes a knowledge of fundamental single-phase and polyphase circuit theory and the operating characteristics of the more important types of alternating-current apparatus. It is, therefore, intended for senior students and graduates in electrical engineering. Because of its recent development the practicing engineer frequently will not have had an opportunity to study the subject in formally organized classes. Special attention has therefore been given to working out, as a part of the text, illustrations of the application of the method. In addition, problems have been included in order to provide exercises and a convenient means for testing the amount of information gathered from study of the individual chapters. The problems have been drawn principally from those encountered by the authors in connection with practical engineering work. It is believed, therefore, that they will be of special interest to the practicing engineer.

Chapters I to X cover the more essential phases of the fundamental theory and the principal application of the method, which is the determination of voltages and currents under unbalanced faults. The remainder of the volume takes up more highly specialized problems. For those interested only in the subject of measurements, Chaps. I, II, XIII, XIV, and XV will be of greatest value. The Appendix to this volume contains much convenient reference material. It includes a review of the principal equations upon which the method of symmetrical components is based and also a number of convenient formulas for network simplification and solution. Particular attention is called to the tables of conductor characteristics, since they include not only the ordinary (positive-sequence) constants but also the ground return or zero-sequence constants. It is hoped that the Appendix will provide all of the additional reference material required for ordinary unbalanced fault calculations by the method of symmetrical components.

In presenting this volume the authors wish to acknowledge the inspiration and help which they have received from Dr. C. L. Fortescue, Mr. J. F. Peters, Dr. J. Slepian, and Mr. C. A. Powel; and the use of both published and unpublished work, including particularly an extensive memorandum prepared by Mr. Peters for use in connection with the Westinghouse Design School. The authors wish also to express their appreciation of assistance from their associates, particularly from Dr. W. A. Lewis for many general suggestions and especially for the use of certain tables given in the Appendix; Mr. S. B. Griscom and Mr. E. L. Harder for general suggestions and for assistance in proofreading; and Mr. W. R. Ellis for the preparation of the majority of the illustrations and for the numerous calculations required.

East Pittsburgh, Pa.
May, 1933.

C. F. Wagner.
R. D. Evans.

CONTENTS

APPENDIX

INTRODUCTION

The complex number arose from attempts by early mathematicians to solve algebraic equations, some of which had known geometrical solutions, in terms of algebraic operations. The first to make use of the minus sign under a radical was the Italian mathematician, Cardan, and the dual numbers he obtained were the same as our present complex numbers. Although many geometric solutions of algebraic equations soluble algebraically only by such numbers had been obtained in the past, it was not until the nineteenth century that the correspondence between the system of real and complex numbers to points of the two-dimensional Cartesian continuum was realized. Until this time such numbers were considered impossible or imaginary numbers as stated in Euler's algebra as late as 1770. It is peculiar that mathematicians before this time did not realize that the system of positive and negative real numbers was not a closed system, since in that domain the square root of a negative quantity had no meaning. With the inclusion of the complex and so-called imaginary numbers the system of numbers was closed for all algebraic operations.

In spite of the fact that until the nineteenth century no interpretation had been given to the complex number, its use had become quite general among mathematicians. So we find DeMoivre introducing it into trigonometry through his famous theorem

$$(\cos \theta + i \sin \theta)^n = (\cos n\theta + i \sin n\theta)$$

while Euler introduced the identity

$$e^{i\theta} = \cos \theta + i \sin \theta$$

and

$$e^{i\pi} = -1$$

It is remarkable that these mathematicians shot so close without hitting the bull's-eye. Apparently, in spite of the development of trigonometry and its relations with algebra, the concept of a vector to define a point in a plane had not yet been realized and the symbol $e^{i\theta}$ was still regarded as a rather absurd algebraic deduction which nevertheless gave useful results. The remarkable beauty of the conception of $e^{i\theta}$ as a unit vector or coordinate oriented at the angle θ from the reference line or vector had not yet dawned on them and therefore the full meaning of DeMoivre's theorem had not been realized. Otherwise Euler from his identity

$$e^{i\pi} = -1$$

would have at once deduced the geometric meaning of $i = \sqrt{-1}$ by the relation

$$e^{i\frac{\pi}{2}} = \sqrt{-1}$$

which is a unit vector at right angles to the line of reference.

The validity of these interpretations and the correspondence of the system of real and complex numbers to points on the two-dimensional Cartesian continuum were established by Gauss, Abel, Cauchy, and Weirstrasse, and the representation of complex numbers by vectors in a plane was introduced by Argand, and diagrams showing the composition and resolution of such numbers in a plane are known as Argand diagrams. These discoveries were the basis of the theory of functions of a complex variable established by Cauchy, Weirstrasse, and Riemann.

Gauss, Abel, Cauchy, and Weirstrasse showed that the system of numbers extended to embrace complex as well as real numbers is closed to infinite processes of analysis as well as to purely algebraic operations. Consequently, all linear operations performed on a complex magnitude will give a number in this system. Two complex numbers are said to be conjugate to each other when their real parts are equal in magnitude and sign and their imaginary parts are equal in magnitude but opposite in sign. If a linear operation is performed on a complex number and its conjugate, the resulting numbers will be conjugate to each other.

The result of this extension of the domain of number was far-reaching; it brought plane geometry under the domain of algebra, and trigonometry became merely a particular branch of algebra dealing with the relation of the component parts of a complex unit number to its direction with respect to some fixed direction in the complex number plane.

The use of complex numbers in the theory of alternating currents was first introduced by Dr. A. E. Kennelly and the late Dr. C. P. Steinmetz. The treatment of periodically varying functions in terms of vectors very greatly simplified the solution of all of the alternating-current problems.

In balanced polyphase problems the phase displacement between currents in the various phases and the voltages in the various phases are equal, and because of this symmetry it is possible to reduce the solution of the problem to that of an equivalent single-phase problem. Unbalanced current and voltage conditions, even in a polyphase system in which the constants are symmetrical, offer very considerable difficulty when attacked by classical methods. In such cases it is necessary to solve the phase currents and voltages simultaneously introducing self and mutual constants between phases. In addition, where rotating machines are involved, it is necessary to introduce impedances relating the rotor and stator circuits. In my early investigation of phase balancers I observed that certain symmetrical relations between phase currents and also between phase voltages recurred frequently, which led me to the investigation of the general problem of unbalance. This investigation finally led to the discovery of the fundamental principles of the Method of Symmetrical Coordinates which I published in 1918. In this paper it is shown that unbalanced problems can be solved by the resolution of the currents and voltages into certain symmetrical relations. When the constants are symmetrical, that is, when the system viewed from any phase is similar, then the symmetrical components of currents do not react upon each other so that it becomes possible to eliminate the mutual relations with their attendant complication in the solution of the problems. These relations are discussed in detail in this volume.

In many respects the method of symmetrical coordinates bears the same relation to the solution of unbalance in polyphase alternating-current problems that the complex variable bears to the solution of single-phase and balanced polyphase problems.

The method of symmetrical coordinates is analogous in some respects to the resolution of a periodic function into its fundamental and higher harmonics by Fourier series. By the method of symmetrical coordinates a set of unbalanced currents may be resolved into systems of balanced currents equal in number to the number of phases involved.

The expression "symmetrical coordinates" is preferable to the more commonly used expression "symmetrical components." The former is more logical from a mathematical sense in that it refers to the axes of reference, but the latter name is probably more descriptive of the process to the laymen.

I have had the pleasure of close association with the authors in their work, and the technical development of the system of symmetrical components owes a great deal to their initiative, enthusiasm, and untiring work during the past several years. This book will be welcomed by engineers as it is the first textbook brought out entirely devoted to the subject of symmetrical components.

C. L. Fortescue.

SYMMETRICAL COMPONENTS

CHAPTER I

INTRODUCTION AND HISTORICAL DEVELOPMENT

The solution of *balanced* polyphase circuits is usually accomplished by converting the constants and the applied voltages to per phase values and solving for one of the phases in a manner similar to single-phase circuits. The mutual reactions between phases may be represented by equivalent self impedances because the symmetry of the problem determines the magnitude and phase position of the other phase currents. The currents and voltages for the other phases are equal in magnitude to those of the first but displaced symmetrically in phase position.

The solution of *unbalanced* polyphase circuits or *balanced* circuits with *unbalanced* terminal conditions does not permit of the same simplification. By the older methods of analysis it was necessary to assign symbols to the quantities in all the phases and carry through the phase solutions simultaneously. This complicates the problem enormously. The modern method of analysis presented in this volume reduces the solution of such problems to a systematic form and results in considerable simplification, especially for balanced systems operating under some kind of unbalance such as a line-to-line, single line-to-ground, or double line-to-ground fault. This new analytical tool is called the **method of symmetrical components.**

1. What Is "Symmetrical Components"?

The fundamentals of the method of symmetrical components are in reality quite simple. No very great amount of mathematics is required for its understanding. It is on the contrary characterized by a large number of relatively simple steps, and once the fundamentals are mastered the application of the principles becomes relatively simple. One can in a very short time indicate these fundamentals and point out their field of application.

Consider for a moment the balanced vectors in Fig. 1(a) of which E_{a1}, E_{b1}, and E_{c1} are the line-to-neutral voltages of phases a, b, and c, respectively, of a three-phase system. The instantaneous values of these vectors are represented by the projection of the vectors upon the X-axis. With the conven-

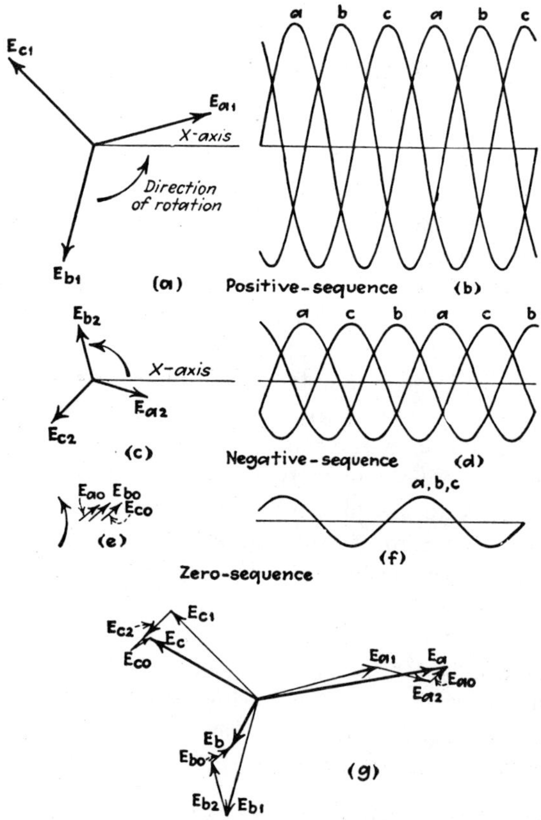

FIG. 1.—Sequence components and their combinations to form unbalanced phase quantities.

tional rotation of vectors in the counter-clockwise direction the instantaneous values of the voltages can be developed as shown in Fig. 1(b). It will be observed that the order of maxima occur in the sequence *abcabc* and for this reason they are said to be a "positive-sequence" set of vectors. Note that this sequence has no reference to the direction of rotation of the vectors themselves.

In Fig. 1(*c*) is shown another system of balanced voltages. Again assuming the conventional direction of rotation of the vectors the instantaneous values of the voltages may be developed as shown in Fig. 1(*d*). For this case the sequence of the maxima of the instantaneous values occurs in the order *acbacb* or *cba* which, since it is opposite to the previous one, is called the "negative-sequence" set of vectors. Note again that the sequence of maxima has no relation to the arbitrarily chosen direction of rotation of the vectors.

Consider lastly another type of three balanced vectors, such as that shown in Fig. 1(*e*), in which all three vectors are in phase. The maxima of the instantaneous values as shown in Fig. 1(*f*) must of course all occur simultaneously, and this set of vectors has been given the name "zero-sequence."

These voltages may exist in separate systems or simultaneously in the same system. In the latter case phase *a* will be composed of E_{a1} from the positive-sequence set of vectors, E_{a2} from the negative-sequence set of vectors, and E_{a0} from the zero-sequence set of vectors. The resultant value for this phase is shown in Fig. 1(*g*). Phase *b* will consist of the corresponding *b* components from the three sequences, and the summation for this phase is also shown in Fig. 1(*g*). The total voltage for phase *c* is obtained in a similar manner. It may be seen therefore that the simultaneous presence of three sets of *balanced* voltages of the type described results in a set of *unbalanced* voltages. Currents can of course be analyzed in a similar manner.

One purpose of the method of symmetrical components is to show how the three unbalanced voltages can be built up in an analytical manner from the three fundamental sequences. Another purpose is to show how any three unbalanced voltages can in turn be resolved into three sets of three balanced or symmetrical components. The question naturally arises, why should the resolution of three vectors into nine vectors necessarily simplify the solution. The answer lies in the fact that the resolution results in three symmetrical systems, each one of which in a balanced system can be treated separately just as the balanced polyphase problems in the past have been solved by reducing the constants and voltages to per phase values and solving on a single-phase basis. In symmetrical circuits, currents and voltages of different sequences do not react upon each other; currents of the positive-sequence produce only voltage drops of positive-sequence, currents of the negative-sequence only voltage drops of negative-sequence, and similarly, currents of the zero-sequence only voltage drops of zero-sequence. This fortunate circumstance results in considerable simplification of all kinds of problems involving asymmetry such as that introduced by short-circuiting conductors of a system either together or to ground, singly or in pairs, or by the open-circuiting of a conductor.

The resolution of the problem into the sequence components has a further advantage in that it isolates the quantities into components which represent a better criterion of the controlling factor or factors in certain phenomena. In stability investigations the synchronizing force between machines is affected principally by the positive-sequence quantities. The demagnetizing factor of armature current in synchronous machines is measured by the positive-sequence component of current only so that it is this component of current which determines excitation requirements. The performance of damper windings, as to both heating and torque, is responsive to the negative-sequence component only. Ground relays and grounding phenomena in general are very closely associated with zero-sequence components. Power quantities can be resolved into components associated with the different sequences. These and many other subjects attest to the value of resolving the problem into the sequence components; the solution of the problem itself is not only simpler, but the results are resolved into factors which can be investigated separately. While the foregoing considered only three-phase systems, the fundamental conceptions are applicable to systems of any number of phases.

Most of the apparatus used in practice such as generators and condensers, induction motors and transmission lines and cables (to within a sufficient degree for practical purposes) are of the symmetrical type. In fact special precautions are taken to insure that they are symmetrical, otherwise undesirable features, such as telephone interference and extra losses, may be introduced. It may be seen, therefore, that this simplification is applicable to the vast majority of systems in commercial use.

In *unsymmetrical* systems currents of one sequence produce not only voltage drops of the same sequence but may produce voltage drops of the other sequences as well. In the method of symmetrical components, rules are developed for solving the fundamental relation between the sequences in such systems.

2. Short Historical Review.

The germ of the idea may be traced to the early analysis of single-phase motors by Ferraris, Lamme, and others, about 1895. As a part of this analysis it is shown that the field set up in the single-phase motor can be resolved into two revolving flux fields rotating in opposite directions. Somewhat later, unbalanced currents from three-phase machines were resolved into the two sets of components that are now known as the positive- and negative-sequence components and their effects were analyzed in terms of the positively and negatively rotating fields in the machine associated with the positive- and negative-sequence components, respectively. This conception was utilized by E. F. W. Alexanderson in connection with his work on phase-balancers,(1)* a qualitative treatment of which was published in 1913, and by L. G. Stokvis in the determination of generator-voltage regulation in terms of the phase currents, a mathematical analysis of which appeared in 1912(101) and 1915.(2)

Stokvis appears to have approached his analysis from the machine point of view. He was thus limited in his resolution of a system of vectors into sets which produced certain effects within the machine and chose for his purpose:

1. A set which produced a positively rotating field.
2. A set which produced a negatively rotating field.
3. A set which produced a pulsating field.

He failed to recognize that what was needed was a new kind of component (the zero-sequence component) which produced neither a *rotating* nor a *pulsating field* in a symmetrical machine. He thus lacked the essential element required to set up components which did not react upon each other in symmetrical parts of systems.

It remained for C. L. Fortescue approaching the problem from a different point of view to perceive the beautiful generality of the method as applied to all kinds of polyphase systems. The general concept was developed by Fortescue and his associates, R. E. Gilman, J. F. Peters, J. Slepian, and others, in studying problems of unbalanced circuits and in analyzing the characteristics of single-phase motors, polyphase motors with unbalanced voltages, and synchronous motor-generator sets and phase-balancers for single-phase railway electrification. In

* Superior numerals in parentheses refer to items in the Bibliography, page 391.

this way a new, simple, and complete method was developed for handling the problem of unbalanced circuits, which was published by C. L. Fortescue in his classic paper[5] in 1918.

3. Application of the Method.

In studying the method developed by Dr. Fortescue, it is convenient to consider its application, first, to commercial three-phase systems which are symmetrical except for the unbalance at a particular point, such as the point of fault; and, second, to the more general case of a system which may be unsymmetrical throughout. As applied to the first type of system, three principal developments may be recognized:

1. Introduction of zero-sequence. This device makes it possible to resolve *any* three unbalanced vectors into three sequence components, namely, positive-, negative-, and zero-sequence.

2. Demonstration that in those parts of a system which are symmetrical, the currents and voltages of one sequence have no influence upon those of another sequence. Recognition of this fact is important as it forms the basis for all the subsequent work on the determination of short-circuit currents of unbalanced faults on commercial systems.

3. Assignment to lines and apparatus of distinct impedances for each sequence that are fixed quantities independent of each other and of the character or amount of the unbalance.

The general method of analysis is also applicable to systems which are unsymmetrical throughout. This case requires additional constants and somewhat more complicated relations to define the reaction of one sequence upon another in the unsymmetrical parts of the system. Dr. Fortescue showed how such calculations could be simplified and systematized by means of the "sequence operator."

He further demonstrated the generality of the method by showing that it could be applied to systems of any number of phases, with similar results. The same general method has even been used to solve the cubic equation and to reduce the degree of a higher order equation.[151] It is this generality of the method which caused Dr. Fortescue to select for the unit sequence vectors the more general term "coordinates" instead of the term "components" which is commonly used in electrical engineering.

Since the presentation of Dr. Fortescue's paper in 1918, the work along this line has been directed principally toward the application of the method to the solution of problems commonly encountered in commercial systems, the most important of these being the determination of system currents and voltages under unbalanced fault conditions. This work has been much facilitated by the concepts of sequence networks and the use of equivalent circuits connecting these networks to represent fault conditions. These short-circuit studies may make use of the direct-current or alternating-current calculating boards, the first application of which was described by R. D. Evans in 1925. The method was next applied to system-stability studies by R. D. Evans and C. F. Wagner in 1926. Also in 1926, A. P. Mackerras collected and presented in two articles the most important features of the method as applied to the determination of single-phase short-circuit currents on three-phase systems.

Another important phase of the subsequent work on symmetrical components has been the determination by calculation and test of the sequence impedance constants of lines and apparatus. This work has given considerable impetus to the further studies of the short-circuits of synchronous machines as evidenced by articles and papers by Bekku, Wagner and Dovjikov, Park and Robertson, Doherty and Nickle, Kilgore, and Wright. Intensive study has also been given to the zero-sequence impedance of transmission lines and cables which is of particular importance in connection with the problem of inductive coordination. The fundamental work of Carson and the theoretical and experimental work of the Joint Sub-committee on Development and Research of the N.E.L.A. and Bell Telephone System have been very valuable.

A knowledge of the method of symmetrical components is essential for an adequate understanding of application problems involving the magnitude and phase relation of voltages and currents under unbalanced conditions and is therefore necessary for the design of power systems from the standpoint of circuit-breaker application, relay protection, and stresses in electrical machinery. The method is particularly suitable for the analysis of the performance of rotating machinery for single-phase or unbalanced polyphase operation and has been used quite extensively in connection with the design of machinery on single-phase railway systems, particularly from the standpoint of phase-converting apparatus. A number of new schemes have appeared commercially which make use of conceptions arising from the method of symmetrical components, such as the negative-sequence protective relay and the positive-sequence voltage regulator. Symmetrical components is also advantageous in analyzing metering means, particularly polyphase instruments on unbalanced loads.

In recent years stability of power systems has been recognized as an important problem, and it has been shown that the limiting condition arises at times of faults, hence the method of symmetrical components has been of very considerable assistance in the analysis of stability problems.

Problems*

1. Given the self and mutual reactances between three long parallel conductors located at the apexes of an equilateral triangle, show that for balanced operation the reactances can be represented by an equivalent self reactance expressed in terms of ohms to neutral per phase.

2. A 625-kva. 2,300–110-volt single-phase transformer has a reactance of 3 per cent. Determine the reactance in ohms on both the high-voltage and low-voltage bases.

3. A 20,000-kva. bank of transformers is connected to a 132,000-volt system. The reactance of the transformers is 9 per cent. Convert to per cent on 100,000-kva. base. Determine the reactance in ohms per phase on the 132,000-volt base and a 220,000-volt base.

4. A 10,000-kva., 132,000–13,200-volt transformer bank having a reactance of 10 per cent is connected on the low-voltage side to a 20,000-kva. generator having a reactance of 30 per cent. A three-phase short-circuit occurs on the high-voltage side of the transformer. Assuming a generator internal voltage equal to rated volts, determine: (*a*) the short-circuit current on the high-voltage side of the transformer, (*b*) the short-circuit current on the low-voltage side of the transformer, and (*c*) the voltage to neutral on the low-voltage side.

* A knowledge of the methods for handling this group of problems dealing with symmetrical three-phase systems is essential for the ready understanding of the treatment presented in subsequent chapters. Deficiency in previous training in this regard may be disclosed at the outset and promptly corrected.

CHAPTER II

FUNDAMENTAL PRINCIPLES

The method of symmetrical components requires no additional knowledge of mathematics beyond that commonly used in the solution of the ordinary alternating-current problems. A working knowledge of the fundamentals of the algebra of complex numbers is, however, essential, and consequently this subject will be reviewed briefly.

4. Laws of Complex Numbers.

Any complex number, such as $a + jb$, may be represented by a single point, p, plotted on Cartesian coordinates, in which a is the abscissa and b the ordinate. This is illustrated in Fig. 2. It is often convenient, however, to represent complex numbers by another method. Referring to Fig. 2, let r represent the length of the line connecting the point with the origin and θ the angle measured from the X-axis to the line r. It is apparent that

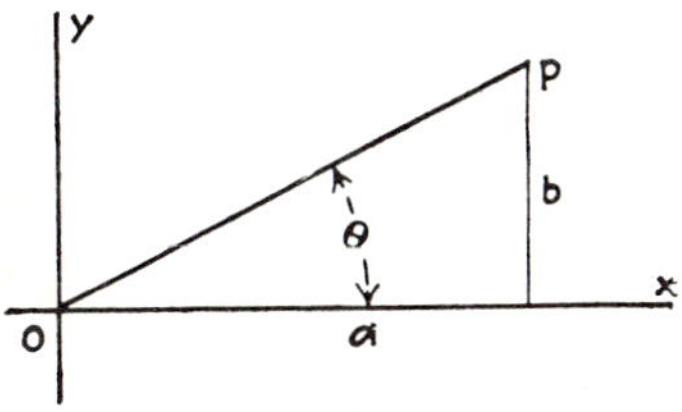

FIG. 2.—A complex number.

$$a = r \cos \theta$$
$$b = r \sin \theta$$

or

$$a + jb = r(\cos \theta + j \sin \theta) \tag{1}$$

This is called the polar form, in which θ is the *argument* or *amplitude* and r the *modulus* or *absolute value.* Increasing θ in a positive sense rotates *op* counter-clockwise. Similarly

$$a - jb = r(\cos \theta - j \sin \theta) \tag{2}$$

Increasing θ in a positive sense in this case produces clockwise rotation.

If ϵ^{θ} be expanded into an infinite series,

$$\epsilon^{\theta} = 1 + \theta + \frac{\theta^2}{2!} + \frac{\theta^3}{3!} + \frac{\theta^4}{4!} + \frac{\theta^5}{5!} + \frac{\theta^6}{6!} + \cdots$$

Similarly,

$$\epsilon^{j\theta} = 1 + j\theta - \frac{\theta^2}{2!} - j\frac{\theta^3}{3!} + \frac{\theta^4}{4!} + j\frac{\theta^5}{5!} - \frac{\theta^6}{6!} \cdots$$

Separating the real and imaginary components,

$$\epsilon^{j\theta} = \left(1 - \frac{\theta^2}{2!} + \frac{\theta^4}{4!} - \frac{\theta^6}{6!} + \cdots\right) + j\left(\theta - \frac{\theta^3}{3!} + \frac{\theta^5}{5!} - \frac{\theta^7}{7!} + \cdots\right)$$

These expansions* are equal to $\cos\theta$ and $\sin\theta$ respectively, so that,

$$\epsilon^{j\theta} = \cos\theta + j\sin\theta \tag{3}$$

Similarly,

$$\epsilon^{-j\theta} = \cos\theta - j\sin\theta \tag{4}$$

This suggests another form to represent a complex number. Substituting (3) in (1) and (4) in (2),

$$a + jb = r\epsilon^{j\theta} \tag{5}$$
$$a - jb = r\epsilon^{-j\theta} \tag{6}$$

These forms will be found valuable for operations involving multiplication and division. Unless otherwise stated this angle will be expressed in degrees throughout this volume.

Returning to equations (1) and (2) and (5) and (6), it will be noted that the real components are equal, but that the imaginary components have opposite signs. These numbers are called conjugates.

In the notation used in this volume an ordinary complex number will be represented by a capital letter and its conjugate by the same letter with a circumflex, for example, I and $\hat{I}$, respectively. If

$$I = a + jb$$

then its conjugate is

$$\hat{I} = a - jb$$

* These relations may be verified by reference to any standard college algebra.

The absolute value of the vector will be designated by a bar above the capital letter, thus

$$\bar{I} = \sqrt{a^2 + b^2}$$

The real and imaginary components may be expressed in terms of the complex number and its conjugate as follows:

$$a = \tfrac{1}{2}(I + \hat{I}) \tag{7}$$
$$jb = \tfrac{1}{2}(I - \hat{I}) \tag{8}$$

The law for the addition of exponents holds equally well for both real and imaginary quantities. Just as $10^3 \times 10^4 = 10^7$ so

$$\epsilon^{j\theta_1} \times \epsilon^{j\theta_2} = \epsilon^{j(\theta_1+\theta_2)}$$

from which it follows that the product of two complex numbers,

$$E = \bar{E}\epsilon^{j\theta_1} \text{ and } I = \bar{I}\epsilon^{j\theta_2}$$

is equal to

$$EI = \bar{E}\bar{I}\epsilon^{j(\theta_1+\theta_2)} \tag{9}$$

It will be noted that the absolute value of the product is the product of the absolute values of the components and that the argument is the sum of the arguments.

Multiplying by $\epsilon^{j\theta}$ merely rotates the modulus through an angle θ in a counter-clockwise direction. This point is important as it forms the basis for the proof of quite a number of circle diagrams.

If a complex number be multiplied by the conjugate of another, the argument of the product is the difference in arguments of the components.

$$E\hat{I} = \bar{E}\epsilon^{j\theta_1}\bar{I}\epsilon^{j-\theta_2}$$
$$= \bar{E}\bar{I}\epsilon^{j(\theta_1-\theta_2)} \tag{10}$$

Resolved into real and imaginary components,

$$E\hat{I} = E\bar{I}[\cos(\theta_1 - \theta_2) + j\sin(\theta_1 - \theta_2)] \tag{11}$$

Squaring a complex number merely increases the modulus as the square and rotates it through double the angle.

$$(\bar{I}\epsilon^{j\theta})^2 = \bar{I}^2\epsilon^{j2\theta} \tag{12}$$

Division, the inverse of multiplication, can be accomplished in a similar manner.

$$\frac{E}{I} = \frac{\bar{E}\epsilon^{j\theta_1}}{\bar{I}\epsilon^{j\theta_2}} = \frac{\bar{E}}{\bar{I}}\epsilon^{j(\theta_1-\theta_2)} \tag{13}$$

The absolute value of the resultant is equal to the quotient of the absolute values, and the argument is the difference of the arguments of the components.

As squaring a number merely increases the modulus to the square of the modulus and doubles the argument, extracting the square root, the inverse, is accomplished by taking the square root of the modulus and halving the argument. Generalizing, the modulus of the nth root of a number is the nth root of the modulus of the number and the argument is $\frac{1}{n}$ of the argument of the number.

$$\sqrt[n]{\bar{I}\epsilon^{j\theta}} = \sqrt[n]{\bar{I}}\epsilon^{j\frac{\theta}{n}} \tag{14}$$

If a complex number be multiplied by its conjugate, the product is a real number of magnitude equal to the square of the absolute value of the number, for example.

$$\begin{aligned} I\hat{I} &= \bar{I}\epsilon^{j\theta}\bar{I}\epsilon^{-j\theta} \\ &= \bar{I}^2\epsilon^{j(\theta-\theta)} \\ &= \bar{I}^2 \end{aligned} \tag{15}$$

5. Vector Representation of Alternating Quantities.

The instantaneous value of a simple harmonic function, such as an alternating electromotive force, may be represented by the equation

$$e = \sqrt{2}\bar{E} \cos (\omega t + \alpha)$$

where

$$\begin{aligned} \omega &= 2\pi f. \\ f &= \text{frequency}. \\ \bar{E} &= \text{r.m.s. magnitude}. \\ \sqrt{2}\bar{E} &= \text{crest value}. \\ \alpha &= \text{angle which determines the value of } e \text{ for } t = 0. \end{aligned}$$

By the addition of equations (3) and (4) it may be seen that $\cos \theta = \frac{1}{2}(\epsilon^{j\theta} + \epsilon^{j\theta})$ and substituting $\omega t + \alpha$ for θ, gives

$$\begin{aligned} \sqrt{2}\bar{E} \cos (\omega t + \alpha) &= \sqrt{2}\bar{E}\left[\frac{\epsilon^{j(\omega t+\alpha)} + \epsilon^{-j(\omega t+\alpha)}}{2}\right] \\ &= \sqrt{2}\bar{E}\left[\frac{\epsilon^{j\alpha}\epsilon^{j\omega t} + \epsilon^{-j\alpha}\epsilon^{-j\omega t}}{2}\right] \end{aligned}$$

so that, if we let $\sqrt{2}\bar{E}\epsilon^{j\alpha} = \sqrt{2}E$ and $\sqrt{2}\bar{E}\epsilon^{-j\alpha} = \sqrt{2}\hat{E}$

$$e = \frac{\sqrt{2}}{2}(E\epsilon^{j\omega t} + \hat{E}\epsilon^{-j\omega t}) \tag{16}$$

It will be noted from this equation that the instantaneous value can be represented as the sum of two oppositely rotating vectors which are conjugate at any instant, the imaginary components canceling out. This relation is shown in Fig. 3, in which the full lines indicate the positions of E and $\hat{E}$ for $t = 0$ and the dotted lines at an instant later. It will be noted that the two vectors are always symmetrically disposed relative to the X-axis so that their sum is always real and does not contain an imaginary component. Furthermore, the projection of either vector on the X-axis is always equal to one-half the sum of the vectors, and the instantaneous value can be represented by the real component of either vector. The same relations apply equally well to the instantaneous values of current.

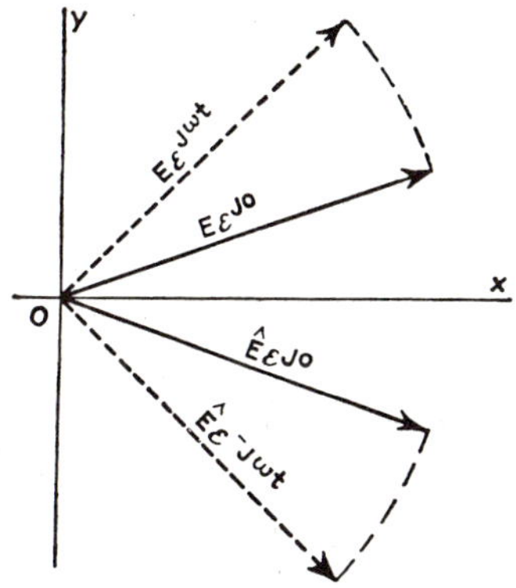

FIG. 3.—Simple harmonic function.

6. Symmetrical Component Systems.

Any point on a line is completely defined by the distance from a given reference point, in either the positive direction or the negative. Similarly a point on a plane is completely defined by the distance from two reference axes and a point in space by the distance from three reference axes. A point on a line, being restrained to lie in the line, is said to possess but one degree of freedom; the point in the plane two degrees of freedom; and the point in space three degrees of freedom. So also with a co-planar vector which, since it is determined by the position of its terminal, is said to possess two degrees of freedom.

Similar considerations apply to systems of vectors. For example, a system of three co-planar vectors is completely defined by six parameters; the system possesses six degrees of freedom. When, however, one applies the restriction that the system be symmetrical, the added restraints reduce the system to one of two degrees of freedom. Going a step further, it is quite conceivable that a system of three co-planar vectors with six degrees of freedom can be defined in terms of three symmetrical systems of vectors each having two degrees of freedom.

Positive-sequence System. Consider first the symmetrical system of vectors E_{a1}, E_{b1} and E_{c1} in Fig. 4(a). Being balanced, the vectors have equal amplitudes, are displaced 120 deg. relative to each other, and

$$\left.\begin{aligned} E_{a1} &= E_{a1} \\ E_{b1} &= \epsilon^{j240}E_{a1} = a^2E_{a1} \\ E_{c1} &= \epsilon^{j120}E_{a1} = aE_{a1} \end{aligned}\right\} \quad (17)$$

in which

$$\left.\begin{aligned} &a \text{ is the unit vector, } \epsilon^{j120} = -0.5 + j0.866 \\ \text{and}\quad & \\ &a^2 \text{ is the unit vector, } \epsilon^{j240} = -0.5 - j0.866 \end{aligned}\right\} \quad (18)$$

Because of the frequent use of this vector it will always be designated by the small letter a. Some of the properties of this vector are given in Table I.

TABLE I.—PROPERTIES OF THE VECTOR a

$$\begin{aligned} 1 &= \epsilon^{j0} = 1.0 + j0.0 \\ a &= \epsilon^{j120} = -0.5 + j0.866 \\ a^2 &= \epsilon^{j240} = -0.5 - j0.866 \\ a^3 &= \epsilon^{j360} = \epsilon^{j0} = 1.0 + j0.0 \\ a^4 &= \epsilon^{j480} = \epsilon^{j120} = a \\ a^5 &= \epsilon^{j600} = \epsilon^{j240} = a^2 \end{aligned}$$

$$\begin{aligned} 1 + a^2 + a &= 0 \\ a - a^2 &= \sqrt{3}\epsilon^{j90} = j\sqrt{3} \\ a^2 - a &= \sqrt{3}\epsilon^{-j90} = -j\sqrt{3} \\ 1 - a &= \sqrt{3}\epsilon^{-j30} = 1.5 - j0.866 \\ 1 - a^2 &= \sqrt{3}\epsilon^{j30} = 1.5 + j0.866 \end{aligned}$$

This system of vectors is called the positive-sequence system, because as explained in Chap. I the order of the sequence of their maxima occur abc. The system must always be treated as a unit. In a three-phase balanced electrical system, E_{a1} would represent the voltage of phase a; a^2E_{a1} the voltage of phase b; and aE_{a1} the voltage of phase c. Fixing the phase position and magnitude of E_{a1} immediately determines E_{b1} and E_{c1}.

Negative-sequence System. Likewise, consider the balanced system of vectors E_{a2}, E_{b2}, and E_{c2} in Fig. 4(b).

$$\left.\begin{aligned} E_{a2} &= E_{a2} \\ E_{b2} &= \epsilon^{j120}E_{a2} = aE_{a2} \\ E_{c2} &= \epsilon^{j240}E_{a2} = a^2E_{a2} \end{aligned}\right\} \quad (19)$$

This system of vectors is called the negative-sequence system.

A physical picture of the significance of these systems may be obtained by considering the fields which result when voltages of the two systems are applied to a three-phase machine with distributed windings, such as an induction motor. If the a, b,

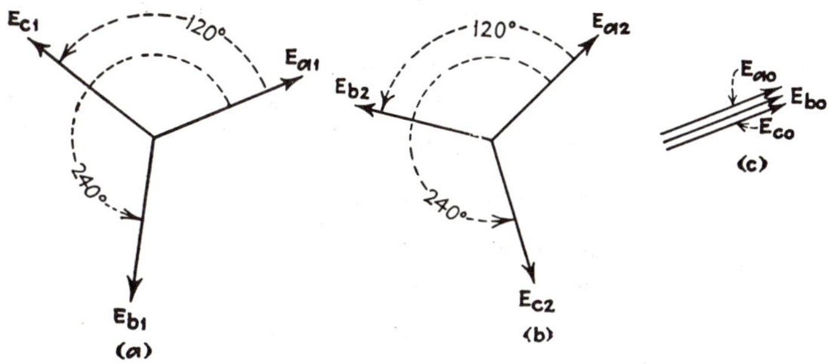

FIG. 4.—Symmetrical vector systems.

and c phases of the positive-sequence voltages be applied to the terminals a, b, and c, respectively, a magnetic field will be produced which will revolve in a certain direction. If now the voltages to terminals b and c are changed by interchanging the leads to terminals b and c, it is well known from induction-motor theory that a magnetic field will be produced which rotates in the opposite direction. A little consideration will show that for this condition the relative phase positions of the voltages applied to the motor are the same as for the negative-sequence system. It follows, therefore, that the negative-sequence set of voltages produces a field rotating in an opposite direction to that of the positive-sequence.

Phase sequence should not be confused with the rotation of the vectors. In this volume the standard convention of counter-clockwise rotation has been followed for all vectors; both positive- and negative-sequence vectors will rotate in the same direction, but the resultant fields in machines will have opposite rotations.

Zero-sequence System. Finally, consider the system E_{a0}, E_{b0}, and E_{c0} in Fig. 4(c) which represents three equal vectors.

$$\left.\begin{aligned} E_{a0} &= E_{a0} \\ E_{b0} &= E_{a0} \\ E_{c0} &= E_{a0} \end{aligned}\right\} \quad (20)$$

This system is called the zero-sequence system.

Combination of Sequence Quantities to Form Phase Quantities. In all three systems of the symmetrical components, the subscripts denote the components in the different phases. The total voltage of any phase is then equal to the sum of the corresponding components of the different sequences in that phase. Therefore, any three arbitrary vectors E_a, E_b, and E_c may be equated.

$$E_a = E_{a0} + E_{a1} + E_{a2} \quad (21)$$
$$E_b = E_{b0} + E_{b1} + E_{b2} \quad (22)$$
$$E_c = E_{c0} + E_{c1} + E_{c2} \quad (23)$$

Or substituting their equivalent values

$$E_a = E_{a0} + E_{a1} + E_{a2} \quad (24)$$
$$E_b = E_{a0} + a^2E_{a1} + aE_{a2} \quad (25)$$
$$E_c = E_{a0} + aE_{a1} + a^2E_{a2} \quad (26)$$

The arbitrary system is thus defined in terms of three balanced systems.

7. Resolution of Three Vectors into Their Symmetrical Components.

The individual components may be obtained as follows.

Zero-sequence. Adding the three equations,

$$\begin{aligned} E_a + E_b + E_c = 3E_{a0} & \\ + (1 + a^2 + a)E_{a1} & \\ + (1 + a + a^2)E_{a2} & \end{aligned}$$

But since $(1 + a^2 + a) = 0$,

$$E_{a0} = \tfrac{1}{3}(E_a + E_b + E_c) \quad (27)$$

Positive-sequence. Multiplying equation (25) by a and equation (26) by a^2 and adding these two equations to (24), there results

$$E_a + aE_b + a^2E_c = (1 + a + a^2)E_{a0} + 3E_{a1} + (1 + a^2 + a)E_{a2}$$

or

$$E_{a1} = \tfrac{1}{3}(E_a + aE_b + a^2E_c) \quad (28)$$

Negative-sequence. Multiplying equation (25) by a^2 and equation (26) by a and adding these two equations to (24), there results

$$E_a + a^2E_b + aE_c = (1 + a^2 + a)E_{a0} + (1 + a + a^2)E_{a1} + 3E_{a2}$$

or

$$E_{a2} = \tfrac{1}{3}(E_a + a^2E_b + aE_c) \tag{29}$$

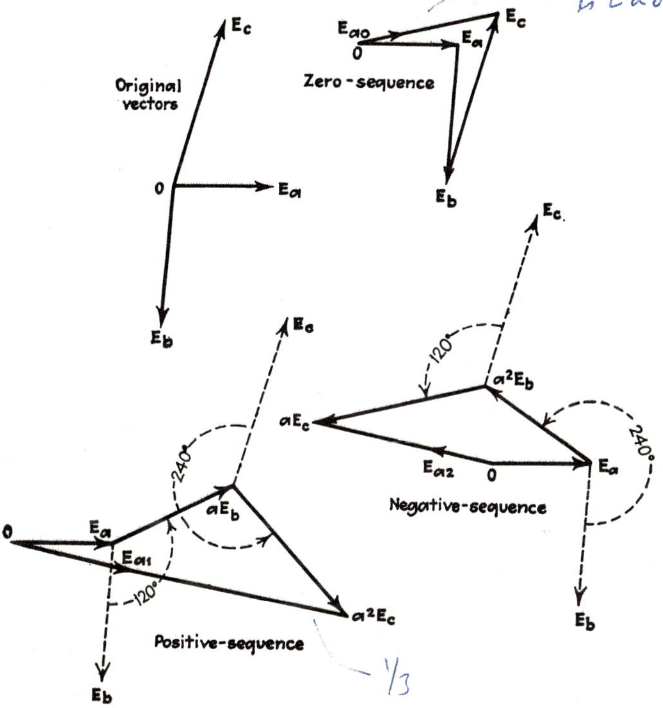

FIG. 5.—Graphical construction for the determination of sequence components.

Thus these three fundamental equations defining the sequence quantities in terms of the phase quantities are grouped together in Fig. 5.

Illustration. Suppose the given voltages are balanced, for example,

$$\begin{aligned} E_a &= E_a \\ E_b &= a^2E_a \\ E_c &= aE_a \end{aligned}$$

then

$$E_{a0} = \frac{(1 + a^2 + a)}{3}E_a = 0$$

$$E_{a1} = \frac{(1 + a^3 + a^3)}{3}E_a = E_a$$

$$E_{a2} = \frac{1 + a^4 + a^2}{3}E_a = 0$$

The zero- and negative-sequence components of voltage in a balanced three-phase system disappear, leaving only the positive-sequence.

The graphical construction for the determination of the sequence components will aid further in elucidating the operations. In Fig. 5 are shown three vectors E_a, E_b, E_c. The zero-sequence component E_{a0} is obtained by direct addition. The positive-sequence component E_{a1} is obtained by adding to E_a, E_b rotated counter-clockwise 120 deg., and E_c rotated counter-clockwise 240 deg., and taking one-third of the sum. Similarly E_{a2} is determined by adding to E_a, E_b rotated counter-clockwise 240 deg., and E_c rotated counter-clockwise 120 deg., and taking one-third of the sum.

Numerical Example. The method may be further illustrated by means of the analytical calculation of the components. Consider the three vectors

$$E_a = 60 + j0$$
$$E_b = 45 - j75$$
$$E_c = -21 + j120$$

From equations (27), (28), and (29),

$$\begin{aligned}
E_{a0} &= \tfrac{1}{3}[E_a + E_b + E_c] \\
&= \tfrac{1}{3}[(60 + j0) + (45 - j75) + (-21 + j120)] \\
&= 28 + j15 \\
E_{a1} &= \tfrac{1}{3}[E_a + aE_b + a^2E_c] \\
&= \tfrac{1}{3}[(60 + j0) + (-0.5 + j0.866)(+45 - j75) + \\
&\qquad (-0.5 - j0.866)(-21 + j120)] \\
&= 72.2 + j11.5 \\
E_{a2} &= \tfrac{1}{3}[E_a + a^2E_b + aE_c] \\
&= \tfrac{1}{3}[(60 + j0) + (-0.5 - j0.866)(45 - j75) + \\
&\qquad (-0.5 + j0.866)(-21 + j120)] \\
&= (-40.2 - j26.5)
\end{aligned}$$

These three values give the symmetrical components of the three original vectors. Using these same values of components the determination of the original vectors in terms of their components will be illustrated.

From equations (24), (25), and (26),

$$\begin{aligned}
E_a &= E_{a0} + E_{a1} + E_{a2} \\
&= (28 + j15) + (72.2 + j11.5) + (-40.2 - j26.5) \\
&= 60 + j0 \\
E_b &= E_{a0} + a^2E_{a1} + aE_{a2} \\
&= (28 + j15) + (-0.5 - j0.866)(72.2 + j11.5) + \\
&\qquad (-0.5 + j0.866)(-40.2 - j26.5) \\
&= 45 - j75
\end{aligned}$$

$$
\begin{aligned}
E_c &= E_{a0} + aE_{a1} + a^2E_{a2} \\
&= (28 + j15) + (-0.5 + j0.866)(72.2 + j11.5) + \\
&\qquad\qquad (-0.5 - j0.866)(-40.2 - j26.5) \\
&= -21 + j120
\end{aligned}
$$

8. Star-delta Transformations.*

Problems sometimes arise involving both star and delta currents and voltages. Some conditions are such that the equations are set up more easily with some star and some delta quantities. It is convenient for the solution of such problems to be able to convert readily from star to delta and *vice versa* in terms of the symmetrical components.

Current Relations. Let I_A, I_B, and I_C, be the delta currents in the delta-connected windings of a machine and I_a, I_b, and I_c, the currents in the lines as shown in Fig. 6. At the junctions the following relations are satisfied:

$$
\left.
\begin{aligned}
I_a &= I_B - I_C \\
I_b &= I_C - I_A \\
I_c &= I_A - I_B
\end{aligned}
\right\} \quad (30)
$$

The zero-sequence component of the star currents, I_{a0}, is, by equation (27),

$$
\begin{aligned}
I_{a0} &= \tfrac{1}{3}(I_a + I_b + I_c) \\
&= \tfrac{1}{3}[(I_B + I_C + I_A) - (I_C + I_A + I_B)] = 0 \qquad (31)
\end{aligned}
$$

This result shows that the zero-sequence current of a polyphase circuit feeding into a delta connection is *always* zero, which may be verified physically by the fact that for zero-sequence currents to flow in the line a neutral return circuit must be available, a condition impossible of fulfillment in the delta connection. The converse of determining I_{A0} in terms of I_{a0} is indeterminate. Currents of zero-sequence *may* circulate within the delta without getting out into the line.

For the positive-sequence, by equation (28),

$$
\begin{aligned}
I_{a1} &= \tfrac{1}{3}(I_a + aI_b + a^2I_c) \\
&= \tfrac{1}{3}[(I_B + aI_C + a^2I_A) - (I_C + aI_A + a^2I_B)] \\
&= \tfrac{1}{3}[(a^2I_A + I_B + aI_C) - (aI_A + a^2I_B + I_C)]
\end{aligned}
$$

By factoring a^2 out of the first term and a out of the second term,

$$
I_{a1} = \frac{a^2}{3}(I_A + aI_B + a^2I_C) - \frac{a}{3}(I_A + aI_B + a^2I_C)
$$

* An application of these principles to transformers is given in Sec. 35.

By reference to equation (28), it may be seen that

$$I_{a1} = (a^2 - a)I_{A1}$$
$$= -j\sqrt{3}I_{A1} \qquad (32)$$

or

$$I_{A1} = \frac{j}{\sqrt{3}}I_{a1} \qquad (33)$$

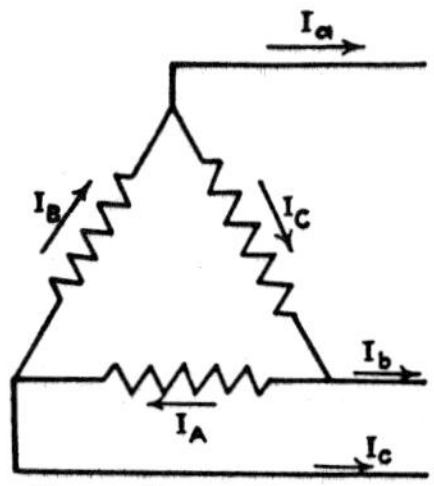

$$I_a = I_B - I_C$$
$$I_b = I_C - I_A$$
$$I_c = I_A - I_B$$

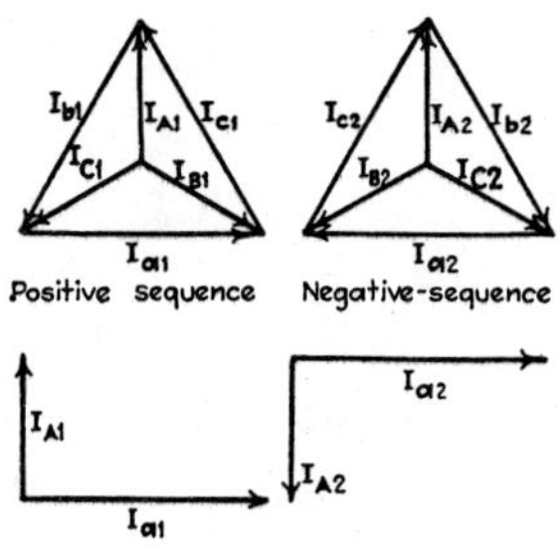

$$I_{A0} = \text{indeterminate}$$
$$I_{A1} = \frac{j}{\sqrt{3}}I_{a1}$$
$$I_{A2} = -\frac{j}{\sqrt{3}}I_{a2}$$
$$I_{a0} = 0$$
$$I_{a1} = -j\sqrt{3}I_{A1}$$
$$I_{a2} = +j\sqrt{3}I_{A2}$$

FIG. 6.—Star-delta current transformations.

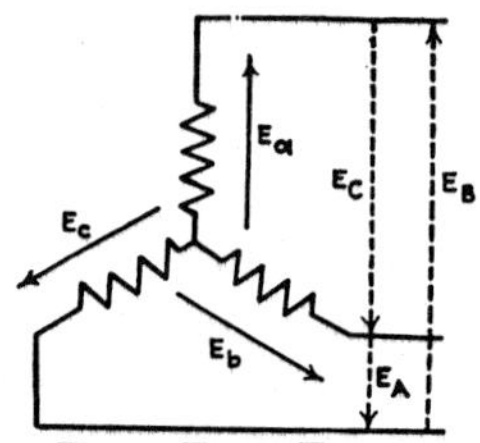

$$E_A = E_c - E_b$$
$$E_B = E_a - E_c$$
$$E_C = E_b - E_a$$

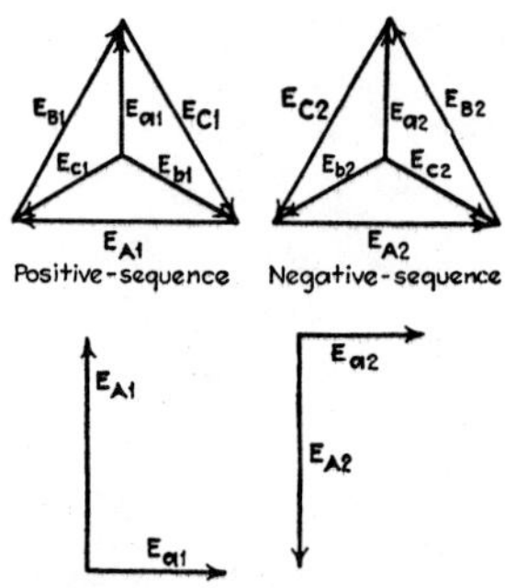

$$E_{A0} = 0$$
$$E_{A1} = j\sqrt{3}E_{a1}$$
$$E_{A2} = -j\sqrt{3}E_{a2}$$
$$E_{a0} = \text{indeterminate}$$
$$E_{a1} = -\frac{j}{\sqrt{3}}E_{A1}$$
$$E_{a2} = +\frac{j}{\sqrt{3}}E_{A2}$$

FIG. 7.—Star-delta voltage transformations.

Similarly it may be shown that

$$I_{a2} = j\sqrt{3}I_{A2} \qquad (34)$$

and

$$I_{A2} = \frac{-j}{\sqrt{3}}I_{a2} \qquad (35)$$

These relations are illustrated in Fig. 6. They could have been obtained more simply by the following reasoning. Assume that positive-sequence currents alone flow in the delta. Positive-sequence currents alone flow in the line, and the values of these line currents can be obtained by applying the relations expressed in equations (30), obtaining thereby the current triangle of Fig. 6 directly. The relations between the I_{a1} and I_{A1} can then be obtained from this figure.

Voltage Relations. Now let E_A, E_B, and E_C of Fig. 7 be the delta voltages and E_a, E_b, and E_c, the star voltages. What are the corresponding equivalences relating the star and delta sequence voltages? By definition:

$$\left.\begin{aligned} E_A &= E_c - E_b \\ E_B &= E_a - E_c \\ E_C &= E_b - E_a \end{aligned}\right\} \quad (36)$$

Since the delta voltages must by their very nature form a closed triangle, $E_A + E_B + E_C$ must always equal zero and $E_{A0} = 0$. The delta voltages, therefore, *can never* contain a zero-sequence component. The star voltages, on the other hand, *may* contain a zero-sequence component. It follows then that E_{a0} cannot be determined from E_{A0} but must be determined by some other relation in the problem.

The relation between the star and delta voltages of the positive- and negative-sequence voltages are shown in Fig. 7. These relations may be verified by applying equations (36). The *a*-phase components of the different sequences are related by the expressions shown also in Fig. 7.

Nomenclature and Convention of Current Flow. A word of warning may be injected at this point concerning the conventions used in the nomenclature and in the direction of current flow. There is no generally accepted convention relating the designation of line conductors and delta phases. The authors have therefore standardized upon that given because of the rather simple relation to remember; conductors and windings of like designation are opposite, thus the *a* conductor is opposite to the *A* winding, etc. The choice of positive sense of current flow is entirely arbitrary. If the opposite directions had been chosen in one set the signs must be reversed in the equations relating the quantities of the two sets. Care must be taken to insure that the directions of current flow are consistent with the foregoing assumptions.

9. Independence of Sequences in Symmetrical Systems.*

In symmetrical systems the different sequences do not react upon each other; positive-sequence currents produce only positive-sequence voltages, negative-sequence currents produce only negative-sequence voltages, and zero-sequence currents produce only zero-sequence voltages. While Fortescue gave a rigorous proof of this statement as applied to both static devices

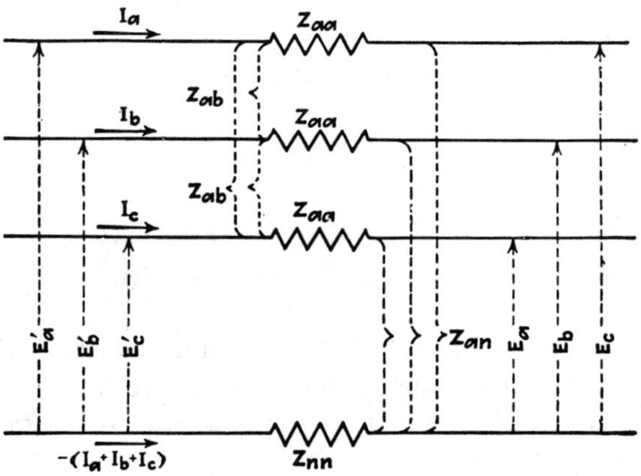

FIG. 8.—Symmetrical static network element.

and rotating machines, the present discussion merely undertakes to show that these conclusions are reasonable.

Static Networks. Consider first the static network shown in Fig. 8 which may represent a perfectly transposed transmission or distribution line in which capacity effects may be neglected. If *only* balanced positive-sequence currents be made to flow through the line conductors, it follows that *no* current flows through the neutral and *no* voltages are induced in the line conductors or impedances due to the mutual coupling between the line conductors and the neutral circuit. Since the mutual coupling between the neutral circuit and the different line conductors are all equal, the total voltage induced in the neutral circuit is equal to zero. Otherwise, the induced voltages in each phase are equal except for the phase displacement of 120 and 240 deg.

* A more extended discussion is given in Chap. XVIII, Sec. 167.

The positive-sequence currents thus produce only positive-sequence voltage drops. It can similarly be shown that negative-sequence currents produce only negative-sequence voltage drops. If only zero-sequence currents flow, equal currents flow in each line conductor and the combined currents of the three line conductors return through the neutral. In this case equal voltages will be induced in all three conductors, and including the drop in the neutral impedance the drops will be the same in all three phases. These equal voltages constitute a zero-sequence from which it may be concluded that zero-sequence currents produce only drops of zero-sequence.

This discussion has considered only series impedances. It should be apparent that similar conclusions may be drawn for the characteristics of symmetrical shunt impedances. Since the steady-state characteristics of lines with distributed constants can be represented by the well-known equivalent π or T, the conclusion may be further extended that the three sequences may be considered independently for static networks with distributed constants.

Rotating Machines. As remarked previously, positive-sequence currents in the stator of a symmetrical machine produce a rotating field which rotates in the same direction as the rotor. This field naturally produces *only* voltages of the same sequence across the terminals. The direct-currents in the rotor of synchronous machines and the slip frequency currents in the rotor of induction machines likewise produce only synchronously rotating fields in the same direction as the rotation of the rotor so that only positive-sequence voltages are produced on the terminals of the stator due to these currents. It is apparent then that under normal operating conditions with positive-sequence voltages applied to the stator of rotating machines *only* positive-sequence currents are produced.

If negative-sequence voltages only are applied to the stator of synchronous or induction machines, a synchronously rotating field is produced which rotates in a direction opposite to the rotation of the rotor. This field induces currents in the rotor, which in turn produces a synchronously rotating field in a direction opposite to that of the rotor. Thus all the currents and voltages in the stator would be of negative-sequence.

Because of the 120-deg. space displacement of the windings, the zero-sequence currents, which are in phase with each other in the three phases, produce no flux in the air gap. Hence, these currents can produce only voltage drops of the zero-sequence.

This analysis indicates that in symmetrically wound machines currents of the different sequences do not react upon each other.

10. Sequence Impedances.

It has been shown that in symmetrical networks the components of current of the different sequences do not react upon each other. When voltage of a given sequence is applied to a piece of apparatus, a very definite current of the same sequence flows. The apparatus may be characterized as having a definite impedance to this sequence. Special names have been given to these impedances, namely: the impedance to positive-sequence currents, the impedance to negative-sequence currents, and the impedance to zero-sequence currents but have been contracted to the simple expressions, *positive-sequence impedance, negative-sequence impedance,* and *zero-sequence impedance.*

The impedances of symmetrical static networks are the same for the positive- and the negative-sequences but may be different for the zero-sequence. For rotating machines the impedances will in general be different for all three sequences. Methods for determining these impedances by calculation and test will be given in subsequent chapters.

Problems

1. With a defined as $(-0.5 + j0.866)$ or ϵ^{j120}, evaluate the following: $(1 - a)(1 - a^2)$, $(a - a^2)(a^2 - 1)$, $\dfrac{1 - a}{1 - a^2}$, $\dfrac{a - a^2}{a^2 - 1}$, $(1 + a)^2$, $(1 + a)(1 + a^2)$, $\dfrac{1 + a}{1 + a^2}$, $\dfrac{1 - a}{1 + a}$, $\dfrac{1 + a^2}{1 - a}$.

2. Determine the six roots of $\sqrt[6]{1}$.

3. Given: $I_a = 8 - j6$; $I_b = -13 - j10$; $I_c = 2 + j10$. Find: I_1, I_2, and I_0, and check.

4. Given: $E_a = 150.4$; $E_b = 220\epsilon^{j250}$; $E_c = 220\epsilon^{-j110}$. Find: E_1, E_2, and E_0, and check.

5. Given: $I_a = 0 + j20$; $I_b = 20 + j0$; $I_c = 0$. Find: I_1, I_2, and I_0, and check.

6. Given: $I_a = 0 + j100$; $I_b = -86.6 - j50$; $I_c = +86.6 - j50$. Find: I_1, I_2, and I_0.

7. The a phase of a three-phase circuit is open-circuited, and currents $I_b = I$ and $I_c = -I$ flow in the two other phases. Determine I_1, I_2, and I_0.

8. The b phase of a three-phase circuit is open-circuited, and currents $I_c = I$ and $I_a = -I$ flow in the two other phases. Determine I_1, I_2, and I_0, and compare with Prob. **7.**

9. Assume that $I_a = 100$ amp., and $I_b = I_c = -50$ amp. What are the sequence currents?

10. The line-to-line voltages of a three-phase system measure 100, 150, and 200 volts. Find the magnitudes of the positive- and negative-sequence components of the delta voltages and the star voltages.

11. The loss in three resistances of R in each phase is equal to R multiplied by the sum of the squares of the currents in each of the phases. Show that the total power loss P is $P = R(\bar{I}_a^2 + \bar{I}_b^2 + \bar{I}_c^2) = 3R(\bar{I}_1^2 + \bar{I}_2^2 + \bar{I}_0^2)$.

12. A grounded-neutral system has a positive-sequence voltage of E_{a1}; show that if the ground be removed and one phase wire be grounded, the positive-sequence voltage remains unchanged.

13. A 13,200-volt delta-connected distribution system is grounded on one phase. What is the magnitude of the zero-sequence voltage?

14. Three resistors of 5, 10 and 20 ohms are connected in delta across phases A, B and C, respectively, of a balanced 100-volt system. What are the sequence components of currents in the resistors and in the supply lines?

CHAPTER III

CALCULATION OF UNBALANCED FAULTS

Probably the most important application of symmetrical components is the calculation of unbalanced faults on commercial systems. These systems are usually symmetrical except for the fault itself, so that the sequences do not react upon each other except as related to each other by the particular characteristics of the fault. The calculation of the currents and voltages under unbalanced conditions requires a knowledge of the generated voltages and the impedances of the network elements to the different sequences. For the present, it will be assumed that these constants are known. Their determination will be taken up in detail in subsequent chapters. The present chapter extends the *single-phase* method of calculating polyphase systems to the calculation of unbalanced faults utilizing the principles of symmetrical components.

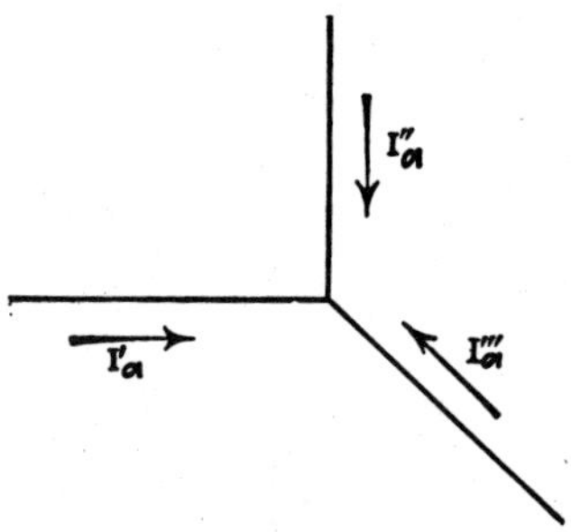

FIG. 9.—Phase *a* currents at a junction point.

11. Kirchhoff's First Law.

The analysis so far considered only one branch or network element. In this element the voltage drops of the various sequences are dependent only upon the currents of the individual sequences and the corresponding impedances. What will be the result when a group or network is considered?

At each junction point of a network, for example, the junction in Fig. 9 in which three circuits converge, by Kirchhoff's first law the sum of the currents in all of the conductors equals zero:

$$\left.\begin{aligned} I_a' + I_a'' + I_a''' &= 0 \\ I_b' + I_b'' + I_b''' &= 0 \\ I_c' + I_c'' + I_c''' &= 0 \end{aligned}\right\} \quad (37)$$

These equations may be expanded by substituting the sequence components for the phase values in accordance with equations (24), (25), and (26), with the result that

$$(I_{a0}' + I_{a0}'' + I_{a0}''') + (I_{a1}' + I_{a1}'' + I_{a1}''') + (I_{a2}' + I_{a2}'' + I_{a2}''') = 0 \quad (38)$$

$$(I_{a0}' + I_{a0}'' + I_{a0}''') + a^2(I_{a1}' + I_{a1}'' + I_{a1}''') + a(I_{a2}' + I_{a2}'' + I_{a2}''') = 0 \quad (39)$$

$$(I_{a0}' + I_{a0}'' + I_{a0}''') + a(I_{a1}' + I_{a1}'' + I_{a1}''') + a^2(I_{a2}' + I_{a2}'' + I_{a2}''') = 0 \quad (40)$$

The addition of equations (38), (39), and (40), remembering that $1 + a^2 + a = 0$, gives

$$I_{a0}' + I_{a0}'' + I_{a0}''' = 0 \quad (41)$$

Upon multiplying equation (39) by a and equation (40) by a^2 and adding them to equation (38) one obtains the following result

$$I_{a1}' + I_{a1}'' + I_{a1}''' = 0 \quad (42)$$

Similarly, multiplying (39) by a^2, (40) by a and adding to (38), we find

$$I_{a2}' + I_{a2}'' + I_{a2}''' = 0 \quad (43)$$

The equations (41), (42), and (43) show that the various sequence components at any junction must individually add up to zero; each sequence of currents obey Kirchhoff's first law separately. This proof has been given for three circuits, but it is apparent that it is a perfectly general relation which must always hold at any junction and may be extended to any number of conductors.

12. Kirchhoff's Second Law.

Consider a general symmetrical network. Since the voltage drops of each sequence, as has been shown in Chap. II, may be considered separately, the sum of the voltage drops of any sequence around any closed circuit must be equal to the e.m.fs. of that same sequence. The network phenomena therefore obey Kirchhoff's second law. This holds true for all points in the system including the point of fault if this point be considered as a source of zero-, negative-, and positive-sequence voltage. The system may therefore be considered as three separate and distinct networks: one for positive-sequence, one for negative-sequence, and the third for zero-sequence. The tie between these systems will be the terminal conditions at the point of fault.

13. Sequence Networks.

It has been shown that each of the individual sequences may be considered independently and since each of the sequence networks involves symmetrical currents and voltages, and impedances in the three phases, each of the sequence networks may be solved by the single-phase method that is commonly used for balanced three-phase systems. The conception of the sequence networks thus makes it possible to solve unbalanced three-phase problems by methods similar to the single-phase solution of balanced problems. These sequence networks may be represented by single-line diagrams of the conventional type in which the currents represent line currents, the voltages line-to-neutral voltages, and the impedances the corresponding star impedances. Each network will thus have its own line conductors and neutral point between which the line-to-neutral voltages are measured and shunt loads are connected. Lines with distributed constants may be represented by series impedances in the lines and shunt impedances between the lines and the neutral of the individual networks.

The positive-sequence network is in all respects identical with the usual networks considered; the resistances and reactances are the values usually given to calculate line regulation. Each synchronous machine must be considered as a source of e.m.f which may vary in magnitude and phase position, depending upon the distribution of power and reactive volt-amperes just previous to the application of the fault. The positive-sequence voltage at the point of fault will drop, the amount being conditioned upon the type of fault; for three-phase faults, it will be zero; for double line-to-ground faults, line-to-line faults, and single line-to-ground faults, it will be higher in the order stated The exact value is calculable and will be determined.

The negative-sequence network is in general quite similar to the positive-sequence network, in fact the number of branches is the same. Because positive-sequence voltages only are generated in the synchronous machines, this network will contain no sources of e.m.f. except the fictitious one at the point of fault.

The zero-sequence network will likewise be free of internal voltages, the flow of current resulting from the voltage at the point of fault. The impedances to this sequence are radically different from those of either the positive- or negative-sequences. The line impedances are those obtained by imagining the three conductors connected together, the ground forming the return conductor. In the zero-sequence network, part of the impedance may be in the outgoing conductor and part in the earth return. However, the total impedance is the only factor affecting current distribution and line-to-ground voltages at the terminals of the network. For the calculation of these quantities the total zero-sequence impedance may be inserted in one conductor of a single-line diagram which becomes similar to the corresponding diagrams for the positive- and negative-sequence networks, the ground constituting a bus of zero potential. The transformer and generator impedances will depend upon the type of connection, whether delta- or star-connected; if star, whether grounded or not. The calculation of the impedances to negative- and zero-sequences will be considered in a subsequent chapter. The number of branches which require consideration in the zero-sequence network is usually less than for the positive- and negative-sequence networks, because of the presence of delta-connected transformers. For the time being, let it be assumed that these quantities are known.

14. Sequence Networks of a Particular System Including Shunt Loads.

As an illustration of the sequence networks, consider the system shown in Fig. 10(*a*). Two generators *A* and *B* at the sending end of a line are connected to two delta-star transformers (only one of which is grounded) which are bussed on the high side. At the receiving end there are two star-star transformer banks (both grounded on the high side, but only one on the low side) connected to separate generators *C* and *D*, one of which is grounded.*

The load is assumed concentrated at the receiving end and is connected as an equivalent star with the neutral point ungrounded. It is represented in the positive-sequence network of (*b*) and the negative-sequence network of (*c*) by shunt impedances connected to the respective neutral buses, but since the

* The unusual connections shown in this diagram are used merely to illustrate the method as applied to different transformer connections. The mutual impedances between lines in the zero-sequence network are neglected.

neutral is ungrounded, the load impedance does not appear in the zero-sequence network of (*d*). If the neutral point of the load at *D* were connected to ground, it would be necessary to add a shunt branch connected to the neutral bus in the zero-sequence network.

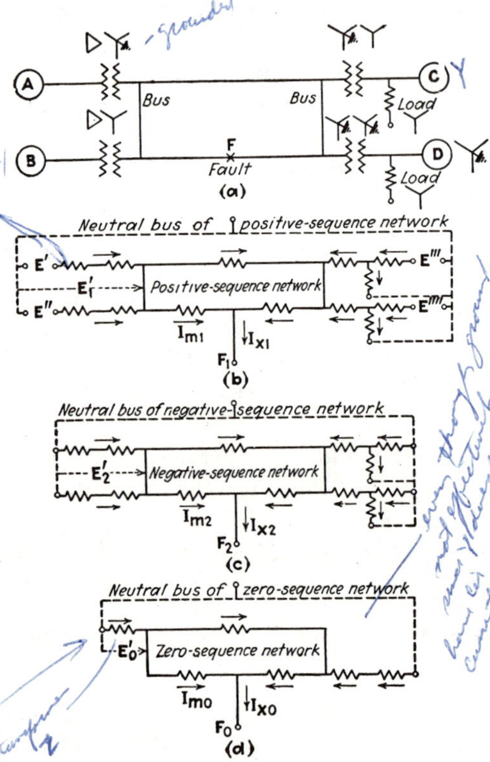

FIG. 10.—Diagrammatic representation of the three sequence networks for a three-phase system.

Assume the fault to occur at the point *F* midway on one of the lines. The single-line impedance diagram for the positive-sequence is shown in Fig. 10(*b*) and those for the negative- and zero-sequences in Fig. 10(*c*) and (*d*). The synthesis of these respective networks will be discussed later. **It is important to retain the positive flow of current, as indicated by the arrows, the same in any branch for all three networks,** as the fundamental equations postulate this condition. E', E'', E''', and E'''' are the internal voltages of the four generators *A*, *B*, *C*, and *D*, respectively. Since positive-sequence voltages only are generated within the machines, the corresponding internal voltages for the negative- and zero-sequence networks are zero. The points associated with the internal voltages may therefore be connected to the neutral buses of the corresponding sequences.

15. Connection of Networks to Represent Faults.

The only tie between these otherwise independent networks is the terminal conditions at the point of fault. The nature of this tie will vary with the character of the fault. Four cases will be considered at this time:

1. Single line-to-ground fault.
2. Double line-to-ground fault.
3. Line-to-line fault.
4. Three-phase fault.

Imagine three short conductors of zero impedance connected to the three line conductors at the point of fault. The terminal conditions imposed by the different types of faults will be applied to these imaginary leads, the potentials to ground of which will be E_x, E_y, and E_z, respectively, and the currents I_x, I_y, and I_z. These imaginary conductors are shown in Fig. 11.

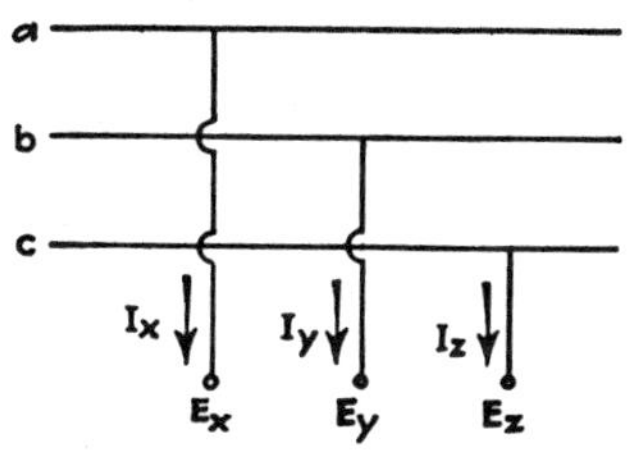

FIG. 11.—Imaginary leads brought out at point of fault.

16. Single Line-to-ground Fault. For this case suppose the a phase to be grounded as indicated in Fig. 12(a). The terminal conditions in this case are:

$$\left.\begin{aligned} E_x &= 0 \\ I_y &= 0 \\ I_z &= 0 \end{aligned}\right\} \quad (44)$$

From equations (27), (28), and (29) of Sec. 6, the symmetrical components of current are:

$$I_{x0} = \tfrac{1}{3}(I_x + I_y + I_z) = \tfrac{1}{3}(I_x) \quad (45)$$
$$I_{x1} = \tfrac{1}{3}(I_x + aI_y + a^2I_z) = \tfrac{1}{3}(I_x) \quad (46)$$
$$I_{x2} = \tfrac{1}{3}(I_x + a^2I_y + aI_z) = \tfrac{1}{3}(I_x) \quad (47)$$

or

$$I_{x0} = I_{x2} = I_{x1} \quad (48)$$

Also

$$E_x = E_{x0} + E_{x1} + E_{x2} = 0 \quad (49)$$

Now from equation (48) it can be seen that the currents at F_1, F_2, and F_0 (Fig. 10) in the three networks for the three sequences are equal, and if the point F_1 of the positive-sequence network be connected to the neutral bus of the negative-sequence network, and the point F_2 in the negative-sequence network be connected to the neutral bus of the zero-sequence network, as shown in Fig. 12, the three networks will be connected in series and the same total current must flow through the three networks. Since the current distribution within each of the networks is determined solely by the impedances in the individual networks, this combination of the networks enables one to determine the current distribution in any branch. However, for this connection, to determine the current completely, the relation expressed in equation (49) must also be satisfied.

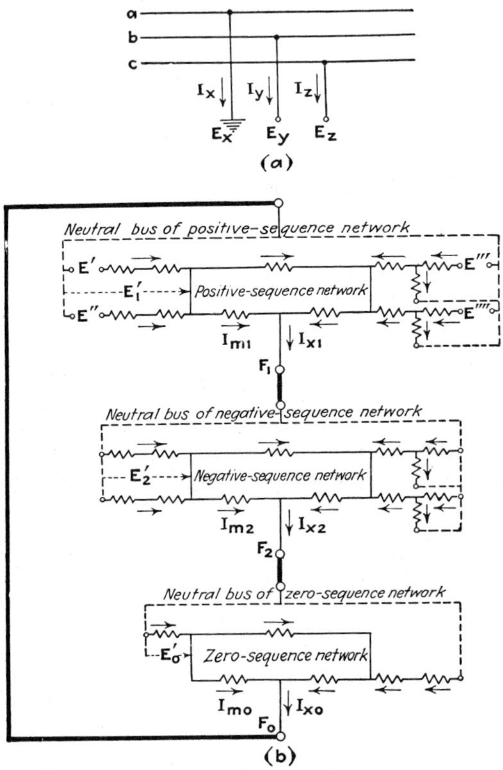

FIG. 12.—Interconnection of the three sequence networks for a line-to-ground fault on one conductor.

Now starting from any one of the internal voltages of Fig. 12, say E', and tracing one's way through the network by any path to F_1, it can be seen that the positive-sequence voltage E_{x1} at F_1 is equal to

$$E_{x1} = E' - \Sigma \text{ positive-sequence voltage drops from } E' \text{ to the point } F_1 \quad (50)$$

Similarly with the negative-sequence network, starting at any point of the neutral bus and tracing one's way to F_2 it will be found that

$$E_{x2} = 0 - \Sigma \text{ negative-sequence voltage drops to the point } F_2 \quad (51)$$

and for the zero-sequence network

$$E_{x0} = 0 - \Sigma \text{ zero-sequence voltage drops to the point } F_0 \quad (52)$$

Adding these three voltages and equating to zero as indicated by equation (49):

$$\left.\begin{aligned} 0 = E' &- \Sigma \text{ voltage drops from } E' \text{ to } F_1 \text{ in positive-sequence network} \\ &- \Sigma \text{ voltage drops from neutral bus to } F_2 \text{ in negative-sequence network} \\ &- \Sigma \text{ voltage drops from neutral bus to } F_0 \text{ in zero-sequence network} \end{aligned}\right\} \quad (53)$$

or

$$\left.\begin{aligned} E' &= \Sigma \text{ voltage drops from } E' \text{ to } F_1 \text{ in positive-sequence network} \\ &+ \Sigma \text{ voltage drops from neutral bus to } F_2 \text{ in negative-sequence network} \\ &+ \Sigma \text{ voltage drops from neutral bus to } F_0 \text{ in zero-sequence network} \end{aligned}\right\} \quad (54)$$

If the point F_0 be connected to the neutral bus of the positive-sequence network, it can be seen that equation (54), and consequently also equation (49), will be satisfied, and the connection of Fig. 12 will completely determine the flow of current. The positive-, negative-, and zero-sequence currents in any branch are equal to the actual currents flowing in the respective branches in the three networks.

The determination of the sequence and phase voltages and currents at any point in the system will be discussed in connection with the application of the calculating board. This will insure a better picture of the correlation of the different networks and provide a better conception of the physical significance of the various quantities.

Since the only e.m.fs. generated in the synchronous machines* are those of the positive-sequence, the question is frequently raised as to the origin of the negative- and zero-sequence voltages that appear throughout the network. Any unbalanced condition such as a line-to-ground fault requires that positive-sequence currents and currents of other sequences flow. The negative- and zero-sequence currents then distribute themselves throughout the system within their respective networks. In each network, as for example the negative-sequence, the impedance drops produced by these currents give rise to voltages of the correspond-

* For special applications such as certain types of phase-balancers, a negative-sequence e.m.f. may be generated. Such cases may be solved by including the negative-sequence generated e.m.f. in the negative-sequence network as is done for the positive-sequence network.

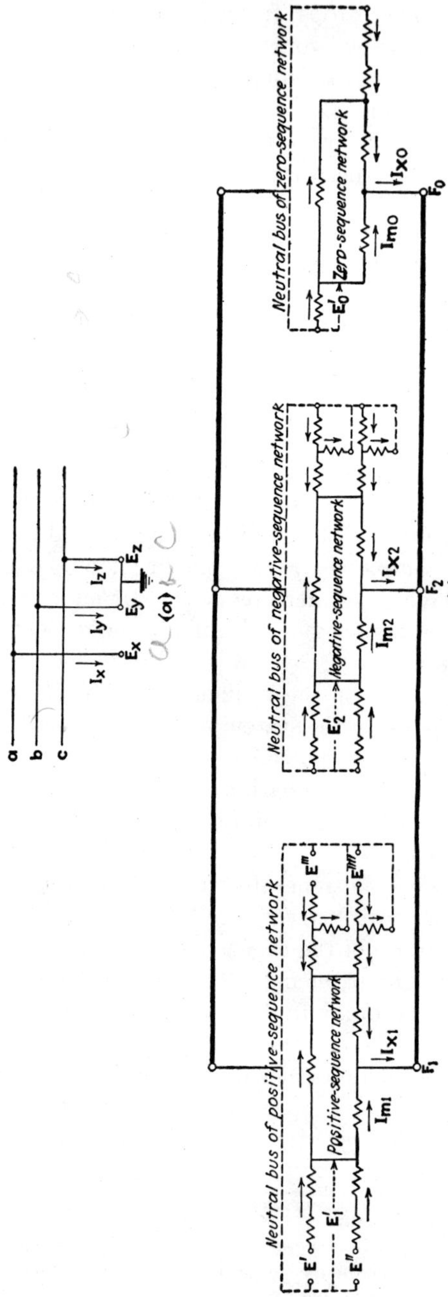

FIG. 13.—Interconnection of the three sequence networks for a double line-to-ground fault.

ing sequence. These voltages are in general a maximum at the fault point and decrease as the neutral bus is approached.

17. Double Line-to-ground Fault.

For this case assume both phases b and c faulted to ground simultaneously as shown in Fig. 13(a). The terminal conditions for this case are then

$$\left.\begin{aligned} E_y &= 0 \\ E_z &= 0 \\ I_x &= 0 \end{aligned}\right\} \tag{55}$$

From equations (27), (28), and (29),

$$\left.\begin{aligned} E_{x0} &= \tfrac{1}{3}(E_x + E_y + E_z) = \tfrac{1}{3}E_x \\ E_{x1} &= \tfrac{1}{3}(E_x + aE_y + a^2E_z) = \tfrac{1}{3}E_x \\ E_{x2} &= \tfrac{1}{3}(E_x + a^2E_y + aE_z) = \tfrac{1}{3}E_x \end{aligned}\right\} \tag{56}$$

or

$$E_{x0} = E_{x1} = E_{x2} \tag{57}$$

Also

$$I_x = I_{x0} + I_{x1} + I_{x2} = 0 \tag{58}$$

Equations (57) and (58) define the conditions which must be fulfilled at the terminals for this case.

Connecting the three sequence networks of Fig. 10, as shown in Fig. 13(b), it will be seen that the two above conditions are fulfilled; equation (58) is fulfilled at the junction of the three networks, and equation (57) is fulfilled because the three voltages are measured across the same terminals and must therefore be equal to each other.

18. Line-to-line Fault.

Assume the fault to occur between phases b and c. The terminal conditions for Fig. 14(a) are then

$$\left.\begin{aligned} I_x &= 0 \\ I_y &= -I_z \\ E_y &= E_z \end{aligned}\right\} \tag{59}$$

The sequence components are then

$$I_{x0} = \tfrac{1}{3}(I_x + I_y + I_z) = \tfrac{1}{3}(0 - I_z + I_z) = 0 \tag{60}$$

$$\begin{aligned} I_{x1} &= \tfrac{1}{3}(I_x + aI_y + a^2I_z) = \tfrac{1}{3}(0 - aI_z + a^2I_z) \\ &= \frac{-a + a^2}{3} I_z \end{aligned} \tag{61}$$

$$\begin{aligned} I_{x2} &= \tfrac{1}{3}(I_x + a^2I_y + aI_z) = \tfrac{1}{3}(0 - a^2I_z + aI_z) = \\ &= \frac{+a - a^2}{3} I_z = -I_{x1} \end{aligned} \tag{62}$$

Since $I_{x0} = 0$, it follows that E_{x0} must also be equal to zero. Furthermore,

$$\left.\begin{aligned} E_{x1} &= \tfrac{1}{3}[E_x + aE_y + a^2E_z] \\ \text{and} \qquad & \\ E_{x2} &= \tfrac{1}{3}[E_x + a^2E_y + aE_z] \end{aligned}\right\} \quad (63)$$

Then, substituting in these equations the relations, $E_y = E_z$ and $E_{x0} = 0$:

$$E_{x1} = \tfrac{1}{3}[E_x + (a + a^2)E_z] \quad (64)$$

$$E_{x2} = \tfrac{1}{3}[E_x + (a + a^2)E_z] \quad (65)$$

$$= E_{x1} \quad (66)$$

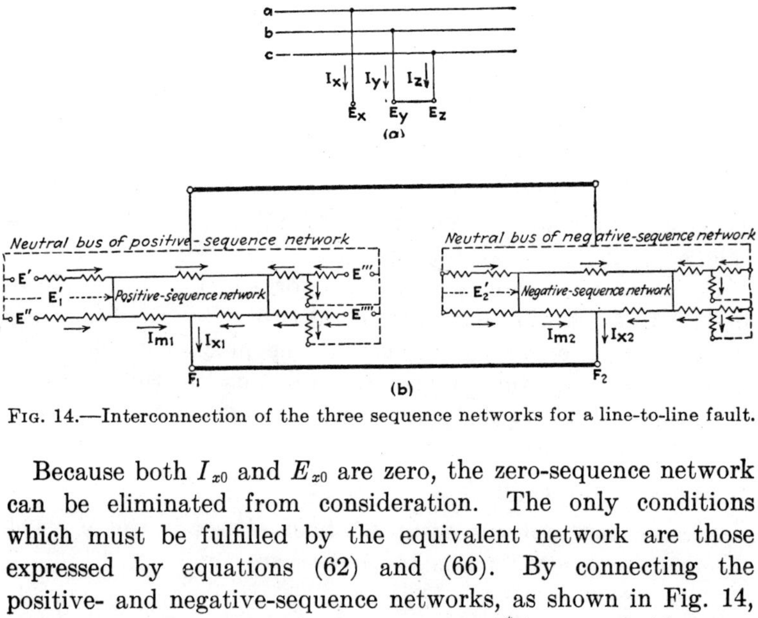

FIG. 14.—Interconnection of the three sequence networks for a line-to-line fault.

Because both I_{x0} and E_{x0} are zero, the zero-sequence network can be eliminated from consideration. The only conditions which must be fulfilled by the equivalent network are those expressed by equations (62) and (66). By connecting the positive- and negative-sequence networks, as shown in Fig. 14, it may be seen immediately that equation (62) is satisfied because I_{x1} must always equal $-I_{x2}$. Equation (66) is satisfied because the equal voltages expressed thereby are measured between the same points and hence must be equal.

19. Three-phase Fault.

This case is the simplest and the one most familiar to persons calculating short-circuits for the application of circuit-breakers and relays. The equivalent diagram reduces to that shown in Fig. 15. Since the system remains balanced, the negative- and zero-sequence networks do not enter the problem.

20. Application to Calculating Boards.

The calculation of system faults by the methods just described may be greatly facilitated by the use of calculating boards.* Figure 16 shows a typical alternating-current calculating board used for power-system studies. This board is made up of a large number of adjustable resistance, reactance, and capacitance branches to

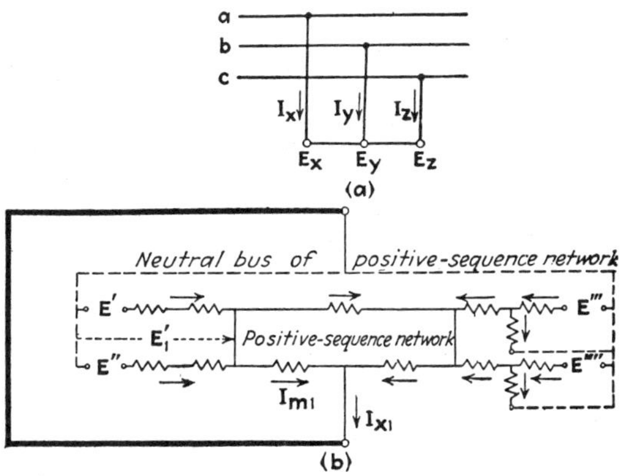

FIG. 15.—Interconnection of the three sequence networks for a three-phase fault.

FIG. 16.—Alternating-current Calculating Board: 179 circuit board of the Commonwealth Edison Company as built by the Westinghouse Electric and Manufacturing Company.

represent the corresponding elements of the actual power system. Special branches are designed to represent inductive coupling

* TRAVERS, H. A., and W. W. PARKER, An Alternating-current Calculating Board, *Elec. Jour.*, pp. 266–270, May, 1930.

between branches. Several sources of e.m.f. are provided with means for adjusting their magnitudes and phase relations. Metering equipment is provided which is suitable for measuring the magnitude and relative phase position of voltage and current vectors as well as power and reactive volt-ampere quantities. The switching arrangements make possible the convenient connection of network branches to represent a network or the combination of the three sequence networks.

By appropriate connection, the alternating-current calculating board may be used to represent the different types of faults. Regardless of the manner in which the sequence networks are connected, the current in phase a at any point can always be obtained by adding the currents of the three networks at the corresponding points. Suppose it is desired to determine the current in the left-hand section of the faulty line in Fig. 10. The current in phase a is determined by adding directly the current I_{m1} from the positive-sequence network, I_{m2} from the negative-sequence network, and I_{m0} from the zero-sequence network. This can be accomplished most conveniently by plugging into the three networks by means of current transformers and totalizing on the secondaries. For phases b and c, networks can be devised which shift the positive- and negative-sequence components 120 and 240 deg.; the shifted currents can then be added to the zero-sequence current. In the absence of such a device, the phase currents can be obtained by adding the quantities analytically after operating, by the unit vectors a and a^2, using the relations

$$I_b = I_{m0} + a^2 I_{m1} + a I_{m2}$$
$$I_c = I_{m0} + a I_{m1} + a^2 I_{m2}$$

The sequence voltage for any point in a network appears only in the corresponding sequence network, so that the manner in which the sequence voltages are measured is independent of the particular way in which the networks are connected together for the representation of the different types of faults. In the positive-sequence network the potential at any point is measured between that point and the neutral bus, indicated by the dotted line in the positive-sequence network of Fig. 10. Expressed differently, it is equal to one of the generated e.m.fs minus the drop from the particular e.m.f. to the point considered. If the value of the particular e.m.f. be zero, the potential at the point will be the negative of the drop from the neutral bus to the point. In the negative-sequence network, since normally no generated e.m.fs. of negative-sequence exist, the negative-sequence potential at any point will be merely the negative of the drop to the point from the negative-sequence neutral bus. Similar considerations apply to the zero-sequence network, the potential at any point being the negative of the drop to the point from the neutral bus of the zero-sequence network. For example, the voltage of phase *a* on the left-hand bus is equal to the vector sum of the voltages E_1', E_2', and E_0', which are indicated in Fig. 10. Note that this voltage is equal to the generated positive-sequence voltage E'' minus the drops in the generator and transformer due to the positive-, negative-, and zero-sequence currents. Thus it may be observed that at the internal voltage of the machine all the drops are zero and the voltage of phase *a* is the generated positive-sequence voltage. Also, for the case of a single line-to-ground fault, as the point under consideration approaches the fault point, the voltage of phase *a* approaches zero. On the alternating-current calculating board the phase *a* voltage can be obtained by direct addition by means of potential transformers, and the voltages of phases *b* and *c* by phase-shifting devices or analytically by means of the fundamental relations connecting phase and sequence voltages.

For short-circuit studies in which only the magnitudes of currents are required it is usually permissible to assume that the generated e.m.fs. are of the same magnitude and in phase so that a single source may be used to represent the generated e.m.f. A further simplification results when it is found sufficiently accurate to assume all the impedances of the same phase angle. In this case the networks may be set up on a direct-current board, the elements of which are pure resistances.

When it is desired to obtain only the zero-sequence currents, or the **residual*** or **ground currents,** the positive- and negative-sequence networks are required only for determining the magni-

* The sum of the zero-sequence currents in the three lines gives $3I_0$; this quantity is sometimes called **residual current.** Part of the total zero-sequence current may flow in a neutral- or ground-wire or cable sheath mounted on the same structure with the line conductors, and part in the ground or remote conductors. The latter is commonly referred to as **ground current** and occasionally as residual current. **Zero-sequence** is invariably used in fault-current calculations. **Residual** current in the past has been used principally in inductive coordination work.

tude of the fault current; the distribution of ground currents is determined of course only by the zero-sequence network.

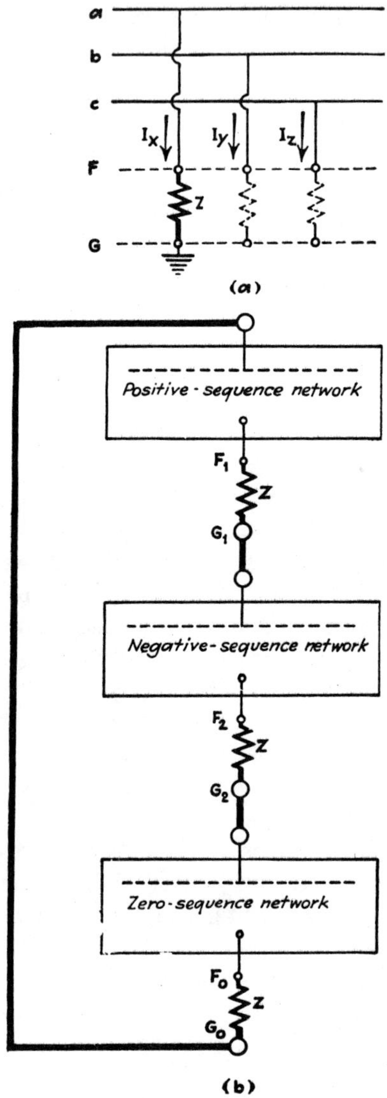

FIG. 17.—Method of connecting fault impedance into network for line-to-ground fault.

Fault impedances may be introduced in the connections of the sequence networks for different types of faults and therefore, as will be shown in the next section, do not affect the method just described for measuring sequence and phase voltages and currents.

21. Representation of Fault Impedances.

In many cases the fault impedance forms a considerable proportion of the total impedance so that the resultant short-circuit current is much smaller than that which would occur if the fault impedance were zero. In addition, the fault impedance may affect the distribution of the current between the negative- and zero-sequence networks, such as occurs for a double line-to-ground fault. The following discussion shows how the equivalent diagram of the system should be altered to include the fault impedance.

Single Line-to-ground Fault. The boundary conditions for this type of fault are indicated in Fig. 17(a), in which Z is the impedance of the fault including the arc, tower footing impedance, etc.

Equations (44) then become:

$$E_x = ZI_x \tag{67}$$

$$\left.\begin{aligned} I_y &= 0 \\ I_z &= 0 \end{aligned}\right\} \quad (67)$$

in which E_x is the voltage of the line conductor x.

The current conditions have not been altered, so that the relations between the different sequence currents obtained for $Z = 0$ still apply. Equation (49), however, showing the relation between the sequence voltages, becomes

$$E_x = E_{x0} + E_{x1} + E_{x2} = ZI_x$$

and since

$$\begin{aligned} I_x &= 3I_{x1} \\ E_x &= 3ZI_{x1} \end{aligned} \quad (68)$$

Equation (53) becomes

$$\left.\begin{aligned} 3ZI_{x1} = E' &- \Sigma \text{ voltage drops from } E' \text{ to } F_1 \text{ in positive-sequence network} \\ &- \Sigma \text{ voltage drops from neutral bus to } F_2 \text{ in negative-sequence network} \\ &- \Sigma \text{ voltage drops from neutral bus to } F_0 \text{ in zero-sequence network} \end{aligned}\right\} \quad (69)$$

$$\left.\begin{aligned} E' = &\ \Sigma \text{ voltage drops from } E' \text{ to } F_1 \text{ in positive-sequence network} \\ &+ \Sigma \text{ voltage drops from neutral bus to } F_2 \text{ in negative-sequence network} \\ &+ \Sigma \text{ voltage drops from neutral bus to } F_0 \text{ in zero-sequence network} + 3ZI_{x1} \end{aligned}\right\} \quad (70)$$

The three networks must therefore be connected to satisfy equations (48) and (70). This will be accomplished by connecting the networks as shown in Fig. 17(*b*).

The three impedances Z can be lumped into one impedance, but by so doing the identity of the points F and G which represent the conductors and the ground is lost. Since no current flows in conductors y and z, the fiction of imaginary impedances Z in these conductors may be carried out to extend the limits to which the system may be considered symmetrical. When this is done each sequence network may be extended by the value of the impedance Z which will be common to each of the networks. The voltages of F and G in each network may then be obtained by merely determining the voltage to the neutral bus in the respective networks.

Double Line-to-ground Fault. The boundary conditions and position of fault impedance for this case are indicated in Fig. 18(a). Equations (55) expressing these boundary conditions become

$$\left.\begin{aligned} E_y &= Z(I_y + I_z) \\ E_z &= Z(I_y + I_z) = E_y \\ I_x &= 0 \end{aligned}\right\} \quad (71)$$

Applying these relations, equations (56) become

$$\left.\begin{aligned} E_{x0} &= \tfrac{1}{3}[E_x + 2E_y] \\ E_{x1} &= \tfrac{1}{3}[E_x + (a^2 + a)E_y] = \tfrac{1}{3}(E_x - E_y) \\ E_{x2} &= \tfrac{1}{3}[E_x + (a + a^2)E_y] = \tfrac{1}{3}(E_x - E_y) \end{aligned}\right\} \quad (72)$$

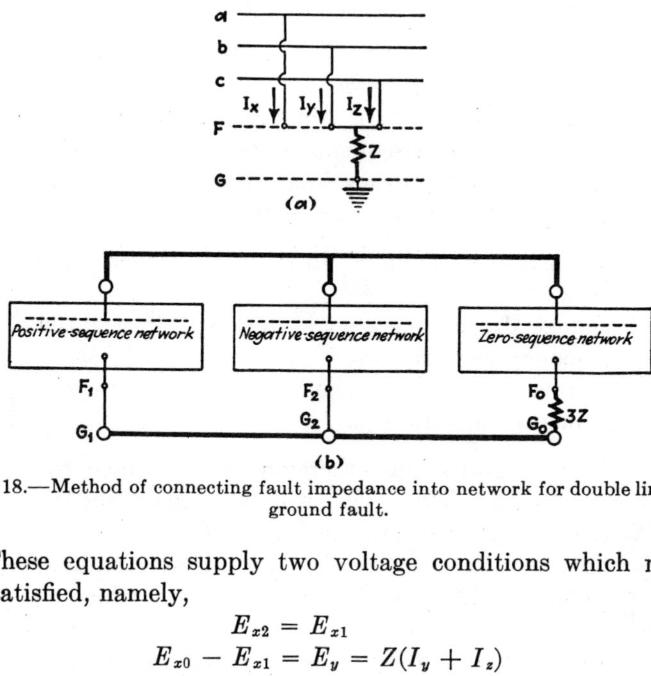

FIG. 18.—Method of connecting fault impedance into network for double line-to-ground fault.

These equations supply two voltage conditions which must be satisfied, namely,

$$E_{x2} = E_{x1} \quad (73)$$

$$E_{x0} - E_{x1} = E_y = Z(I_y + I_z) \quad (74)$$

The voltage E_y from equation (74) may be evaluated in terms of sequence components of currents as follows:

$$I_y = I_{x0} + a^2 I_{x1} + aI_{x2}$$

$$I_z = I_{x0} + aI_{x1} + a^2 I_{x2}$$

Adding

$$I_y + I_z = 2I_{x0} + (a^2 + a)(I_{x1} + I_{x2}) \quad (75)$$

Now from the last of equations (71)

$$I_x = I_{x0} + I_{x1} + I_{x2} = 0 \tag{76}$$

or

$$I_{x1} + I_{x2} = -I_{x0}$$

Substituting this relation in (75)

$$I_y + I_z = 3I_{x0} \tag{77}$$

and (74) becomes

$$E_{x0} - E_{x1} = 3ZI_{x0} \tag{78}$$

Now, since the three networks are sufficient in themselves, but their manner of connection is dependent upon the boundary condition, the necessary and sufficient conditions to be met by the equivalent networks are those expressed by equations (73), (76), and (78). These conditions are satisfied when the networks are connected as shown in Fig. 18(*b*).

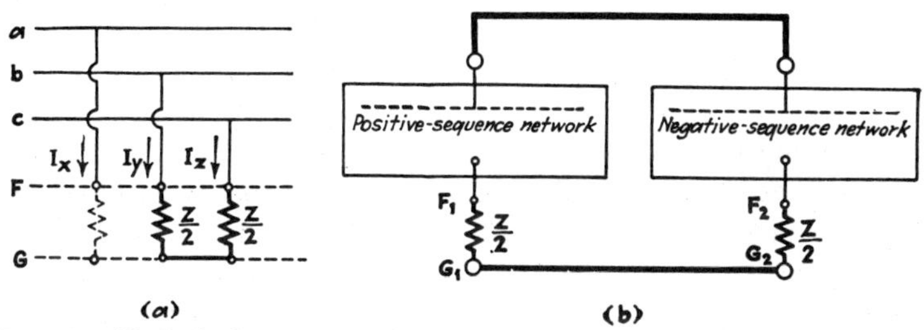

FIG. 19.—Method of connecting fault impedance into network for a line-to-line fault.

Line-to-line Fault. The terminal conditions for this particular fault are indicated in Fig. 19(*a*). Expressed mathematically they are

$$\left.\begin{aligned} I_x &= 0 \\ I_y &= -I_z \\ E_y &= E_z - ZI_z \end{aligned}\right\} \tag{79}$$

in which E_y and E_z are the potentials of the conductors y and z.

Since the current relations are unaltered, equations (60), (61), and (62) still apply. From equation (61)

$$I_z = \frac{3}{-a + a^2} I_{x1}$$

so that

$$E_y = E_z - \frac{3}{-a + a^2} I_{x1} Z \tag{80}$$

From the fundamental relations expressed in equations (63) which connect the phase and the sequence quantities

$$E_{x1} - E_{x2} = \tfrac{1}{3}(a - a^2)E_y + \tfrac{1}{3}(a^2 - a)E_z$$

Substituting E_y from equation (80)

$$\begin{aligned} E_{x1} - E_{x2} &= \frac{1}{3}(a - a^2)E_z - \frac{(a - a^2)Z}{-a + a^2}I_{x1} + \frac{1}{3}(a^2 - a)E_z \\ &= ZI_{x1} \end{aligned} \tag{81}$$

From equations (62) and (81) the two necessary and sufficient conditions which must be met are

$$\left.\begin{aligned} I_{x2} &= -I_{x1} \\ E_{x1} &= E_{x2} + ZI_{x1} \end{aligned}\right\} \tag{82}$$

These are satisfied by the connection shown in Fig. 19(b).

22. Short-circuit of Generators.

To illustrate the application of the more general relations which have been developed, a simple analytical example will be considered. For this purpose let it be desired to determine the voltages and currents during the short-circuit at the terminals of a symmetrical star-connected generator under different types of unbalanced faults. The impedances to positive-, negative-, and zero-sequence currents will be Z_1, Z_2, and Z_0, respectively. This particular example was chosen not only because it is an interesting problem in itself, but also because it furnishes the simplest or elemental form of network which still retains unequal and finite impedances to the three sequences.

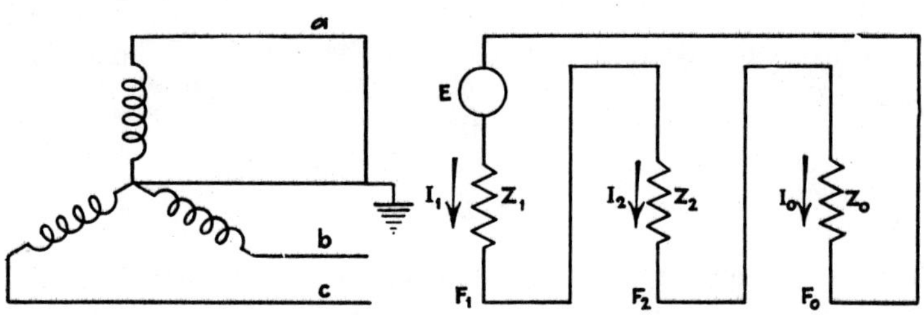

FIG. 20.—Line-to-ground fault on a generator.

Single Line-to-ground Fault. The equivalent diagram for this condition is illustrated in Fig. 20. As can be seen, the currents of the three sequences are all equal to

$$I_1 = I_2 = I_0 = \frac{E}{Z_1 + Z_2 + Z_0} \tag{83}$$

The current in phase a (*i.e.*, the short-circuit current) is

$$\begin{aligned} I_a &= (I_0 + I_1 + I_2) \\ &= \frac{3E}{Z_1 + Z_2 + Z_0} \qquad (84) \\ I_b &= I_0 + a^2 I_1 + a I_2 \\ &= (1 + a^2 + a) I_{a1} = 0 \\ I_c &= I_0 + a I_1 + a^2 I_2 \\ &= (1 + a + a^2) I_{a1} = 0 \end{aligned}$$

The positive-sequence component of the terminal voltage is the voltage between F_1 and F_0 (Figs. 12 and 20), namely,

$$E_1 = E - Z_1 I_1$$

After substituting the value of I_1 from (83),

$$E_1 = \left[1 - \frac{Z_1}{Z_1 + Z_2 + Z_0}\right] E$$

or

$$= \frac{Z_2 + Z_0}{Z_1 + Z_2 + Z_0} E \qquad (85)$$

Similarly the negative- and zero-sequence voltages are

$$\begin{aligned} E_2 &= -Z_2 I_2 \\ &= -\frac{Z_2}{Z_1 + Z_2 + Z_0} E \qquad (86) \end{aligned}$$

and

$$\begin{aligned} E_0 &= -Z_0 J_0 \\ &= -\frac{Z_0}{Z_1 + Z_2 + Z_0} E \qquad (87) \end{aligned}$$

The voltage of phase a is then

$$\begin{aligned} E_a &= E_0 + E_1 + E_2 \\ &= E \frac{Z_2 + Z_0 - Z_2 - Z_0}{Z_1 + Z_2 + Z_0} = 0 \end{aligned}$$

and of phases b and c

$$\begin{aligned} E_b &= E_0 + a^2 E_1 + a E_2 \\ &= a^2 E - \frac{E}{Z_1 + Z_2 + Z_0}(Z_0 + a^2 Z_1 + a Z_2) \\ &= \frac{E}{Z_1 + Z_2 + Z_0}[(a^2 - a) Z_2 + (a^2 - 1) Z_0] \qquad (88) \\ E_c &= E_0 + a E_1 + a^2 E_2 \\ &= a E - \frac{E}{Z_1 + Z_2 + Z_0}(Z_0 + a Z_1 + a^2 Z_2) \end{aligned}$$

$$= \frac{E}{Z_1 + Z_2 + Z_0}[(a - a^2)Z_2 + (a - 1)Z_0] \tag{89}$$

Double Line-to-ground Fault. The equivalent diagram for this case reduces to that shown in Fig. 21. The solution of this network gives

$$I_1 = \frac{E}{Z_1 + \dfrac{Z_0Z_2}{Z_0 + Z_2}} = \frac{(Z_0 + Z_2)}{Z_0Z_1 + Z_1Z_2 + Z_2Z_0}E \tag{90}$$

$$I_2 = -\frac{Z_0}{Z_0 + Z_2}I_1$$

$$= -\frac{Z_0}{Z_0Z_1 + Z_1Z_2 + Z_2Z_0}E \tag{91}$$

$$I_0 = -\frac{Z_2}{Z_0Z_1 + Z_1Z_2 + Z_2Z_0}E \tag{92}$$

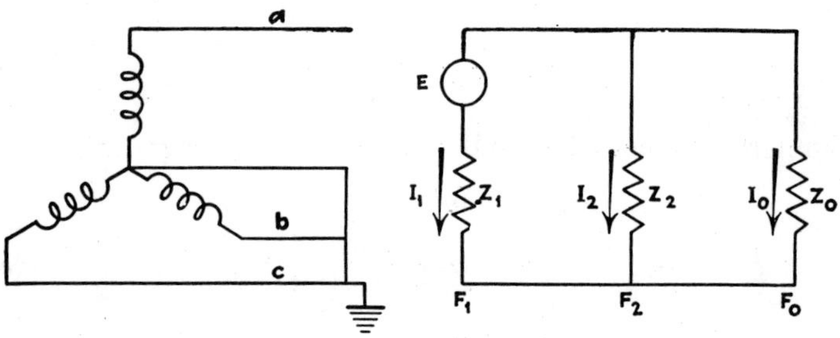

FIG. 21.—Double line-to-ground fault on a generator.

from which

$$I_a = I_0 + I_1 + I_2$$

$$= \frac{E}{Z_0Z_1 + Z_1Z_2 + Z_2Z_0}[-Z_2 + Z_0 + Z_2 - Z_0] = 0$$

$$I_b = I_0 + a^2I_1 + aI_2$$

$$= \frac{E}{Z_0Z_1 + Z_1Z_2 + Z_2Z_0}[-Z_2 + a^2(Z_0 + Z_2) - aZ_0]$$

$$= \frac{E}{Z_0Z_1 + Z_1Z_2 + Z_2Z_0}[(a^2 - 1)Z_2 + (a^2 - a)Z_0] \tag{93}$$

$$I_c = I_0 + aI_1 + a^2I_2$$

$$= \frac{E}{Z_0Z_1 + Z_1Z_2 + Z_2Z_0}[-Z_2 + a(Z_0 + Z_2) - a^2Z_0]$$

$$= \frac{E}{Z_0Z_1 + Z_1Z_2 + Z_2Z_0}[(a - 1)Z_2 + (a - a^2)Z_0] \tag{94}$$

Since the three sequence components of terminal voltage are equal, *i.e.*,

$$E_1 = E_2 = E_0 \tag{95}$$

and

$$E_2 = -I_2Z_2$$
$$= E\frac{Z_0Z_2}{Z_0Z_1 + Z_1Z_2 + Z_2Z_0} \tag{96}$$

then

$$E_a = 3E_2$$
$$= 3E\frac{Z_0Z_2}{Z_0Z_1 + Z_1Z_2 + Z_2Z_0} \tag{97}$$

Line-to-line Fault. The circuit condition to which this case reduces is shown in Fig. 22.

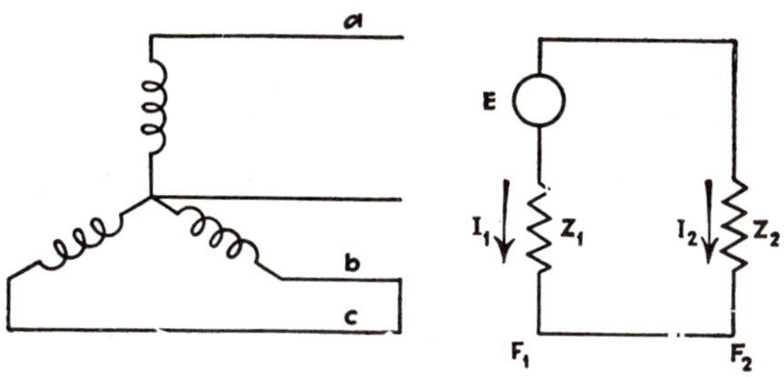

FIG. 22.—Line-to-line fault on a generator.

$$I_1 = \frac{E}{Z_1 + Z_2} \tag{98}$$

$$I_2 = -\frac{E}{Z_1 + Z_2} \tag{99}$$

$$I_0 = 0$$

Therefore

$$I_a = I_0 + I_1 + I_2$$
$$= 0 + \frac{E}{Z_1 + Z_2} - \frac{E}{Z_1 + Z_2} = 0$$
$$I_b = I_0 + a^2I_1 + aI_2$$
$$= 0 + a^2\frac{E}{Z_1 + Z_2} - a\frac{E}{Z_1 + Z_2}$$
$$= (a^2 - a)\frac{E}{Z_1 + Z_2} \tag{100}$$
$$I_c = I_0 + aI_1 + a^2I_2$$
$$= 0 + a\frac{E}{Z_1 + Z_2} - a^2\frac{E}{Z_1 + Z_2}$$
$$= (a - a^2)\frac{E}{Z_1 + Z_2} = -I_b \tag{101}$$

Also

$$E_1 = E_2 = -Z_2 I_2$$

$$= \frac{Z_2}{Z_1 + Z_2} E \qquad (102)$$

Therefore

$$E_a = E_0 + E_1 + E_2$$

$$= \frac{2Z_2}{Z_1 + Z_2} E \qquad (103)$$

$$E_b = E_0 + a^2 E_1 + a E_2$$

$$= \frac{(a^2 + a) Z_2}{Z_1 + Z_2} E$$

$$= -\frac{Z_2}{Z_1 + Z_2} E \qquad (104)$$

$$E_c = E_0 + a E_1 + a^2 E_2$$

$$= \frac{(a + a^2) Z_2}{Z_1 + Z_2} E = E_b \qquad (105)$$

23. Numerical Example of Generator-fault Calculations.

To aid in a more thorough understanding of the different types of generator faults, calculations will be made of an assumed machine for sustained short-circuits. Consider a machine in which

$Z_1 = 0 + j1.0$ (synchronous impedance)
$Z_2 = 0.1 + j0.3$ (impedance to negative-sequence current)
$Z_0 = 0 + j0.1$ (impedance to zero-sequence current)
$E = 100 + j0$

Single Line-to-ground Fault. Referring to Fig. 20:

$$I_0 = I_1 = I_2 = \frac{100}{Z_0 + Z_1 + Z_2} = \frac{100}{0.1 + j1.4}$$

$$= 5.08 - j71.1$$

$$I_a = I_0 + I_1 + I_2 = 3(5.08 - j71.1)$$

$$= 15.24 - j213.3$$

$$E_1 = E - I_1 Z_1$$

$$= 100 - j1.0(5.08 - j71.1)$$

$$= 28.9 - j5.08$$

$$E_2 = -I_2 Z_2$$

$$= -(5.08 - j71.1)(0.1 + j0.3)$$

$$= -21.84 + j5.59$$

$$E_0 = -I_0 Z_0$$

$$= -(5.08 - j71.1)(0 + j0.1)$$

$$= -7.11 - j0.508$$

$$
\begin{aligned}
E_b &= E_0 + a^2E_1 + aE_2 \\
&= (-7.11 - j0.508) + (-0.5 - j0.866)(+28.9 - j5.08) \\
&\qquad + (-0.5 + j0.866)(-21.84 + j5.59) \\
&= (-7.11 - j0.508) + (-18.85 - j22.49) + (6.08 - j21.70) \\
&\qquad = -19.88 - j44.70
\end{aligned}
$$

$$
\begin{aligned}
E_c &= E_0 + aE_1 + a^2E_2 \\
&= (-7.11 - j0.508) + (-0.5 + j0.866)(+28.9 - j5.08) \\
&\qquad + (-0.5 - j0.866)(-21.84 + j5.59) \\
&= (-7.11 - j0.507) + (-10.05 + j27.57) + (15.76 + j16.1) \\
&\qquad = -1.40 + j43.16
\end{aligned}
$$

Double Line-to-ground Fault. For this case, referring to Fig. 21,

$$
\begin{aligned}
\frac{Z_0Z_2}{Z_0 + Z_2} &= \frac{j0.1(0.1 + j0.3)}{j0.1 + 0.1 + j0.3} \\
&= \frac{-0.03 + j0.01}{0.1 + j0.4} \\
&= 0.00588 + j0.0765
\end{aligned}
$$

$$
\begin{aligned}
I_1 &= \frac{E}{Z_1 + \dfrac{Z_0Z_2}{Z_0 + Z_2}} \\
&= \frac{100}{j1.0 + 0.00588 + j0.0765} \\
&= \frac{100}{0.00588 + j1.0765} \\
&= 0.50 - j92.8
\end{aligned}
$$

$$
\begin{aligned}
I_2 &= -\frac{\dfrac{Z_0Z_2}{Z_0 + Z_2}}{Z_2}I_1 \\
&= \frac{-(0.00588 + j0.0765)}{0.1 + j0.3}(0.50 - j92.8) \\
&= -5.58 + j21.81
\end{aligned}
$$

$$
\begin{aligned}
I_0 &= -\frac{\dfrac{Z_0Z_2}{Z_0 + Z_2}}{Z_0}I_1 \\
&= \frac{-(0.00588 + j0.0765)}{+j0.1}(0.50 - j92.8) \\
&= 5.07 + j71.02
\end{aligned}
$$

Combining the sequence components, the following phase currents are obtained:

$$
\begin{aligned}
I_a &= 0 \\
I_b &= I_0 + a^2 I_1 + a I_2 \\
&= (5.07 + j71.02) + (-0.5 - j.866)(0.50 - j92.8) \\
&\qquad + (-0.5 + j.866)(-5.58 + j21.81) \\
&= -91.8 + j101.3 \\
I_c &= I_0 + a I_1 + a^2 I_2 \\
&= (5.07 + j71.02) + (-0.5 + j.866)(0.5 - j92.8) \\
&\qquad + (-0.5 - j.866)(-5.58 + j21.81) \\
&= 106.9 + j111.8
\end{aligned}
$$

The sequence voltages and phase a voltage are

$$
\begin{aligned}
E_1 &= E_2 = E_0 \\
E_2 &= -Z_0 I_0 \\
&= -j0.1(5.07 + j71.02) \\
&= 7.10 - j0.507 \\
E_a &= 3E_2 \\
&= 3(7.10 - j0.507) \\
&= 21.3 - j1.521
\end{aligned}
$$

Line-to-line Fault. Referring to Fig. 22:

$$
\begin{aligned}
I_1 &= \frac{E}{Z_1 + Z_2} \\
&= \frac{100}{j1 + 0.1 + j0.3} \\
&= 5.88 - j76.5 \\
I_2 &= -I_1 \\
&= -5.88 + j76.5 \\
I_0 &= 0 \\
I_a &= I_0 + I_1 + I_2 = 0 \\
I_b &= I_0 + a^2 I_1 + a I_2 = 0 + (a^2 - a) I_1 \\
&= -j1.732 \frac{E}{Z_1 + Z_2} = -j1.732(5.88 - j76.5) \\
&= -132.5 - j10.2 \\
I_c &= I_0 + a I_1 + a^2 I_2 = 0 + (a - a^2) I_1 = -I_b \\
&= 132.5 + j10.2 \\
E_1 &= E_2 = -Z_2 I_2 \\
&= -(0.1 + j0.3)(-5.88 + j76.5) \\
&= 23.54 - j5.89 \\
E_0 &= 0
\end{aligned}
$$

$$\begin{aligned} E_a &= E_0 + E_1 + E_2 \\ &= 0 + 2(23.54 - j5.89) \\ &= 47.08 - j11.78 \\ E_b &= E_0 + a^2E_1 + aE_2 \\ &= 0 + (a^2 + a)E_1 = -E_1 \\ &= -23.54 + j5.89 \\ E_c &= E_0 + aE_1 + a^2E_2 \\ &= 0 + (a + a^2)E_1 = -E_1 \\ &= -23.54 + j5.89 \end{aligned}$$

Problems

1. Assume a system with a sustained supply voltage of 2,300 volts from line-to-neutral and with line impedances $Z_1 = +j10$, $Z_2 = +j8$, and $Z_0 = 10 + j12$. Find the currents in the different phases for different types of faults as follows: (*a*) single line-to-ground; (*b*) double line-to-ground; (*c*) line-to-line; (*d*) three-phase.

2. Assume that a generator with constant internal voltage is subjected to different types of faults and that all resistances are negligible. Also assume that the absolute values of fault current for the different types are as follows: (*a*) three-phase, 1,000 amp.; (*b*) line-to-line, 1,400 amp.; (*c*) line-to-ground, 2,220 amp. If the positive-sequence voltage to neutral is 2,000 volts, what are the positive-, negative-, and zero-sequence impedances?

3. A generator is subjected to a sustained fault with phases *b* and *c* short-circuited together and to neutral. Show that the zero-sequence impedance may be determined from the following relation: $Z_0 = \frac{E_a}{I_n}$ where E_a is the voltage from the unfaulted conductor to neutral and I_n is the neutral current, *i.e.*, the sum of the currents I_b and I_c.

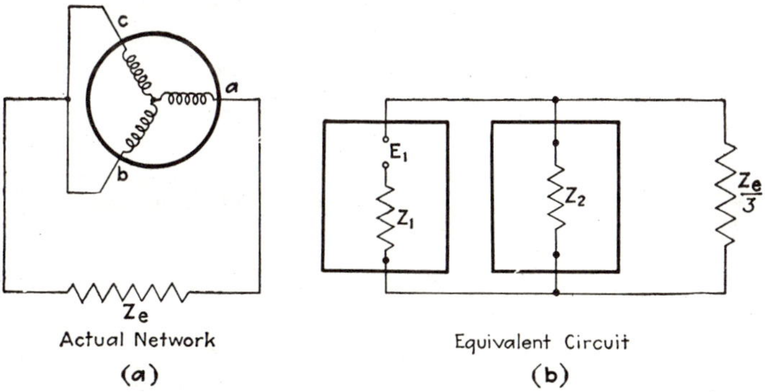

4. Assume a generator in which phases *b* and *c* are short-circuited and connected through impedance Z_e to phase *a* as illustrated in (*a*) of the accompanying figure. If the generator impedances are Z_1 and Z_2, and the internal voltage is E_1, show that the equivalent circuit is given in (*b*).

CHAPTER IV

ILLUSTRATION OF UNBALANCED FAULT CALCULATIONS

The calculation of system performance at times of unsymmetrical faults is one of the most important applications of the method of symmetrical components. Such calculations are important from the standpoint of circuit-breaker and relay application, mechanical and thermal limitations of apparatus, and system stability.

In Chaps. II and III, a method was developed for the calculation of unsymmetrical faults on systems in which the impedances may be considered as symmetrical in the three phases. It was shown that the three sequences could be represented individually by separate networks, the connection between these networks being determined by the particular type of fault considered. An example has already been given showing the application of these methods to the simplest type of network, namely a generator, which was completely solved analytically and then numerically by substitution in the derived formulas. Such analytical expressions for practical networks, however, become so unwieldy that recourse is usually taken to the solution of the network with numerical circuit constants. The present chapter presents the solution of a network of sufficient complexity to illustrate some of the necessary steps involved in the simplification of the network and also the manipulation of the sequence networks to obtain the current and voltage distribution in the different phases.

24. Conventional Simplifying Assumptions in Short-circuit Calculations.

A large percentage of the problems, within the accuracy ordinarily required, can be solved by making the simplifying assumptions enumerated below. These simplifying assumptions are those which are normally employed for the calculation of short-circuit currents by means of the direct-current calculating board.

1. All generated e.m.fs. are of equal magnitude and in phase.*

2. All resistances are neglected, only the reactance of impedances or an equivalent reactance being used.

3. All shunt-impedance branches, such as the equivalent circuits for representing magnetizing currents of transformers and charging current of transmission lines, as well as equivalent impedance branches representing normal loads, are neglected.

4. Mutual† impedance between lines is neglected.

Consideration will be given subsequently to methods for avoiding the limitations imposed by the above assumptions.

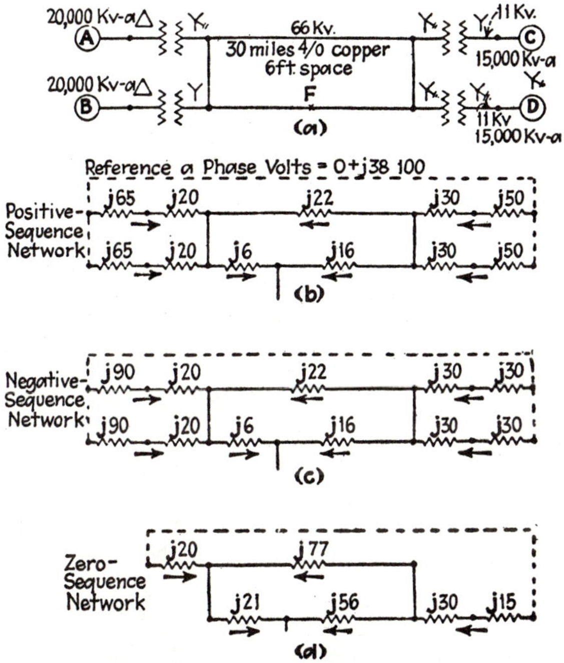

FIG. 23.—Single-line diagram of a network and the corresponding positive-, negative-, and zero-sequence networks.

25. Layout of Assumed System.

To illustrate the method for the calculation of faults in which these assumptions are justified, a numerical example will be worked out. For this purpose the network‡ shown in Fig. 23

* The case of non-identical generated e.m.fs. is treated in Sec. 113.

† The zero-sequence mutual impedance between lines, if the fault occurs between bussing points, must be taken into consideration for faults to ground.

‡ This network corresponds to Fig. 10 of Chap. III, with the omission of the shunt loads.

is selected as having a degree of complexity sufficient to illustrate some of the necessary steps of network simplification. This network consists essentially of two generating stations connected through two 30-mile, 66-kv. transmission lines. The lines may be considered as on separate rights-of-way, to justify the assumption that mutual impedance may be neglected. Faults of different characters will be considered as occurring at the point F, about eight miles from the left-hand bus. The unusual transformer connections are employed merely to illustrate the manipulation of the zero-sequence currents.

26. Single-line Diagram of the System and the Three Sequence Networks.

The first step is to set up the network impedances in the form of a single-line diagram for the three separate sequences. All the impedances should be reduced to a common voltage, which in this case is the 66-kv. transmission voltage. It will be assumed that these impedances are known and of the values indicated in Fig. 23(*b*), (*c*), and (*d*). The negative-sequence network is identical with the positive-sequence network except for the impedance of the synchronous machines. With the exception of the transformer impedances, the zero-sequence impedances are, in general, different from the impedances of either the positive- or negative-sequence. Also, the zero-sequence diagram generally possesses a smaller number of branches. From what has already been said, branches which do not permit the flow of neutral or ground current can be neglected in the consideration of the zero-sequence network. Also, in the solution of all network problems, one is at liberty to choose an arbitrary direction of current flow in the individual branches in setting up the problem. This is a freedom permissible in these problems with the limitations that the *current flow must be the same in the corresponding branches of all three sequences.* This point was discussed in a previous chapter, but it is again introduced because of the ease with which this very important point might be overlooked.

27. Outline of Short-circuit Calculations.

The general outline of the calculations is somewhat as follows:

1. The determination of the single equivalent impedance of the positive-sequence network and the distribution factors

giving the current in the individual branches for unit positive-sequence current at the point of fault.

2. The corresponding equivalent impedance and distribution factors for the negative-sequence network.

3. The corresponding equivalent impedance and distribution factors for the zero-sequence network.

The above calculations will be common to all types of faults. The remainder of the calculations are different for the different types of faults, but in all cases the necessary steps are as follows:

4. Grouping of equivalent impedances and the determination of the sequence currents at the fault.

5. Determination of the sequence-current distribution throughout the three networks by the application of the distribution factors to the sequence currents at the fault.

6. Synthesis of phase currents from the sequence components.

7. Determination of sequence voltages throughout the three networks from the current distribution and branch impedances.

8. Synthesis of phase voltages and line-to-line voltages from the sequence components.

9. The conversion of the equivalent voltages and currents from the common voltage base to the actual voltage corresponding to the actual transformer connections.

28. Equivalent Impedances and Distribution Factors.

The first step is the determination of the single **equivalent impedance** and the current **distribution factors** for each sequence network. These two calculations are conveniently made together taking the three sequences in turn.

The calculation of the **positive-sequence equivalent impedance** from the sources to the fault will now be undertaken. Figure 24(*a*) shows the positive-sequence network of Fig. 23(*b*) redrawn and simplified by combining the generator and transformer impedances. Figure 24(*a*) may be simplified to (*b*) by reducing by conventional methods;* the two parallel impedances at the extreme right to a single impedance, the two parallel impedances at the extreme left to a single impedance, and the remaining impedances, which are connected in delta, to an equivalent group of star-connected impedances. By combining impedances in series or in parallel successively, it is possible to reduce the network of Fig. 24(*b*) to the form shown in (*c*) and (*d*), giving finally

* See Appendix, Section VI.

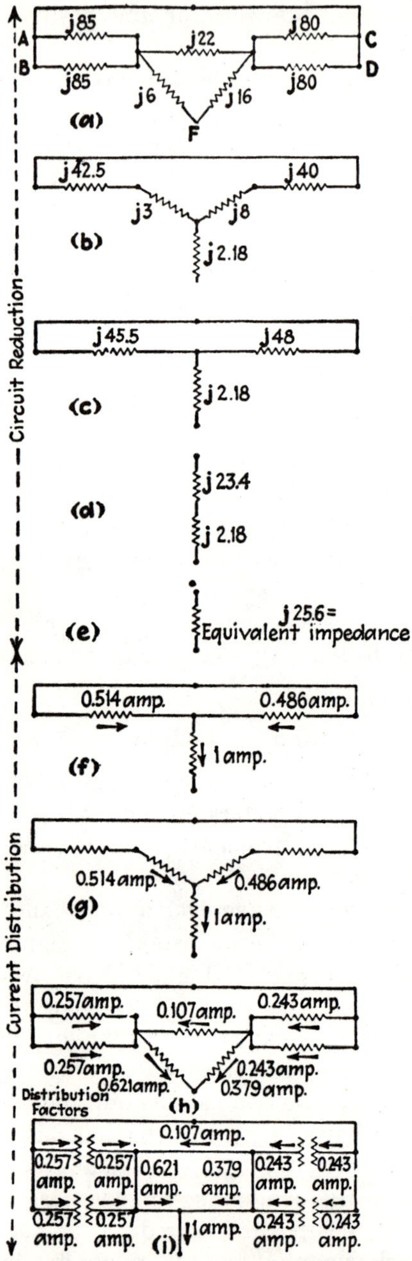

FIG. 24.—Circuit reduction and distribution factors for the positive-sequence network.

the **equivalent impedance** of the positive-sequence network shown in (*e*) and having a value of $j25.6$.

The distribution factor of any branch for the positive-sequence network is defined as the ratio of the positive-sequence current in that branch to the positive-sequence current at the fault. The most convenient way to obtain these factors is to assume unit positive-sequence current at the point of fault and determine the distribution of current by working back through the network. The various steps in this procedure are indicated in Fig. 24(*e*) to (*i*). The division of current in (*f*) and (*g*) can be obtained from the familiar fact that the current divides inversely as the impedances.

In the transition from Fig. 24(*g*) to (*h*), it is necessary to find the current distribution in a delta group of impedances from the known current distribution in the equivalent star group. This may be accomplished by making use of the fact that the potential difference between any two terminals of the delta group is the potential difference between the two corresponding terminals of the star group. The potential difference between any two terminals in the star group can be calculated from the impedances and the known current distribution, after which, by dividing the potential difference between any two terminals by the impedance of the delta group connected across these terminals, the current in the particular impedance branch can be determined. The remaining current divisions may be computed in a similar manner or by equating the currents at various junction points. For example, the potential difference between the left-hand bus and the point F, from Fig. 24(b) and (g), is

$$(1.0)(j2.18) + (0.514)(j3.0) = j3.722 \text{ volts}$$

from which the current in the delta branch between these two points in Fig. 24(h) is

$$\frac{j3.722}{j6.0} = +0.621 \text{ amp.}$$

For clearness, the distribution of the positive-sequence current throughout the system for unit positive-sequence current at the fault, or the distribution factor, is shown in Fig. 24(i).

The equivalent impedance and the distribution factors for the **negative-sequence** network, shown in Fig. 23(c), may be obtained in a similar manner. The principal steps involved are indicated in Fig. 25, of which (e) gives the equivalent impedance and (f) gives the distribution factors. Similar data for the **zero-sequence** network are shown in Fig. 26.

The distribution in all three sequences is based on a system voltage of 66 kv. and therefore does not take into consideration transformer ratios or phase shift in star-delta transformers. These are discussed subsequently.

29. Single Line-to-ground Fault Calculations.

Having obtained the equivalent impedance and the distribution factors for the different sequences, the next step is to determine the value of the sequence current at the fault. This will vary with the particular type of fault under consideration. Probably the most important type of unbalanced faults on transmission systems is the line-to-ground fault. As this fault involves all three sequences, the calculations will be carried out in detail in order to illustrate the steps essential to the general method.

30. Determination of Sequence Currents.

For a line-to-ground fault the three sequence networks must be connected in series,* and as corollary of this connection the

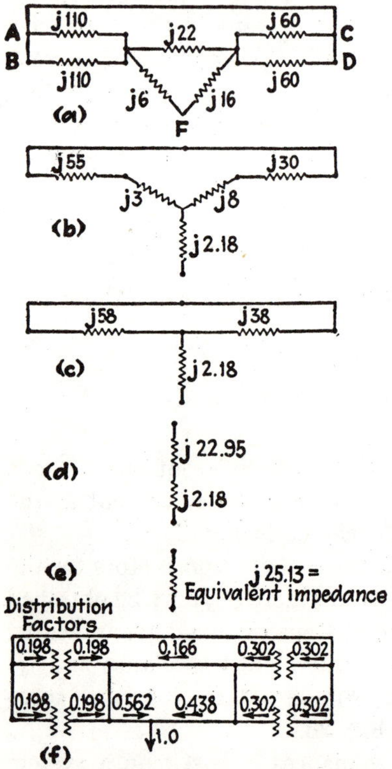

FIG. 25.—Circuit reduction and distribution factors for the negative-sequence network.

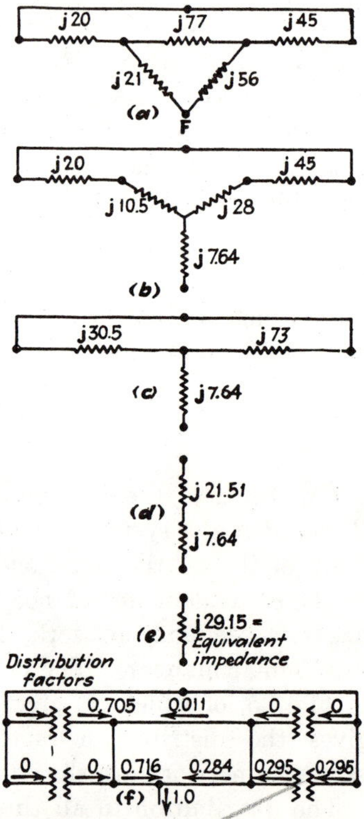

FIG. 26.—Circuit reduction and distribution factors for the zero-sequence network.

three sequence currents at the fault are necessarily equal. The value of these currents, since only the terminal currents are required, may be obtained by replacing the three sequence networks by their equivalent impedances. The equivalent network for the determination of the sequence currents at the fault, therefore, reduces to that shown in Fig. 27, from which

$$I_1 = I_2 = I_0 = \frac{E}{Z_1 + Z_2 + Z_0} \tag{106}$$

Now for the particular case under consideration, in which all the impedances are purely reactive, the equivalent impedances

* This was developed in Chap. III and is illustrated in Fig. 12.

will be purely imaginary. If the phase of the voltage E be taken as purely imaginary so that $E = j\bar{E}$, the values of I_1, I_2, and I_0 are all real and the currents in all branches are real. This arbitrary choice of phase of the reference voltage is taken to avoid the awkward negative imaginary terms for currents that result if the reference voltage were located along the real axis. Using this assumption gives identically the type of solution obtained with the direct-current calculating board, in which the system reactances are replaced by resistances in the board. With this voltage reference

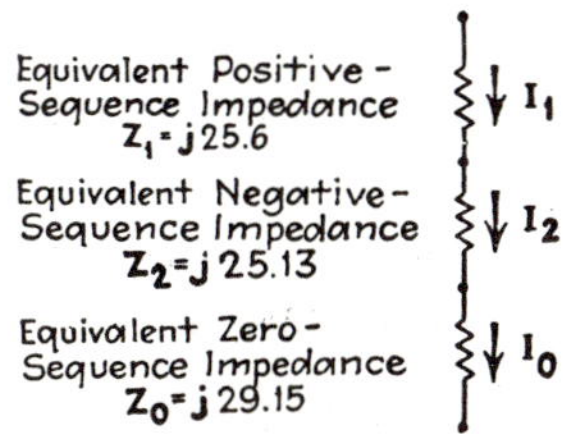

FIG. 27.—Combination of equivalent impedances for a single line-to-ground fault.

$$I_1 = I_2 = I_0 = \frac{j38{,}100}{j(25.6 + 25.13 + 29.15)} = 477 \text{ amp.}$$

The current distribution in the individual networks may now be obtained by the application of the distribution factor to the sequence current at the fault. For example, the positive-sequence current in the sound line is equal to the positive-sequence current at the point of fault (477 amp.) times the distribution factor for the positive-sequence current [0.107 from Fig. 24(*i*)], which is equal to 51.1 amp. Following this method the distribution of current for the three sequences is obtained, the results of which are shown in Fig. 28.

31. Phase Currents from Sequence Currents.

In the development of the general method in the preceding chapters, the a phase was used as the reference phase and the line-to-ground fault was applied to this phase. The currents in the different phases may therefore be obtained by application of the following formulas:

$$\left.\begin{aligned} I_a &= I_0 + I_1 + I_2 \\ I_b &= I_0 + a^2 I_1 + a I_2 \\ I_c &= I_0 + a I_1 + a^2 I_2 \end{aligned}\right\} \quad (107)$$

These formulas give correct results only *if the arbitrarily chosen directions of positive current flow are identical for the three sequence networks.*

As an illustration of the procedure which may be employed for each branch, the currents flowing in the individual conductors of the faulty line from the right-hand bus will be determined. From Fig. 28 the sequence currents in this line are

$$I_0 = 135.3 \text{ amp.}$$
$$I_1 = 180.7 \text{ amp.}$$
$$I_2 = 209.0 \text{ amp.}$$

These currents are all in phase and, because of the choice of the phase position of the generated voltage, are all real. The current in phase a is then

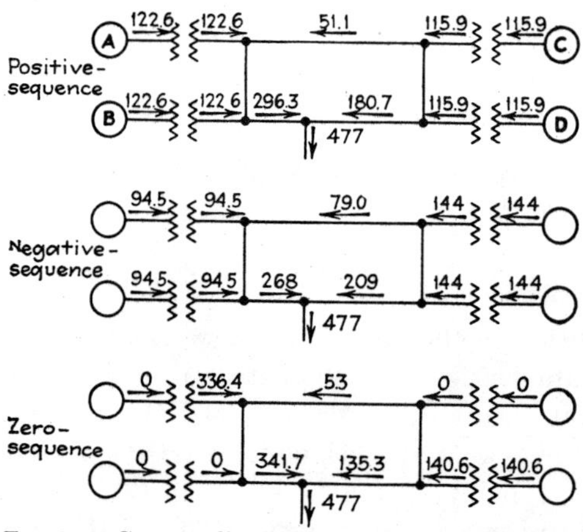

FIG. 28.—Current distribution in the three sequence networks.

$$I_a = 135.3 + 180.7 + 209.0 = 525 \text{ amp.}$$

The currents in phases b and c are obtained in a similar manner by applying the above relations, but, since the sequence components of current are all real, a simpler form may be used to advantage. Rewriting the expression for I_b and I_c and remembering that I_0, I_1, and I_2 are all real, there follows that

$$I_b = I_0 + (-0.5 - j0.866)I_1 + (-0.5 + j0.866)I_2$$
$$I_c = I_0 + (-0.5 + j0.866)I_1 + (-0.5 - j0.866)I_2$$

and

$$I_b = I_0 - 0.5(I_1 + I_2) - j0.866(I_1 - I_2) \quad (108)$$
$$I_c = I_0 - 0.5(I_1 + I_2) + j0.866(I_1 - I_2) \quad (109)$$

It will be observed that I_b and I_c are conjugate, *i.e.*, that their real parts are identical and their imaginary parts are equal but of opposite sign. Substituting the numerical values given

$$I_b = 135.3 - 0.5(180.7 + 209.0) - j0.866(180.7 - 209.0)$$
$$= -59.6 + j24.5$$
$$I_c = -59.6 - j24.5$$

In this manner the phase currents throughout the entire system are determined. The results may be conveniently tabulated in the form shown in Fig. 29. The currents shown are expressed in amperes on the 66-kv. base throughout. They must be modified by the introduction of transformer ratios to show the actual phase currents.

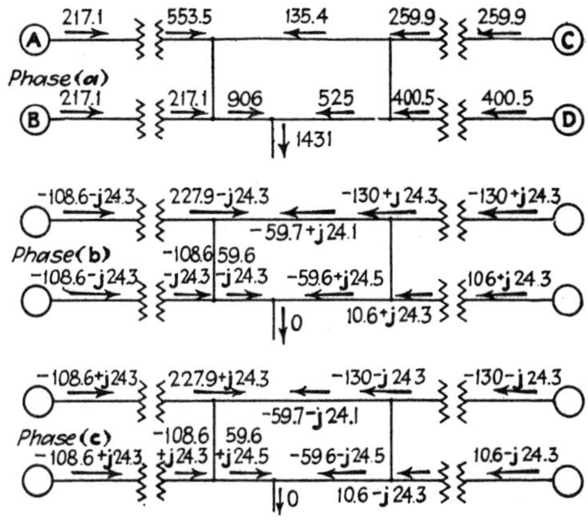

FIG. 29.—Phase current distribution.

32. Determination of Sequence Voltages.

The sequence component of line-to-neutral voltage may be obtained from a knowledge of the sequence-current distribution and the branch impedances. In particular, the positive-sequence voltage at a point is equal to the generated voltage minus the positive-sequence drop from the source to the point under consideration. For example, the positive-sequence voltage of the left-hand bus is equal to the generated voltage ($j38,100$) minus the drop due to the positive-sequence current (122.6 amp.) in the generator and transformer ($j85 \times 122.6$), which is equal to $j27,680$ volts.

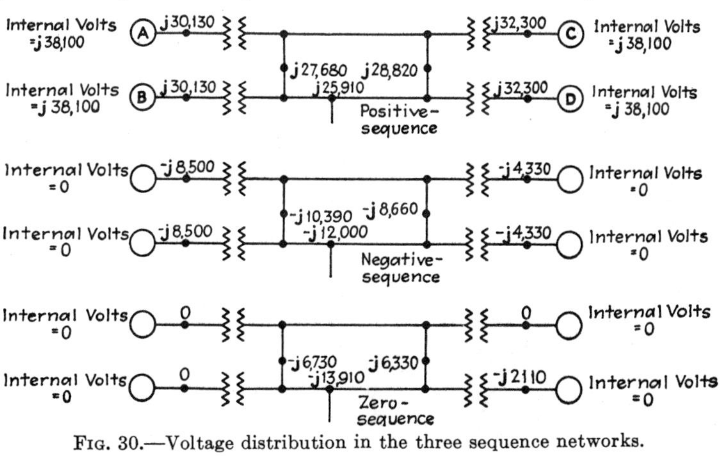

FIG. 30.—Voltage distribution in the three sequence networks.

Similarly, the negative-sequence voltage at any point is equal to the generated negative-sequence voltage (which in this case is zero) minus the negative-sequence drop from the source to the point under consideration. For the left-hand bus this becomes

$$0 - j110 \times 94.5 = -j10{,}390 \text{ volts}$$

Similarly, for the zero-sequence network the voltage of the left-hand bus is

$$0 - j20 \times 33.64 = -j6{,}730 \text{ volts}$$

The voltage distribution for the three sequences for the entire system is obtained in this manner, the results being shown in Fig. 30.

For cases in which extensive voltage calculations are involved, the voltage distribution factor may be obtained for different points throughout the system for each of the three sequence networks. These may be obtained and applied in the manner analogous to that described for currents.

33. Line-to-neutral Voltages from Sequence Voltages.

The line-to-neutral voltages are obtained from the sequence voltages by applying the fundamental conversion formulas (24), (25), and (26), which for ready reference are repeated below

$$\begin{aligned} E_a &= E_{a0} + E_{a1} + E_{a2} \\ E_b &= E_{a0} + a^2E_{a1} + aE_{a2} \\ E_c &= E_{a0} + aE_{a1} + a^2E_{a2} \end{aligned}$$

These expressions may be simplified by rewriting in the following form

$$E_a = E_{a0} + E_{a1} + E_{a2} \qquad (110)$$

$$E_b = -j0.866(E_1 - E_2) + E_0 - 0.5(E_1 + E_2) \qquad (111)$$

$$E_c = +j0.866(E_1 - E_2) + E_0 - 0.5(E_1 + E_2) \qquad (112)$$

Since, for this particular case, all the voltages are purely imaginary, it will be observed that the first terms of E_b and E_c are real and the negative of each other, and that the second terms of E_b and E_c are purely imaginary and equal. Voltages E_b and E_c are then mirror images of each other about the imaginary axis.

Inserting the values of positive-, negative-, and zero-sequence voltages for the left-hand bus, the following values of line-to-neutral voltages are obtained.

$$
\begin{aligned}
E_a &= j(-6{,}730 + 27{,}680 - 10{,}390)\\
&= j10{,}560 \text{ volts}\\
E_b &= -j0.866\,(j27{,}680 + j10{,}390)\\
&\qquad + [-j6{,}730 - 0.5(j27{,}680 - j10{,}390)]\\
&= 32{,}970 - j15{,}370 \text{ volts}\\
E_c &= -32{,}970 - j15{,}370 \text{ volts}
\end{aligned}
$$

The line-to-neutral voltages for the entire system are given in Fig. 31.

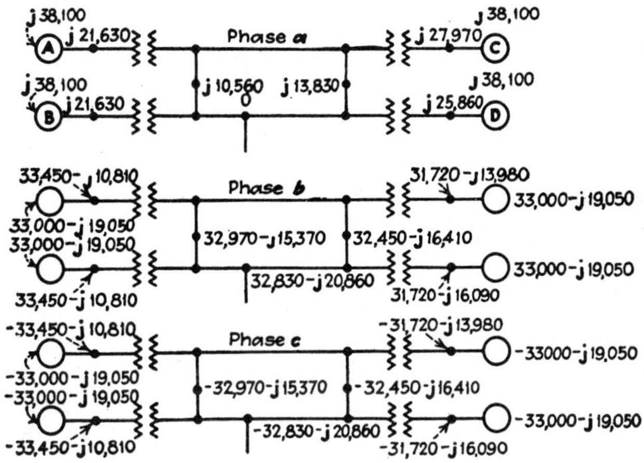

FIG. 31.—Line-to-neutral voltage distribution.

34. Line-to-line Voltages.

The line-to-line voltages may be obtained from a knowledge of the line-to-neutral voltages by applying the relations of equations (36) and Fig. 7. These equations are repeated for convenience.

$$
\begin{aligned}
E_A &= E_c - E_b\\
E_B &= E_a - E_c\\
E_C &= E_b - E_a
\end{aligned}
$$

Applied to the voltages at the left-hand bus

$$
\begin{aligned}
E_A &= (-32{,}970 - j15{,}370) - (32{,}970 - j15{,}370)\\
&= -65{,}940\\
E_B &= j10{,}560 - (-32{,}970 - j15{,}370)\\
&= 32{,}970 + j25{,}930\\
E_C &= (32{,}970 - j15{,}370) - (j10{,}560)\\
&= 32{,}970 - j25{,}930
\end{aligned}
$$

When the line-to-neutral voltages are not desired, the line-to-line voltages may be obtained directly from the sequence components of line-to-neutral voltages by means of the following relations:

$$\left.\begin{aligned} E_A &= j\sqrt{3}(E_{a1} - E_{a2}) \\ E_B &= 1.5(E_{a1} + E_{a2}) - j0.866(E_{a1} - E_{a2}) \\ E_C &= -1.5(E_{a1} + E_{a2}) - j0.866(E_{a1} - E_{a2}) \end{aligned}\right\} \quad (113)$$

The foregoing equations, it must be remembered, are valid only for the particular assumptions which for the sake of emphasis will be repeated. The resistances of the network branches are negligible and the generated voltages are of the positive-sequence only. In addition, the voltage of the reference phase (phase a of the star group) is purely imaginary.

35. The Conversion of Currents and Voltages Due to Transformers.

To this point all calculations have been made on the equivalent star basis corresponding to the transmission voltage of 66 kv. For the transmission line itself this gave the actual currents and voltages of the three conductors, but for the low-voltage portion of the system certain transformation constants must be applied. Hence, the determination of the actual currents and voltages where transformations have taken place requires conversion in terms of the new base. The phenomenon is complicated to a certain extent by the fact that associated with every transformation there must necessarily be some impedance and magnetizing current. In most cases the magnetizing current is so small in comparison with the load current under short-circuit conditions that it may be neglected, as has been done in this particular case. The problem may be further simplified by divorcing the impedance and conversion aspects by considering the actual transformer replaced by a perfect transformer with an impedance equal to the transformer impedance in series with it. In this case the impedance will be considered as being in series on the high side, because the voltage of the high side was chosen as the voltage base. It can be shown that the impedance could be inserted on either side.

Determination of Actual Currents. For **star-star** transformers, the currents on the low side are equal to the actual turns ratio times the currents on the high side. For example, the actual phase currents in generator C are from Fig. 29

$$I_a = \frac{66{,}000}{11{,}000} 259.9 = 1{,}559 \text{ amp.}$$

$$I_b = \frac{66{,}000}{11{,}000}(-130 + j24.3) = -780 + j146 \text{ amp.}$$

$$I_c = \frac{66{,}000}{11{,}000}(-130 - j24.3) = -780 - j146 \text{ amp.}$$

Star-delta transformers introduce not only a change in magnitude of voltages and currents but also a change in phase angle. For this reason it is necessary that some convention be adopted

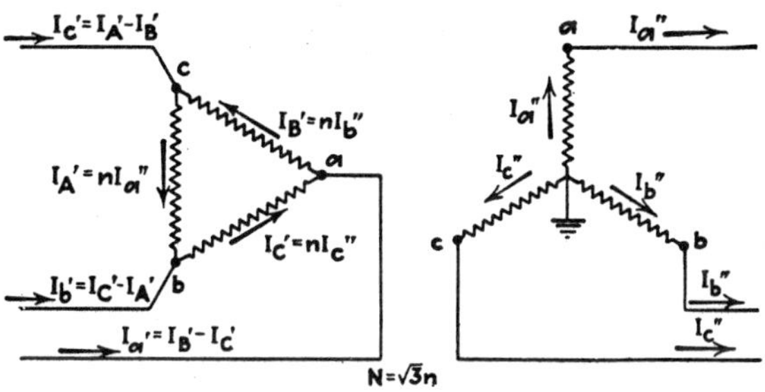

$$I_{a1}' = -jNI_{a1}'' \qquad I_{a1}'' = \frac{j}{N} I_{a1}'$$

$$I_{a2}' = jNI_{a2}'' \qquad I_{a2}'' = -\frac{j}{N} I_{a2}'$$

$$I_{a0}' = 0 \qquad I_{a0}'' = \text{indeterminate}$$

FIG. 32.—Relation between star currents in star-delta or delta-star transformations.

for designating the phases on the delta side with reference to those on the star side. Figure 32 shows the convention that has been chosen. It will be noted that on the delta side the subscript for the current in any terminal is the same letter as the subscript of the current in the winding opposite. In the notation small subscripts refer to line currents, large subscripts to delta currents, and single prime and double primes to assumed line currents on the delta and star sides, respectively. This notation conforms to that chosen in Sec. 8.

Calling n the turns ratio of the transformer and N the line voltage ratio ($N = \sqrt{3}n$), the following relations may be derived by inspection of Fig. 32:

$$\left.\begin{aligned} I_a' &= I_B' - I_C' = n(I_b'' - I_c'') = \frac{N}{\sqrt{3}}(I_b'' - I_c'') \\ I_b' &= I_C' - I_A' = n(I_c'' - I_a'') = \frac{N}{\sqrt{3}}(I_c'' - I_a'') \\ I_c' &= I_A' - I_B' = n(I_a'' - I_b'') = \frac{N}{\sqrt{3}}(I_a'' - I_b'') \end{aligned}\right\} \quad (114)$$

Following this method the actual currents in all parts of the system can be determined, taking both turn ratio and transformer connections into account.

Illustration. To illustrate these relations, let it be desired to determine the currents in generator A. The star currents on the 66-kv. base may be obtained from Fig. 29, from which

$$I_a' = \frac{66,000}{11,000}\frac{1}{\sqrt{3}}[(-108.6 - j24.3) - (-108.6 + j24.3)] = -j168.4$$

$$I_b' = \frac{66,000}{11,000}\frac{1}{\sqrt{3}}[(-108.6 + j24.3) - (217.1)] = -1,128 + j84.2$$

$$I_c' = \frac{66,000}{11,000}\frac{1}{\sqrt{3}}[(217.1) - (-108.6 - j24.3)] = 1,128 + j84.2$$

An alternative method for determining the phase currents, when star-delta transformations are involved, is to consider the effect of the transformation upon the sequence component of currents individually. Since the equations (114) hold for any values of current, then for the positive-sequence component alone, from the first equation

$$I_{a1}' = \frac{N}{\sqrt{3}}(a^2 - a)I_{a1}'' = -jNI_{a1}'' \quad (115)$$

Similarly, for the negative-sequence component alone

$$I_{a2}' = \frac{N}{\sqrt{3}}(a - a^2)I_{a2}'' = jNI_{a2}'' \quad (116)$$

and for the zero-sequence alone

$$I_{a0}' = \frac{N}{\sqrt{3}}(1 - 1)I_{a0}'' = 0 \quad (117)$$

Conversely, the line currents on the delta side may be obtained in terms of the line currents on the star side.

$$\left.\begin{aligned} I_{a1}'' &= \frac{j}{N}I_{a1}' \\ I_{a2}'' &= -\frac{j}{N}I_{a2}' \\ I_{a0}'' &= \text{indeterminate} \end{aligned}\right\} \quad (118)$$

The above relations hold whether the transformer is grounded or not. In general, n and N may be greater or less than unity.

With **cascade transformers** either of the methods just described may be applied to each transformation successively. However, in general, it will be found more convenient to follow the effect of the various transformations upon each of the phase-sequence components of current. Proceeding from any point at which the actual line currents are known, the rule is as follows.

In passing through a star-delta or a delta-star transformation the positive-sequence components of line or phase currents on the star side *lead* the corresponding quantities on the delta side by 90 deg.; similarly the negative-sequence components on the star side *lag* the corresponding quantities on the delta side by 90 deg. In addition, the ratio of transformation must be taken into account. The zero-sequence current will be blocked at the first star-delta transformation. Applying this rule to the transformer

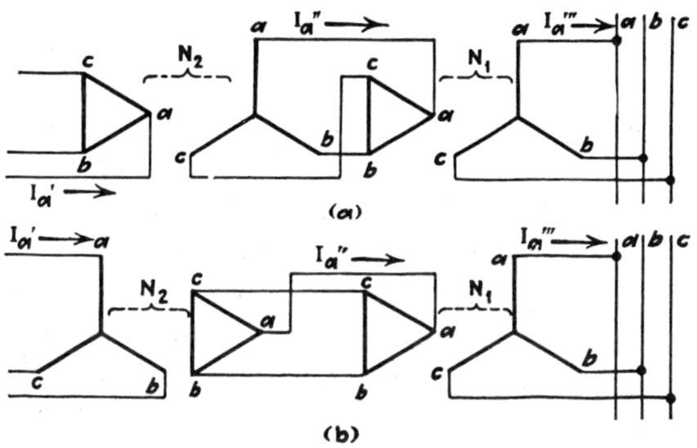

FIG. 33.—Cascade transformer connections.

combination in Fig. 33(*a*), it will be observed that the following relations hold.

$$I_{a1}'' = -jN_1I_{a1}'''$$
$$I_{a2}'' = jN_1I_{a2}'''$$

and

$$I_{a1}' = -jN_2I_{a1}'' = -N_1N_2I_{a1}'''$$
$$I_{a2}' = jN_2I_{a2}'' = -N_1N_2I_{a2}'''$$

and to Fig. 33(*b*)

$$I_{a1}' = +jN_2I_{a1}'' = N_1N_2I_{a1}'''$$
$$I_{a2}' = -jN_2I_{a2}'' = N_1N_2I_{a2}'''$$

A double star-delta transformation reverses both the positive- and negative-sequence components of current and consequently the phase currents, but a star-delta, delta-star transformation retains the original sense of the sequence components and phase currents without change in phase relation.

It is important that the notation indicated (see also Fig. 32) be strictly adhered to, for otherwise the reference phases become confused.

Determination of Actual Voltages. The voltages on the low side of a **star-star** transformer are equal to the voltages on the high side less the drop through the transformer branch on the high-voltage base divided by the ratio of transformation.

The effect of **star-delta transformers** upon the phase position of the voltages to neutral must be the same as that upon the phase position of the line currents, for otherwise the values of power and reactive volt-amperes on both sides of the ideal transformer would not be equal. Therefore, the positive-sequence star voltage in going through a star-delta transformation is rotated backward 90 deg., and in passing through a delta-star transformation forward 90 deg. Similarly, the negative-sequence voltage-to-neutral in passing through a star-delta transformation is rotated forward 90 deg. and, passing through a delta-star transformation, backward 90 deg. These relations, with the notation corresponding to that used in Fig. 32 are expressed in Fig. 34.

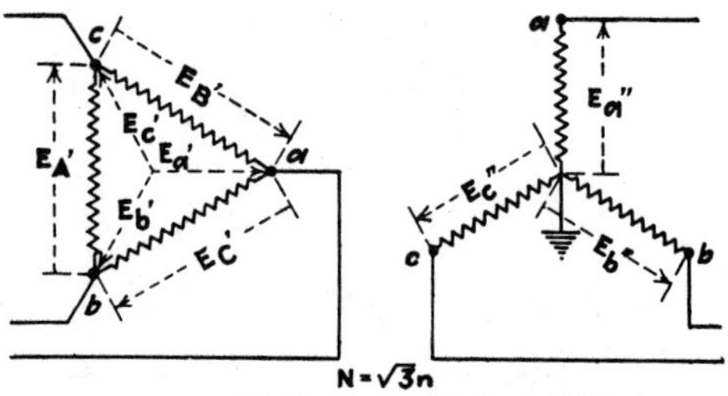

$$E_{a1}' = -j\frac{E_{a1}''}{N} \qquad E_{a1}'' = jNE_{a1}'$$

$$E_{a2}' = j\frac{E_{a2}''}{N} \qquad E_{a2}'' = -jNE_{a2}'$$

$$E_{a0}' = 0 \qquad E_{a0}'' = \text{indeterminate}$$

FIG. 34.—Relation between star voltages in star-delta or delta-star transformations.

Since

$$E_{A1}' = j\sqrt{3}E_{a1}'$$

it follows that

$$E_{A1}' = j\sqrt{3}\left(-j\frac{E_{a1}''}{N}\right) = \frac{\sqrt{3}}{N}E_{a1}'' = \frac{E_{a1}''}{n}$$

Also, since

$$E_{A2}' = -j\sqrt{3}E_{a2}'$$

it follows that

$$E_{A2}' = -j\sqrt{3}\left(j\frac{E_{a2}''}{N}\right) = \frac{\sqrt{3}}{N}E_{a2}'' = \frac{E_{a2}''}{n}$$

Cascaded transformers may be handled by considering the transformations successively just as outlined in the treatment of the current relations.

36. Double Line-to-ground Fault Calculations.

The calculation of double line-to-ground faults differs from that for single line-to-ground faults only in the arrangement of the equivalent impedances. As developed in Sec. 17, the impedances should be grouped as shown in Fig. 35. It is important as emphasized and explained before, that the *positive direction of current flow* be maintained as indicated. Solving, it is found that

$$I_1 = \frac{j38{,}100}{Z_1 + \dfrac{Z_0 Z_2}{Z_0 + Z_2}}$$
$$= \frac{j38{,}100}{j25.6 + j\dfrac{29.15 \times 25.13}{54.28}}$$
$$= 975 \text{ amp.}$$
$$I_2 = \frac{Z_0}{Z_0 + Z_2}(-I_1)$$
$$= \frac{j29.15}{j54.28}(-975)$$
$$= -523.5 \text{ amp.}$$
$$I_0 = -451.5 \text{ amp.}$$

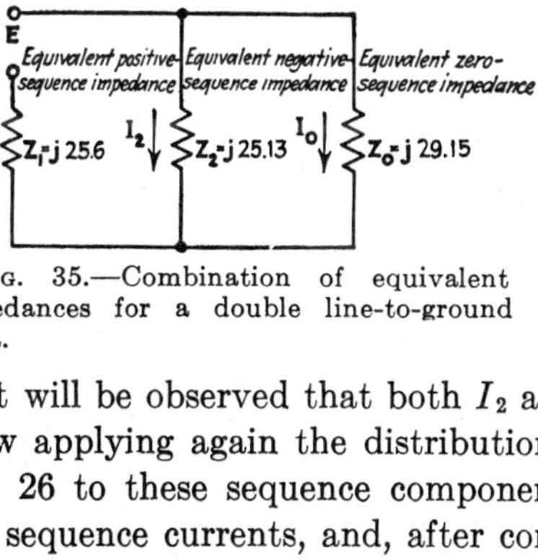

FIG. 35.—Combination of equivalent impedances for a double line-to-ground fault.

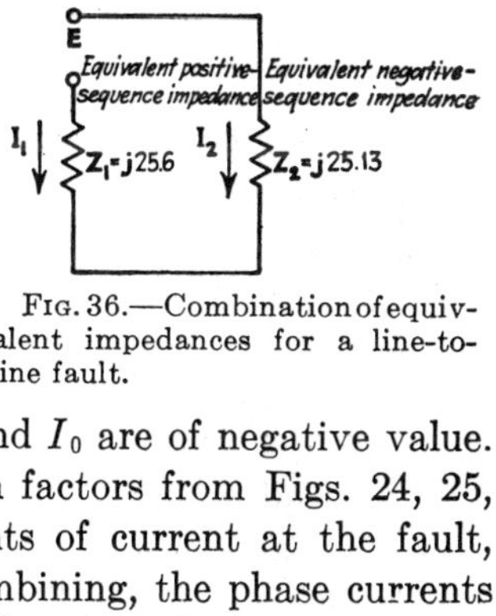

FIG. 36.—Combination of equivalent impedances for a line-to-line fault.

It will be observed that both I_2 and I_0 are of negative value. Now applying again the distribution factors from Figs. 24, 25, and 26 to these sequence components of current at the fault, the sequence currents, and, after combining, the phase currents at all points in the system may be determined. Voltages may be calculated in the same manner as described for single line-to-ground faults.

37. Line-to-line Fault Calculations.

The zero-sequence network does not enter into the calculation of the line-to-line fault. This is to be expected, since the boundary conditions at the fault require only positive- and negative-sequence voltages to define them. The arrangement of the equivalent network for this type of fault was developed in Sec. 18 and is as indicated in Fig. 36. Solving the network,

$$I_1 = \frac{j38{,}100}{j(25.6 + 25.13)} = 751 \text{ amp.}$$

$$I_2 = -I_1 = -751 \text{ amp.}$$

From this point the procedure is the same as for the calculation of the two other types of faults.

38. Comparison of Short-circuit Calculations and Test Results

It is of interest to compare the results of calculations by the method of symmetrical components of fault currents and voltages on a complicated system with the results obtained by actual tests. In 1925, during an extensive stability investigation, an opportunity was afforded to obtain a very complete record of power-system voltages and currents at different points in the system. In addition the system characteristics were accurately known because of the stability investigation, and calculations were made for the conditions existing at the time of the fault.

The power system is shown diagrammatically in Fig. 37.* The transmission circuits included a 220-kv. transmission line 202 miles long paralleled by a similar line operated at 110 kv. The short-circuit calculations require consideration of

1. Distributed capacity of the 202-mile transmission line.
2. Inductive coupling to zero-sequence between the 110-kv. and 220-kv. lines.
3. Unequal transformer ratios in the different circuits.

The results of calculation are shown by means of voltage and current vectors at different points in the system diagram of Fig. 37. All the vectors are given on the common voltage base of 110 kv., and to obtain the actual values of the currents and voltages in any particular part of the circuit it is necessary to take into consideration the transformer ratio and phase shifts due to star-delta transformations.

Figure 37 shows the radical distortion of voltages and currents that obtain under unbalanced faults. The distortion is accentuated by the relatively small value of load current in comparison with the fault current and even with the charging current on

* Figure 37 and Table II are taken from "Studies of Transmission Stability."[(28)]

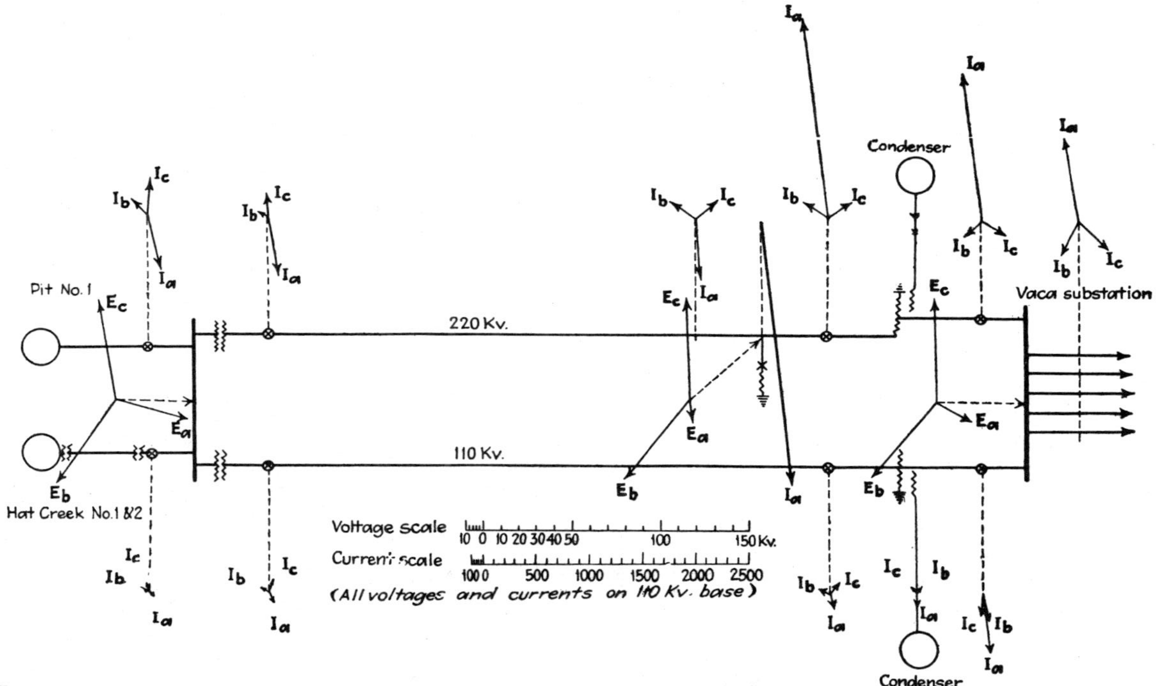

FIG. 37.—Voltage and current distribution on the transmission system of the Pacific Gas and Electric Company with a single line-to-ground fault on the 220-kv. line near Vaca substation—calculated.

the 220-kv. line. The effect of the charging current is brought out by the change in the currents between the sending and fault ends of the 220-kv. line. It may be pointed out that the vector diagrams show that the proportion of zero-sequence current is relatively large in the 220-kv. line and relatively small in the 110-kv. line. This is due to the fact that the zero-sequence current induced in the 110-kv. line by the current in the 220-kv. line is opposite in direction and approximately equal to the zero-sequence current that would tend to flow conductively through the 110-kv. line to the Vaca bus and through the autotransformer to the fault.

In Table II is given the comparison of the calculated voltages and currents and the corresponding quantities observed during

TABLE II.—COMPARISON OF MEASURED AND CALCULATED INSTANTANEOUS SYMMETRICAL VOLTAGES AND CURRENTS FOR LINE-TO-GROUND FAULT AT THE VACA-DIXON SUBSTATION

Currents and voltages	Measured	Calculated
Zero-sequence current in 220-kv. circuit-breaker at Vaca	1,020	1,130
Voltage, c phase at Vaca	60,000	59,300
Zero-sequence voltage, 110-kv. bus at Vaca	14,200	13,300
Zero-sequence current, 220-kv. line at Pit 1	140	138
Current a phase, 220-kv. line at Pit 1	235	280
Voltage A phase, Pit 1 generator	4,840	4,996
Current A phase, 110-kv. line on low voltage side at Pit 1	1,070	1,030

the test. The calculated values are based on the instantaneous symmetrical value of the short-circuit currents and the test results are based on the values during the fourth cycle after the application of the fault, this cycle being chosen to minimize the effect of the rapidly decaying subtransient components of current. The close agreement obtained, considering the factors involved, serves as a satisfactory check on the general method.

Problems

1. For the system of Fig. 23(a), determine the actual currents and voltages of the generators C and D for the single line-to-ground fault at F.

2. Assume that faults of different types occur on the right-hand bus of Fig. 23(a). Determine the fault currents for (a) single line-to-ground fault; (b) double line-to-ground fault; (c) line-to-line fault; (d) three-phase fault.

3. For the same system and faults as Prob. 2 determine the corresponding fault voltages.

4. In Fig. 29 it is shown that for a single line-to-ground fault at F the currents in the b and c conductors of the transmission line are not zero. Explain why this condition exists. State the conditions in regard to network impedances which are necessary in order to eliminate the phase b and c currents in generator C and in generator D.

5. Consider the system of Fig. 23(a) and assume that a single line-to-ground fault occurs on phase a through a resistance of 115 ohms. Determine the voltages and currents for the point F, the fault location.

6. Determine the ground currents of all transmission circuits of the system of Prob. 5.

7. For the system of Fig. 23(a), if generator A is disconnected, how will the fault currents and ground currents be affected? In case generator D is disconnected? How are the current distribution factors affected in both cases?

8. Assume a generator with an internal voltage E_{a1} of 7,000 volts line-to-neutral under sustained single-phase load conditions on phases a and b, If the generator resistances are negligible and the reactances are $X_1 = 1.0$. $X_2 = 0.15$, and the load impedance $Z_L = 0.8 + j0.6$, what are the delta terminal voltages E_A, E_B, and E_C?

9. Assume a generating station of n similar units, each with negligible resistance and with reactances of X_1, X_2, and X_0, and that each machine is operating with an internal voltage of E_{a1}. Calculations are to be made to determine the relative current stress in the generators for different methods of grounding the generator neutrals. Find the analytical expression for the phase current in each machine for (a) the neutral grounded on all machines; (b) the neutral grounded on only one machine. If the number of ungrounded machines, in parallel with the grounded neutral machine, approaches infinity, what value does the current in the grounded neutral machine approach? Compare this expression with that for the currents on a three-phase fault.

CHAPTER V

CONSTANTS OF SYNCHRONOUS MACHINES*

Only positive-sequence voltages are generated in most commercial systems. System short-circuit phenomena are so closely related to the characteristics of synchronous generators, motors and condensers that their performance under short-circuit conditions will be reviewed at some length.

39. Three-phase Short-circuits of Generators.

Assume that an instantaneous three-phase short-circuit is applied across the terminals of a synchronous machine operating

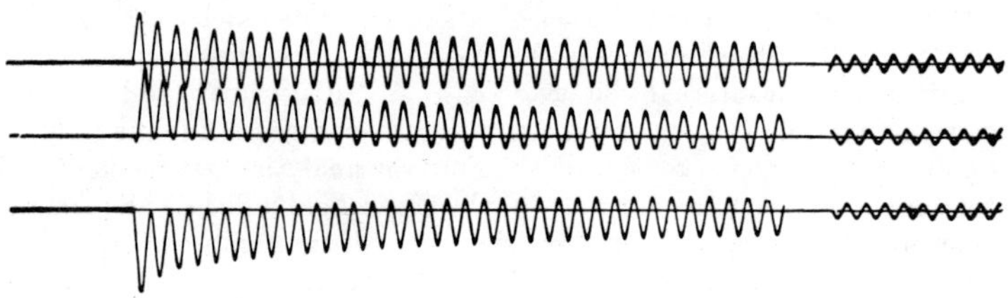

FIG. 38.—Three-phase short-circuit on a 75,000-kva. 11,600-volt 60-cycle 1,800 r.p.m. turbo-generator.

with *constant excitation* and no load. The characteristic current waves for the three phases are shown in Fig. 38. It will be observed that in general each current wave consists of two kinds of components, namely,

1. Alternating-current components.
2. Unidirectional or direct-current components.

The former are equal in all three phases. The latter are dependent upon the particular point of the cycle at which the short-circuit occurs.

* The treatment in this chapter presupposes an elementary knowledge of synchronous-machine theory as well as simple transients. For the latter the reader is referred to "Electric Transients" by C. E. Magnusson, McGraw-Hill Book Company, Inc., New York, 1926.

40. Alternating-current Components.

Figure 39 shows an enlarged view of the alternating-current component from which it may be observed that this wave may in turn be resolved into three components.

a. Sustained Current. For a prolonged short-circuit the armature current finally attains the sustained value I indicated by the cycle shown in dotted lines. The magnitude of this

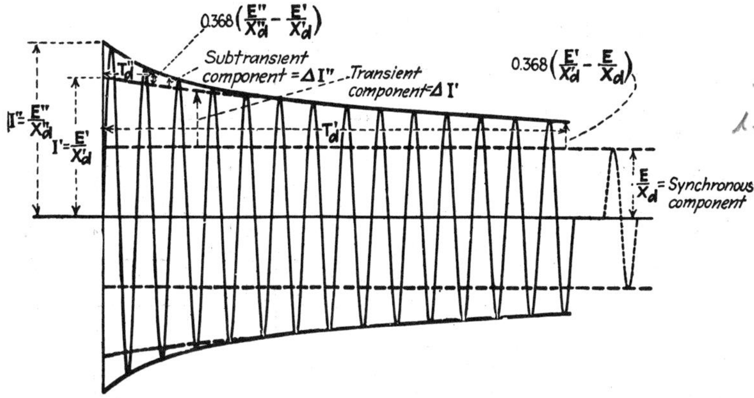

FIG. 39.—Curve of symmetrical (alternating-current) component of short-circuit current from no load.

quantity is determined by the expression $\frac{E}{x_d}$, where E is the synchronous internal voltage which in this case is equal to the air-gap voltage corresponding to the no-load excitation, and x_d* is the constant of the machine known as the **synchronous reactance.** This reactance is the value that is commonly used for all problems involving steady-state calculations of machine performance and includes armature-leakage reactance and the reactance equivalent to armature reaction. It is usually obtained from the saturation curve and the three-phase short-circuit curve (see Fig. 40) and is defined as the *ratio of the field current at rated armature current on sustained symmetrical short-circuit to the field current at normal*

* The notation used is based on Blondel's two-reaction theory of synchronous machines in which x_d and x_q represent reactances in direct and quadrature axes, respectively. For most problems the quadrature-axis reactances may be ignored and the direct-axis quantities only are required. For a discussion of reactances reference may be made to A.I.E.E. papers by R. H. Park and B. L. Robertson,(41) L. A. Kilgore,(72) and S. H. Wright.(73)

*open-circuit voltage on the air-gap line (that is, the extended straight part of the magnetization curve).** The definition gives the result in the per unit† quantity which from Fig. 40 may be seen to be

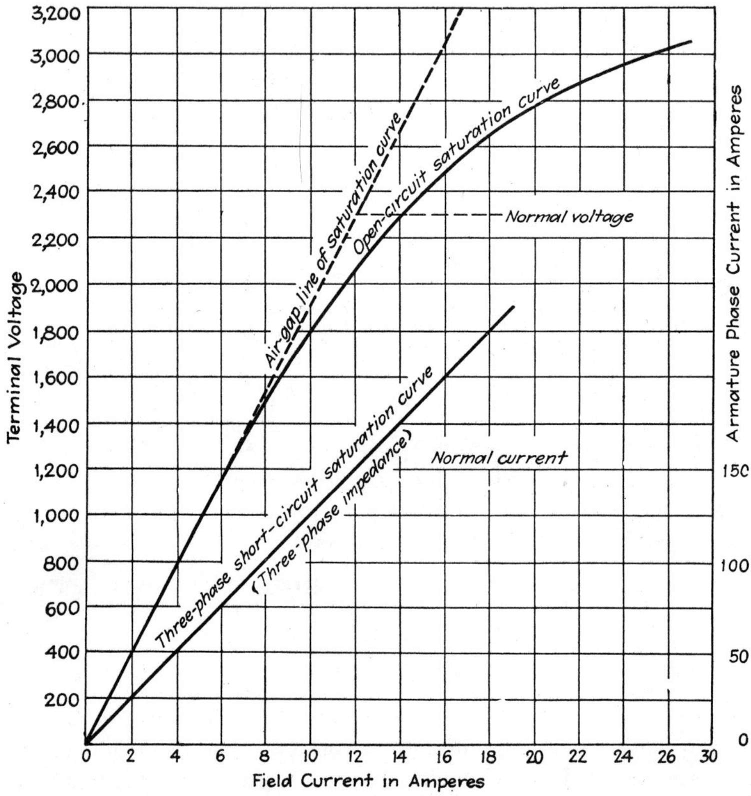

FIG. 40.—Saturation and synchronous impedance.

$\frac{12}{12}$ = unity. This unit may be easily converted to per cent values or to actual ohms at any voltage base. If E is given as a r.m.s. value, I will also be expressed as a r.m.s. value.

* Excerpt from *Report of Subject Committee on Definitions of Terms Used in Power System Studies.*[78]

† The per unit quantities are similar to per cent quantities except that 100 per cent is equal to one per unit; rated armature current is 100 per cent armature current or one per unit armature current. This general system has most of the advantages of the per cent system in comparing performance of machines or lines of different rating plus the advantage that it is unnecessary to carry through awkward figures that result from multiplying an impedance by a current expressed in per cent to obtain a voltage drop in per cent. Thus 100 per cent reactance times 100 per cent current is equal to 10,000 per cent voltage which, of course, must be corrected to 100 per cent voltage.

b. Transient Current. If the envelope of the current wave be projected back to zero time *neglecting the first few cycles having a very rapid decrement,* the value thus obtained for zero time is called the **transient current** I'. This value is determined by dividing the **transient internal voltage** E', in this case the open-circuit voltage before the short-circuit, by the **transient reactance** x_d'. The portion of the current to neglect in obtaining the envelope may be made more evident by plotting the envelope of

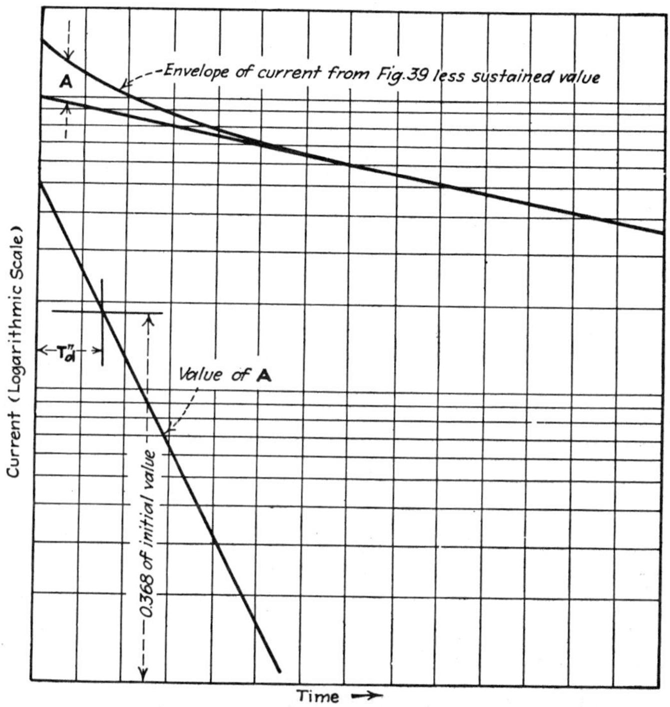

FIG. 41.—Determination of T_d''.

the excess of current over the sustained value on semilogarithmic paper, as shown in Fig. 41.

At the instant of sudden short-circuit the demagnetizing effect of the armature current tends to decrease the flux linkages with the field winding. However, the flux linkages with any inductive circuit cannot be changed instantly but tend to remain constant. This results in an induced current in the field winding of such magnitude as to just annul the demagnetizing effect of the armature current. If the leakage flux associated with the field winding were zero, the air-gap flux would remain constant and the initial armature current would be determined by the armature-leakage reactance alone, and the transient reactance would be equal to the leakage reactance. Actually, however, the field winding does possess a certain amount of leakage, and with the increase of field current the leakage flux also increases. In order to maintain constant flux linkages with the field winding the air-gap flux must decrease. Since the armature current is equal to the air-gap voltage divided by the armature-leakage reactance, the initial armature current will be smaller than if the field leakage were zero. This effect may be included by increasing the transient reactance by an amount proportional to the field leakage. The transient reactance thus includes the effect of both armature and field leakages.*

c. Subtransient Current. There remains yet to consider the **subtransient component,** the value in Fig. 39 between the actual current and the transient component. The total initial current is determined by dividing the **subtransient internal voltage** E'', which in this case is equal to the open-circuit voltage, by the **subtransient reactance** x_d''. The increment of armature current over that of the sum of the transient component and the sustained value, namely, the subtransient component, is due to the induced currents in the damper windings or any similar circuit. If only the phenomenon associated with the field winding were considered, the air-gap flux would decrease upon the application of a short-circuit, but actually the close proximity of the damper winding imbedded in the surface of the pole piece prevents the air-gap flux from changing instantaneously. In the damper winding currents are induced which tend to maintain the air-gap flux and thus increase the initial value of armature short-circuit current. For a perfect damper winding, in the sense that it does not possess any leakage flux, the transient reactance of the machine is therefore equal to the armature-leakage reactance. Actually, however, damper windings possess some leakage so that the subtransient reactance is somewhat greater than the armature-leakage reactance.

* A more complete discussion of the physical significance of transient reactance in terms of the armature and field leakages is given in a paper by C. F. Wagner, Effect of Armature Resistance upon Hunting, *Trans A.I.E.E.*, vol. 49, p. 1011, July, 1930.

Subtransient and transient reactances can be determined from short-circuit tests. An alternative method for determining the subtransient reactance is to apply voltage across any two terminals (excluding the neutral) with the rotor at rest and short-circuited on itself through an ammeter. One half the voltage required to circulate rated current is equal to x_d'' in per unit values. For this test the rotor must be in the position of maximum induced field current (the direct axis). If the rotor is in the position of minimum induced field current the **quadrature subtransient reactance** x_q'' is obtained.

41. Time Constants.

The initial value of the transient component of armature current is produced by the induced current in the field winding. The field winding and the coupled armature circuits act substantially as a simple resistance and inductance. The induced portion of the field current and consequently the transient component of armature current therefore decay in an exponential manner. Any exponential curve may be expressed in the from $I\epsilon^{-\frac{t}{T}}$ in which I is the magnitude for $t = 0$ and T is the time constant. These curves have the property of decaying in any interval of time t to $\frac{1}{\epsilon}$ or 0.368 of its value at the beginning of the interval. The time T_d' in Fig. 39 required for the transient component to decay to 0.368 of its initial value is thus the time constant of the transient component and is called the **transient short-circuit time constant.**

The **transient open-circuit time constant** of a synchronous machine is the time constant of the exponential curve of armature voltage obtained when the exciting voltage is suddenly removed from a machine operating at no load, the field circuit of course being maintained. The transient short-circuit time constant T_d' may be obtained in terms of the transient open-circuit time constant T'_{d0} by the expression[34]

$$T_d' = \frac{x_d'}{x_d} T'_{d0} \tag{119}$$

The subtransient current is maintained by currents flowing in the damper windings or any other similarly located circuit. This circuit is similar in nature to the field circuit except for the relative values of inductance and resistance. Because of the relatively high value of resistance the time constant is usually very small, which accounts for the rapid decay. In general, therefore, it will be found that the **subtransient short-circuit time constant** T_d'' is very small and is of the order of 0.1 or 0.2 sec. or smaller. Figure 39 shows the significance of the time constant T_d'' and Fig. 41 indicates how it may be obtained from the alternating-current component of short-circuit current.

42. Direct-current Components.

Considering only the alternating-current components of current in the individual phases of the armature requires that the armature currents in at least two, and possibly three, of the phases increase instantly from zero at times of three-phase short-circuit. Because of the presence of the inductance in the individual phases of the armature the currents cannot change instantly from zero to a finite value. A direct-current or equalizing current therefore flows in each phase of such magnitude at $t = 0$ to bring the current equal to zero. Thereafter the individual direct-current components decay in an exponential manner corresponding to the **direct-current time constant** T_a which is equal for all three phases. The value of this time constant is dependent upon the ratio of inductance to resistance in the armature circuit. As will be shown later the negative-sequence reactance of the synchronous machine x_2 is a sort of average of the armature subtransient reactances so that the direct-current time constant may be written[(41)]

$$T_a = \frac{x_2}{2\pi f r_a} \tag{120}$$

where r_a is the armature resistance, and x_2 and r_a must be expressed in the same units, either per unit or ohms per phase.

43. R.M.S. Total Current.

The r.m.s. total current at any instant is equal to the square root of the sum of the squares of the alternating-current and direct-current components; the minimum currents thus occur on that phase in which the direct-current component is equal to zero and the maximum currents on that phase in which the direct-current component is a maximum. Since the maximum value that the direct-current component can attain is $\sqrt{2}I''$, the maximum value of the r.m.s. total current is

$$\sqrt{I''^2 + (\sqrt{2}I'')^2} = \sqrt{3}I''.$$

44. Short-circuit of Loaded Generator.

The foregoing assumed that the generator under consideration had been operating at no load just previous to the short-circuit. Such is not usually the case. It is necessary therefore to be able to calculate short-circuits for the more usual case of the loaded machine.(73) This involves principally the determination

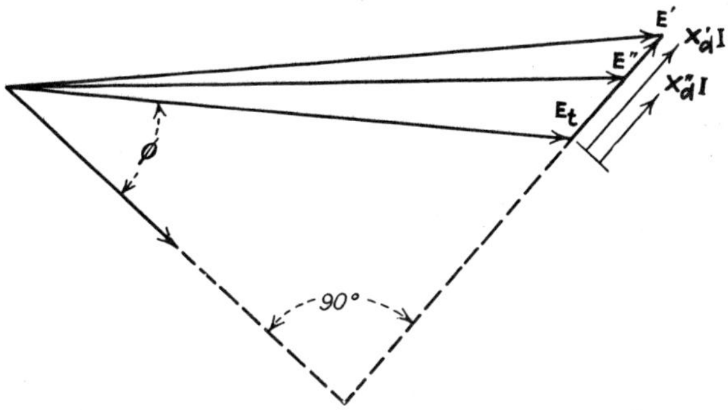

Fig. 42.—Construction to determine E' and E''.

of the synchronous, transient. and subtransient internal voltages. The latter two may be obtained by merely adding vectorially the transient and subtransient reactance drops, respectively, to the terminal voltage E_t. An example of this construction* is shown in Fig. 42. In determining the sustained value of short-circuit current the internal voltage used is that fictitious value corresponding to the point on the air-gap line for the particular excitation. This is permissible because armature reaction decreases the air-gap flux below the saturation point. Saturation is an important factor, however, in determining the excitation for the particular load carried by the machine before the short-circuit. A convenient method for obtaining this excitation is the A.I.E.E. method, which utilizes the no-load and the zero power-factor saturation curves shown in Fig. 43. To obtain the excitation at any other power factor for rated current the distance ab between the no-load saturation curve and the zero

* This construction is not strictly correct for salient pole machines but is sufficiently accurate for practical purposes.

power-factor curve for the excitation 0*c* is laid off along the line *bd* (Fig. 43) which makes an angle equal to the power-factor angle with the *X*-axis. A line equal to the distance *ac* (the no-load voltage corresponding to the excitation 0*c*) is then scribed from the point *a* intercepting the *Y*-axis. The horizontal line

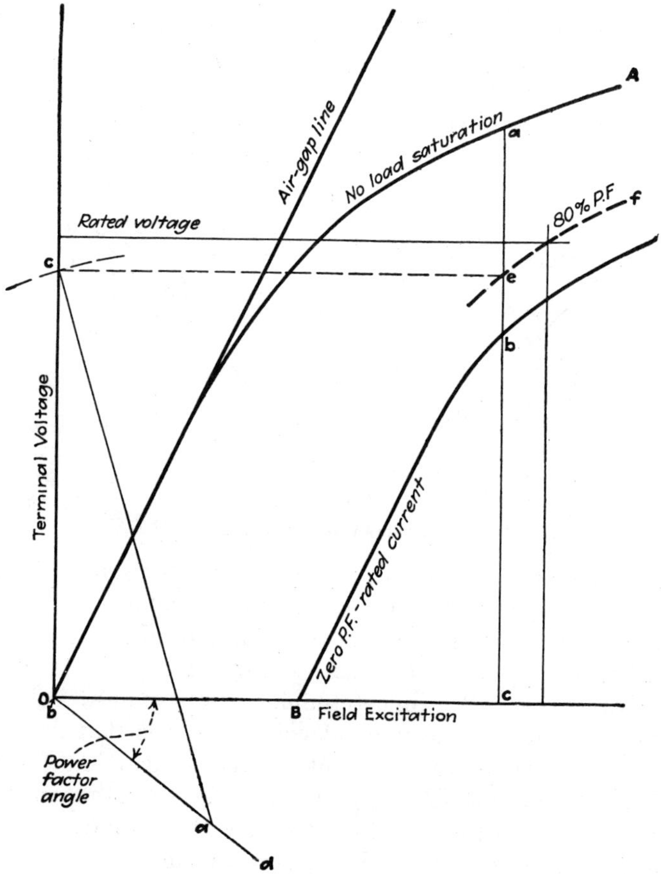

FIG. 43.—Determination of excitation for load power factors other than zero by the A.I.E.E. method.

ce is then the terminal voltage for the particular excitation. Following this procedure another excitation is chosen and the construction repeated from which the dotted line *ef*, shown in Fig. 43, is obtained. The intersection of this line with the normal voltage gives the excitation for the particular power factor at rated load. If the machine is not operating at rated load, the zero power-factor curve corresponding to the particular current should be used. The synchronous internal voltage expressed in per unit is the ratio of the excitation under consideration to the excitation required to generate rated voltage at no load, measured on the air-gap line.

The initial values of I' and I'' are then equal, respectively, to $\frac{E'}{x_d'}$ and $\frac{E''}{x_d''}$. The maximum value of the direct-current component I_{dc} is obtained by taking $\sqrt{2}$ times the negative of the

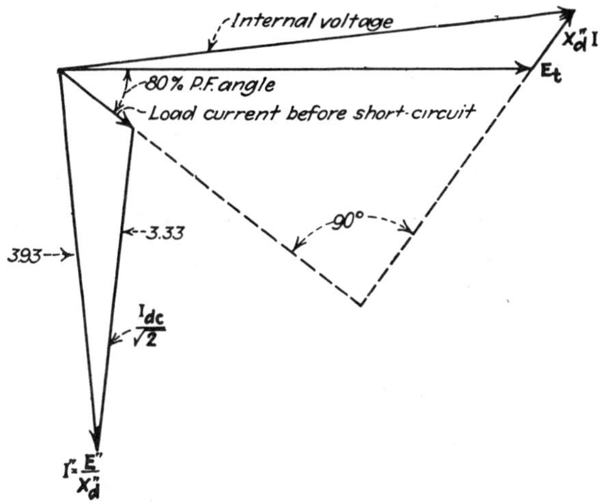

FIG. 44.—Construction to obtain the maximum value of I_{dc} for a three-phase short-circuit on a loaded machine.

vector difference between I'' and the load current before the short-circuit. This construction is shown in Fig. 44.

45. External Reactance.

Machine short-circuits with external reactances may be treated after determining the internal voltages in all respects as though the machine reactances were increased corresponding amounts; thus calling x_e the external reactance, $x_d + x_e$, $x_d' + x_e$, and $x_d'' + x_e$ should be used as the synchronous, transient, and subtransient reactances, respectively. The time constants vary in a similar manner, thus

$$T_d' = \frac{x_d' + x_e}{x_d + x_e} T'_{d0}. \tag{121}$$

46. Example of Decrement Calculations.

To illustrate the foregoing discussion, let it be desired to calculate the three-phase short-circuit of a large typical turbine generator whose characteristics are given in Fig. 45. An external or feeder reactance of 0.145 will be assumed. This odd value

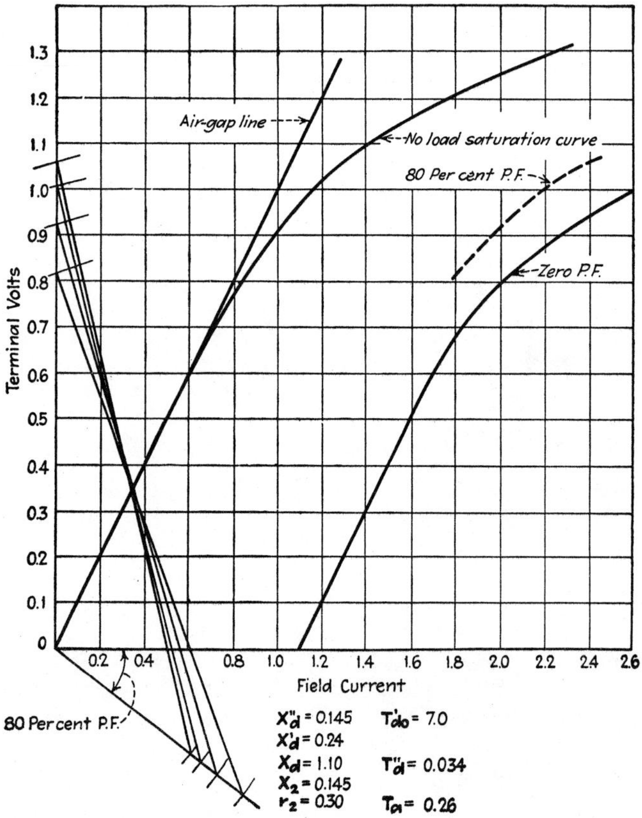

FIG. 45.—Characteristics of turbine generator.

is chosen as it is equal to the negative-sequence reactance; the results of this calculation can therefore be used later in illustrating unbalanced short-circuit calculations. The machine is assumed to be fully loaded at 80 per cent power factor before the short-circuit and the short-circuit to occur on an unloaded feeder.

Internal Voltages. The construction of the 80 per cent power-factor regulation curve is shown in Fig. 45. From this curve it will be seen that the excitation for rated voltage is 2.19 in per unit terms using the excitation for normal voltage for the air-gap line as unity. Therefore, $E_i = 2.19$. Using the values $x_d' = 0.24$ and $x_d'' = 0.145$, the corresponding internal voltages are $E_i' = 1.165$ and $E_i'' = 1.095$.

Initial Alternating-current Components. For the short-circuit condition, the load will be represented by a reactance which takes the same reactive kilovolt-amperes as the load before the fault. This reactance is $j1.67$. The external or feeder reactance is by assumption $j0.145$. During the fault condition these reactances will be in parallel, so that the net external reactance is

$$x_e = \frac{(0.145)(1.67)}{0.145 + 1.67} = 0.134$$

The initial machine currents then become

$$I'' = \frac{E_i''}{x_d'' + x_e} = \frac{1.095}{0.145 + 0.134} = 3.93$$

$$I' = \frac{E_i'}{x_d' + x_e} = \frac{1.165}{0.24 + 0.134} = 3.12$$

$$I = \frac{E_i}{x_d + x_e} = \frac{2.19}{1.1 + 0.134} = 1.77$$

A portion of the above currents is by-passed through the load reactance, so that the initial feeder currents are

$$I'' = \frac{0.134}{0.145}3.93 = (0.924)(3.93) = 3.63$$

$$I' = (0.924)(3.12) = 2.88$$

$$I = (0.924)(1.77) = 1.64.$$

It will be observed from the above that the currents taken by the feeder and the machine vary only about 8 per cent.

Direct-current Components. The maximum direct-current component of the machine current is obtained by taking the vector difference between I'' and the load current. Figure 44 has been laid out to scale for the particular case under consideration, from which the difference is 3.33. The direct-current component is then

$$I_{dc} = \sqrt{2} \times 3.33 = 4.72$$

For the feeder, since it had been unloaded,

$$I_{dc} = \sqrt{2} \times 3.63 = 5.1$$

Total Initial Current. For the machine

$$I_{total} = \sqrt{(4.72)^2 + (3.93)^2} = 6.15$$

For the feeder

$$I_{total} = \sqrt{3} \times 3.63 = 6.29$$

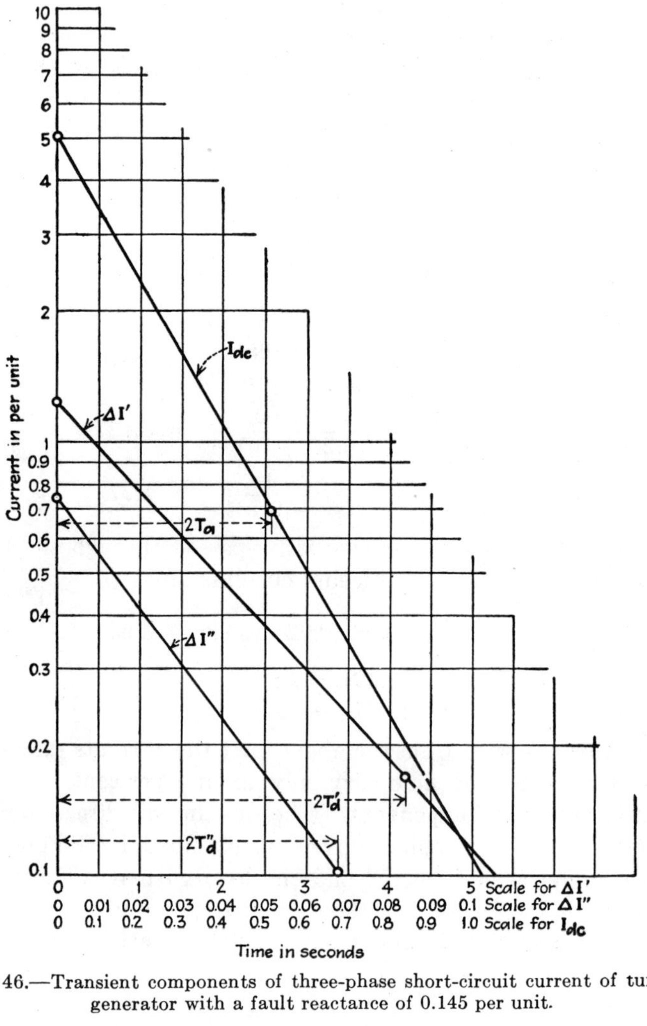

FIG. 46.—Transient components of three-phase short-circuit current of turbine generator with a fault reactance of 0.145 per unit.

Instantaneous Values. The instantaneous values of the feeder currents only will be determined. The initial subtransient component $\Delta I''$ (see Fig. 39) for the feeder is

$$\Delta I'' = I'' - I' = 3.63 - 2.88 = 0.75$$
$$\Delta I' = I' - I = 2.88 - 1.64 = 1.24$$

The transient time constant is, from equation (121)

$$T_d' = \frac{0.24 + 0.134}{1.1 + 0.134}7 = 2.1$$

and from Fig. 45,

$$T_d'' = 0.034$$

It will be assumed that the ratio of the negative-sequence reactance of the external impedance to its resistance is the same as

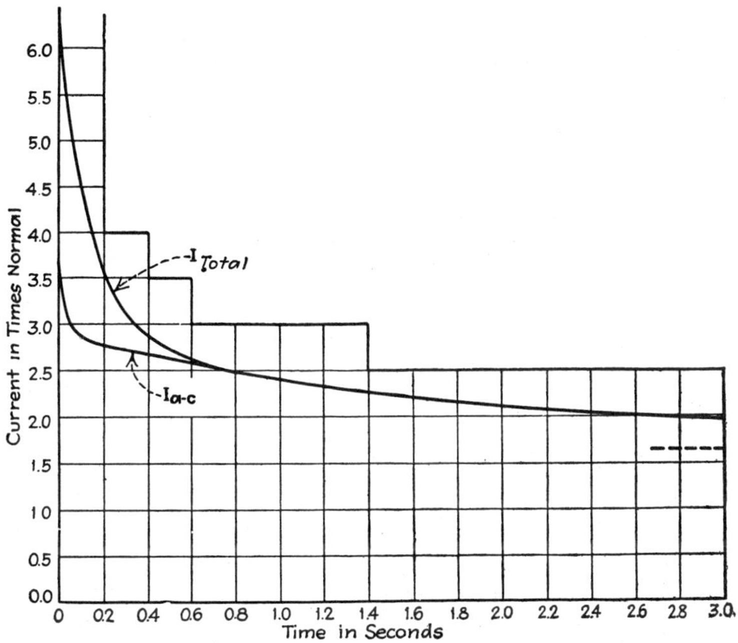

FIG. 47.—Curves illustrating example of decrement current calculation.

that of the generator so that the direct-current time constant remains the same as that of the generator, namely,

$$T_a = 0.26$$

The instantaneous values of the exponentials may be obtained with the assistance of the single logarithmic plots of Fig. 46. These may be drawn as follows. Plot the initial value for $t = 0$, and $(0.368)^2$ or 0.135 of the initial value for t equal to two times the time constant. The straight line connecting these points gives the value of the particular component as a function of time. Adding $\Delta I'$, $\Delta I''$, and the sustained value, 1.64, gives the total alternating-current component; and combining the alternating-current and direct-current components by taking the square root of the sum of the squares gives the total current. Both of these are plotted on Fig. 47. Thus, at 0.05 sec., the currents in the feeder are

$$
\begin{aligned}
I &= 1.64 \\
\Delta I'' &= 0.17 \\
\Delta I' &= \underline{1.21} \\
I_{ac} &= 3.02 \\
I_{dc} &= 4.20 \\
\text{R.m.s. total current} &= \sqrt{3.02^2 + 4.20^2} = 5.18
\end{aligned}
$$

47. Negative-sequence Reactance.

The negative-sequence impedance of a machine is the impedance offered to the flow of negative-sequence current, *i.e.*, it is the negative-sequence voltage across the machine when one unit of negative-sequence current flows through the machine. As stated previously, when negative-sequence voltage is applied to the armature, a field is set up which rotates with synchronous velocity in a direction opposite to that of the rotor. This field sets up currents of double system frequency in the rotor. The subtransient reactances may be measured by blocking the rotor, with the field winding short-circuited, and applying a single-phase alternating voltage across two terminals of the armature. The reactance per phase measured in this manner varies with the position of the rotor. If the machine is a salient pole machine without damper windings the variation is very great, for when the axis of the rotor coincides with the axis of the pulsating field, the field winding constitutes a short-circuited secondary producing a low impedance, but when the two axes are in quadrature, the impedance is merely that determined by the exciting current and is quite high. The upper curve in Fig. 48 shows these subtransient reactances plotted as a function of the angular position of the rotor. For the negative-sequence measurement a similar phenomenon is involved except that the rotor is rotating with double frequency with relation to the field set up by the applied voltage and is taking successively all the possible positions used in determining the subtransient reactance. One would expect, therefore, that the imaginary component of z_2, namely, the negative-sequence reactance x_2, be some kind of a mean between the maximum value of subtransient reactance x_q''

and the minimum value x_d''. Park and Robertson(41) give for this value, when the circuit has a large value of reactance in series, *i.e.*, when the current wave form is essentially sinusoidal,

$$x_2 = \tfrac{1}{2}(x_d'' + x_q'') \tag{122}$$

Distortion arising from saliency effects results in somewhat different values depending upon whether sinusoidal voltage is applied or sinusoidal current is circulated. The dotted line

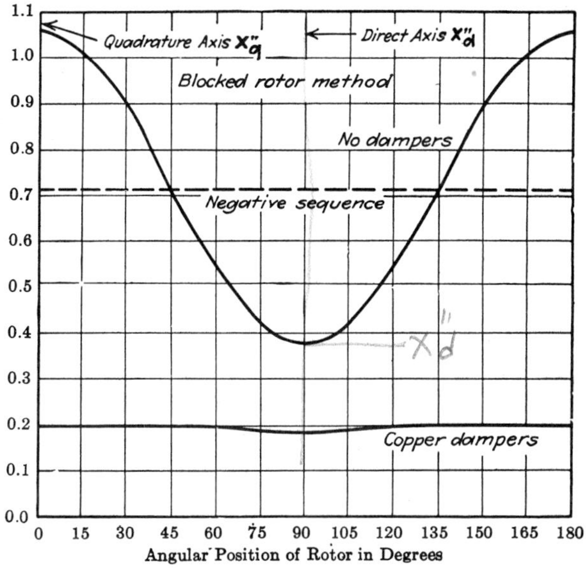

FIG. 48.—Relation between subtransient and negative-sequence reactance.

midway between the maximum and minimum values gives the test result for the negative-sequence reactance of the same salient pole machine without damper windings.

The addition of low-resistance damper windings has different effects upon the reactances in the two axes. In the direct axis the close proximity of the damper windings to the air gap and hence to the armature winding results in a smaller leakage and thus x_d'' is reduced somewhat. A machine without damper windings has no windings on the field structure in the quadrature axis and thus x_q'' is determined by the exciting reactance in this axis. A damper winding would, however, constitute a short-circuited turn and reduce the reactance from one of an exciting reactance to one approaching a leakage reactance. Actually, the reactances in the two axes for a machine with damper windings are very nearly equal, as is shown by the curve for this case in Fig. 48. It follows, therefore, that the negative-sequence reactance is also very nearly equal to these values.

Method of Test. In addition to the method outlined whereby x_2 is defined as the arithmetic mean of x_d'' and x_q'', x_2 may be determined directly either by applying negative-sequence voltage or by the method[(34),(73)] shown in Fig. 49. With the machine driven at rated speed, and with a single-phase short-circuit

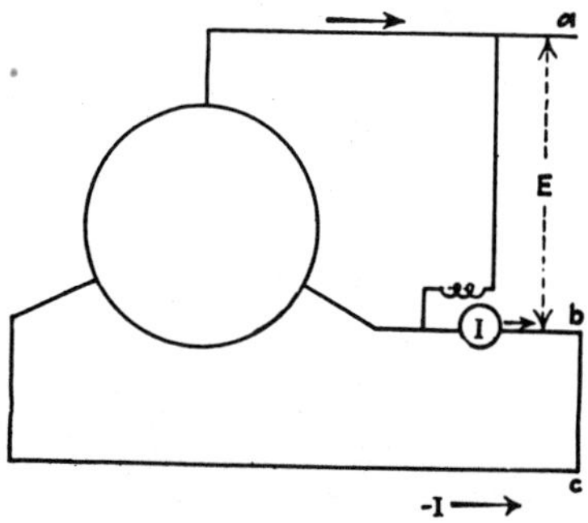

$$I_a = o \quad I_b = I \quad I_c = -I \qquad I_{a2} = \tfrac{1}{3}(o + a^2 I - aI) = \frac{a^2 - a}{3} I$$

$$E_A = o \quad E_B = E \quad E_C = -E \qquad E_{A2} = \tfrac{1}{3}(o + a^2 E - aE) = \frac{a^2 - a}{3} E$$

$$E_{a2} = j\frac{E_{A2}}{\sqrt{3}} = j\frac{a^2 - a}{3\sqrt{3}} E \qquad z_2 = \frac{E_{a2}}{I_{a2}} = \frac{jE}{\sqrt{3}I}$$

If $\phi = \cos^{-1} \frac{P}{\bar{E}\bar{I}}$, where P = wattmeter reading,

then, $$z_2 = \frac{\bar{E}}{\sqrt{3}\bar{I}}(\sin \phi + j \cos \phi) = r_2 + jx_2$$

FIG. 49.—Determination of the negative-sequence impedance of symmetrically wound machines.

applied between two of its terminals (neutral excluded), the sustained armature current and the voltage between the terminal of the free phase and either of the short-circuited phases are measured. The reading of a single-phase wattmeter with its current coil in the short-circuited phases and with the above mentioned voltage across its potential coil is also recorded. The negative-sequence impedance equals the ratio of the voltage to the current so measured, divided by the square root of three. The per unit negative-sequence reactance equals this value of impedance multiplied by the ratio of power to the product of voltage and current.

48. Negative-sequence Resistance.

The power associated with the negative-sequence current may be expressed as a resistance times the square of the current. This resistance is designated the *negative-sequence resistance.* For a machine with no dampers the only source of loss is in the armature and field resistances, eddy currents, and iron loss. The copper loss in the armature and field is very small as is also the iron and eddy loss in the armature, but the iron and eddy loss in the rotor may reach quite high values. Copper damper windings provide a lower impedance path for the eddy currents and hinder the penetration of flux into the pole structure. The relatively low resistance of this path results in a smaller negative-

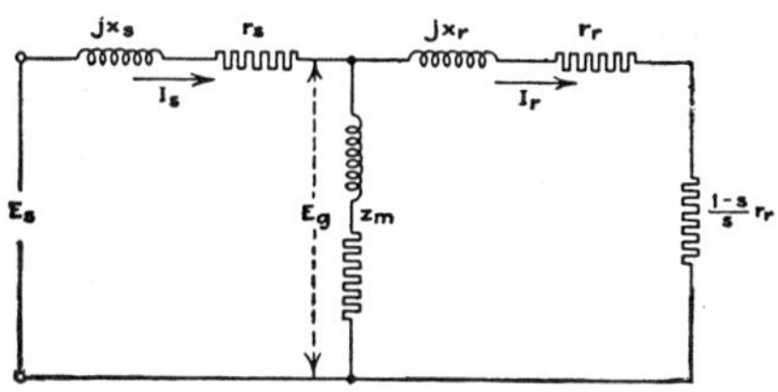

FIG. 50.—Equivalent circuit of induction motor.

sequence resistance. For higher resistance dampers, the negative-sequence resistance increases to a point beyond which the larger resistance diminishes the current in the rotor circuits sufficiently to decrease the loss.

Induction Motor Diagram. The nature of the negative-sequence resistance is best visualized by analyzing the phenomena occurring in induction motors. Figure 50 shows the usual equivalent circuit* of an induction motor in which

r_s = stator resistance.
x_s = stator-leakage reactance at rated frequency.
r_r = rotor resistance.
x_r = rotor-leakage reactance at rated frequency.
z_m = shunt impedance to include the effect of magnetizing current and no-load losses.
E_s = applied voltage.
I_s = stator current.
I_r = rotor current.
s = slip.

* For more detailed description of this circuit see any standard text book such as Ralph R. Lawrence, "Principles of Alternating Current Machinery," McGraw-Hill Book Company, Inc., New York.

All of the foregoing quantities are assumed to be given in per unit quantities.

The justification for this diagram may be seen briefly as follows: The air-gap flux due to the currents I_s and I_r induce the voltage E_g in the stator and sE_g in the rotor. In the rotor the impedance drop is

$$r_r I_r + jsx_r I_r$$

since the reactance varies with the frequency of the currents in the rotor. The rotor current is therefore determined by the equation

$$sE_g = r_r I_r + jsx_r I_r$$

or

$$E_g = \frac{r_r}{s} I_r + jx_r I_r \tag{123}$$

It follows from this equation that the rotor circuit can be completely represented by placing a circuit of impedance $\frac{r_r}{s} + jx_r$ across the voltage E_g. The total power absorbed by $\frac{r_r}{s}$ must be the sum of the rotor losses and the useful shaft power, so that, resolving $\frac{r_r}{s}$ into the resistances r_r and $\frac{1-s}{s} r_r$, the power absorbed by r_r represents the rotor copper loss and the power absorbed by $\frac{1-s}{s} r_r$ represents the useful shaft power.

Neglecting r_s and the real part of z_m, the only real power is that concerned in the rotor circuit. Now assume that the induction motor is loaded by means of a direct-current generator connected to the shaft. At small slips the electrical input into the stator is equal to the copper loss, *i.e.*, the $I_r^2 r_r$ of the rotor plus the shaft load. With the rotor locked the shaft load is zero, and the total electrical input into the stator is equal to the rotor copper loss. At 200 per cent slip, *i.e.*, with the rotor rotating at synchronous speed in the reverse direction, the copper loss is $I_r^2 r_r$, the electrical input into the stator is $\frac{I_r^2 r_r}{2}$, and the shaft load $\frac{1-2}{2} r_r I_r^2$ or $-\frac{I_r^2 r_r}{2}$. A negative shaft load signifies that the direct-current machine instead of functioning as a generator is now a motor. Physically that is just what would be expected, for as the slip increases from zero the shaft power increases to a maximum and then decreases to zero for 100 per cent slip. A further increase in slip necessitates motion in the opposite direction, which requires a driving torque. It will be observed, therefore, that at 200 per cent slip the electrical input into the stator is equal to the mechanical input through the shaft; half of the copper loss is supplied from the stator and half through the shaft. This is the condition obtaining with respect to the negative-sequence in which the rotor is rotating at a slip of 200 per cent relative to the synchronously rotating negative-sequence field in the stator.* **Half of the machine loss associated with the negative-sequence current is supplied from the stator and half by shaft torque through the rotor.**

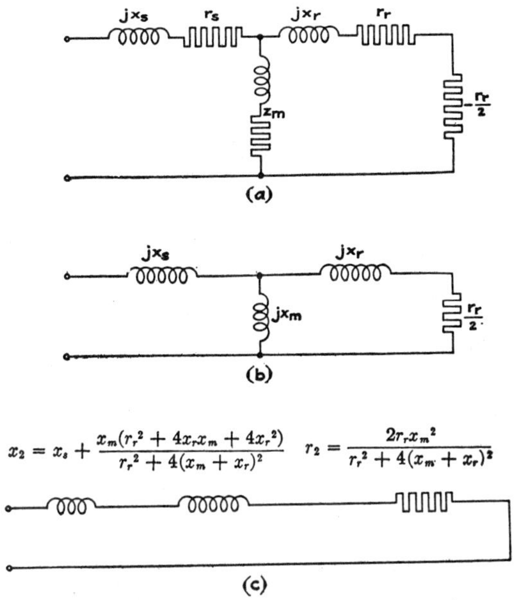

FIG. 51.—Development of negative-sequence resistance and reactance from equivalent circuit of induction motor. (*a*) Negative-sequence diagram for induction motor; (*b*) neglecting armature and no load losses; (*c*) simplified network-negative-sequence resistance and reactance.

The factor of fundamental importance is the power supplied by the stator and through the shaft, which can always be determined by solving the equivalent circuit involving the stator and rotor constants and the magnetizing-current constants. A more convenient device, since s is constant and equal to 2 for the negative-sequence, is to reduce the equivalent network to a simple series impedance as shown in Fig. 51(*c*). The components of this impedance will be called the negative-sequence resistance

* For a more complete discussion of induction-motor performance with the application of negative-sequence voltage refer to Chap. XVII.

r_2 and the negative-sequence reactance x_2. The values of these constants are also given in Fig. 51. The current flowing through the negative-sequence impedance is the current flowing through the stator of the machine, and the power loss in r_2 is equal to the loss supplied from the stator of the machine and the equal loss supplied through the shaft.

The total electrical effect of the negative-sequence resistance is obtained by inserting the negative-sequence resistance in the negative-sequence network and solving the network in the usual manner. All three of the sequence currents are thus affected by a change in the negative-sequence resistance. The total electrical output of a generator is equal to the total terminal power output plus the losses in the machine. However, since the negative- and zero-sequence power outputs are merely the negative of their losses, the contribution to the electrical output by the negative- and zero-sequences is zero. The total electrical output is therefore that due to the positive-sequence and to include the positive-sequence armature-resistance loss it is only necessary to use the positive-sequence internal voltage in the calculations. Or viewed differently, since there are no internal generated voltages of the negative- or zero-sequence, the corresponding internal power must be zero. In addition to this electrical output which produces a torque tending to decelerate the rotor, there also exists the negative-sequence shaft power supplied through the rotor. It was shown that the value of this power tending to decelerate the rotor is numerically equal to the negative-sequence power supplied to the stator which is equal to the loss absorbed by the negative-sequence resistance. Therefore, the total decelerating power is equal to the positive-sequence power output plus the loss in the negative-sequence resistance.

The assumption was made that the stator resistance and the losses in the magnetizing branch were neglected. For greater refinements, the stator resistance and the losses in the magnetizing branch can be taken into consideration by substituting them in the equivalent circuit and reducing that circuit to a simple series resistance and reactance, wherein the resistance becomes the negative-sequence resistance and the reactance the negative-sequence reactance. The ratio of the negative-sequence shaft power to the negative-sequence stator power is then equal to the ratio of the power loss in $\frac{r_r}{2}$ for unit negative-sequence current in the stator to r_2. This ratio can be obtained very easily by test by measuring the shaft torque and the negative-sequence input when negative-sequence voltages only are applied to the stator.

While this analysis has premised induction-motor construction, the conclusions may also be applied to synchronous machines.

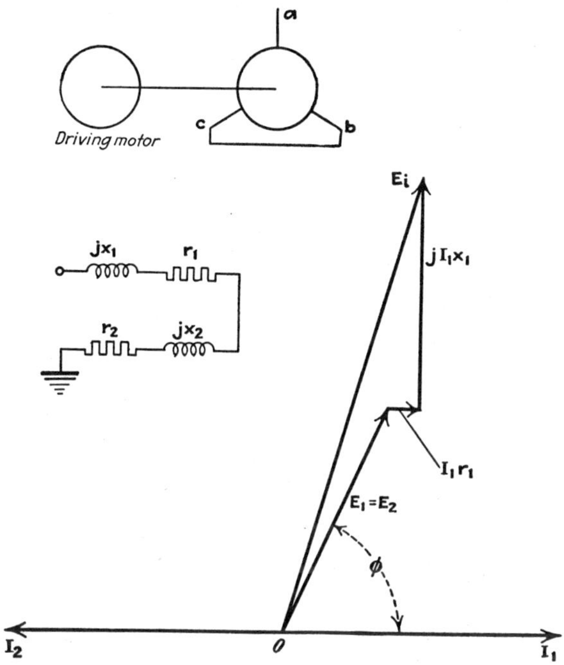

FIG. 52.—Negative-sequence resistance of a synchronous machine.

$$r_2 = \frac{3(P_s - P_{F+W})}{2I^2}$$

in which

P_s = shaft input.
P_{F+W} = friction and windage loss.

Method of Test. While r_2 and x_2 can be determined by applying negative-sequence voltage from another source of supply to the armature, the following method[58] has the advantage that the machine supplies its own negative-sequence voltage. Two terminals of the machine under test are short-circuited and the machine driven at rated frequency by means of a direct-current motor. The equivalent circuit and vector diagram for this connection are shown in Fig. 52. The positive-sequence power at the terminals is equal to the product of $\bar{E}_1$ and $\bar{I}_1$ and the cosine of the angle ϕ. It will be observed that this power is positive. However, the negative-sequence power output is equal to the product of $\bar{E}_2$, $\bar{I}_2$, and the cosine of the angle between E_2 and I_2, and since $I_2 = -I_1$, and $E_1 = E_2$, the negative-sequence power output is the negative of the positive-sequence power output, which, of course, must follow since the output of the machine is zero. A negative output is equivalent to a positive input. This input is equal to $r_2\bar{I}_2^2$. Therefore, the positive-sequence terminal output is $r_2\bar{I}_2^2$, and adding to this the copper loss due to I_1, gives the total shaft power due to the positive-sequence as $r_2\bar{I}_2^2 + r_1\bar{I}_1^2$. Now from Fig. 51($a$), if z_m be neglected, the negative-sequence input is equal to

$$\left(r_r + r_s - \frac{r_r}{2}\right)\bar{I}_2^2 \text{ or } \left(\frac{r_r}{2} + r_s\right)\bar{I}_2^2,$$

from which it follows that

$$r_2 = \frac{r_r}{2} + r_s \tag{124}$$

As shown previously the negative-sequence shaft power is equal to $\frac{r_r}{2}\bar{I}_2^2$, which on substituting $\frac{r_r}{2}$ from (124) reduces to $(r_2 - r_s)\bar{I}_2^2$. But since $r_s = r_1$, the expression for the negative-sequence shaft power may also be written $(r_2 - r_1)\bar{I}_2^2$. Therefore the total shaft input into the alternating-current machine is equal to $r_2\bar{I}_2^2 + r_1\bar{I}_1^2 + (r_2 - r_1)\bar{I}_2^2$ and, since $I_1 = I_2$, reduces to $2r_2\bar{I}_2^2$.

Including the effect of friction and windage, $P_{(F+W)}$, and calling P_S the total input into the alternating-current machine from the driving tool,

$$r_2 = \frac{P_S - P_{(F+W)}}{2\bar{I}_2^2} \tag{125}$$

and, since $I_2 = \frac{I}{\sqrt{3}}$ where I is the actual measured phase current,

$$r_2 = \frac{3[P_S - P_{(F+W)}]}{2\bar{I}^2} \tag{126}$$

49. Zero-sequence Impedance.

The zero-sequence impedance is the impedance offered to the flow of unit zero-sequence current, *i.e.*, the voltage drop across any one phase (star-connected) for unit current in each of the phases. The machine must, of course, be star-connected for otherwise the term zero-sequence impedance has no significance as no zero-sequence current can flow.

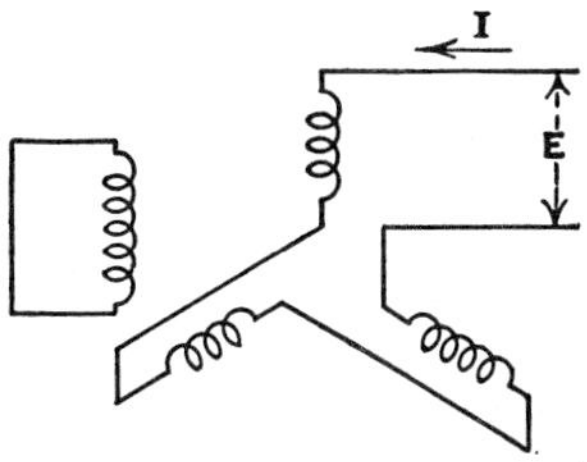

Rotor at synchronous speed (or locked)
Zero-sequence impedance,

$$z_0 = \frac{E}{3I}$$

FIG. 53.—Connection for measuring zero-sequence impedance.

The zero-sequence impedance of synchronous machines is quite variable and depends largely upon pitch and breadth factors. In general, however, the values are much smaller than the positive- and negative-sequences. The nature of the impedance may be suggested by considering that, if the armature windings were infinitely distributed so that each phase produced a sinusoidal distribution of the m.m.f., then the superposition of the three phases with equal instantaneous currents cancel each other and produce zero field and consequently zero reactance except for slot and end-connection fluxes. The departure from this ideal condition introduced by chording and the breadth of the phase belt determines the zero-sequence impedances.

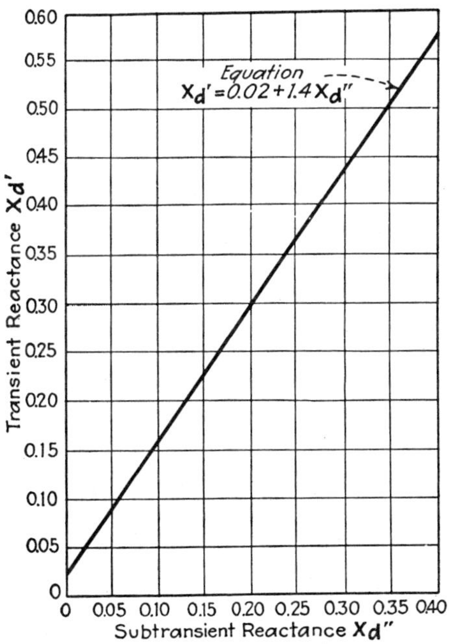

FIG. 54.—Relation between x_d' and x_d'' (saturated values) for three-phase synchronous machines in per unit quantities.

The zero-sequence resistance is equal to, or somewhat larger than, the positive-sequence resistance. In general, however, it is neglected in most calculations.

Method of Test. The most convenient method for test is to connect the three phases together, as shown in Fig. 53, with the

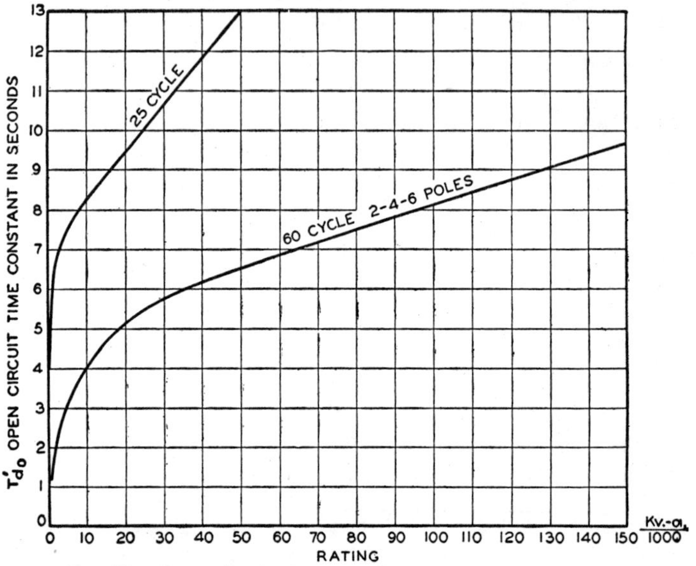

FIG. 55.—Open-circuit time constants of turbine generators.

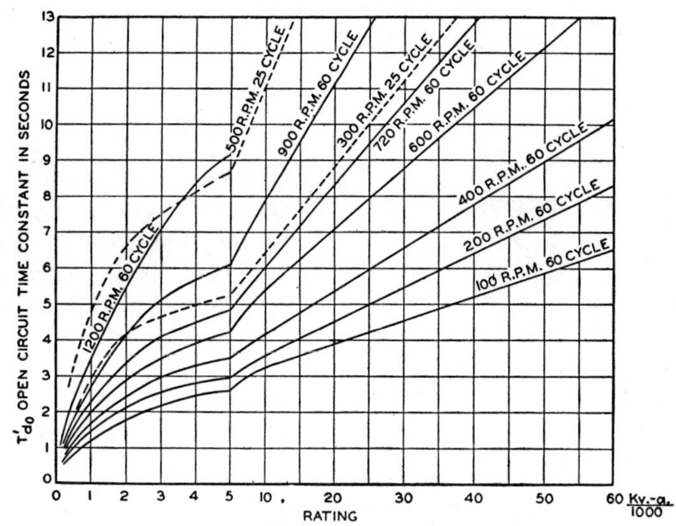

FIG. 56.—Open-circuit time constants of alternating-current generators and motors.

field short-circuited on itself. This connection insures equal distribution of current between the three phases and for this reason is preferable to connecting the three phases in parallel. The zero-sequence impedance is then equal to $Z_0 = \frac{E}{3I}$ as indicated in the figure.

50. Typical Values of Synchronous Machine Constants.

The constants of synchronous machines vary over a wide range with the particular type of machine, rating, and speed. As a result of averaging the data on a large number of machines the straight-line relation between x_d' and x_d'' in Fig. 54 was found to hold remarkably well. The relation between the time constants and the rating can also be represented by curves when segregated into 25- and 60-cycle turbine generators and salient pole generators and motors of different speeds. These are shown in Figs. 55 and 56. Other constants classified as to

TABLE III.—TYPICAL THREE-PHASE 60-CYCLE SYNCHRONOUS MACHINE CONSTANTS*

Machines	x_d	x_d'	x_d''	x_2	x_0†
Turbine generator	$\frac{1.10}{0.95 - 1.45}$	$\frac{0.19}{0.12 - 0.26}$	$\frac{0.12}{0.07 - 0.17}$	$= x_d''$	$\frac{0.03}{0.01 - 0.14}$
Salient pole motors and generators (with damper windings)	$\frac{1.10}{0.60 - 1.45}$	$\frac{0.33}{0.20 - 0.51}$	$\frac{0.22}{0.13 - 0.35}$	$= x_d''$ (nearly)	$\frac{0.06}{0.02 - 0.20}$
Waterwheel generators (no damper windings)	$\frac{1.10}{0.60 - 1.45}$	$\frac{0.35}{0.20 - 0.45}$ (satur.)	$0.85\,x_d'$	$\frac{0.50}{0.30 - 0.70}$	$\frac{0.07}{0.04 - 0.22}$
Condensers	$\frac{1.80}{1.50 - 2.20}$	$\frac{0.37}{0.27 - 0.55}$	$\frac{0.25}{0.18 - 0.38}$	$\frac{0.24}{0.17 - 0.37}$	$\frac{0.08}{0.02 - 0.15}$

* For test data of specific machines, refer to bibliography item 73.

† x_0 varies so critically with armature winding-pitch that the average values given are not very dependable.

type of machine are given in Table III, the figure above the line representing average values.

51. Unbalanced Faults.

It was shown in Chap. III that unbalanced short-circuits on synchronous machines can be represented by equivalent circuits involving the positive-, negative-, and zero-sequence impedances. From these circuits it may be seen that the relations between positive-, negative-, and zero-sequence currents are dependent only upon the character of the fault and the negative- and zero-sequence impedances. It follows therefore that the negative- and zero-sequence currents vary proportionately with the positive-sequence current, and in determining the negative- and zero-sequence components of current it is necessary to obtain merely the positive-sequence time variation of current. It may further be seen from these diagrams that the effect of the unbalance upon the positive-sequence current may be completely simulated by placing an impedance in series relation with the machine; for a line-to-ground fault the impedance is $Z_2 + Z_0$, for a line-to-line fault Z_2, and for a double line-to-ground fault $\frac{Z_0Z_2}{Z_0 + Z_2}$. The time variation of the positive-sequence current can then be determined in the usual manner after which the negative- and zero-sequence currents are determined by applying the proportionality constants.

The foregoing applies only to the alternating-current component of current. The direct-current component is obtained as for the balanced short-circuits by taking the negative of the difference between the initial instantaneous alternating-current component and the instantaneous current in the different phases before the fault.

52. Decrement Curves.*

The commercial application of relays and circuit-breakers demands a simpler means than that described for calculating the short-circuit current at any time. By making certain assumptions which represent typical operating conditions and machine characteristics it is possible to compute a set of curves to evaluate the short-circuit current which are useful over a wide range of machine constants. These curves are called **standard**

* This section is based on data and curves presented in an A.I.E.E. paper by W. C. Hahn and C. F. Wagner.(75)

decrement curves. They are applicable not only to single machines but to any number of parallel machines whose constants are sufficiently similar. The assumptions upon which they are based are:

1. The generators are assumed to be operating at rated voltage and kilovolt-amperes at 80 per cent power factor immediately preceding the short-circuit.
2. No automatic voltage regulator is used.
3. The actual system subjected to fault may be represented by a single equivalent generator of the same total rating and an external reactance.
4. The load is assumed to be located at the machine terminals, and the machine reactance is taken as 15 per cent unless the total reactance is less than 15 per cent. In this event all the reactance is assumed to be in the machine.
5. The short-circuit is assumed to occur on an unloaded feeder.
6. The short-circuit is assumed to occur at the point of the voltage wave which corresponds to maximum possible instantaneous current.
7. All resistance in the circuit including the resistance of the fault is neglected.
8. All the machine e.m.fs. are assumed to be in phase.
9. The machine reactances and time constants are taken as representative of modern machines. Of particular importance is the relation between transient and subtransient reactance of machines (see Fig. 54).

Description of Decrement Curves. The decrement curves of Figs. 57 and 58 give the r.m.s. total current expressed in terms of *times normal current* with total connected synchronous capacity in kilovolt-amperes as a base. In Fig. 57, the elapsed time is the abscissa and reactance is the parameter. In Fig. 58, reactance is plotted as the abscissa with elapsed time from the beginning of the fault as a parameter for the curves.

The reactance to be used with the curves for any kind of a fault may be obtained either by means of analytical calculations or by use of the calculating board. In order to choose the proper decrement curve, it is essential that the reactance used to select a curve in Fig. 57 or a line in Fig. 58 be expressed as a per cent of the **total connected synchronous capacity in kilovolt-amperes** rather than, as is common practice, an arbitrary value conveniently chosen to expedite system calculations.

If the system open-circuit time constant referred to the point of fault is known to be other than 5.0 sec., the time scale of the curves may be corrected to conform to it by means of the formula given on the curve sheets.

Three-phase Short-circuit. The reactance to be used to select the proper decrement curve for a three-phase fault is

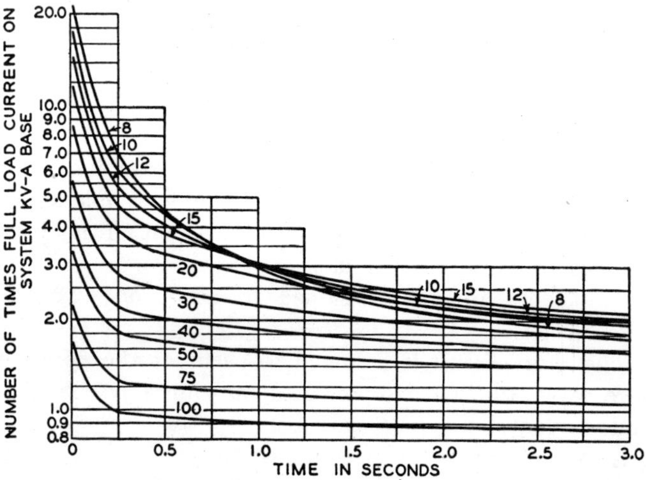

FIG. 57.—Short-circuit decrement curves.

System reactance *must be* based on system kilovolt-amperes and *not* on the particular kilovolt-amperes chosen as a base for calculations.

System kilovolt-amperes is connected synchronous capacity in kilovolt-amperes.

Note: These curves are based on $T'_{d0} = 5$ sec.

For other values of T'_{d0}, multiply actual times by $\frac{5}{T'_{d0}}$, to get equivalent times to use in curves.

How to Use Curves:

For three-phase short-circuit:

Reactance: Use system reactance to the point of fault.

Times full load (normal) scale: Use scale reading.

For line-to-line short-circuit:

Reactance: Use two times the system reactance for the three-phase fault.

Times full-load (normal) scale: Multiply scale reading by $\sqrt{3}$.

For line-to-ground short-circuit:

Reactance: Use $(2X_1 + X_0)$

Times full-load (normal) scale: Multiply scale reading by 3.

what is commonly spoken of as the *system reactance* referred to or viewed from the point of fault. It is the reactance used with the decrement curves formerly published. When determining this reactance it is important that synchronous machines be represented by their subtransient reactances, and also that **all loads other than synchronous be neglected.** Proper account has been taken of these loads for the average system in deriving the decrement curves.

The system reactance referred to the point of fault enables the proper decrement curve to be selected which gives the three-phase r.m.s. short-circuit current or kilovolt-amperes in **times normal** at any time after the occurrence of the short circuit, subject to the assumptions previously given.

Line-to-line Short-circuit. For a line-to-line short-circuit, the fault current is determined by the use of the standard decre-

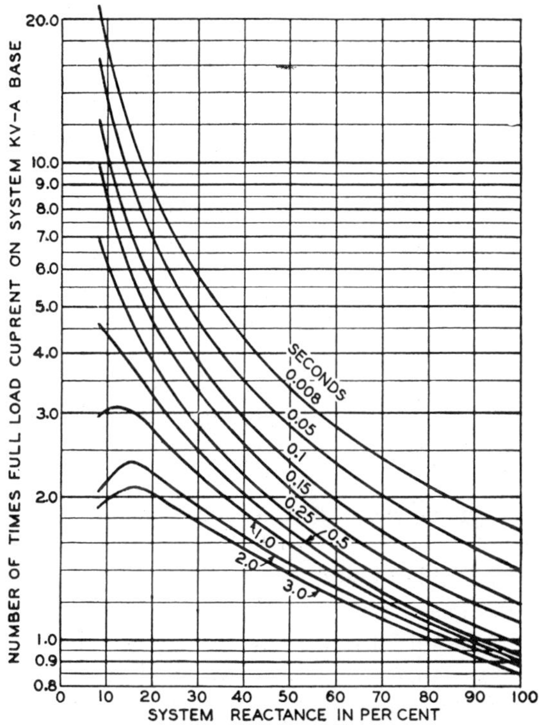

Fig. 58.—Short-circuit decrement curves.
For way to use curves, see Fig. 57.

ment curves of Fig. 57 or Fig. 58 in the same manner as the three-phase short-circuit current by using a reactance equal to twice the system reactance (the value used for three-phase short-circuit calculation) and multiplying the resulting current in times normal read from the curves by $\sqrt{3}$.

Where the negative-sequence reactance of the system is known, more accurate results may be obtained by using a reactance equal to the sum of the system reactance x_1 and the negative-sequence system reactance x_2 instead of twice the system reactance x_1. The negative-sequence reactance of the system is obtained in the same manner as the system reactance x_1 except that the negative-sequence reactance of the machines is substituted for the subtransient reactance of machines.

Line-to-ground Fault. The value of reactance to be used with the curves for a line-to-ground short-circuit is the sum of the positive-, negative-, and zero-sequence reactances of the system $(x_1 + x_2 + x_0)$ as measured from the point of fault. For ordinary calculations it is sufficiently accurate to use a reactance equal to $(2x_1 + x_0)$, where x_1 is the system reactance used for three-phase short-circuit calculations and x_0 is the zero-sequence reactance of the system. The line-to-ground current at any time will be equal to three times the value of *times normal current* read from the decrement curves using the above values for system reactances.

Minimum Fault Current. For certain relay applications it is sometimes necessary to know the alternating component of the fault current, *i.e.*, with no direct-current component, rather than the maximum current at any time as given by the standard decrement curves which are based on an asymmetrical fault. The initial r.m.s. alternating (symmetrical) currents for values of system reactance are given in Table IV. As the asymmetrical component is negligible after 0.3 sec., the r.m.s. alternating (symmetrical) current for any time less than 0.3 sec. may be obtained by interpolating between the initial value obtained from Table IV and the curve values at 0.3 sec. Satisfactory results can readily be obtained by sketching in a curve starting from the initial r.m.s. alternating (symmetrical) current and making it tangent to the corresponding standard decrement curve at 0.3 sec.

TABLE IV.—INITIAL R.M.S., ALTERNATING (SYMMETRICAL) CURRENT

Reactance on decrement curve	8	10	12	15	20	30	40	50	75	100
Initial symmetrical current	12.3	9.96	8.38	6.88	5.11	3.37	2.52	2.01	1.33	1.0

R.M.S. Total Current for Salient Pole Machines without Dampers. Where salient pole machines without damper windings supply the short-circuit current the standard decrement curves give good results after 0.2 sec., provided that for these machines the equivalent subtransient reactances are obtained from the known transient reactances by means of the relation given in Fig. 54. The initial values indicated by the standard decrement curves may be 30 per cent too high. If more accurate values of initial current are desired, it is necessary to make a separate calculation of the system reactance using the actual value of the subtransient reactance of the salient pole machines without damper windings instead of the equivalent value obtained from the transient reactance by means of the relation given in Fig. 54. A curve may readily be sketched between the new initial point and the 0.2-sec. point on the standard decrement curve corresponding to the equivalent subtransient reactance of Fig. 54.

The subtransient reactance of machines without damper windings has generally been taken as equal to the transient reactance. However, there is an appreciable subtransient component in the short-circuit current of such a machine due to the damping action of the pole rivets and other closed circuits and also due to the rapidly disappearing effects of saturation. The test data available indicate that the subtransient reactance of a machine without damper windings should be taken as about 85 per cent of the transient reactance.

53. Internal Voltage Method.

The internal voltage method* of calculating short-circuit current variation was developed to allow solution of short-circuit problems of systems that cannot be replaced by a single equivalent generator, *i.e.*, systems in which machines are located unsymmetrically with respects to the fault or have different time constants and reactances. The method is adapted to the use of the calculating board, which use reduces considerably the time required for solution. It is also easily extended to include the effect of variation in exciter voltage.

It may be seen from Fig. 47 that the contribution of the subtransient component is practically negligible, 0.1 sec. after a fault occurs, and since one is usually interested in a breaker opening time of either one cycle (practically instantaneous) or something over 0.1 sec., the subtransient component may be neglected except as it influences the evaluation of the direct-current component at zero time.

* For a more detailed discussion including an illustrative example of this method refer to bibliography item 76.

Fundamental Assumption. The variation of the transient component of the alternating current is determined from the variation of internal transient voltage. The assumption is made that the internal transient voltages of the individual machines vary in an exponential manner between their initial values and their final values. The initial values are obtained from the loading of the machines previous to the inception of the fault, and the final values are obtained by using the synchronous internal voltages of all the machines, calculating the sustained current in all of the machines and then subtracting from the synchronous internal voltages of the individual machines

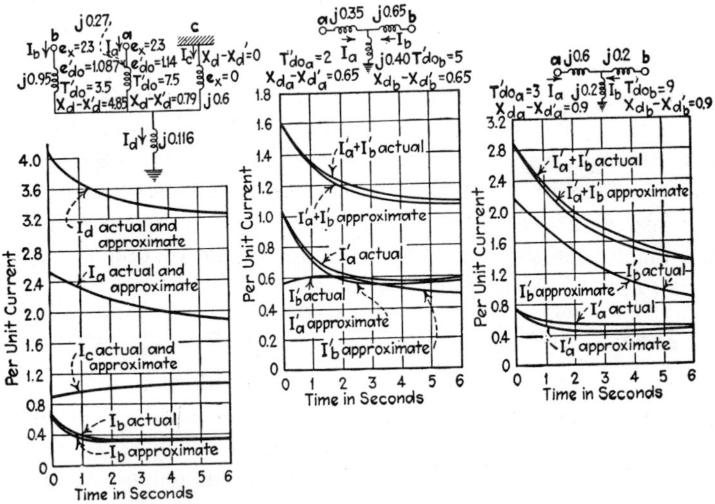

FIG. 59.—Comparison between actual short-circuit currents and those obtained by using "internal voltage" method of calculation. Subtransient and direct-current components have been neglected.

the difference in the synchronous and the transient reactance drops. This will give the sustained transient internal voltage back of the transient reactance for the individual machines. The time constant for the exponential curve connecting the initial and final values of transient internal voltage is determined as though the particular machine were the only one connected to the system, the other machines being replaced by their respective synchronous reactances. The accuracy of this assumption may be gauged from Fig. 59 which gives the transient alternating-current component for systems shown in the inserts as determined in two ways: (1) by the internal voltage method, and (2) by an accurate differential-equation solution. Two of these systems consist of two generators each having different time constants. The shunt reactances may represent either the reactance of a high-reactance three-phase fault or the equivalent negative-sequence reactance of a line-to-line fault. The error involved may be seen to be very small and becomes still smaller as the shunt reactance is decreased. The left-hand figure represents the current resulting when a line-to-line fault is applied to the system shown, which consists of two generators connected to an infinite system. It is evident that the accuracy of the internal voltage method is quite sufficient for most practical purposes.

Alternating-current Component. Having given the instantaneous value of the transient internal voltage for the individual machines, the positive-sequence component of the transient component of current in any branch of the network can be obtained for any instant by inserting the particular values of transient internal voltages obtained from the exponential curves for the particular instant. For the different kinds of faults the negative- and zero-sequence components can be obtained by calculating the currents in the particular branches of the negative- and zero-sequence networks. For line-to-line faults the maximum positive- and negative-sequence components of current in the reference phase a are in phase opposition, so that the maximum current occurs in phases b and c and the sequence components are 60 deg. out of phase. If I_1' and I_2' be the positive- and the negative-sequence currents in any branch, then the total current is given by

$$\bar{I}' = \sqrt{(\bar{I}_1')^2 + (\bar{I}_2')^2 + \bar{I}_1\bar{I}_2}. \tag{127}$$

For a line-to-ground fault the maximum current occurs in the particular phase upon which the fault occurs and the value of this current is given by

$$I' = I_1' + I_2' + I_0' \tag{128}$$

Direct-current Component. The initial value of the subtransient current I'' is necessary in the determination of I_{dc} and is obtained by computation or measurement on a calculating board by using the initial value of subtransient internal voltages and the subtransient reactances in the various machines. The initial value of I_{dc} is obtained from the vector difference between the initial value of I'' and the load current in a manner analogous to the case shown in Fig. 44.

Calculations have shown that the direct-current component will increase the r.m.s. value of short-circuit current less than 5 per cent after 0.1 sec. if the direct-current time constant of the circuit is 0.05 sec. Hence for such machines the contribution of the direct-current component may be neglected after 0.1 sec. For faults involving only synchronous machines and transformers the time constants are usually greater than 0.05 sec., and therefore the direct-current component will be of importance beyond 0.1 sec.; but for circuits involving transmission or distribution lines the time constant is less than 0.05 sec., and therefore the contribution of the direct-current component is negligible after 0.1 sec.

Summary. To summarize the method, the subtransient component because of its rapid attenuation is negligible after 0.1 sec., and therefore the total alternating-current component is equal to the transient component which is obtained by calculating the time variation of the transient internal voltage of the individual machines and using these voltages in connection with the network which is set up with the transient reactances of the machines. The direct-current component is obtained from a knowledge of the initial subtransient alternating-current component and the load current before the fault. The initial value of total current is then obtained by taking the square root of the sum of the squares of the initial direct-current and the initial subtransient current. The total current, when the fault involves transmission or distribution lines, is equal to the transient alternating-current component after 0.1 sec., and it is only when the fault involves only transformers and generators that the direct-current component must be taken into consideration after 0.1 sec.

54. Effect of Varying Exciter Voltage.

The internal voltage method is easily extended to include the effect of alteration in exciter voltage,[76] in fact this is one of the principal advantages of the method. The direct-current and subtransient alternating-current components are affected only slightly by exciter-voltage variation and may therefore be computed in the usual manner neglecting the variation in exciter voltage. The problem then reduces to the determination of the instantaneous values of transient internal voltage of each machine, after which the procedure is identical with that for constant exciter voltage as discussed previously.

In brief, the method consists of the following. The sustained values of transient internal voltages corresponding to the instantaneous values of exciter voltage are computed. In this way curves are obtained of the values the transient internal voltage would attain if the time constant of the machine were zero. The transient time constants of the individual machinesare then obtained as described under the internal voltage method. The actual instantaneous transient internal voltage for a particular machine is then obtained by plotting these sustained values displaced a distance equal to the transient time constant from the point of zero time as shown in Fig. 60. Starting with the initial value of transient internal voltage which is given by the point a, the actual locus is built up in small increments by following along the lines ab, cd, etc., as shown in the figure. This must be repeated for each machine, after which the instantaneous values of transient component of current are easily obtained.

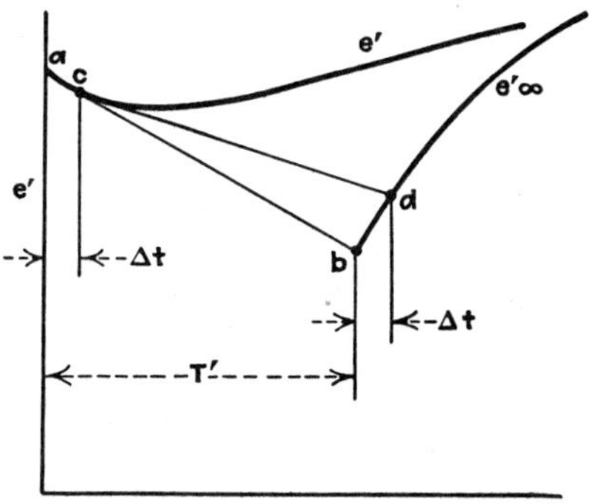

FIG. 60.—Determination of e' with variable excitation.

Problems

1. A three-phase generator operating at normal voltage no load is subjected to a three-phase short-circuit. From the oscillograms the current in per unit values after eliminating the direct-current component is found to be as follows:

Time	I	Time	I	Time	I
0.00	4.48	0.25	2.47	1.50	1.27
0.05	3.41	0.40	2.19	2.00	1.13
0.10	2.96	0.50	2.04	2.50	1.05
0.15	2.72	0.70	1.80	3.00	1.01
0.20	2.56	1.00	1.55	....	0.96

Find the values of X_d', X_d'', X_d, T_d', and T_d''.

2. The direct-current component of current in the three phases of a short-circuited generator are 1,060, 1,932, and 516 amp. In this test the short-circuit occurred at such an instant as not to produce the maximum dissymmetry. Determine (*a*) the maximum possible value of the direct-current component; (*b*) the peak value of the alternating-current component; (*c*) the maximum possible r.m.s. total current.

3. A three-phase 50,000-kva., 11,000-volt, 60-cycle generator is subjected to different kinds of short-circuits while operating at rated voltage no load. The sustained short-circuit currents are found to be: three-phase fault, 2,000 amp.; line-to-line fault, 1,800 amp.; line-to-neutral fault, 2,200 amp. The instantaneous symmetrical three-phase short-circuit current is found to be 20,000 amp. Determine the values of X_d'', X_d, X_2, and X_0 in per cent and in per unit values.

4. Compute the value of the fault current at the point F of Fig. 23(*a*), Chap. IV, at 0.2 sec. after the inception of the fault for different types of faults as follows: (*a*) line-to-ground; (*b*) line-to-line; (*c*) three-phase. Use the standard decrement curves assuming that the reactances given include the subtransient reactance X_d'' for each machine and neglect the effect of arc resistance.

5. Assume the same system as in Prob. 4 but that the generators are all of the salient pole construction without damper windings. Determine the value of the instantaneous symmetrical short-circuit current at 0.05 sec. after the inception of the fault for (*a*) line-to-ground fault; (*b*) line-to-line fault; (*c*) three-phase fault.

6. Assume that an exciter builds up its voltage immediately upon the occurrence of a fault to a magnitude and at a rate given by the expression

$$E_x = 3 - 2\epsilon^{-\frac{t}{T}}$$

where E_x is expressed in per unit excitation, t is the time as measured from the application of the fault, and T is the time constant of the exciter build-up curve. Compute the alternating-current component of the armature current as a function of time for the case of the generator subjected to a three-phase short-circuit from no-load normal voltage. Neglect the subtransient effects and assume the constants

$$X_d' = 0.40 \qquad T = 0.3$$
$$X_d = 1.00 \qquad T_{d0}' = 2.0$$

CHAPTER VI

CONSTANTS OF TRANSFORMERS

The sequence impedances of transformers and similar devices, including series and shunt impedances, two- and three-winding transformers, and autotransformers, will now be considered. Since this apparatus is non-rotative, and since symmetry between the different phases has been assumed, it follows that the impedance of these devices is independent of the phase rotation of the e.m.fs. applied to the terminals, and that the positive- and negative-sequence impedances are identical. The zero-sequence impedances of these devices are, in general, different from the positive- or negative-sequence impedances. While the impedance of the device itself may be represented by the same equivalent circuit, the actual impedance is dependent upon the external connections, which may be different for the zero-sequence.

55. Two-winding Transformers, Series and Shunt Impedances.

Series-impedance branches which are symmetrical in the different phases have the same value for positive-, negative-, and zero-sequence, if there is a neutral return as illustrated in Fig. 61(*a*). If there is no path for zero-sequence current, the corresponding zero-sequence impedance is infinite. Each device may be represented in the zero-sequence network by a suitable connection, as shown in the right-hand column of Fig. 61 with the impedance the same as for positive- or negative-sequence.

Shunt-impedance branches which are symmetrical in the different phases have the same value for positive-, negative-, and zero-sequence, if there is a neutral connection as illustrated in Fig. 61(*c*). However, if the shunt branches are connected in delta or in star without neutral connection, as illustrated in Fig. 61(*d*), the impedance to zero-sequence is infinite and is represented by an open-circuit.

Neutral impedance in a neutral-wire or ground connection is. of course, wholly zero-sequence. The value of this impedance should be *multiplied by three* to obtain the equivalent value for phase-to-neutral. This is illustrated in Fig. 61(*e*), which shows a star-connected group of shunt impedances with a neutral impedance in the ground connection. It will be noted that the equivalent circuit for zero-sequence includes two impedances in series, the first impedance having the same value as the actual impedance per phase in the group of star-connected impedances, whereas the second impedance has three times the value of the impedance in the neutral connection.

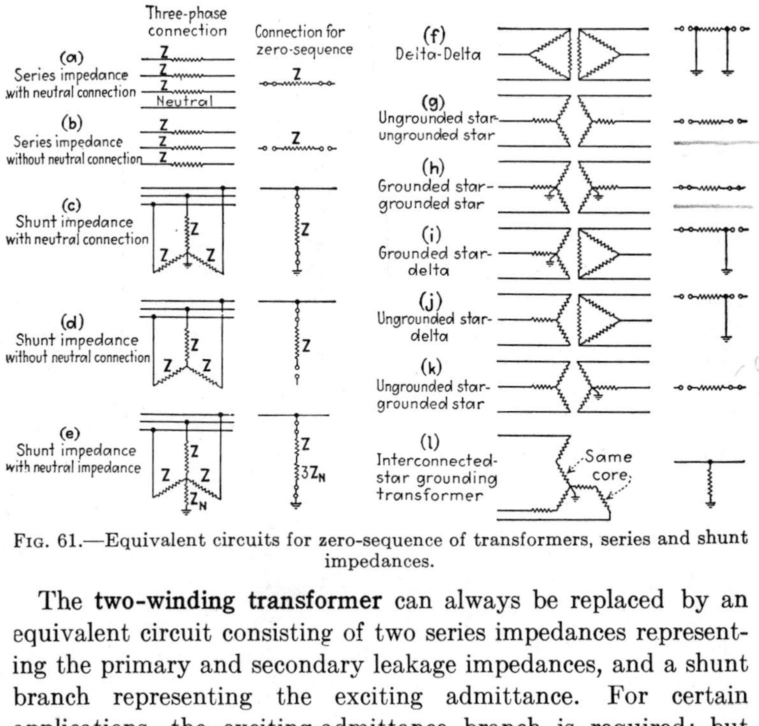

FIG. 61.—Equivalent circuits for zero-sequence of transformers, series and shunt impedances.

The **two-winding transformer** can always be replaced by an equivalent circuit consisting of two series impedances representing the primary and secondary leakage impedances, and a shunt branch representing the exciting admittance. For certain applications, the exciting-admittance branch is required; but for short-circuit calculations it may be neglected and the equivalent circuit reduced to a single series impedance. Typical values of transformer impedance are given in Table V, these values representing the **impedances to positive- or negative-sequence.** These impedances are usually expressed in per cent on a given kilovolt-ampere base, or in ohms on a given voltage base. In any event, they must be reduced to the common base chosen for the calculations.

TABLE V.—TYPICAL VALUES OF TRANSFORMER REACTANCES

	Per Cent
Distribution	3
Network	5
Power	
Up to 66 kv	5–7
88 and 110 kv	6–9
132 and 154 kv	8–10
187 and 220 kv	10–14

The direct-current resistance varies from 0.35 to 0.50 per cent.

The **impedance to zero-sequence** of a bank of three two-winding transformers* is either the same as for positive- or negative-sequence, or is infinite depending upon the type of winding. The connections of the **equivalent circuit** of a two-winding transformer to the external circuit for zero-sequence are illustrated in Fig. 61(*f* to *l*). The most common connections are delta, star, and grounded star. In the case of any delta winding, it may be seen (1) that no zero-sequence current can flow from the delta winding to the external circuit or *vice versa;* and (2) that zero-sequence current can circulate in the delta winding, without flowing through the external circuit. The zero-sequence connections for the equivalent circuit of a delta winding are therefore an open-circuit between the delta winding and the external circuit, and connection to ground of the delta side of the equivalent impedance, as illustrated in (*f*). In the case of a star winding with free neutral, no path is provided for the flow of zero-sequence current either from the external circuit or in the transformer winding itself. Hence, for the zero-sequence connection, a star winding with a free neutral is represented by an open-circuit of the star end of the branch representing the transformer for both the external circuit and to ground, as shown in (*g*). In the case of the grounded star winding, zero-sequence current can flow through the transformer winding only when the external circuit provides a completing circuit. Hence, the connections of the equivalent circuit for grounded star winding are as shown in (*h*). Other combinations of ungrounded star, grounded star and delta windings are shown in (*i*) (*j*) and (*k*).

The **interconnected star** or zigzag grounding **transformer,** illustrated in (*l*), is the only other form of two-winding trans-

* For three-phase transformers see Sec. 57.

former commonly encountered in a three-phase bank. This transformer has two separate windings on a common core, which are interconnected between phases, the windings on the same core being represented by parallel branches in the diagram. This transformer has an open-circuit or, more accurately, an exciting admittance for positive- or negative-sequence e.m.f. supplied to the terminals. For zero-sequence, however, the currents in all line terminals are of the same value. Hence, the impedance per phase to zero-sequence is the leakage impedance from one winding to the other winding on the same core. The zero-sequence connection is therefore a simple shunt impedance to ground, as shown in (*l*). The interconnected star-delta transformer is discussed with the three-winding transformers.

56. Three-winding Transformers.

For stability or short-circuit calculations the exciting current of three-winding transformers may be neglected and only the

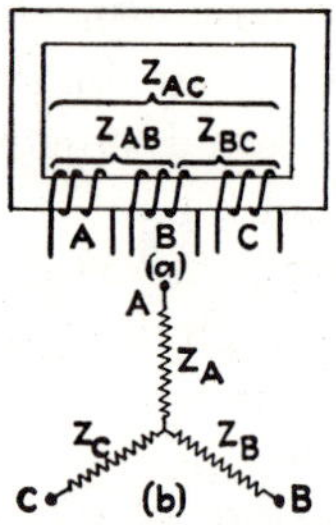

$$Z_{AB} = Z_A + Z_B$$
$$Z_{BC} = Z_B + Z_C$$
$$Z_{CA} = Z_C + Z_A$$
$$Z_A = \tfrac{1}{2}(Z_{AB} + Z_{AC} - Z_{BC})$$
$$Z_B = \tfrac{1}{2}(Z_{AB} + Z_{BC} - Z_{AC})$$
$$Z_C = \tfrac{1}{2}(Z_{CA} + Z_{BC} - Z_{AB})$$

FIG. 62.—Equivalent star representation of three-winding transformer in terms of impedance between windings. Impedances Z_{AB}, etc., are the short-circuit impedances measured between the windings corresponding to the subscripts, with the other winding open. All impedances must be expressed in the same kilovolt-ampere and voltage base.

series drop characteristics taken into consideration. The characteristics of such transformers are usually given in terms of the reactance between two windings taken at a time, the third being open-circuited. While in two-winding transformers both windings have the same rating, the difference in rating of the three windings in three-winding transformers requires that special care be exercised to specify the particular base to which the impedance refers. It has been shown* that when the magnetizing current is neglected, the performance of the transformer can

* PETERS, J. F., and M. E. SKINNER, Transformers for Interconnecting High Voltage Systems, *Trans. A.I.E.E.*, Vol. 40, p. 1181, 1921.

be represented by an **equivalent circuit** consisting of three star-connected impedances. In Fig. 62(*a*), let the impedance between windings *A* and *B* be designated Z_{AB}; that between *B* and *C*, Z_{BC}; and that between *A* and *C*, Z_{AC}. The equivalent star impedances are shown in Fig. 62(*b*). For winding *C* open-circuited, it follows that

$$Z_{AB} = Z_A + Z_B \tag{129}$$

and similarly for the other windings

$$Z_{BC} = Z_B + Z_C \tag{130}$$
$$Z_{AC} = Z_C + Z_A \tag{131}$$

It also follows from the equivalent diagram that the voltage across winding *B* with voltage applied to *A* and with *C* short-circuited is the drop across the impedance Z_C. Expressions for converting these impedances from one form to another are given in Fig. 62. It should perhaps be pointed out that the star point of the equivalent circuit in Fig. 62(*b*) is a fictitious point and does not represent the system neutral and that loads or short-circuits can be applied only to terminals.

Example. As an example of these relations consider a three-phase transformer bank having the following characteristics:

Winding *A*—7,500 kva. (2,500 kva. per phase) star-connected for 66,000-volt line
Winding *B*—10,000 kva. (3,333 kva. per phase) star-connected for 11,000-volt line
Winding *C*—5,000 kva. (1,667 kva. per phase) delta-connected for 2,300-volt line

The per cent resistance drop for this transformer may be assumed to be as follows:

Winding *A*—0.75 per cent resistance drop on 7,500 kva.
Winding *B*—1 per cent resistance drop on 10,000 kva.
Winding *C*—1 per cent resistance drop on 5,000 kva.

The per cent reactance drop between the various windings may be assumed to be as follows:

Winding *A-B*—10 per cent reactance drop on 10,000 kva.
Winding *B-C*— 5 per cent reactance drop on 10,000 kva.
Winding *C-A*—10 per cent reactance drop on 5,000 kva.

In obtaining the data for making the substitution of the star group of impedances for the three-winding transformer, it is important to express the impedances in terms of the same voltage and kilovolt-amperes per winding in order to avoid error. For this purpose express the transformer impedances in terms of a 10,000-kva. bank for a system voltage of 11,000 volts. The phase voltage is $\frac{11,000}{\sqrt{3}} = 6,350$ volts, and the phase current is $\frac{10,000,000}{3 \times 6,350} = 525$ amp. Hence the equivalent winding resistances in ohms per phase are as follows:

$r_a = 0.75$ per cent on 7.500 kva. = 1 per cent on 10,000 kva. $= \frac{0.01 \times 6,350}{525}$
$= 0.12$ ohm

$r_b = 1$ per cent on 10,000 kva. $= 0.12$ ohm

$r_c = 1$ per cent on 5,000 kva. = 2 per cent on 10,000 kva. $= \frac{0.02 \times 6,350}{525}$
$= 0.24$ ohm

Similarly, the winding reactances in ohms per phase are as follows:

$x_{ab} = 10$ per cent on 10,000 kva. $= \frac{0.10 \times 6,350}{525} = 1.21$ ohms

$x_{bc} = 5$ per cent on 10,000 kva. $= \frac{0.05 \times 6,350}{525} = 0.60$ ohm

$x_{ca} = 10$ per cent on 5,000 kva. = 20 per cent on 10,000 kva. $= \frac{0.20 \times 6,350}{525}$
$= 2.42$ ohms

The transformer impedance between windings A, B, and C may now be written

$$
\begin{aligned}
Z_{AB} &= r_a + r_b + jx_{ab} = 0.24 + j1.21 \text{ ohms} \\
Z_{BC} &= r_b + r_c + jx_{bc} = 0.36 + j0.60 \text{ ohms} \\
Z_{CA} &= r_a + r_c + jx_{ac} = 0.36 + j2.42 \text{ ohms}
\end{aligned}
$$

The above values of transformer impedance may be substituted in the equations of Fig. 62 to obtain the equivalent star-connected impedances with the results.

$$
\begin{aligned}
Z_A &= \frac{Z_{AB} + Z_{AC} - Z_{BC}}{2} \\
&= \frac{(0.24 + j1.21) + (0.36 + j2.42) - (0.36 + j0.60)}{2} \\
&= \frac{0.24 + j3.03}{2} = 0.12 + j1.51 \text{ ohms} \\
Z_B &= \frac{Z_{AB} + Z_{BC} - Z_{CA}}{2} \\
&= \frac{(0.24 + j1.21) + (0.36 + j0.60) - (0.36 + j2.42)}{2} \\
&= \frac{0.24 - j0.61}{2} = 0.12 - j0.305 \text{ ohms} \\
Z_C &= \frac{Z_{CA} + Z_{CB} - Z_{AB}}{2} \\
&= \frac{(0.36 + j2.42) + (0.36 + j0.60) - (0.24 + j1.21)}{2} \\
&= \frac{0.48 + j1.81}{2} = 0.24 + j0.905 \text{ ohms}
\end{aligned}
$$

NOTE. The resistances of the branches Z_A, Z_B, and Z_C, are identical with the values of r_a, r_b, and r_c, respectively. This will always be the case for transformers with separate windings, so that the resistance terms of Z_A, Z_B, and Z_C may be determined in this manner.

The **impedance of three-winding transformers** for the **positive- and negative-sequences** may be represented by the groups of three star-connected impedances just described.

The **equivalent circuit of three-winding transformers** to **zero-sequence** involves the same complication due to the connection of windings, whether they are delta-, or grounded- or ungrounded-star, as arises in the case of the two-winding transformer. The same conceptions with respect to the connection of the impedances to the external circuit or to ground as were used in connection with the two-winding transformer diagram in Fig. 61, may be applied to any type of three-winding transformer. The results for three common types are shown in Fig. 63. The impedances in these diagrams are, of course, the per phase values. For the delta-connected winding they may be visualized best by regarding the winding as a fictitious star-connected winding of the same capacity and same percentage impedance.

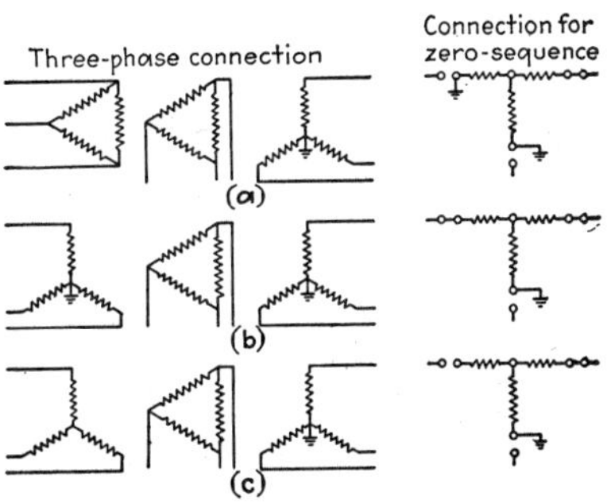

FIG. 63.—Common three-winding transformer connections with equivalent circuits for zero-sequence.

57. Three-phase Transformers.

The preceding discussion applied directly to three-phase banks made up of single-phase units. Three-phase transformers require further consideration because of the fact that flux paths for the several windings may be common to each other.

The iron circuit of a common form of the three-phase **shell-type transformer** is illustrated in Fig. 64(*b*). The middle winding is reversed to decrease the flux in the section between any two windings. For convenience in comparing the iron circuits the three-phase bank of single-phase transformers is illustrated in Fig. 64(*a*). For positive- or negative-sequence voltages applied to terminals *a*, *b*, and *c*, of either the single-phase bank or the three-phase transformer, the impedance is very high since it is that corresponding to the exciting impedance. In the case of zero-sequence voltages applied, it will be observed that the outer legs provide an iron return path for the flux set up by the zero-sequence currents. Thus the zero-sequence exciting impedance of the single-phase transformer bank of Fig. 64(*a*) or the three-phase shell-type transformer is the same as for positive- or negative-sequence voltage of the same value across each phase. The presence of secondary windings has the same effect upon the short-circuit impedance for either the single-phase bank or the three-phase shell transformer.

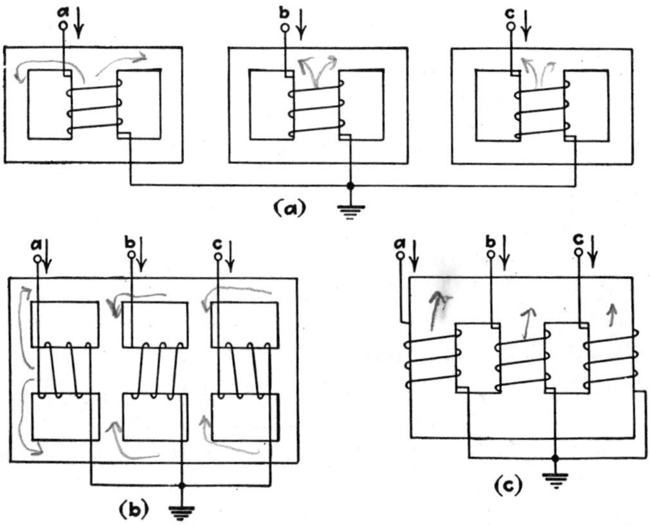

FIG. 64.—Diagrams to illustrate the zero-sequence impedances of three-phase transformer banks. (*a*) Three-phase bank of single-phase transformers; (*b*) three-phase shell-type transformer; (*c*) three-phase core-type transformer.

In the case of the three-phase **core-type transformer** with three legs as illustrated in Fig. 64(*c*), it will be observed that for positive- or negative-sequence currents a return circuit is provided through the other legs and the exciting impedance is thus very high. For the case of zero-sequence voltages, however, it will be noted that there is no iron return path for the flux set up by the currents in the windings on the three legs. Consequently, zero-sequence currents will produce relatively little flux within the transformer and the reactance to zero-sequence will be low compared with the ordinary exciting impedances, though the value will be high when compared with the ordinary series leakage reactance of the transformer. In commercial three-phase core-type transformers the zero-sequence exciting impedance of a grounded-star winding varies from 30 to 300 per cent, the higher values applying for the larger size units. The presence of the secondary windings has practically the same effect upon the short-circuit impedance as for a bank of single-phase transformers. However, because of its lower exciting impedance, the zero-sequence impedance of the core-type transformer will vary from 90 to 100 per cent of its positive-sequence impedance.

The core-type three-phase transformer may be considered in the same manner as the other three-phase banks if the effect of the magnetic circuit be replaced by a fictitious delta tertiary winding of very high impedance. This tertiary winding tends to reduce the flux in the transformer due to the flow of zero-sequence currents from the value which obtains for positive-sequence currents and makes it possible to obtain the intermediate value of impedance as described in the preceding paragraph. This tertiary winding should be considered in addition to any other delta-connected windings that may exist within the transformer, for example, a three-phase core-type transformer with the primary winding connected in grounded-star and a secondary connected in delta would have an impedance to zero-sequence corresponding to Fig. 63(*a*). Thus it appears that the zero-sequence impedance of grounded star-delta transformers of the three-phase core construction is somewhat less than the positive-sequence impedance.

58. Autotransformers.

Autotransformers encountered in power systems usually consist of two windings, parts of which are common, and a third independent winding called the tertiary winding. The common windings of three-phase banks are connected in star, and the tertiary in delta. Exceptions to this type of autotransformer are to be found in applications of single-phase transformers to railway work. The present analysis was developed formally for the three-winding autotransformer, permitting the two-winding autotransformer to be evaluated as a special case.

When the magnetizing current is neglected, it is possible to select three pairs of terminals and completely represent the characteristics of the transformer by a set of three star-connected impedances between these terminal pairs. A choice exists as to the appropriate pairs of terminals involved in the common winding. Figure 65(*a*) and (*b*) shows the two combinations that are usually specified. For convenience, the left-hand combination will be referred to as on the **winding basis** and the right-hand combination, as on the **circuit basis.**

The **equivalent circuits of the autotransformer for the positive- and negative-sequence** are shown in Fig. 65(*c*) and (*d*) for the

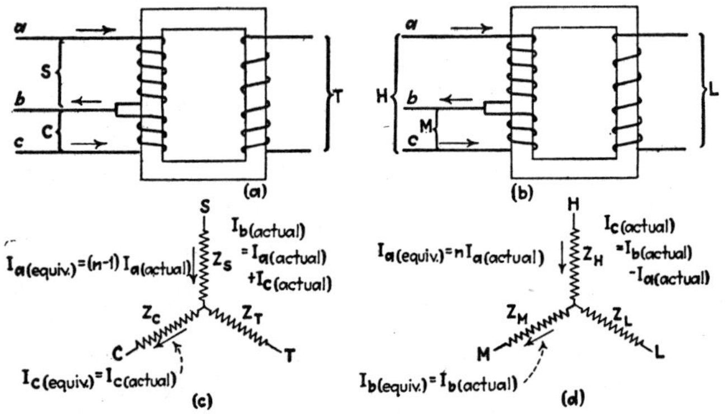

$$Z_T = \tfrac{1}{2}(Z_{TC} + Z_{TS} - Z_{CS})$$
$$Z_C = \tfrac{1}{2}(Z_{CT} + Z_{CS} - Z_{ST})$$
$$Z_S = \tfrac{1}{2}(Z_{ST} + Z_{SC} - Z_{CT})$$
(*e*)

$$Z_L = \tfrac{1}{2}(Z_{LM} + Z_{LH} - Z_{MH})$$
$$Z_M = \tfrac{1}{2}(Z_{ML} + Z_{MH} - Z_{LH})$$
$$Z_H = \tfrac{1}{2}(Z_{HL} + Z_{HM} - Z_{LM})$$
(*f*)

$$Z_{ML} = Z_{CT}$$
$$Z_{HM} = \left(\frac{n-1}{n}\right)^2 Z_{SC}$$
$$Z_{HL} = \frac{Z_{TC}}{n} + \frac{n-1}{n} Z_{TS} - \frac{n-1}{n^2} Z_{CS}$$
(*g*)

$$Z_{CT} = Z_{ML}$$
$$Z_{SC} = \left(\frac{n}{n-1}\right)^2 Z_{HM}$$
$$Z_{TS} = \frac{n}{n-1} Z_{HL} - \frac{1}{n-1} Z_{ML} + \frac{n}{(n-1)^2} Z_{HM}$$
(*h*)

$$Z_C = \frac{n}{n-1} Z_M$$
$$Z_T = Z_L - \frac{Z_M}{n-1}$$
$$Z_S = \left(\frac{n}{n-1}\right)^2 Z_H + \frac{n}{(n-1)^2} Z_M$$
(*i*)

$$Z_M = \frac{n-1}{n} Z_C$$
$$Z_L = Z_T + \frac{Z_C}{n}$$
$$Z_H = \left(\frac{n-1}{n}\right)^2 Z_S - \frac{n-1}{n^2} Z_C$$
(*j*)

FIG. 65.—Conversion formulas for autotransformer with tertiary winding. C and M windings used as base. Ratio $H/M = n$.

winding and circuit bases, respectively. It will be noted that the form is identical with that of the transformer with separate windings.

The **equivalent circuit to zero-sequence** for autotransformers involves the same connections of the equivalent circuits as for transformers with three separate windings. This is illustrated in the example given in a later paragraph of this section.

Conversion of Autotransformer Impedances from Winding Basis to Circuit Basis. Frequently the autotransformer impedances are given on one basis and it is desired to use impedances in the network on the other basis. It therefore becomes necessary to derive expressions whereby the impedances on one basis may be converted to the impedances on the other basis.

Before going directly to these derivations it may be well to specify more clearly the currents in the different circuits. The impedances shown in Fig. 65 and consequently the currents flowing therein are all based on a common voltage. The actual currents in the various circuits can be obtained from the currents in the equivalent circuits by converting the currents from the common voltage base used in the calculations to the actual voltage base of the particular circuit. The currents in branches C and S of Fig. 65(c) represent, after converting to their respective voltage bases, the actual currents in conductors c and a, respectively, of Fig. 65(a). The actual current flowing in the common conductor b of Fig. 65(a) cannot be represented in the equivalent diagram, but is the sum of the actual currents in conductors a and c. Similarly, currents in branches H and M of Fig. 65(d), after reducing to the proper voltage base, represent the actual currents in conductors a and b, respectively, and the actual current in conductor c is the difference of the actual currents in conductors a and b.

For the purpose of this analysis the winding M will be used as the reference winding, and the impedances will be expressed in terms of this voltage as the base. A ratio n will be taken to give the turns ratio of the high-voltage to the medium-voltage winding, that is

$$n = \frac{E_H}{E_M} = \frac{E_C + E_S}{E_C} \tag{132}$$

and

$$\frac{E_S}{E_C} = (n - 1) \tag{133}$$

In addition, windings T and L may be assumed for the present, without limiting the general application of the formulas, to have the same voltage base as windings C and M.

Let it be assumed that all impedances appearing in the equations of Fig. 65(e) are given on the same kilovolt-ampere base and the voltage base of the C winding, and it is desired to convert these impedances to the terms appearing in the equations of Fig. 65(*f*), all of which are expressed on the same kilovolt-ampere base and voltage base of the *M* winding.

Determination of Z_{HL}. Apply voltage to the *H* winding and circulate unit current through this winding with the *L*

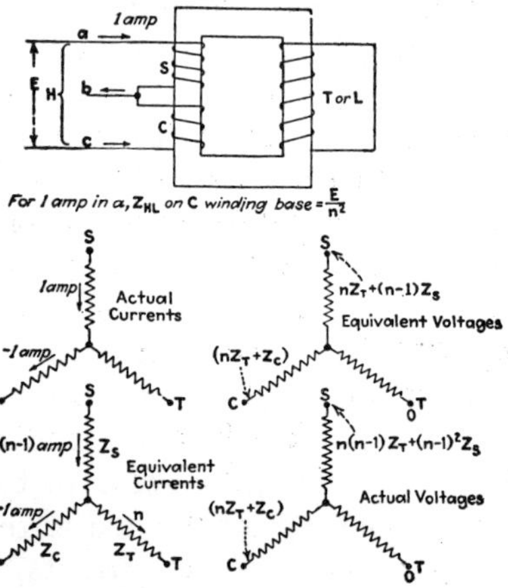

$$E = E_{S(actual)} + E_{C(actual)}$$
$$= n^2 Z_T + (n-1)^2 Z_S + Z_C$$
$$Z_{HL} = \frac{E}{n^2} = Z_T + \left(\frac{n-1}{n}\right)^2 Z_S + \frac{Z_C}{n^2}$$
$$= \frac{1}{n} Z_{TC} + \frac{n-1}{n} Z_{TS} - \frac{n-1}{n^2} Z_{CS}$$

FIG. 66.—Determination of Z_{HL} in terms of *S*, *C* and *T* constants of transformer.

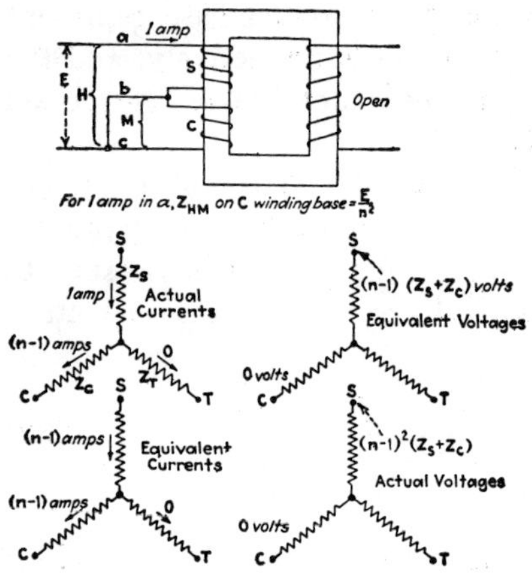

$$E = (n-1)^2(Z_S + Z_C)$$
$$= (n-1)^2 Z_{SC} \text{ volts}$$
$$Z_{HM} = \frac{E}{n^2} = \left(\frac{n-1}{n}\right)^2 Z_{SC}$$

FIG. 67.—Determination of Z_{HM} in terms of *S*, *C*, and *T* constants of transformer.

winding short-circuited as shown in Fig. 66. For this condition the actual currents flowing in circuits *S* and *C* are the actual currents in the terminals *a* and *c*, namely, unity in both terminals. These currents in the equivalent circuit correspond to $(n-1)$ in circuit *S* and unity in circuit *C*, the former flowing into and the latter flowing out of the star point. The equivalence of these currents in the actual and in the equivalent windings may easily be verified by consideration of the corresponding magnetic effect as influenced by the relative number of turns. The current in the T winding must equal the difference of these, namely, $(n - 1) - (-1) = n$ flowing from the star point. Since the T winding is short-circuited, its voltage is zero. The equivalent voltage across the S winding is then the sum of the drops in the T and S impedances, thus $nZ_T + (n - 1)Z_S$, and the equivalent voltage across the C winding is the difference in drops in the T and C windings, thus $nZ_T - (-1)Z_C = nZ_T + Z_C$. The actual voltages are obtained by multiplying the equivalent voltage of S by $(n - 1)$, the ratio of the turns of S to those of C, and the equivalent voltage of C by unity. The actual voltage E across H is the sum of these two voltages. Since by assumption unit current is flowing in these windings, the value of E is equal to the impedance Z_{HL} in terms of the voltage base of H. Dividing by n^2 converts this impedance to the common voltage of the C winding with the final result

$$Z_{HL} = Z_T + \left(\frac{n-1}{n}\right)^2 Z_S + \frac{Z_C}{n^2} \tag{134}$$

Inserting the values of Z_T, Z_C, and Z_S from Fig. 65(e) gives

$$Z_{HL} = \frac{Z_{TC}}{n} + \frac{n-1}{n}Z_{TS} - \frac{n-1}{n^2}Z_{CS} \tag{135}$$

Determination of Z_{HM}. The impedance Z_{HM} between the high- and medium-voltage windings in terms of the impedances of the other combination may be obtained in a similar manner. The details of the steps involved are indicated in Fig. 67, giving finally

$$Z_{HM} = \left(\frac{n-1}{n}\right)^2 Z_{SC} \tag{136}$$

Determination of Z_{ML}. It is evident that

$$Z_{ML} = Z_{CT} \tag{137}$$

as it involves merely a change in notation.

The results of these transformations are tabulated in Fig. 65(g) for ready reference. The converse equations defining Z_{CT}, Z_{SC}, and Z_{TS} in terms of Z_{ML}, Z_{HM}, and Z_{HL} are also given in Fig. 65(h). To make the conversions complete, the star impedances are also defined for the two combinations. These are tabulated under (i) and (j) of Fig. 65 and were obtained by substituting equations (*g*) and (*h*) in (*e*) and (*f*), remembering that $Z_{TS} = Z_T + Z_S$, etc.

For the above development, the voltage of the *C* and *M* windings has been chosen as the base, and all impedances must of course be expressed in terms of this base. When any of the other windings are chosen as base, it is necessary to multiply all the impedances given by the ratio squared of the respective voltage bases.

Example. To illustrate the use of these relations consider a three-phase autotransformer bank whose *H*, *M*, and *L* windings are connected for 220, 150, and 11 kv. respectively. The reactances between the various combinations of windings are

Z_{HM} – 9.5 per cent on 35,700 kva.
Z_{ML} – 9.2 per cent on 11,000 kva., or 29.85 per cent on 35,700 kva.
Z_{HL} – 14.0 per cent on 11,000 kva., or 45.4 per cent on 35,700 kva.

or reduced to ohms on the 150-kv. base

$$Z_{HM} = j\frac{(0.095) \times 150{,}000 \times 150{,}000}{35{,}700{,}000} = j59.9 \text{ ohms}$$
$$Z_{ML} = j188 \text{ ohms}$$
$$Z_{HL} = j287 \text{ ohms}$$

The resistance of the windings is neglected.

The star impedances, by application of the equations, Fig. 65(*f*), are

$$\left.\begin{aligned} Z_L &= \frac{j}{2}(188 + 287 - 59.9) = j207.6 \\ Z_M &= \frac{j}{2}(188 + 59.9 - 287) = -j19.6 \\ Z_H &= \frac{j}{2}(287 + 59.9 - 188) = j79.5 \end{aligned}\right\} \text{ohms on 150-kv. base}$$

If the nature of the problem requires the determination of the star impedances in the *S*, *C*, and *T* combination, they may be obtained by substituting the above numerical values of Z_L, Z_M, and Z_H in equations Fig. 65(*i*), remembering that $n = \frac{220}{150} = 1.467$.

$$\left.\begin{aligned} Z_C &= \frac{1.467}{1.467 - 1}(-j19.6) = -j61.6 \\ Z_T &= j207.6 - \frac{(-j19.6)}{1.467 - 1} = j249.6 \\ Z_S &= \left(\frac{1.467}{1.467 - 1}\right)^2 (j79.5) + \\ &\quad \frac{1.467}{(1.467 - 1)^2}(-j19.6) = j652.6 \end{aligned}\right\} \text{ohms on 150-kv. base}$$

Similarly, given the leakage reactances between two windings at a time in the *H*, *M*, and *L* combination, the leakage reactances between windings for the *S*, *C*, and *T* combination may be obtained by substituting the numerical values of Z_{HM}, Z_{ML}, and Z_{HL}, in equations Fig. 65(*h*).

$$\left.\begin{aligned}Z_{CT} &= j188\\ Z_{SC} &= \left(\frac{1.467}{1.467-1}\right)^2(j59.9) = j591\\ Z_{TS} &= \left(\frac{1.467}{1.467-1}\right)(j287) - \\ &\quad \frac{1}{1.467-1}(j188) + \frac{1.467}{(1.467-1)^2}(j59.9)\\ &= j902\end{aligned}\right\}\text{ohms on 150-kv. base}$$

59. Inclusion of Series Impedance in Autotransformer Circuits.

In some cases it is desirable to include series impedances into the equivalent network. It will be assumed that these impedances are expressed in terms of the C and M winding voltage base. To be more specific, impedances in T or L windings are reduced to the C and M base by dividing the actual impedance in ohms by the square of the ratio of turns; impedances in the a conductor are reduced to the C and M winding base by dividing the actual ohms by n^2; and impedances in the b and c conductors are already on the M and C winding turns base. Series impedances in either the T or L windings or the a conductor offer no difficulty, because they are merely added externally to their respective windings. Impedances in conductors b and c may require some manipulation depending upon the combination of windings chosen. It is more convenient, for transmission work, to use the combination H, M, and L, Fig. 65(b). Assume then the impedances S_a, S_b, and S_c in conductors a, b, and c, respectively, Fig. 68(a). It should be evident that, expressing the transformer impedances in the H, M, and L combination, S_a and S_b can be treated as an impedance in series with the H and M windings, respectively. The impedance S_c should be treated as an impedance in series with the C winding of the S, C, and T combination, Fig. 65(a). Therefore, the same coefficients should be applied to this impedance as was applied to Z_C in the equations in Fig. 65(j), in obtaining the star impedances for the H, M, and L combination, namely, $\frac{n-1}{n}$, $\frac{1}{n}$ and $-\frac{n-1}{n^2}$ for Z_M, Z_L, and Z_H, respectively. The result of this conversion is shown in Fig. 68(b).

The most important application of these considerations is to the determination of the zero-sequence impedance. A form which the problem sometimes takes is shown in Fig. 69, in which a star-delta connected machine, which may be a transformer or synchronous machine, and whose neutral is grounded through an impedance, is connected to the low-voltage side of the common winding of the autotransformer bank. The autotransformer itself may be connected to ground through an impedance. The tertiary of the autotransformer will be delta-connected. The problem is to determine the zero-sequence impedance to ground of this combination. The notation and the current flow assumed are shown in Fig. 69(*a*). The single-line diagram for this combination reduces to that shown in Fig. 69(*b*). In particular, the problem reduces to that of finding the impedance of the *H* winding with the two other windings short-circuited as indicated. Applying the relations indicated in Fig. 68(*b*), the equivalent circuit, using the *M* winding as base, becomes that shown in Fig. 69(*c*). Now the impedance desired is merely the impedance of the *H* winding branch in series with the *M* and *L* branches in parallel, which is

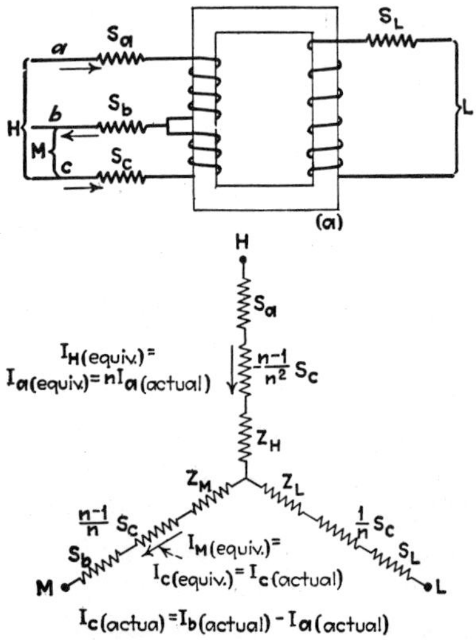

FIG. 68.—Equivalent circuit of autotransformer with series impedance. Winding *M* used as base. Ratio $H/M = n$.

$$Z_0 = Z_H - \frac{3(n-1)}{n^2}Z_{n2} + \frac{\left(Z_L + \frac{3Zn_2}{n}\right)\left[Z_M + Z_g + 3Z_{n1} + \frac{3(n-1)}{n}Z_{n2}\right]}{\left(Z_L + \frac{3Z_{n2}}{n}\right) + \left[Z_M + Z_g + 3Z_{n1} + \frac{3(n-1)}{n}Z_{n2}\right]} \quad (138)$$

This is the value of impedance to use in the zero-sequence network when using winding *M* as base. When using the *H* winding as base, multiply this impedance by n^2.

The actual currents in the different windings in terms of the equivalent current in the *H* winding (the current which one will obtain from the solution of the complete network) are developed in Fig. 69. The important idea in this development is to keep in mind the conversion coefficients from equivalent to actual currents. It is interesting to observe in passing, as inspection of the equation for the neutral current in Fig. 69 discloses, that the current in the neutral under some conditions is of the same sign as the high-tension current, under other

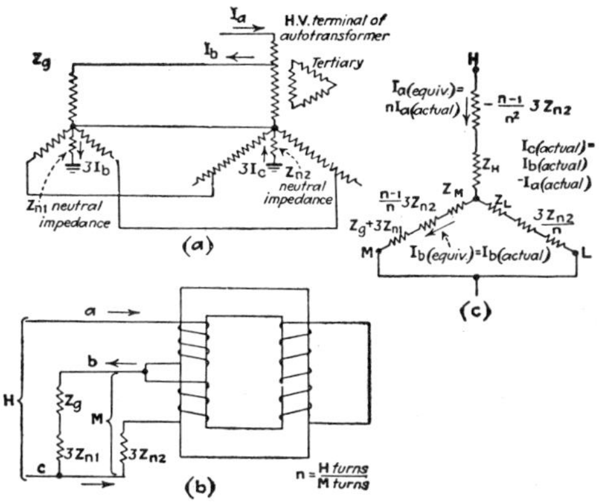

$$Z_0 = Z_H - \frac{3(n-1)}{n^2}Z_{n2} + \frac{\left(Z_L + \frac{3Z_{n2}}{n}\right)\left[Z_M + Z_g + 3Z_{n1} + \frac{3(n-1)}{n}Z_{n2}\right]}{\Delta}$$

in which

$$\Delta = \left(Z_L + \frac{3Z_{n2}}{n}\right) + \left[Z_M + Z_g + 3Z_{n1} + \frac{3(n-1)}{n}Z_{n2}\right]$$

$$I_{a(actual)} = I_{H(actual)} = \frac{I_{H(equiv.)}}{n}$$

$$I_{n1(actual)} = 3I_{b(actual)} = 3I_{M(equiv.)} = \frac{3\left(Z_L + \frac{3Z_{n2}}{n}\right)}{\Delta}I_{H(equiv.)}$$

$$I_{n2(actual)} = 3I_{c(actual)} = 3I_{M(equiv.)} - 3\frac{I_{H(equiv.)}}{n}$$

$$= \frac{3[(n-1)Z_L - Z_M - Z_g - 3Z_{n1}]}{n\Delta}I_{H(equiv.)}$$

FIG. 69.—Zero-sequence impedance and current distribution for autotransformer with M winding connected to star-connected grounded machine. Neutrals of stars grounded through different impedances. All impedances expressed in terms of M winding.

conditions is of opposite sign, and under certain other conditions is zero. This is, of course, due to the fact that two paths are provided in the transformer for the flow of current, these paths causing current of opposite signs to flow through the transformer neutral connection. Consequently, the actual current will have a magnitude and sign which depend upon the relative impedance of the different paths. This is an important point in connection with certain relay schemes.

Ungrounded Autotransformer with Tertiary Winding. The ungrounded autotransformer with delta-connected tertiary winding may be considered as a special case of the foregoing in which

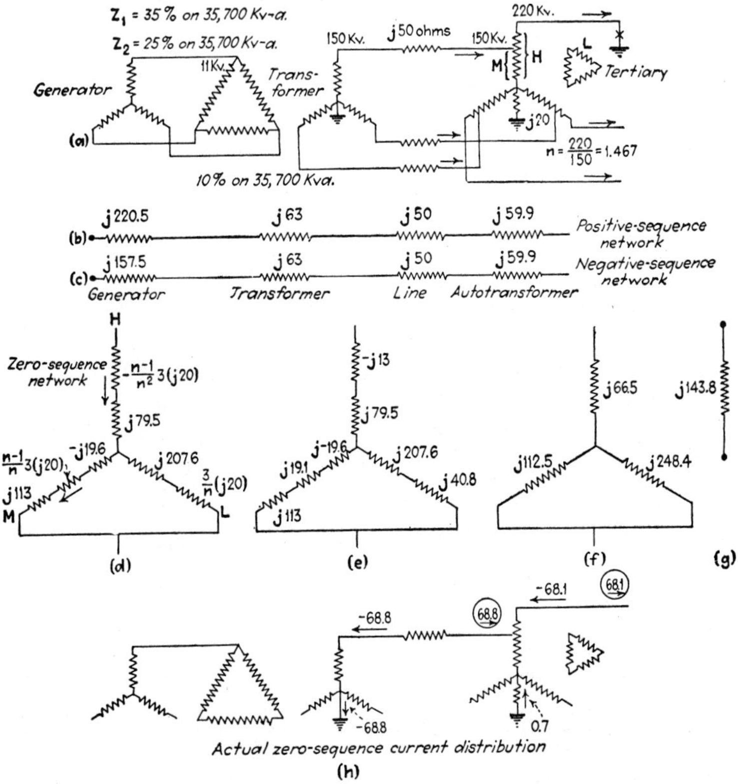

FIG. 70.—Single line-to-ground fault calculation. Quantities in the circles represent the same positive sense of current flow as in positive- and negative-sequence networks.

Z_{n2} becomes infinitely large. The value of Z_0 for this case may be obtained by converting the expression for Z_0 to a simple fraction and then letting $Z_{n2} = \infty$, thus obtaining

$$Z_0 = Z_H + \left(\frac{n-1}{n}\right)^2 Z_L + \frac{Z_M + Z_g + 3Z_{n1}}{n^2} \tag{139}$$

Using the relations given in Fig. 65 this can be reduced to

$$Z_0 = \left(\frac{n-1}{n}\right)^2 Z_{ST} + \frac{Z_g + 3Z_{n1}}{n^2} \tag{140}$$

It will be recalled that the impedances are expressed in terms of the M winding. To transform to the H winding turns it is necessary to multiply through by n^2, after which it will be observed that the impedance consists of two parts, that due to the impedance between the S and T windings and that due to the generator and the neutral impedance, the latter impedances being their actual impedances since the same current flows through the generator and neutral impedance as flows through the high-voltage line.

Example. To illustrate the application of these relations to a practical problem, let it be desired to calculate the current distribution for a single line-to-ground fault on the 220-kv. circuit for the system shown in Fig. 70, consisting of a generator and two transformers connected by a section of 150-kv. line. The autotransformers will be assumed to have the same characteristics as those illustrated by the example under Autotransformers, namely,

$$\left.\begin{aligned} Z_L &= j207.6 \\ Z_M &= -j19.6 \\ Z_H &= j79.5 \end{aligned}\right\} \text{ohms on 150-kv. base}$$

The positive- and negative-sequence networks for this system consist of merely the generator, transformer A, the line and autotransformer impedances, reduced to a 150-kv. base, connected in series. The autotransformer impedance for this case is the impedance Z_{HM} (59.9 ohms). The impedance diagrams for the two networks are the same except for the generator impedances which are different for the positive- and negative-sequences. After adding, these impedances become $j393.4$ and $j330.4$ for the positive- and negative-sequences, respectively.

The zero-sequence diagram reduces to the same kind as has just been under consideration, and which is shown in Fig. 69. The series impedance Z_g of Fig. 69 in this case equals the sum of the line impedance $j50$ ohms (which is already on the 150-kv. base) and the transformer impedance $j63$ ohms. Since transformer A is solidly grounded, Z_{n1} is equal to zero. The neutral impedance Z_{n2} is equal to $j20$ ohms. The equivalent diagram of Fig. 69(c) with these values substituted is shown in Fig. 70(d), which through successive simplification reduces to $j143.8$ ohms.

The sequence components of current at the point of fault (on the 150-kv. base, of course) are then equal to the line-to-neutral voltage divided by the sum of the three impedances, namely,

$$I_0 = I_1 = I_2 = \frac{j150{,}000}{\sqrt{3}j(393.4 + 330.4 + 143.8)} = 99.8 \text{ amp.}$$

A word of caution is necessary at this point. The theory, upon which the method of symmetrical components as applied to single line-to-ground faults was developed, presupposed that the positive direction of current at any point in the system is the same for all three sequences. It will be observed that in Fig. 70(a) the positive direction of current in the 220-kv. line, for example, was taken from left to right, but in Fig. 69 and Fig. 70(d) the positive sense was taken oppositely. This difference in the positive sense of current flow, which was arbitrarily chosen, does not alter the value of the zero-sequence impedance and consequently the value of the three sequence components of current at the point of fault, but it does mean that the current I_H (Fig. 70(*d*)) is the negative of the zero-sequence current at the fault or -99.8 amp. The zero-sequence current in the M winding [terminal *b*, Fig. 69(*b*)] is then

$$I_M = \frac{248.4}{112.5 + 248.4}(-99.8) = -68.8 \text{ amp.}$$

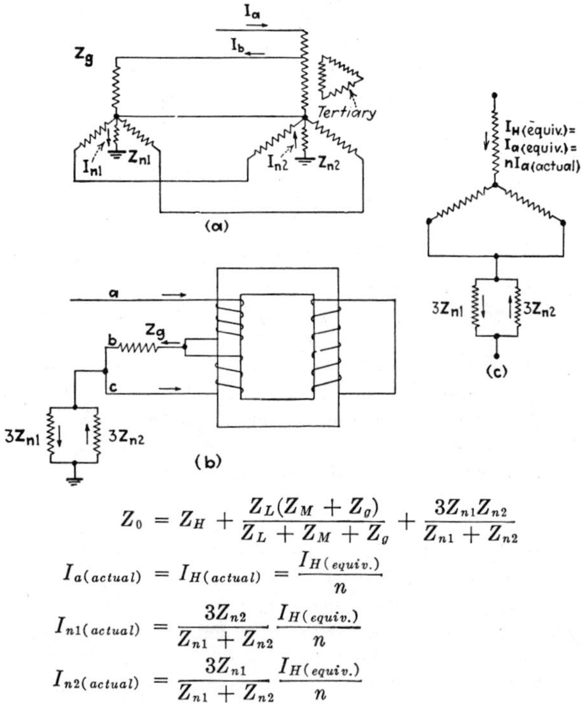

$$Z_0 = Z_H + \frac{Z_L(Z_M + Z_g)}{Z_L + Z_M + Z_g} + \frac{3Z_{n1}Z_{n2}}{Z_{n1} + Z_{n2}}$$

$$I_{a(actual)} = I_{H(actual)} = \frac{I_{H(equiv.)}}{n}$$

$$I_{n1(actual)} = \frac{3Z_{n2}}{Z_{n1} + Z_{n2}} \frac{I_{H(equiv.)}}{n}$$

$$I_{n2(actual)} = \frac{3Z_{n1}}{Z_{n1} + Z_{n2}} \frac{I_{H(equiv.)}}{n}$$

FIG. 71.—Zero-sequence impedance and current distribution for autotransformer with tertiary, connected to a star-connected grounded machine. Neutrals of stars tied together and grounded through different impedances. All impedances given in terms of the low voltage (M) winding of autotransformer. (*a*) System lay-out; (*b*) single-line schematic diagram; (*c*) equivalent circuit.

This is the actual zero-sequence component of current in the 150-kv. line in a direction toward the generator and in the neutral of transformer A in a direction into the ground. The total neutral current in transformer A is then 3×68.8 or 206.4 amp. out (because of the negative sign) of the ground. The actual current in the grounding reactor is

$$3(I_{M(actual)} - I_{H(actual)}) = 3\left(-68.8 - \frac{(-99.8)}{1.467}\right) = 0.7 \text{ amp.}$$

in a direction up from the ground. The actual current distribution in the zero-sequence network is shown in Fig. 70(*h*), the values in the circles corresponding, in the positive sense of current flow, to that of the positive- and negative-sequence networks. The phase currents at any point can be obtained from the sequence components by the usual methods.

In Fig. 71 is shown another combination which is sometimes met in practice. The figure is self-explanatory in depicting the development of the zero-sequence network corresponding to the circuit condition given.

60. Interconnected Star-delta Transformers.

An interesting case involving three-winding transformer considerations is the interconnected star-delta transformer connection. The usual method of designating this connection is shown in Fig. 72(*a*), but to aid in the analysis the complete wiring diagram is shown in (*b*), which also shows the nomenclature which will be used. It will be assumed that equivalent star impedances for the three windings of each transformer are given, designated as Z_A for the delta-connected winding, Z_B for the star portion next to the high-tension side, and Z_C for the star portion next to the ground side. This notation will be clear from Fig. 72(*c*), which shows the diagram for the left-hand transformer. The current flow throughout the entire bank is shown in (*b*), from which it is seen that current I_x flows out of the Z_B branch and $-I_z$ out of the Z_C branch in (*c*). With these currents flowing, the following equations may be written:

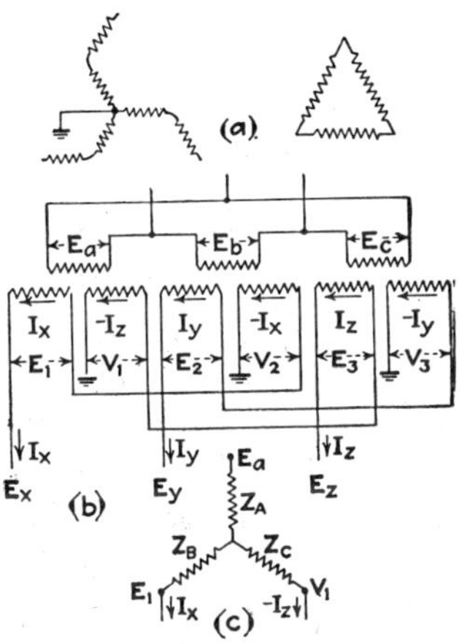

FIG. 72.—Interconnected star-delta transformer.

$$E_1 = E_a - Z_A(I_x - I_z) - Z_B I_x \tag{141}$$

and

$$V_1 = E_a - Z_A(I_x - I_z) + Z_C I_z \tag{142}$$

Similar expressions may be written for the two other transformers giving

$$E_2 = E_b - Z_A(I_y - I_x) - Z_B I_y \tag{143}$$
$$V_2 = E_b - Z_A(I_y - I_x) + Z_C I_x \tag{144}$$

and

$$E_3 = E_c - Z_A(I_z - I_y) - Z_B I_z \tag{145}$$
$$V_3 = E_c - Z_A(I_z - I_y) + Z_C I_y \tag{146}$$

Now from Fig. 72(*b*)

$$E_x = E_1 - V_2 \tag{147}$$

and substituting the above values of E_1 and V_2

$$E_x = E_a - E_b - Z_A(2I_x - I_z - I_y) - Z_B I_x - Z_C I_x \tag{148}$$

Similar expressions can be obtained for the two other phases, but because of symmetry all of the necessary information can be obtained from equation (148) alone.

First let it be desired to obtain the **positive-sequence impedance.** For this case

$$E_b = a^2 E_a \tag{149}$$
$$I_y = a^2 I_x \tag{150}$$
$$I_z = a I_x \tag{151}$$

substituting

$$\begin{aligned} E_x &= (1 - a^2)E_a - Z_A(2 - a - a^2)I_x - Z_B I_x - Z_C I_x \\ &= (1 - a^2)E_a - (3Z_A + Z_B + Z_C)I_x \end{aligned} \tag{152}$$

This equation indicates that on the basis of equal turns on all windings when no current is flowing, namely, when $I_x = 0$, the star voltage on the interconnected star side is $(1 - a^2)$ times the line-to-line voltage on the delta side, and when current flows the star voltage on the star-connected side decreases by the amount $(3Z_A + Z_B + Z_C)$ per ampere flowing in that side. It follows, therefore, that, given the transformer impedances Z_A, Z_B, and Z_C in ohms on the voltage base corresponding to the operating voltage across the individual windings of the two in series, the quantity $(3Z_A + Z_B + Z_C)$ is the series impedance in ohms per phase for the positive-sequence when using the voltage of the interconnected star side as base. A concrete example which will be worked out later will serve to illustrate these relations.

The apparatus being non-rotative the **negative-sequence impedance** is equal to the positive-sequence impedance.

For the **zero-impedance,** all similar voltages and currents in the three phases are equal, so that letting

$$E_b = E_a \tag{153}$$
$$I_y = I_z = I_x \tag{154}$$

and substituting in equation (148) gives

$$\begin{aligned} E_x &= E_a - E_a - Z_A(2I_x - I_x - I_x) - Z_B I_x - Z_C I_x \\ &= -(Z_B + Z_C)I_x \\ &= -Z_{BC} I_x \end{aligned} \tag{155}$$

The impedance Z_{BC} in ohms on the voltage base of each of the individual windings, therefore, represents the impedance per phase for the zero-sequence. It is interesting to observe that the impedance of the delta winding does not enter this relation. This results from the fact that the two sections of the interconnected star winding were assumed to have the same number of turns, and since equal and opposite currents flow through these windings because of the series relationship, the m.m.f. conditions within the transformer are satisfied and no current flows in A branch, Fig. 72(c).

Example. In order to illustrate the calculation of the sequence impedances of an interconnected star-delta transformer, refer to Fig. 72(b) and assume a bank of three 2,000-kva. transformers with E_a as 11 kv. on the delta side, and E_1 and V_1 as 22 kv., each corresponding to a line-to-line voltage of 66 kv. on the interconnected star side. Further, assume that transformer resistances may be neglected and the reactances are as follows:

X_{AB} = 15 per cent on 2,000 kva.
X_{AC} = 15 per cent on 2,000 kva.
X_{BC} = 10 per cent on 1,000 kva. = 20 per cent on 2,000 kva.

The equivalent star impedance diagram for the single-phase transformer may be calculated in the usual manner, using the notation of Fig. 72(c).

$$Z_A = j\left(\frac{15 + 15 - 20}{2}\right) \text{ per cent} = +j5 \text{ per cent}$$

$$Z_B = j\left(\frac{15 + 20 - 15}{2}\right) \text{ per cent} = +j10 \text{ per cent}$$

$$Z_C = j\left(\frac{20 + 15 - 15}{2}\right) \text{ per cent} = +j10 \text{ per cent}$$

The above impedances may be converted into ohms at 22 kv. (line-to-neutral) with the following results:

$$\left.\begin{aligned} Z_A &= j\frac{5}{100}\,\frac{(22{,}000)^2}{2{,}000 \times 1{,}000} = j12.1 \\ Z_B &= j\frac{10}{100}\,\frac{(22{,}000)^2}{2{,}000 \times 1{,}000} = j24.2 \\ Z_C &= j24.2 \end{aligned}\right\}\text{ohms at 22-kv. (line-to-neutral)}$$

The positive- or negative-sequence reactance of this path of three 2,000-kva. transformers may be computed from the reactance of the single-phase transformer by the use of the expression

$$(3Z_A + Z_B + Z_C) = 3(j12.1) + j24.2 + j24.2 = j84.7 \text{ ohms}$$

This is the proper value of reactance for the 38-kv. line-to-neutral voltage corresponding to a 66-kv. line on the interconnected star side of the transformer. The impedance of this transformer on the 11-kv. side can readily be obtained by taking the ratio of transformation into account, with the following result:

$$j84.7\left(\frac{11,000}{66,000}\right)^2 = j2.35 \text{ ohms.}$$

The zero-sequence impedance of the three 2,000-kva. transformers from the interconnected star side is

$$Z_{BC} = j48.4 \text{ ohms}$$

which is, of course, the value to be used with the zero-sequence voltage to neutral of 38 kv., corresponding to a line-to-line voltage of 66 kv.

Problems

1. Determine the equivalent star impedance diagram for a bank of three-winding single-phase transformers whose connections are as follows: high-voltage winding, grounded star, 66 kv. between line conductors; low-voltage winding, grounded star, 11 kv. between line conductors; tertiary winding, delta, 2.2 kv. The reactances between transformer windings for the bank ratings are as follows: *H–L*, 10 per cent on 30,000 kva.; *H–T*, 6 per cent on 10,000 kva.; *L–T*, 14 per cent on 15,000 kva. Also determine the reactances in ohms on: (*a*) 66-kv. base; (*b*) 11-kv. base; (*c*) 2.2-kv. base.

2. If the transformer of Prob. 1 is supplied by a generator having a zero-sequence reactance of 10 per cent on 30,000 kva., what is the zero-sequence reactance of the combination as measured from the 66-kv. side? What is the corresponding reactance when the generator is disconnected?

3. Assume an autotransformer for connection between the 220-kv. and 110-kv. lines, with an 11-kv. delta-connected tertiary. If all the reactances are reduced to a 20,000-kva. base, they may be expressed as follows: *H–M*, 10 per cent; *H–L*, 50 per cent; *L–M*, 20 per cent; as given in Fig. 65(*b*). Determine the reactances for the equivalent star diagram. Find the zero-sequence reactance as viewed from (*a*) the high-voltage winding and (*b*) the medium-voltage winding, assuming transformer neutral is solidly grounded.

4. If the transformer of Prob. 3 is grounded through a resistor whose zero-sequence resistance per phase is 30 per cent on the 20,000-kva. base, find the equivalent circuit for zero-sequence.

5. A power system is supplied by a generator through a bank of delta grounded-star step-up transformers and a three-phase transmission line. At the receiving end, a bank of three two-winding transformers is connected in grounded-star on the transmission side and the secondary windings are kept separate. If one of the transformer secondaries is short-circuited, current will flow through the ground connections. If the secondary windings are connected in delta and a short-circuit is placed on one phase as before, will current flow through the transformer neutrals? Explain your answer in terms of zero-sequence. Also give the explanation in the terms of the single-phase solution.

6. A system is supplied by a grounded generator through a circuit-breaker to a transformer which has a grounded-star low-voltage winding, a delta-connected tertiary winding, and a grounded interconnected-star high-voltage winding. It is assumed that the component windings of the interconnected star are symmetrically coupled with the delta-connected tertiary and the grounded-star low-voltage windings. What is the zero-sequence impedance of the system as viewed from the high-voltage terminals with the circuit-breaker closed and also with the circuit-breaker open? Will current flow through the generator neutral?

7. Two parts of a system are supplied by delta-connected transformers and are connected through a zigzag autotransformer. This autotransformer employs an interconnected-star connection, and the intermediate tap is taken off the junction points of the two windings. Does this autotransformer ground the system on both sides of the transformer? How is the answer affected by transformer construction whether made of single-phase units or three-phase core-type units.

CHAPTER VII

CONSTANTS OF SHORT TRANSMISSION LINES WITHOUT GROUND WIRES

For the purpose of determining their characteristics transmission lines may conveniently be divided into three general groups, (*a*) short lines *without* ground wires, (*b*) short lines *with* ground wires, and (*c*) long lines. The treatment of long lines differs from that of short lines only because of the effects of distributed capacity. The present chapter is concerned with the first classification only, after which the two other classifications will be considered in succeeding chapters.

61. Positive- and Negative-sequence Impedance of Single Circuits.

The impedance of symmetrical non-rotative apparatus is the same for both the positive- and negative-sequences. For short transmission lines, in which the distributed capacitance may be neglected, the impedance is equal to $R + jX$, in which

R = total line resistance = rl.
X = total line reactance = xl.
l = length of line.
r = resistance per unit length of line, single conductor.
x = reactance per unit length of line.

The resistance r and the reactance x may be obtained from tables which take into consideration the effect of stranding, twist, skin effect, and effect of steel cores. Such tables are given in the Appendix.

In determining the reactance of symmetrically disposed three-phase conductors, the flux linking the conductors is integrated from the center of the conductor to the center of the two other conductors. For convenience this integration may be divided into two parts: (1) that including the flux within the conductor and also that external to the conductor to a radius of 1 ft., and (2) that including the flux between a radius of 1 ft. and the center of the two other conductors. W. A. Lewis has made use of this conception in the preparation of tables of reactances by which the reactance of any conductor for any spacing may be obtained by adding the reactance for the particular conductor for 1-ft. spacing and the reactance corresponding to the flux between 1 ft. and the actual spacing.

For properly transposed unsymmetrically spaced conductors the effect of the dissymmetry may be taken into consideration by introducing the concept of the geometric mean of the three spacings. This quantity is usually abbreviated *G.M.D.* and for three-phase circuits is equal to the cube root of the product of the three separations. This conception will be discussed in more detail in connection with the determination of the zero-sequence reactances.

Neglecting "skin effect," the flux within a conductor per unit current is dependent only upon the configuration of the cross section and can therefore be represented by a constant K. The reactance due to the external flux within a radius of 1 ft. is equal in ohms per mile at 60 cycles to $0.2794 \log \frac{1}{a}$, where a is equal to the radius of the conductor in feet. By adding the reactance due to the internal flux, the reactance due to all flux within a radius of 1 ft. is

$$0.2794\left(\log_{10}\frac{1}{a} + K\right)$$

or

$$0.2794\left[\log_{10}\frac{1}{a} + \log_{10}(10)^{K}\right]$$

or

$$0.2794 \log_{10}\frac{10^{K}}{a}$$

The expression $\frac{a}{10^{K}}$ is commonly called the **geometric mean radius** and is designated by the abbreviation *G.M.R.* Observe that it is a fictitious radius which when inserted in the logarithmic term enables one to include the effect of both the internal flux and the external flux within a radius of 1 ft. For an infinitely thin tube the *G.M.R.* is equal to a and for a round homogeneous conductor, to $0.779a$. Table VI gives the *G.M.R.* for the more common strandings and conductor patterns met in practice. Tests have shown that for multilayer A.C.S.R. conductors the

TABLE VI.—GEOMETRIC MEAN RADII* AND DISTANCES

Solid round conductor	$0.779a$
Full stranding	
7	$0.726a$
19	$0.758a$
37	$0.768a$
61	$0.772a$
91	$0.774a$
127	$0.776a$
Hollow stranded conductors and A.C.S.R. (neglecting steel strands)	
30-two layer	$0.826a$
26-two layer	$0.809a$
54-three layer	$0.810a$
Single layer A.C.S.R.	$0.35a$–$0.70a$
Point within circle to circle	a
Point outside circle to circle	distance to center of circle
Rectangular section of sides α and β	$0.2235\ (\alpha + \beta)$
Circular tube	

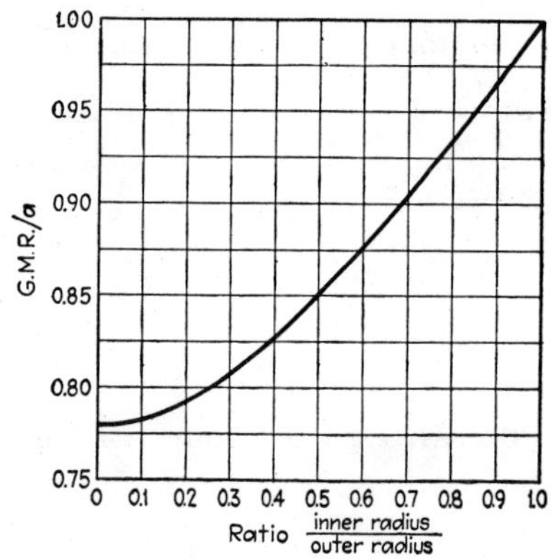

$a = \frac{1}{2}$ outside diameter

* The *G.M.R.* of standard conductors are given in the impedance tables of the Appendix.

steel strands have an inappreciable effect upon the reactance so that the *G.M.R.* is dependent only upon the configuration of the aluminum strands. For single-layer aluminum over steel this is not the case, and the *G.M.R.* varies considerably with the different conductors and also varies somewhat with the current flow.

Thus the total effect of unequal spacings and conductor stranding may be represented conveniently by means of the formula

$$x = 0.2794 \log_{10} \frac{G.M.D.}{G.M.R.} \text{ ohms per mile per phase at 60 cycles} \tag{156}$$

Both of these dimensions must be expressed in the same units.

62. Positive- and Negative-sequence Reactance of Parallel Circuits.

Due to the close proximity of parallel circuits, especially those on the same tower, transposition will not entirely eliminate the effect of mutual inductance between circuits. For the usual transposition and the configuration shown in Fig. 73 the reactance of the paralleled circuits is

$$x = 0.2794\left[\frac{1}{2}\log_{10}\frac{\sqrt[3]{d_{ab}d_{bc}d_{ca}}}{G.M.R.} - \frac{1}{12}\log_{10}\frac{d_{aa'}^{4}d_{bb'}^{2}}{d_{ab'}^{2}d_{ca'}d_{ac'}d_{ba'}^{2}}\right] \text{ ohms per mile per phase at 60 cycles} \tag{157}$$

in which the distances refer to distances between conductors in the first section.

The first term is merely the reactance of the combined circuits neglecting mutual effects in which $\sqrt[3]{d_{ab}d_{bc}d_{ca}}$ is the *G.M.D.* or the geometric mean separation of the conductors. The second represents the correction factor and may reduce the reactance 3 to 5 per cent. The formula assumes transpositions in the order *abc, cab, bca, c'b'a', b'a'c'* and *a'c'b'*, thus permitting each conductor to occupy each of the six positions in turn. The formula also assumes symmetry about the vertical axis but not necessarily about the horizontal axis.

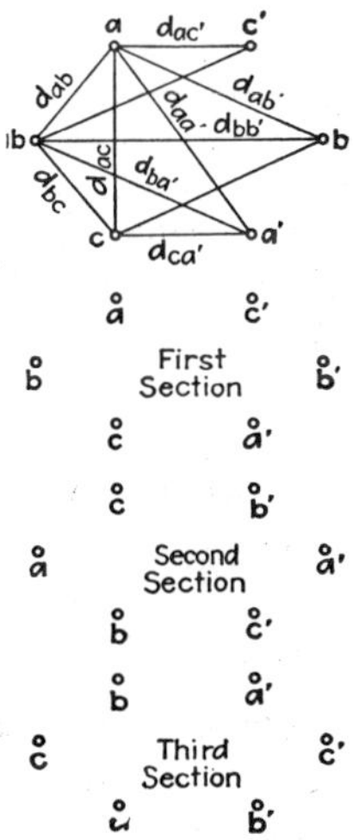

FIG. 73.—Configuration and usual order of transposition of twin circuits.

63. Zero-sequence Impedance.

The zero-sequence impedance of a transmission line is of an entirely different character from either the positive- or the negative-sequence impedance. By its very nature it involves the impedance to currents that are in phase in the three conductors, which necessitates a return path either in the earth or in a neutral or ground wire. For the important case in which the ground constitutes the only return path, the problem involves the determination of the current distribution in the earth. This problem has been attacked, among others, by Rudenberg,[137] Mayr* and Pollaczek† in Europe, and by Carson[32] and Campbell‡ in this country. Carson and Pollaczek consider a conductor parallel to the ground and assume the earth to have uniform resistivity and to be of infinite extent. No other assumption is used. Rudenberg assumes the conductor to be placed at the center of a hollowed cylindrical depression in the earth's surface of radius equal to the height of the conductor above ground and that the earth has a uniform resistivity. Mayr uses a radically different assumption—that the ground current flows in a very thin stratum next to the surface of the earth. His argument for this assumption is that on penetrating the earth's surface solid granite is soon reached. Current cannot penetrate into or beyond this stratum and must therefore flow next to the surface. Carson's formulas will be considered in detail.

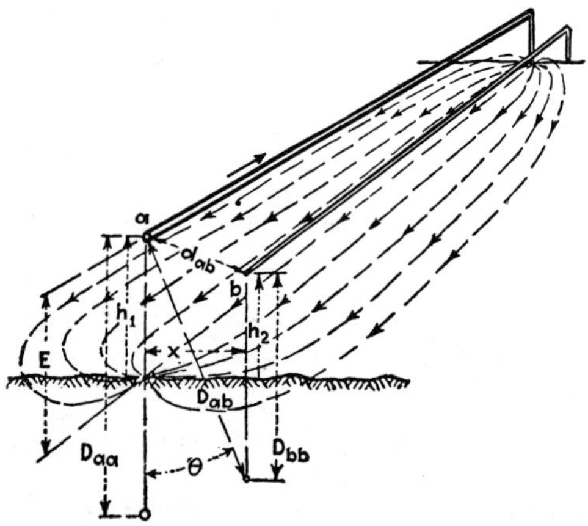

FIG. 74.—Current flow in determination of self and mutual impedances in ground return circuits.

64. Carson's Formulas.

The problem will take two phases: first the determination of the self impedance z_g of conductor a, Fig. 74, with earth return (the voltage between a and earth for unit current in the conductor), and, second, the mutual impedance z_{gm} between conductors a and b with common earth return (the voltage between b and earth for unit current in a and earth return). Reference

* Die Erde als Wechselstromleiter, *E T Z.* vol. 46, pp. 1352–1436, September, 1925.

† *Elekt. Nachrichten-Tech.*, vol. 3, p. 339, 1926.

‡ Mutual Impedance of Grounded Circuits, *Bell System Tech. Jour.*, p. 1. October, 1923.

should be made to Fig. 74 for the geometry of the system. All distances will be measured in feet, including the radius of the conductor, and θ will be measured in radians. The conductor will be considered of such length that the end effects may be neglected. Let

ρ = resistivity of earth in ohms per meter cube.
f = frequency in cycles per second.
r_c = resistance of conductor in ohms per mile.
a = radius of conductor in feet.
$\omega = 2\pi f$.

Carson in his development shows that the self and mutual impedances for ground return circuits are equal to the self and mutual impedances for a perfectly conducting earth plus a term involving $P + jQ$ for both impedances, which is a function of the variables p and θ, which will be defined. This function is the same for both the self and mutual impedances, but the variables p and θ are different. Carson then develops formulas equivalent to the following

1. Self impedance of ground return circuits $= z_g$.

$$\theta = 0.$$

$$p = 1.713 \times 10^{-3} h_1 \sqrt{\frac{f}{\rho}} = 8.565 \times 10^{-4} D_{aa} \sqrt{\frac{f}{\rho}} \tag{158}$$

$$z_g = r_c + j0.004657f \log_{10} \frac{2h_1}{G.M.R.} + 0.004044f(P + jQ) \text{ in ohms per mile} \tag{159}$$

2. Mutual impedance of ground return circuits $= z_{gm}$

$$\theta = \sin^{-1} \frac{x}{D_{ab}} = \tan^{-1} \frac{x}{h_1 + h_2} \tag{160}$$

$$p = 8.565 \times 10^{-4} D_{ab} \sqrt{\frac{f}{\rho}} \tag{161}$$

$$z_{gm} = j0.004657f \log_{10} \frac{D_{ab}}{d_{ab}} + 0.004044f(P + jQ) \text{ in ohms per mile} \tag{162}$$

For the evaluation of P and Q, Carson has derived formulas for three ranges of the values of p as given in Fig. 75. For the intermediate range ($0.25 < p < 6.0$) the curves may be used, and for the highest ($p > 5.0$) and lowest ($p < 0.25$) ranges the series given in the figure may be used. For most power-system calculations, p falls within the lowest range. To form an idea

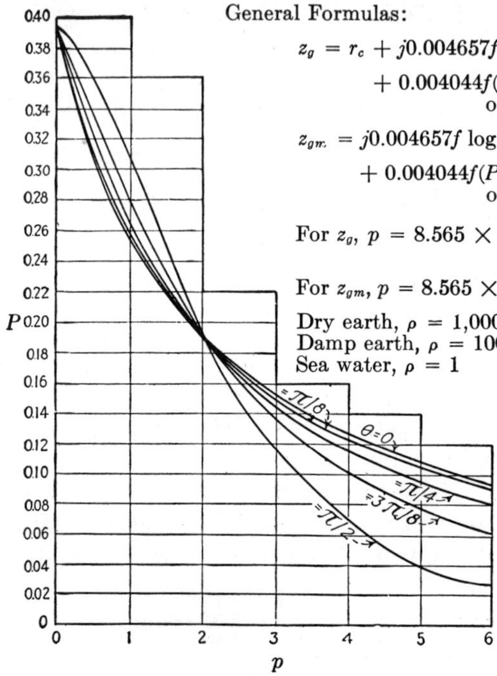

General Formulas:

$$z_g = r_c + j0.004657f \log_{10} \frac{D_{aa}}{G.M.R.} + 0.004044f(P + jQ)$$

ohms per mile

$$z_{gm} = j0.004657f \log_{10} \frac{D_{ab}}{d_{ab}} + 0.004044f(P + jQ)$$

ohms per mile

For z_g, $p = 8.565 \times 10^{-4}\sqrt{\frac{f}{\rho}}D_{aa}$

For z_{gm}, $p = 8.565 \times 10^{-4}\sqrt{\frac{f}{\rho}}D_{ab}$

Dry earth, $\rho = 1{,}000$
Damp earth, $\rho = 100$
Sea water, $\rho = 1$

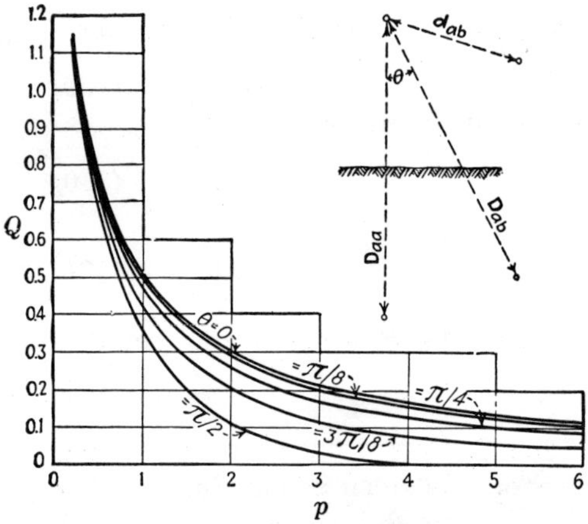

For $p = 0.25$ to 5, use curves.

For $p < 0.25$:

$$P = \frac{\pi}{8} - \frac{1}{3\sqrt{2}}p\cos\theta + \frac{p^2}{16}\cos 2\theta\left(0.6728 + \log_\epsilon \frac{2}{p}\right) + \frac{p^2}{16}\theta \sin 2\theta$$

$$Q = -0.0386 + \frac{1}{2}\log_\epsilon \frac{2}{p} + \frac{1}{3\sqrt{2}}p\cos\theta$$

For $p > 5$:

$$P = \frac{\cos\theta}{\sqrt{2}p} - \frac{\cos 2\theta}{p^2} + \frac{\cos 3\theta}{\sqrt{2}p^3} + \frac{3\cos 5\theta}{\sqrt{2}p^5}$$

$$Q = \frac{\cos\theta}{\sqrt{2}p} - \frac{\cos 3\theta}{\sqrt{2}p^3} + \frac{3\cos 5\theta}{\sqrt{2}p^5}$$

FIG. 75.—Self and mutual impedance of ground return circuits. (*Based on data given by Carson in Bell System Technical Jour., vol.* 5, pp. 539–554, *October*, 1926.) For zero-sequence, multiply by 3.

of the maximum probable value of this quantity consider the case in which $f = 60$, $h = 50$, and $\rho = 10$, for which

$$p = (1.713 \times 10^{-3})(50)\sqrt{\frac{60}{10}}$$
$$= 0.22$$

For lower frequencies, smaller heights, and larger values of earth resistivity, p will be still smaller than the above value.

An important simplifying approximation, which ordinarily introduces but a small error, is to neglect the terms in the series for ($p < 0.25$) which contain θ. The errors introduced by this approximation will be discussed in the next section. Neglecting the θ terms, P and Q become

$$P = \frac{\pi}{8} = 0.3927$$

$$Q = -0.0386 + \frac{1}{2}\log_\epsilon \frac{2}{p}$$

and since (-0.0386) is equal to $1.151 \log_{10} \frac{1}{1.079}$

$$Q = 1.151 \log_{10} \frac{2}{1.079p} = 1.151 \log_{10} \frac{1.853}{p}$$

Substituting these values of $P + jQ$ in the expressions for z_g and z_{gm} (from Fig. 75), there results

$$z_g = r_c + j0.004657f \log_{10} \frac{2h}{G.M.R.}$$
$$+ 0.004044f\left(0.3929 + j1.151 \log_{10} \frac{1.853}{p}\right)$$
$$= r_c + 0.00159f$$
$$+j0.004657f\left[\log_{10} \frac{(2h)}{G.M.R.} \frac{1.853}{p}\right]$$

and substituting the value of p from equation (158), there results

$$z_g = r_c + 0.00159f + j0.004657f \log_{10} \frac{2{,}160\sqrt{\frac{\rho}{f}}}{G.M.R.} \text{ in ohms per mile} \quad (163)$$

$$z_{gm} = 0.00159f + j0.004657f \log_{10} \frac{2{,}160\sqrt{\frac{\rho}{f}}}{d_{gb}} \text{ in ohms per mile} \quad (164)$$

For $p > 0.25$, the self and mutual impedances of earth return circuits may be obtained by the use of the curves in Fig. 76 which give the increases in resistance and reactance due to the

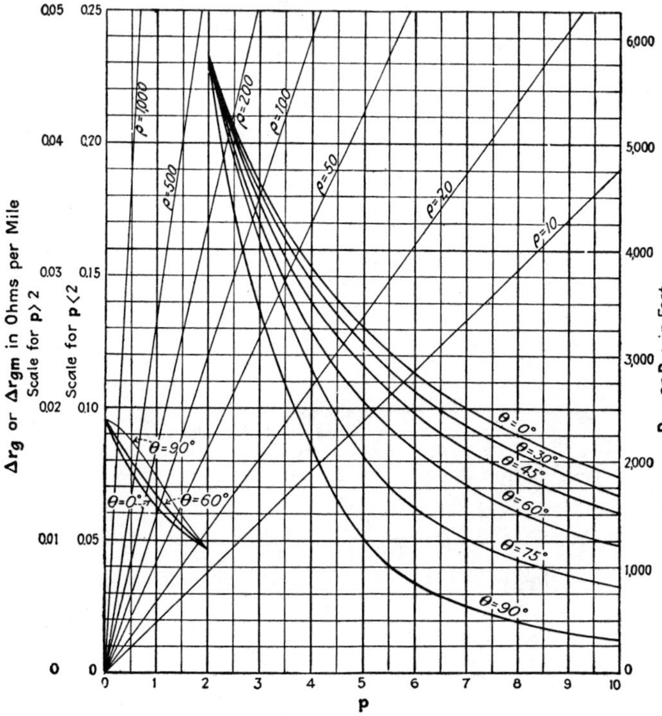

$$Z_g = r_c + j0.2794 \log \frac{D_{aa}}{G.M.R.} + (\Delta r_g + j\Delta x_g)$$

$$Z_{gm} = +j0.2794 \log \frac{D_{ab}}{d_{ab}} + (\Delta r_{gm} + j\Delta x_{gm})$$

where $(\Delta r_g + j\Delta x_g)$ and $(\Delta r_{gm} + j\Delta x_{gm})$ are determined from curves *a* and *b* for 60 cycles. For other frequencies use the fictitious value of $\rho = \frac{60}{f}\rho_{actual}$ and multiply ordinate by $\frac{f}{60}$.

FIG. 76(*a*).—Curves of the increase in earth return circuit resistance and reactance due to the resistivity of the earth.

departure of earth resistivity from zero. The caption gives the reactance for $\rho = 0$, and correction terms are obtained from the curves. These curves are laid out on a 60-cycle basis, so that p can be determined from equations (158) and (161) when ρ and D_{aa} (or D_{ab}) are given. This is accomplished graphically in Fig. 76 by using the right-hand scales and the straight lines labeled for the corresponding values of ρ. After p is determined, Δr_g or Δx_g is read from the curve for $\theta = 0$ and the determined values of p. Similarly, Δr_{gm} and Δx_{gm} are used from the curves labeled for the appropriate angle.

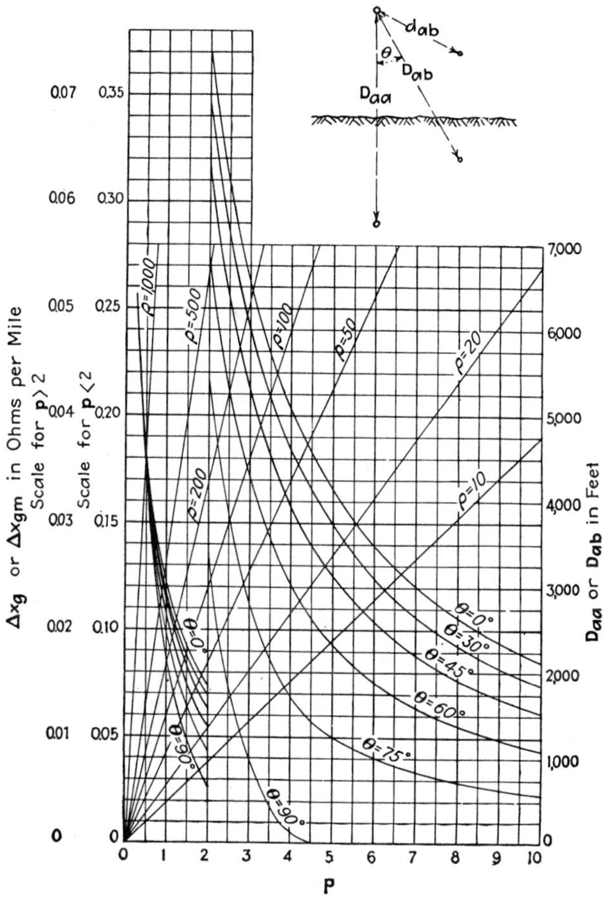

FIG. 76(*b*).

It will be observed from equations (158) and (161) that p varies directly with $\sqrt{f}$ and inversely as $\sqrt{\rho}$. Since the curves are laid out for $f = 60$, they may be used for other frequencies by using a fictitious value of ρ equal to $\frac{60}{f}$ times the actual value. The increment in impedance varies directly with the frequency so that it is necessary to multiply the ordinates by $\frac{f}{60}$.

65. Physical Conception and Factors Affecting Earth Return Circuits.

A very useful physical conception for use in the analysis of earth return circuits is that of concentrating the current in a fictitious return conductor at some considerable depth below

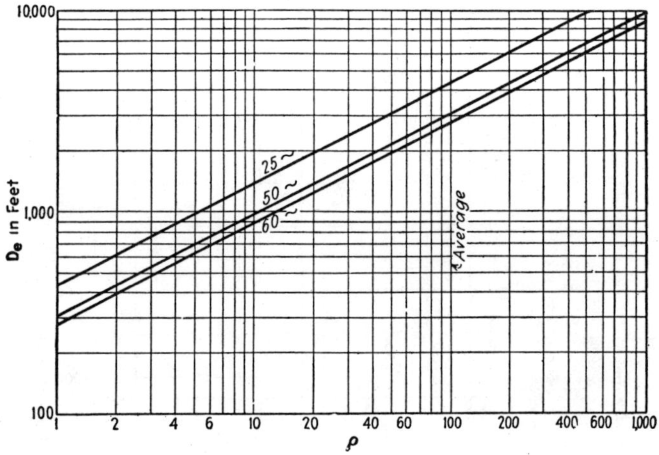

FIG. 77.—Equivalent depth of ground return circuit as a function of the earth resistivity ρ in meter-ohms and frequency.

the outgoing conductor. The **equivalent depth of earth return*** D_e is the distance from the outgoing conductor to the fictitious conductor and is the value to be used in the equation

$$x = 0.004657f \log_{10} \frac{D_e}{G.M.R.} \text{ ohms per mile} \tag{165}$$

This conception has been used for some time with good results because the equation takes the same form as that of the simplified form of Carson's. Thus, by equating the logarithmic expressions of (163) and (165), there results

$$D_e = 2{,}160\sqrt{\frac{\rho}{f}} \tag{166}$$

* The distance to the equivalent earth plane assuming infinite conductivity is half of D_e.

The values of D_e in feet for different values of ρ and f are given in Fig. 77.

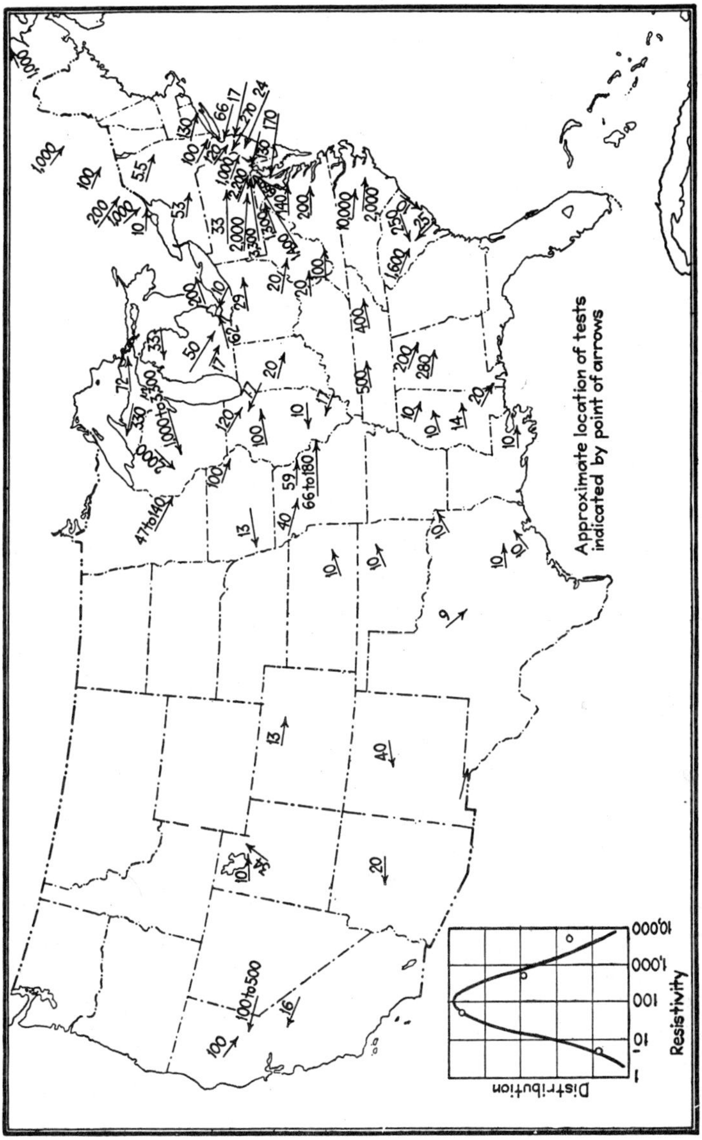

FIG. 78.—Test results for earth resistivity ρ in meter-ohms. The values of earth resistivity given were obtained by the Joint Subcommittee on Development and Research of the National Electric Light Association and the Bell Telephone System.

Earth Resistivity. During the past few years considerable data have been collected regarding the values of ρ met in practice. Most of these data have been obtained by measuring the self impedance of a ground return circuit of known geometrical configuration and determining the equivalent ρ which when used in Carson's formula gives the same impedance as that measured. The map shown in Fig. 78 has been prepared as a result of work carried out by the Joint Research and Development Subcommittee of the N.E.L.A. and the Bell Telephone System.(79) Table VII gives the resistivity for various constituents of the earth's surface. Comparison of the values of ρ from Table VII

TABLE VII.—EARTH RESISTIVITY IN METER-OHMS

Average of large number of determinations*	100
Sea water	0.01–1.0
Swampy ground	10–100
Dry earth	1,000
Pure slate	10^7
Sandstone	10^9

* See insert, Fig. 77.

and the insert of Fig. 78 shows that the average value corresponds to damp earth. This comparison also shows that the resistivity of the earth is not uniform, a result to be expected from the presence in some localities of coal or mineral veins, surface or subsurface streams, railroad tracks, pipe lines or other underground structures. Also it is known that in some mountainous localities the geological formation consists of a solid granite base with merely a relatively shallow layer of soil. This fact forms the basis of an advanced method of ground return circuit analysis in which the earth is assumed to be composed of two homogeneous layers of different resistivities. The condition of the soil as influenced by climatic changes and seasons also affects the conductivity. Some localities such as deserts and prairies become extremely dry and offer high resistance to current flow. Then, again, in some districts the resistivity of the soil may be uniformly low in the winter and spring, but during the hot season the soil may dry out to a considerable depth, increasing its resistivity and, in effect, increasing the effective height of the conductors above true ground. Of course, in such cases it is recognized that the earth cannot be considered homogeneous, but that the conductivity increases with the distance from the surface. The value of the Carson form of the expression is that it enables one to evaluate the influence of peculiar local conditions and thus to estimate these reactances more accurately.

Current Distribution. The current density in the earth is greatest at the surface directly underneath the conductor and decreases gradually as the distance from the conductor increases. The curves in Fig. 79, showing the current density at the earth's surface in terms of the density directly under the conductor, were calculated for the two more common frequencies met in practice and for damp earth. It will be observed that at a distance of 1 mile, the current density has reduced to 6 per cent for the 60-cycle case and to 9.5 per cent for the 25-cycle case. For smaller heights the current is still more concentrated. The significance of these curves is, as Rudenberg[137] points out,

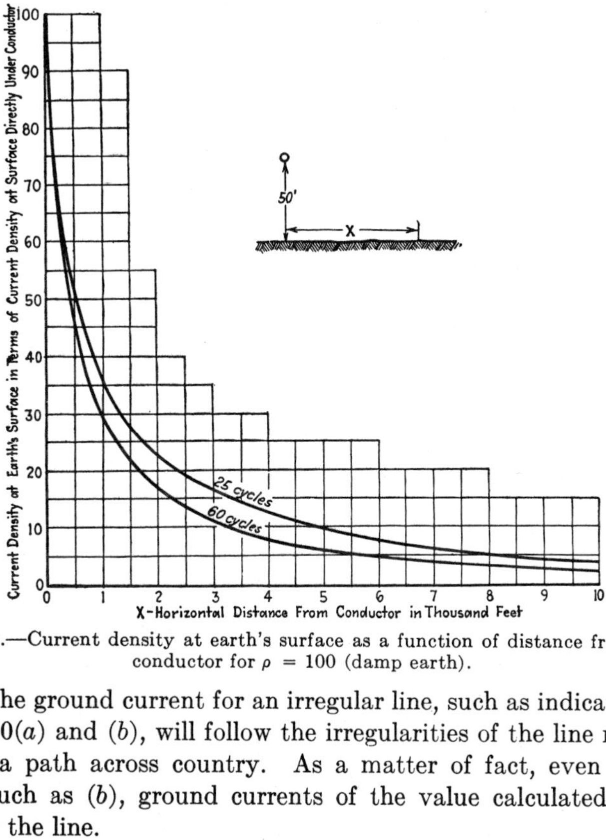

FIG. 79.—Current density at earth's surface as a function of distance from the conductor for $\rho = 100$ (damp earth).

that the ground current for an irregular line, such as indicated in Fig. 80(*a*) and (*b*), will follow the irregularities of the line rather than a path across country. As a matter of fact, even for a line such as (*b*), ground currents of the value calculated exist under the line.

Effect of Resistivity and Height. It may be noted from the approximate formulas, equations (163) and (164), that for the range of earth resistivity met in practice and for commercial frequencies both the self and mutual impedances are independent of height above ground. It is interesting to note that the resistance component of the self impedance is independent of the resistivity of the earth. This paradox is explained by the fact that at high resistivity the current spreads out over a larger area and at low resistivity is restricted to an area near the conductor. Thus the ground resistance is for 25 cycles 0.04, and for 60 cycles 0.095 ohm per mile.

It should be remembered, however, that the approximate formulas neglected certain terms including θ in the expressions for P and Q. These terms include resistivity, frequency, and certain geometrical factors (h or D_{ab}). The effect of these terms will be to decrease the resistance as indicated by the curves of Fig. 81. Except in very rare cases such as river crossings (which may represent a very small portion of the total length of the line under consideration), the highest average height between towers is about 50 ft. For $\rho = 10$, the resistance of the ground return is 0.086 ohm per mile. The use of 0.095 instead of 0.086 therefore represents an error of 10 per cent. The equivalent resistance per mile contributed by the resistance of

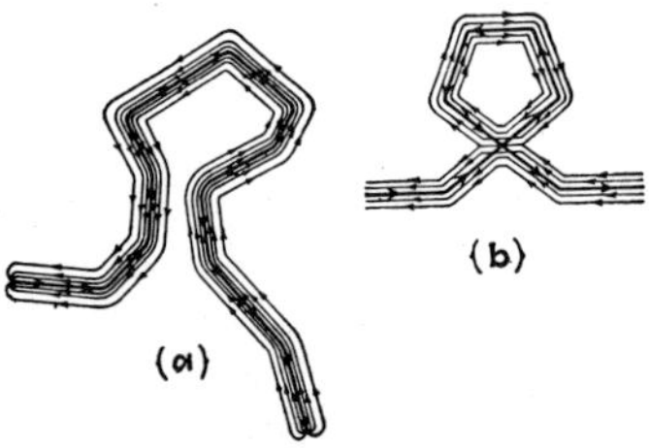

FIG. 80.—Ground current follows irregularities in the line.

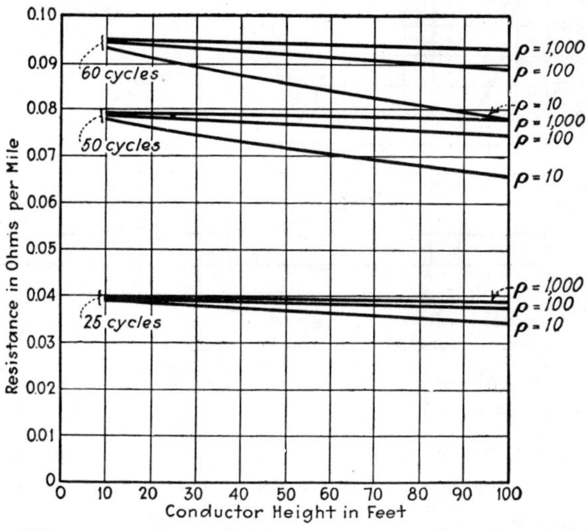

FIG. 81.—Effect of conductor height, frequency, and earth resistivity upon resistance of ground return circuits. Zero-sequence resistance is three times the ordinate.

the conductor r_c, the tower-footing resistance, and the station ground will be of the order of 0.1 ohm per mile, so that the error in terms of the resistance will be about 5 per cent. In addition the reactance is usually larger than the resistance, and since they add at right angles the effect of this error in the resistance term upon the magnitude of the current is still further reduced. If the phase position of the zero-sequence current is desired, the error is of course still effective. In view of the variable character of the tower footing and station grounds, and also the departure of the earth from the assumption of uniform conductivity, this error is quite tolerable.

The effect of neglecting the cos θ term of the Q expression upon the reactance of the ground return circuit may be obtained by multiplying the cos θ term $\left(\frac{1}{3\sqrt{2}}p \cos \theta\right)$ by $0.004044f$ and inserting the value of p, giving the term

$$1.63 \times 10^{-6} hf\sqrt{\frac{f}{\rho}} \tag{167}$$

which must be added to equation (163) to give the self reactance of the ground return circuit and the term

$$0.82 \times 10^{-6} D_{ab} f\sqrt{\frac{f}{\rho}} \cos \theta \tag{168}$$

which must be added to equation (164) to give the mutual reactance of ground return circuits. For the self-reactance calculations this correction is always less than 1 per cent, but for mutual reactances in extreme cases, such as $\rho = 10$, $f = 60$, and $D_{ab} = 200$ ft., the error may become as great as 4 per cent.

Therefore, for practically all power work, expressions (163) and (164) are sufficiently accurate. For inductive coordination work the distances and frequencies involved are greater so that p is frequently larger than 0.25. For this field of application it is necessary to use the curves of Fig. 75 or 76. The Joint Development and Research Subcommittee of the N. E. L. A. and the Bell Telephone System have prepared elaborate curves[79] for this purpose which will save much time if many calculations of induced voltage are to be made.

66. Simplified Formulas for Zero-sequence Impedances.

Using the formulas involving the equivalent depth of earth return current, the expressions for self and mutual impedance become

25 cycles

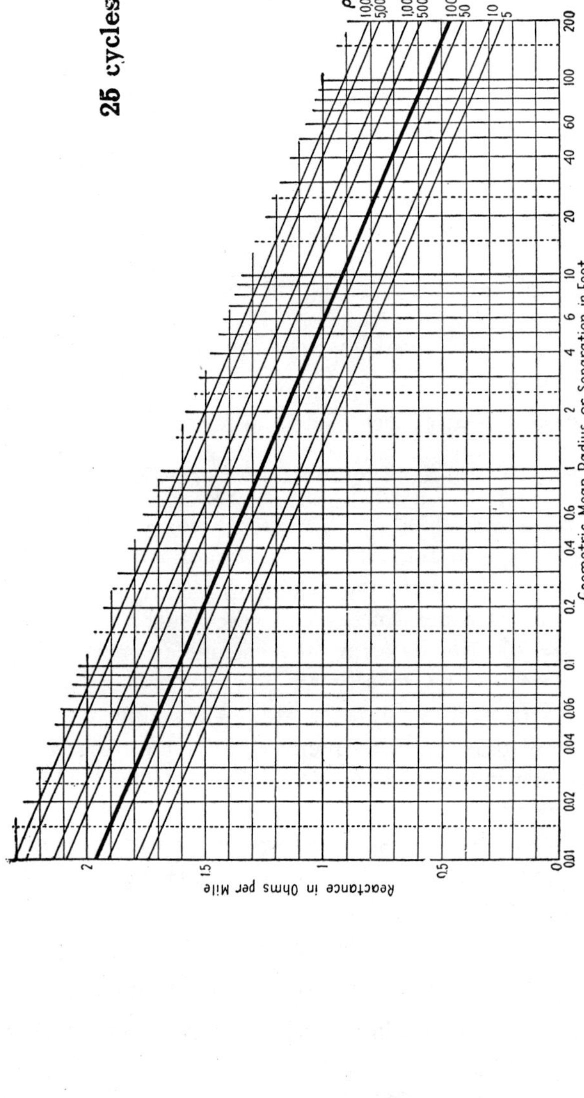

FIG. 82(*a*).—Zero-sequence self and mutual reactance per phase of single equivalent conductor at 25 cycles. Zero-sequence resistance = 3(resistance of conductors considered as a group) + 0.12 ohm per mile per phase

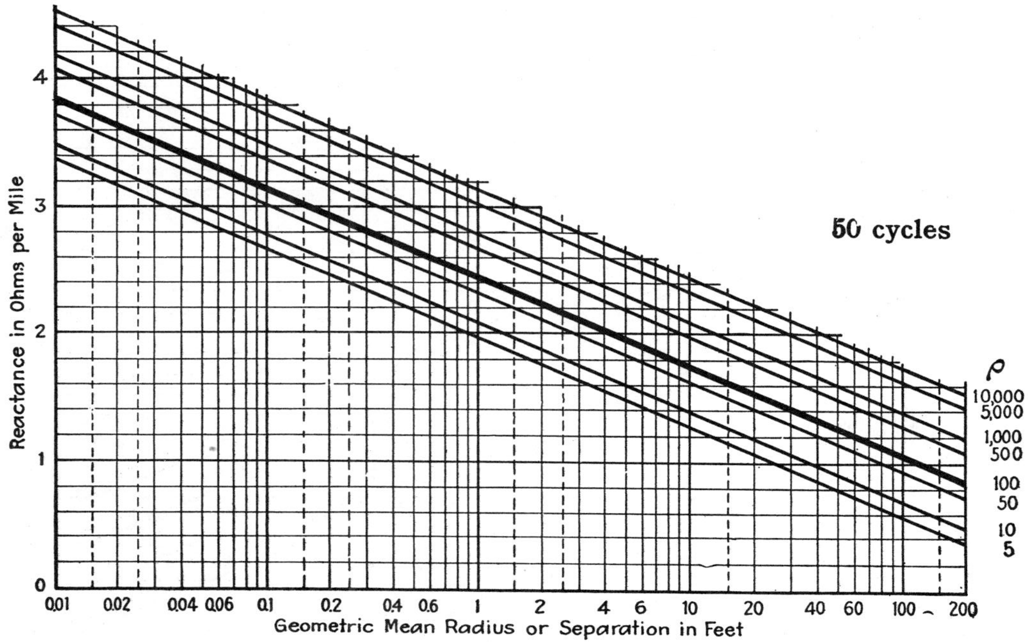

FIG. 82(*b*).—Zero-sequence self and mutual reactance per phase of single equivalent conductor at 50 cycles. Zero-sequence resistance = 3(resistance of conductors considered as a group) + 0.24 ohm per mile per phase

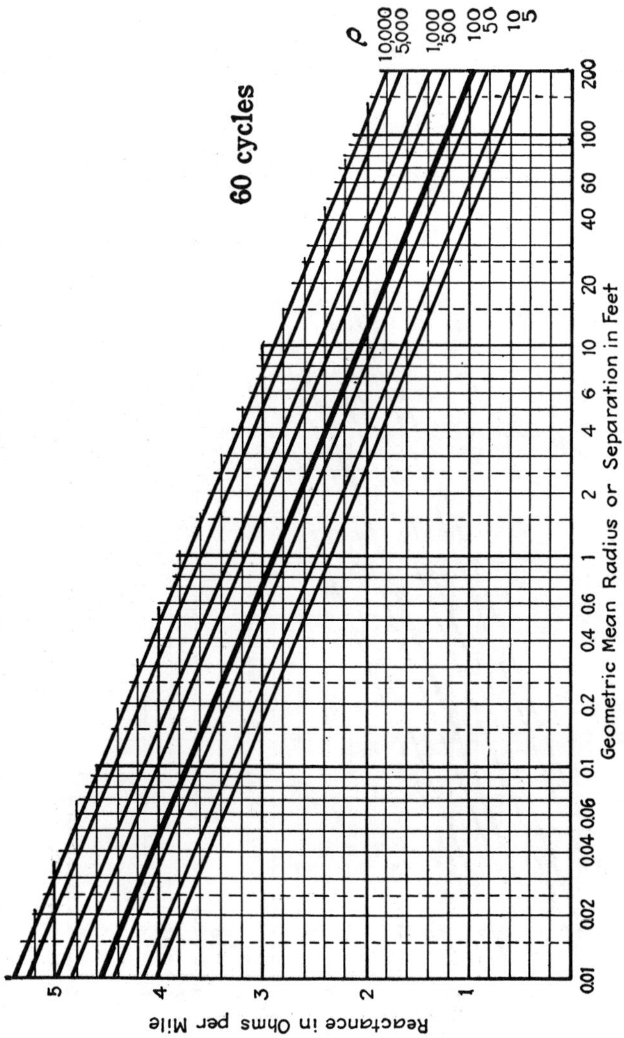

FIG. 82(*c*).—Zero-sequence self and mutual reactance per phase of single equivalent conductor at 60 cycles. Zero-sequence resistance = 3(resistance of conductors considered as a group) + 0.286 ohm per mile per phase

$$z_g = r_c + 0.00159f + j0.004657f \log_{10} \frac{D_e}{G.M.R.} \text{ ohms per mile} \quad (169)$$

$$z_{gm} = 0.00159f + j0.004657f \log_{10} \frac{D_e}{d_{ab}} \text{ ohms per mile} \quad (170)$$

These equations can also be applied to multiple-conductor circuits if r_c, the *G.M.R.*, and d_{ab} refer to the conductor or conductors as a group. This point will be amplified later.

Unit zero-sequence current consists of 1 amp. in each phase conductor and therefore of 3 amp. in the earth. In replacing the three conductors by a single equivalent conductor, 3 amp. flow in the equivalent conductor for every ampere of zero-sequence current. Hence the corresponding zero-sequence impedances are three times the above values. Calling the zero-sequence impedances z_0 and z_{0m},

$$z_0 = 3r_c + 0.00477f + j0.01397f \log_{10} \frac{D_e}{G.M.R.} \text{ ohms per phase per mile} \quad (171)$$

$$z_{0m} = 0.00477f + j0.01397f \log_{10} \frac{D_e}{d_{ab}} \text{ ohms per phase per mile} \quad (172)$$

The resistance r_c in this formula is, of course, the resistance of the equivalent conductor and is equal to one-third the resistance of a single conductor when a single three-phase circuit is being considered; $3r_c$ in this case thus becomes the resistance of one phase conductor.

The substitution of the value of D_e from equation (166) in equations (171) and (172) gives the reactance of earth return circuits. This reactance is plotted for the three frequencies 25, 50, and 60 cycles, in Fig. 82(*a*), (*b*), and (*c*), respectively, as a function of the *G.M.R.* or separation for different earth resistivities. In the absence of definite information regarding the value of ρ in the particular geographical location under consideration, the values corresponding to the heavy lines ($\rho = 100$) may be taken as representative of average conditions.

67. Self Impedance of Parallel Conductors with Earth Return.

So far the self and mutual impedances between single cylindrical conductors with circular cross section only have been considered. In practice, however, the problem usually arises to determine the self impedance of three similar conductors connected in parallel. Consider any three conductors, such as a, b, and c, in Fig. 83, whose geometrical dispositions are as indicated. With conductors transposed, the current distributes uniformly between conductors, so that for a total current equal to unity the current in each conductor is one-third. The voltage drop in conductor a for the position indicated is

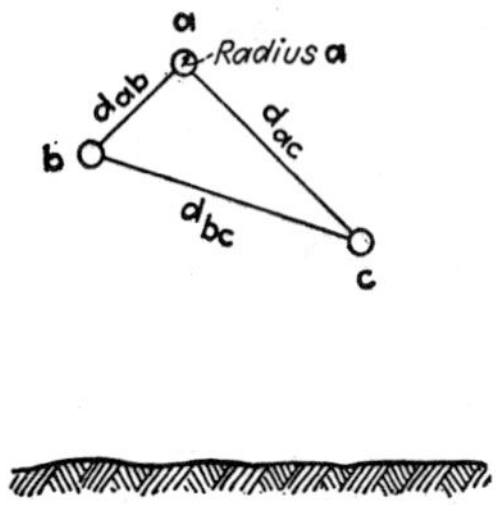

FIG. 83.—Notation for the determination of the zero-sequence reactance.

$$\frac{Z_{aa}}{3} + \frac{Z_{ab}}{3} + \frac{Z_{ac}}{3}$$

for conductor b

$$\frac{Z_{ab}}{3} + \frac{Z_{bb}}{3} + \frac{Z_{bc}}{3}$$

and for conductor c

$$\frac{Z_{ac}}{3} + \frac{Z_{bc}}{3} + \frac{Z_{cc}}{3}$$

in which Z_{ab}, Z_{bc}, and Z_{ac} are the mutual impedances between the conductors represented by the subscripts and Z_{aa}, Z_{bb}, and Z_{cc} are the self impedances of the three conductors with ground return. Since conductor a takes all these positions successively, the drop per conductor is

$$\tfrac{1}{9}(Z_{aa} + Z_{bb} + Z_{cc} + 2Z_{ab} + 2Z_{bc} + 2Z_{ca})$$

After inserting the values of impedances from equations (169) and (170), this drop, or impedance, since unit current is flowing, becomes

$$z_g = \tfrac{1}{9}\left[3r_c + 9(0.00159f) + j0.004657f \left\{3 \log_{10} \frac{D_e}{G.M.R.} + 2 \log_{10} \frac{D_e}{d_{ab}} + 2 \log_{10} \frac{D_e}{d_{bc}} + 2 \log_{10} \frac{D_e}{d_{ca}}\right\}\right] \text{ ohms per mile}$$

Combining terms

$$z_g = \frac{r_c}{3} + 0.00159f + j0.004657f \log_{10} \frac{D_e}{\sqrt[9]{(G.M.R.)^3 d^2_{ab} d^2_{bc} d^2_{ca}}} \text{ ohms per mile} \quad (173)$$

and multiplying by 3, the zero-sequence impedance per phase becomes

$$z_0 = r_c + 0.00477f + j0.01397f \log_{10} \frac{D_e}{\sqrt[9]{(G.M.R.)^3 d^2_{ab} d^2_{bc} d^2_{ca}}}$$

ohms per mile per phase (174)

The ninth root in the denominator of the logarithmic term is the *G.M.R.* and is equal to the radius of an infinitely thin tube which would have the same inductance as the system represented. Equation (173) reduces to (169), r_c in (169) being the resistance of the three conductors in parallel, and r_c in (173) and (174) the resistance of one conductor. In determining the reactance for the positive- and negative-sequence impedances the *G.M.D.* of the conductors was used. This equivalent separation was obtained by taking the cube root of the three separations. From equation (174)

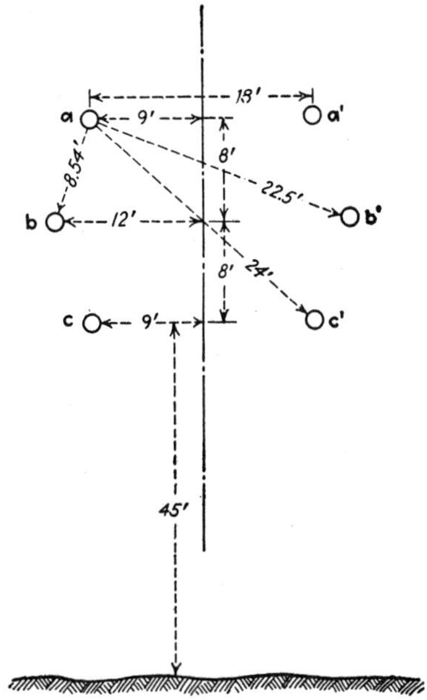

FIG. 84.—Configuration of typical twin-circuit transmission line. 397,500 cir. mils A.C.S.R. – 60 cycles Conductor radius = 0.403 in. Conductor *G.M.R.* (30 strand-two layer) = 0.826 × 0.403 in.

$$G.M.R._{circuit} = \sqrt[9]{(G.M.R.)^3_{conductor} d^2_{ab} d^2_{bc} d^2_{ca}}$$
$$= \sqrt[3]{(G.M.R.)_{conductor} (G.M.D.)^2_{separation}} \quad (175)$$

Example. Let it be desired to determine the zero-sequence impedance of one circuit of the double-circuit line shown in Fig. 84, assuming damp soil (ρ = 100 meter-ohms).

Assume a resistance of each conductor = 0.235 ohm per mile

$$G.M.D. = \sqrt[3]{8.54 \times 8.54 \times 16} = 10.5 \text{ ft.}$$

$$G.M.R. = \sqrt[3]{0.826 \times \frac{0.403}{12} (10.5)^2} = 1.45 \text{ ft.}$$

$$D_e \text{ (from Fig. 77)} = 2{,}800 \text{ ft.}$$

$$z_0 = 0.235 + 0.00477 \times 60 + j0.01397 \times 60 \log_{10} \frac{2{,}800}{1.45}$$
$$= 0.521 + j2.76 \text{ ohms per mile per phase}$$

This same result may be obtained by referring to Fig. 82(*c*) for *G.M.R.* = 1.45 and ρ = 100 and picking off the reactance which is equal to 2.76 and adding the resistance of the conductor to the resistance of the ground return circuit [0.286 also from Fig. 82(*c*)].

68. Mutual Impedance between Two Circuits with Earth Return.

The mutual impedance between two circuits with common earth return may be derived in a similar manner and is found to be

$$z_{m0} = 0.00477f + j0.01397f \log_{10} \frac{D_e}{G.M.D.} \text{ ohms per mile per phase} \quad (176)$$

where the *G.M.D.*, in the case of the three-phase circuit, is the ninth root of the product of the nine possible distances between the conductors in the two circuits.

Example. Let it be desired to calculate the mutual impedance between the two circuits in Fig. 84.

$$\begin{aligned} D_e \text{ (from Fig. 77)} &= 2{,}800 \text{ ft.} \\ G.M.D. &= \sqrt[9]{(18)^2(22.5)^4(24)^3} \\ &= 21.9 \text{ ft.} \\ z_{m0} &= 0.00477 \times 60 + j0.01397 \times 60 \log_{10} \frac{2{,}800}{21.9} \\ &= 0.286 + j1.77 \text{ ohms per mile per phase} \end{aligned}$$

Or using Fig. 82(*c*) for *G.M.D.* = 21.9 ft. and ρ = 100, the mutual reactance is 1.77. The mutual resistance is 0.286.

69. Self Impedance of Two Identical Parallel Polyphase Circuits with Earth Return.

The impedance of two parallel identical circuits is

$$z_0 = \frac{r_c}{2} + 0.00477f + j0.01397f \log_{10} \frac{D_e}{\sqrt{(G.M.R.)(G.M.D.)}} \text{ ohms per mile per phase} \quad (177)$$

in which *G.M.R.* is the geometric mean radius of one set of conductors and *G.M.D.* is the geometric mean distance between the two sets of conductors.

Example. To illustrate the application, calculate the impedance of both circuits in Fig. 84 in parallel. From a previous calculation

$$\begin{aligned} G.M.R &= 1.45 \text{ ft.} \\ G.M.D. &= 21.9 \text{ ft.} \\ \sqrt{(G.M.R.)(G.M.D.)} &= 5.64 \text{ ft.} \end{aligned}$$

Substituting in formula (177)

$$z_0 = 0.4035 + j2.26$$

This should check against $\frac{1}{2}(z_0 + z_{m0})$ from previous calculations.

$$z_0 = \tfrac{1}{2}(0.521 + j2.76 + 0.286 + j1.77)$$
$$= 0.4035 + j2.26 \text{ ohms per mile per phase}$$

For non-identical circuits it is better to compute the mutual and self impedance for the individual circuits.

70. Equivalent Circuits for Parallel Lines with Mutual Inductive Coupling.

Frequently it is required to calculate faults at locations between bussing points on physically parallel circuits. The coupling between these circuits for the positive- and negative-

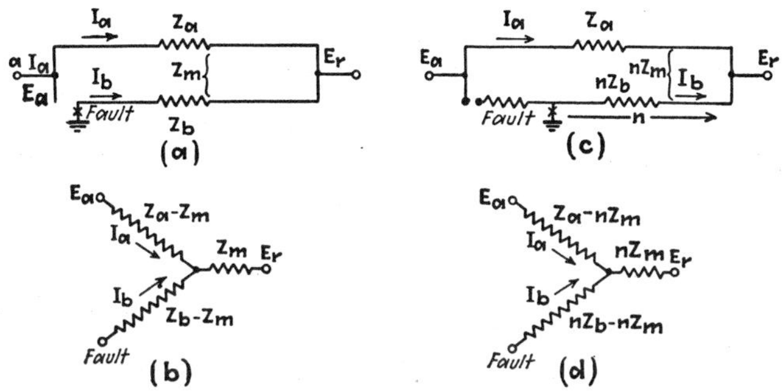

Fault at One End — Fault at an Intermediate Point

FIG. 85.—Zero-sequence equivalent networks for parallel lines after one circuit breaker has opened.

sequences is small, and it is further reduced by transpositions. For the positive- and negative-sequence networks, then, this effect is negligible. Such is not the case for the zero-sequence, and transpositions do not reduce the effect. One important case involving these considerations is illustrated in Fig. 85(*a*), in which it is assumed that the breaker at the left bus has already opened for the fault which is assumed to have occurred just outside of the breaker. If the problem is such that capacitance effects may be neglected, either because of the shortness of the line or because of the accuracy required, the actual network may be replaced by the equivalent shown in (*b*). The values of the constants are given in the figure and the proof is indicated subsequently in the discussion of the more general case involving distributed capacitance. For faults at intermediate points the circuit condition and equivalent circuit are shown in Fig. 85(*c*) and (*d*), respectively.

Another important case is that of a fault on a two-circuit line at some point between buses, illustrated in Fig. 86. The impedances Z_a, Z_b, and Z_m are the zero-sequence impedances for the whole line. It will be observed that if the two circuits are divided into two parts at the fault point, each side can be represented by the equivalent circuits of the type shown in Fig. 85. This has been done in Fig. 86(b). For the usual case $Z_a = Z_b = Z_L$ and, using this assumption, the circuit can be successively simplified as indicated in (c), (d), and (e). The

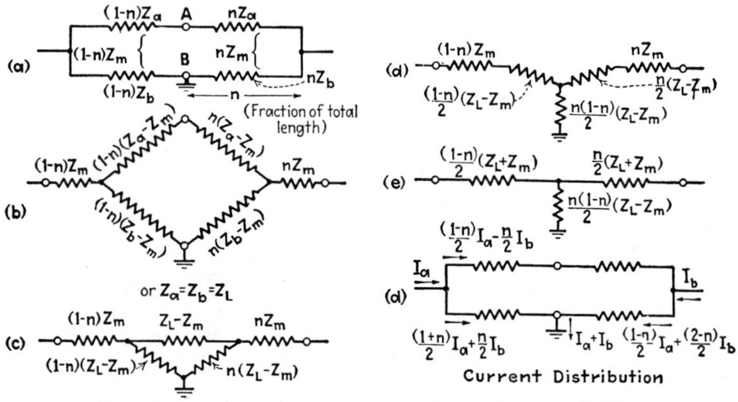

FIG. 86.—Zero-sequence equivalent network for a two-circuit line with a fault at an intermediate point.

current distribution in the faulted and unfaulted lines in terms of the current input into the network is given in (f).

Example. Let it be desired to determine the zero-sequence network for a 25-mile double-circuit line whose configuration is shown in Fig. 84. The self and mutual capacitances will be neglected. Further, let the fault occur at a point 10 miles from the right-hand bus, Fig. 86(a), so that $n = \frac{10}{25} = 0.4$. The self and mutual impedances for the zero-sequence have already been calculated for this configuration and were found to be

$$r_0 + jx_0 = 0.521 + j2.76 \text{ ohms per mile per phase}$$
$$r_{m0} + jz_{m0} = 0.286 + j1.77 \text{ ohms per mile per phase}$$

Therefore

$$Z_L = (r_0 + jx_0)l = 13.02 + j69.0 \text{ ohms}$$
$$Z_M = (r_{m0} + jz_{m0})l = 7.15 + j44.25 \text{ ohms}$$

and

$$\frac{n(1-n)}{2}(Z_L - Z_M) = \frac{(0.4)(1-0.4)}{2}(13.02 + j69.0 - 7.15 - j44.25) = 0.704 + j2.97$$

$$\frac{n}{2}(Z_L + Z_M) = \frac{0.4}{2}(13.02 + j69.0 + 7.15 + j44.25) = 4.034 + j22.65$$

$$\frac{(1 - n)}{2}(Z_L + Z_M) = \frac{1 - 0.4}{2}(13.02 + j69.0 + 7.15 + j44.25) = 6.05 + j34.0$$

These values, substituted in their respective positions in Fig. 86(*e*), thus give the equivalent network. The current distribution is given in (*f*) in terms of the external currents.

Problems

1. Assume that the ground-return-circuit solution given by Carson and the normal values of earth resistivity apply for radio frequency. Determine the resistance of the earth return circuit per mile for 1,000 kilocycles, assuming $\rho = 100$ meter-ohms and $h = 50$ feet.

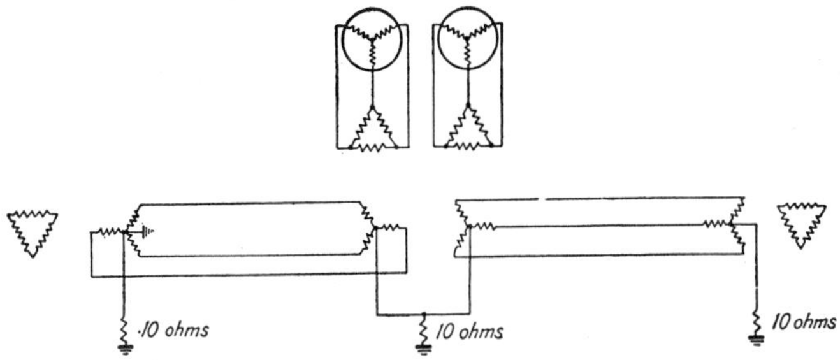

2. Two different power stations and the step-up transformers are located at the same point as indicated in the accompanying figure. Each generator and each transformer has a reactance of 10 per cent on a 20,000-kva base. Each transmission line is 40 miles in length and consists of 4/0 stranded copper conductors spaced triangularly 10 ft. apart and transposed at intervals. The transmission line is operating at its no-load normal voltage of 66 kv. At the generating station the transformer neutrals are connected and then grounded together through a connection that has a resistance of 10 ohms. The ground connections for each receiver transformer has a resistance of 10 ohms. Determine the amount of current that flows in the various transformer neutrals for a line-to-ground fault at the receiver end of one transmission line assuming that the transmission lines do not parallel each other. The system frequency is 60 cycles per second.

3. Assume the same systems as in Prob. 2 except that the ground connections are assumed to be of zero impedance, but that the transmission lines parallel each other throughout their entire length at a separation of 100 ft. Determine the current flowing through each of the four transformer neutrals for the same fault conditions.

4. Assume that a 33-kv. line forming a ten mile square is paralleled at an average separation of 50 feet by a 132-kv. system as shown in the accompanying figure. The 33-kv. circuit is made up of 266,800 A.C.S.R. conductors spaced 6 ft. delta. If the 132-kv. transmission circuit is operated at 60 cycles and is subjected to a line-to-ground fault which draws 1,000 amp. through the parallel, determine the zero-sequence current flowing in the 33-kv. system.

Explain why zero-sequence current can flow in the 33-kv. lines which have no ground connections.

5. Assume a source of negligible impedance to positive-, negative-, and zero-sequence in comparison with the impedance of a 40-mile transmission line made up of 3/0 stranded copper conductors spaced 10 ft. apart and symmetrically transposed. Assume that one conductor is grounded at the far end. Determine the positive-, negative-, and zero-sequence impedances of the line by the method of symmetrical components and the line-to-neutral fault current for a normal voltage of 110 kv. between lines. Determine the impedance of a single-phase circuit with one line conductor and ground as

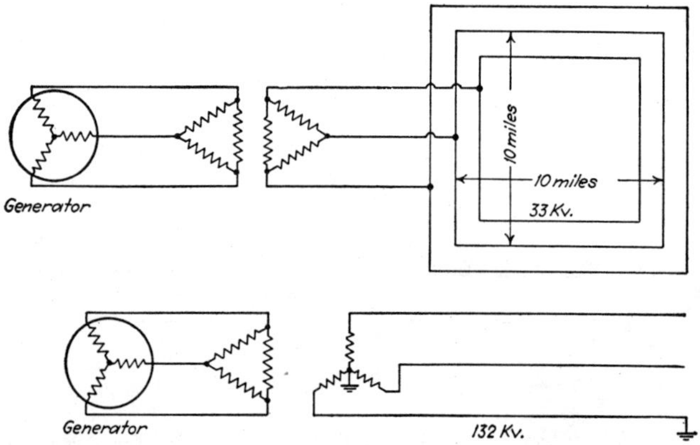

the return and the resultant fault current by the "single-phase" method. Analyzed by the single-phase method, it is necessary merely to determine the supply frequency, the resistivity of the earth, and the conductor material, stranding, and diameter. By the method of symmetrical components it is necessary to determine, in addition, the spacing from the faulted conductor to the other conductors. Show that the single-phase method and symmetrical components will give the same analytical expression for both the resistance and reactance components.

6. Tests were made to determine the resistivity of the earth in a particular area. A long, insulated cable was laid on the surface of the earth and carefully grounded at each end. A test loop was formed by laying an insulated wire parallel to the cable at a distance of 100 ft. and grounding at one end. The power cable is sufficiently long so that end effects may be eliminated from the test loop. A current of 100 amp. at 60 cycles is caused to flow through the grounded conductor. The voltage induced in the test loop is found by test to be $(9.5 + j36)$ volts. Determine the equivalent depth of the earth return current and the resistivity of the earth. Check the induced voltage by the geometry of the circuit and the value of the earth resistivity derived from test results.

CHAPTER VIII

CONSTANTS OF SHORT TRANSMISSION LINES WITH GROUND WIRES

Previous chapters have emphasized that, in the symmetrical portions of a system, the positive-, negative-, and zero-sequence quantities are independent of each other; whereas, in unsymmetrical portions, the sequence quantities are not independent. When the impedances are unsymmetrical, positive-sequence currents may produce negative- and zero-sequence voltages, and similarly for the other sequences. Due to the physical arrangement of conductors, all transmission lines are unsymmetrical to some extent. Fortunately, the error introduced by dissymmetry is negligibly small for the arrangements of conductors and ground wires usually met in practice if the geometric mean distances (*G.M.D.*) are used. This phase of the problem will be illustrated by working out several examples, using both a rigorous method and the simplified method of symmetrical components, which assumes that the phases are balanced, and comparing the results. For extreme cases, investigation of this error should be undertaken before using the simplified method. The detailed assumptions involved will be specified in the discussion of the determination of the different sequence impedances.

71. Positive- and Negative-sequence Impedance.

The ground wires may be neglected in the calculation of the positive- and negative-sequence impedance. This assumption is rigorously accurate for the case of perfectly transposed lines with the ground wires grounded only at barrel points, and line faults occurring at barrel points, because the sum of the voltages induced in the ground wires by positive- and negative-sequence currents is zero. In practice, however, the ground wires are grounded at every tower, so that the induced currents in the ground wires will have a certain definite but negligibly small effect upon the positive- and negative-sequence impedances of the line. The error is quite small also for faults at intermediate points in a transposition barrel.

72. Application of Carson's Formulas to the Calculation of Zero-sequence Impedance.

It was shown in Sec. 65 that, as a result of Carson's work, the self impedance of a conductor and the mutual impedance between two parallel conductors, with earth return (neglecting end effects), can be calculated by equations (169) and (170) and Fig. 75. The corresponding impedances in terms of zero-sequence

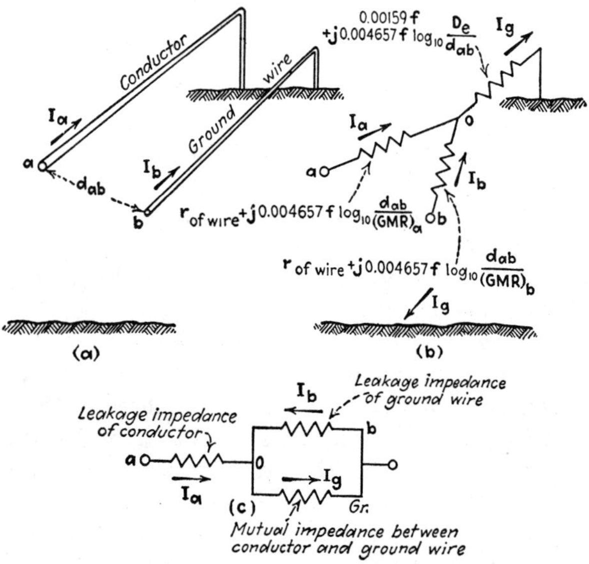

FIG. 87.—Equivalent network for two inductively coupled aerial circuits with ground return.

(*a*) Physical arrangement; (*b*) equivalent circuit of two aerial conductors with ground return; (*c*) equivalent circuit when one conductor is grounded.

D_e = equivalent depth of ground return

G.M.R. = geometric mean radius

For 60 cycles 0.004657f = 0.2794

For zero-sequence constants, multiply by 3.

are given by equations (171) and (172) and Fig. 82(*a*), (*b*), and (*c*) for 25, 50, and 60 cycles, respectively.

The disposition of two such conductors is shown in Fig. 87(*a*). Conductor *a* is grounded at the far end to show the manner of obtaining the simplified network. This would actually be the case only when the circuit is grounded for test through a grounding device having zero impedance to zero-sequence current. The resultant impedance, converted to a zero-sequence base, is the impedance which is to be connected in series relation between the two points in the zero-sequence network of the system under consideration.

If current I_a flows into conductor a and returns through the ground, and I_b flows into conductor b and returns through the ground, then the voltages of a and b, above ground per mile of length may be written:

$$E_a = \left[r_a + 0.00159f + j0.004657f \log_{10} \frac{D_e}{(G.M.R.)_a}\right]I_a + \left[0.00159f + j0.004657f \log_{10} \frac{D_e}{d_{ab}}\right]I_b \quad (178)$$

$$E_b = \left[r_b + 0.00159f + j0.004657f \log_{10} \frac{D_e}{(G.M.R.)_b}\right]I_b + \left[0.00159f + j0.004657f \log_{10} \frac{D_e}{d_{ab}}\right]I_a \quad (179)$$

The **equivalent circuit** for such a combination is shown in Fig. 87(b). It can readily be seen that the branch between o and ground is the mutual impedance, and that the remaining impedances are associated with the circuits a and b and are the differences between the self and mutual impedances (a sort of leakage impedance, such as between the windings of a transformer).

This equivalent circuit is of course independent of the voltages impressed across the terminals. Applied to the ground wire case, one of the conductors, say the b conductor, is at ground potential, which requires that b be connected to ground in the equivalent network shown in Fig. 87(c). The total impedance of the circuit between a and ground is found by solving for the total impedance, composed of ao in series with the parallel circuit composed of the mutual impedance and the leakage impedance of the ground wire.

The current distribution between the ground wire and ground is obtained by solving for the current distribution between the two parallel branches.

Application of the curves of Fig. 82(a), (b); and (c) will reduce the labor involved in the determination of the reactances in the three branches in Fig. 87. The procedure is indicated in Fig. 88, which is largely self-explanatory. The curve gives the zero-sequence reactance per phase as a function of the $G.M.R.$, for a particular value of ρ. The sum of the reactance in the a branch and of the mutual reactance is equal to the reactance read from the curve for an abscissa equal to the $G.M.R.$ of the *a* conductor. The mutual reactance is the value read from the curve for an abscissa equal to the geometric mean separation between the two sets of conductors. It follows therefore that the reactance in the *a* branch is the difference between these two quantities, read from the curve. Similar relations apply to the *b* branch. Since these values have already been reduced to per phase quantities, it is unnecessary to multiply them by 3.

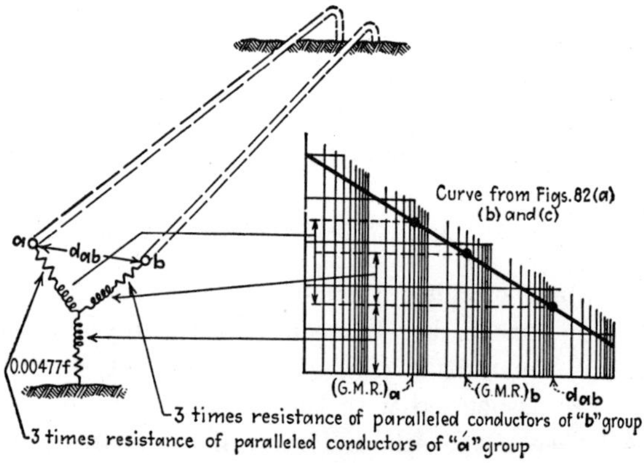

FIG. 88.—Calculation of the equivalent zero-sequence network of two inductively coupled aerial circuits with earth return.

The resistance component for the *a* and *b* branches are, respectively, equal to three times the resistance of each set considered as a group.

73. Practical Calculation of Zero-sequence Impedance.

The foregoing considered only the impedance between one conductor and the combined ground wire and ground. Practically, the zero-sequence circuit necessarily involves three or more conductors. These cases may be solved by resorting again to the idea of the *G.M.D.* of combined conductors. Several different cases will be considered.

Single-circuit and One Ground Wire. The disposition of the conductors for a typical case is shown in Fig. 89 together with the detailed calculations. The *G.M.R.* of the conductors is found by taking the ninth root of the product of the nine possible distances between the three conductors comprising the polyphase circuit, including the *G.M.R.* of each of the three conductors, which may be considered as the *G.M.D.* from the conductor to itself. As stated in the preceding chapter, this is also equal to the cube root of the *G.M.R.* of a conductor times the square of the geometric mean separation of the three conductors. The geometric mean separation of the conductors

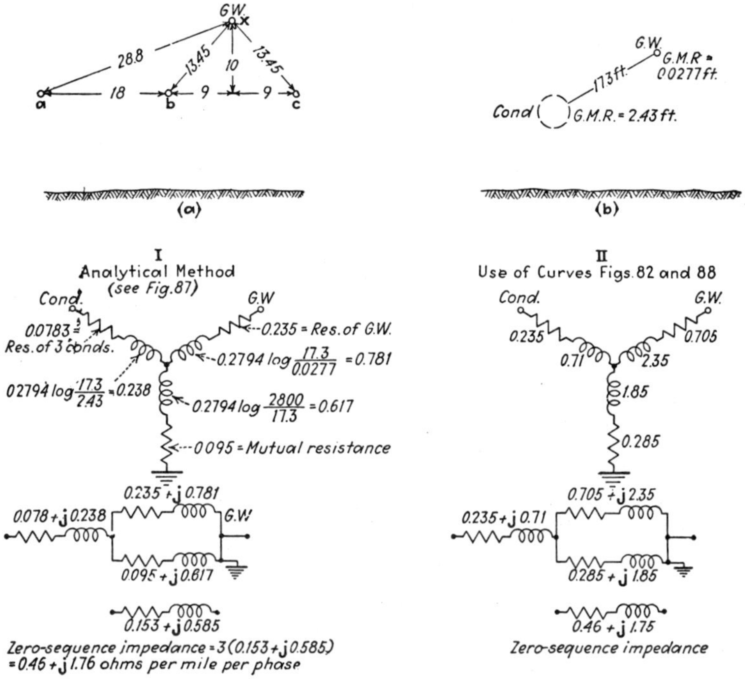

Conductors and ground wire = 397,500 cir. mils. A.C.S.R. (30 × 7)
G.M.R. of conductor and ground wire = 0.0277 ft.
Resistance per mile of conductor = 0.235 ohm
Frequency = 60 cycles
Damp earth—ρ = 100 meter-ohms. D_e = 2,800 ft.
Mutual resistance = $0.0159f$ = 0.095 ohm per mile

$$G.M.R. = \sqrt[9]{(d_{ab})^2(d_{bc})^2(d_{ca})^2(G.M.R.)^3}$$
$$= \sqrt[9]{(18)^2(18)^2(36)^2(0.0277)^3} = 2.43 \text{ ft.}$$
$$G.M.D._{\text{separation}} = \sqrt[3]{(d_{xa})(d_{xb})(d_{xc})}$$
$$= \sqrt{(28.8)(13.45)^2} = 17.3 \text{ ft.}$$

FIG. 89.—Calculation of zero-sequence impedance for a single circuit with a single ground wire. Forty-three per cent of the current returns in the ground wire.

and ground wire is merely the cube root of the three separations. With these quantities obtained, two procedures are available. The left-hand column illustrates the analytical method, which follows the outline indicated in Fig. 87. The treatment is carried through in terms of actual impedance, and to convert to zero-sequence impedance per phase it is necessary to multiply by 3 at the end. The procedure outlined in the right-hand column of Fig. 89 is carried through on the basis of zero-sequence impedance per phase, as in Fig. 88, and the reactances are obtained from the curves of Fig. 82(c).

To give an idea of the error involved for the particular configuration shown, the current was calculated by the above method for a line-to-ground fault at the open-circuited receiving end of the transmission line, assuming the voltage maintained

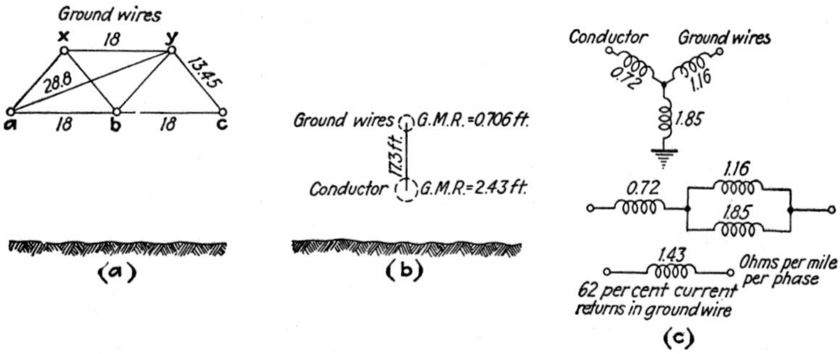

Conductors and ground wire = 397,500 cir. mils. A.C.S.R.
$G.M.R.$ of each conductor and ground wire = 0.0277 ft.
Neglect resistance of wires.
Damp earth, ρ = 100 meter-ohms. D_e = 2800 ft.

$$G.M.R._{\text{ground wire}} = \sqrt{(d_{xy})(G.M.R.)_{\text{ground wire}}} = \sqrt{(18)(0.0277 \text{ ft.})} = 0.706 \text{ ft.}$$

$$G.M.R._{\text{conductors}} = \sqrt[9]{(d_{ab})^2(d_{bc})^2(d_{ca})^2(G.M.R.)^3} = \sqrt[9]{4(18)^6(0.0277 \text{ ft.})^3} = 2.43 \text{ ft.}$$

$$G.M.D._{\text{separation}} = \sqrt[6]{(d_{ax})(b_{bx})(d_{cx})(d_{ay})(d_{by})(d_{cy})} = \sqrt[6]{(13.45)^4(28.80)^2} = 17.3 \text{ ft.}$$

FIG. 90.—Calculation of zero-sequence reactance for a single circuit with two ground wires. (a) Physical configuration of conductors; (b) equivalent conductors; (c) reduction of equivalent network.

at the sending end and the line transposed. These results gave, at 60 cycles, a current which was 0.2 per cent too high.

In general, the effect upon the reactive component of neglecting the resistances of the line conductors, ground wires, and mutual branch is very small. In the particular case just calculated the total reactance, neglecting the resistances in the branches, is $j0.71 + \dfrac{(j2.35)(j1.85)}{j(2.35 + 1.85)} = j1.745$. This value compared with $j1.75$ constitutes a substantial check. For more ground wires, the check should be closer. An important exception to this case is the use of steel ground wires.

In many cases short-circuit currents are calculated by using the reactance alone, and neglecting the resistance. In the example just discussed the magnitude of the zero-sequence impedance is 1.81 ohms as compared with the reactance of 1.75 ohms, giving a difference of 3.5 per cent. Hence the error in current magnitude introduced by neglecting the resistance cannot exceed this value.

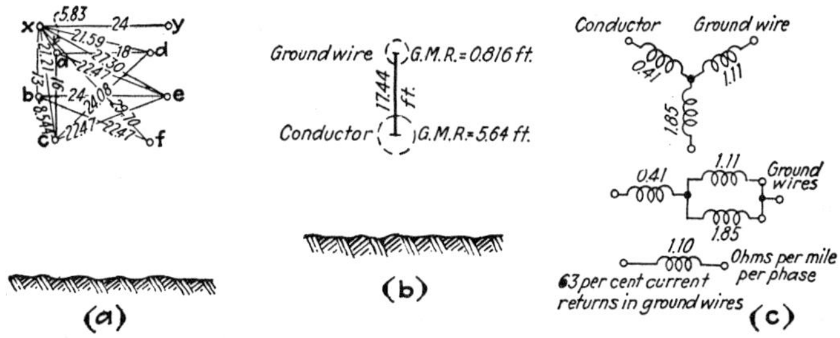

Conductors and ground wires = 397,500 cir. mils. A.C.S.R.
G.M.R. of each conductor and ground wire = 0.0277 ft.
Neglect resistance.
ρ = 100 meter-ohms.

$$G.M.R._{conductors} = \sqrt[36]{[(G.M.R.)_{conductors}(d_{ab})(d_{ac})(d_{ad})(d_{ae})(d_{af})]^4 \, [(G.M.R.)_{conductors}(d_{ba})(d_{bc})(d_{bd})(d_{be})(d_{bf})]^2}$$
$$= \sqrt[18]{(0.0277)^3(8.544)^4(16)^2(24.08)^2(18)^2(22.47)^4(24)} = 5.64 \text{ ft.}$$

$$G.M.R._{ground\ wires} = \sqrt[2]{(G.M.R.)_{ground\ wire}(d_{xy})} = \sqrt[2]{(0.0277)(24)} = 0.816 \text{ ft.}$$

$$G.M.D._{separation} = \sqrt[12]{(d_{xa})(d_{xb})(d_{xc})(d_{xd})(d_{xe})(d_{xf})(d_{ya})(d_{yb})(d_{yc})(d_{yd})(d_{ye})(d_{yf})}$$
$$= \sqrt[6]{(5.83)(13)(21.21)(29.70)(27.30)(21.59)} = 17.44 \text{ ft.}$$

FIG. 91.—Calculation of zero-sequence reactance for a twin circuit with two ground wires. (*a*) Physical configuration of conductors; (*b*) equivalent conductors; (*c*) reduction of equivalent network.

Single-circuit and Two Ground Wires (Resistance Neglected). The configuration of a typical single-circuit line equipped with two ground wires is shown in Fig. 90. Calculating the equivalent radii as indicated gives an equivalent system in which the conductor group has a radius of 2.43 ft. and the ground-wire group 0.706 ft. The equivalent separation of the two groups is 17.3 ft. The remaining steps in the calculation of the zero-sequence reactance are shown in the figure, the resistance being neglected in this case.

To test the order of the error in using the simplified method, the line alone was considered, with the receiving end open-circuited and constant balanced voltages maintained at the sending end. Assuming a transposed line, a comparison between a rigorous solution and the result obtained by the simplified method of symmetrical components showed that the error in the current was but 0.18 per cent for a single line-to-ground fault at the receiving end. For a double line-to-ground fault on phases b and c at the receiving end, the rigorous calculation gave

$$I_b = -0.4194 - j1.0923$$

and

$$I_c = -0.4194 + j1.0923$$

Corresponding quantities obtained by the simplified method are

$$I_b = -0.4155 - j1.0856$$
$$I_c = -0.4155 + j1.0856$$

indicating an error of about 0.6 per cent.

Twin-circuit Line and Two Ground Wires (Resistance Neglected). The configuration for this case is shown in Fig. 91, together with the details of calculation. The general method is essentially the same as for the previous cases. The fault current was calculated for a line-to-ground fault at the receiving end of an open-circuited transposed line, assuming the voltage maintained at the sending end. A comparison of the results obtained by the rigorous method with that obtained by the simplified method of symmetrical components indicates an error of 0.25 per cent.

In the determination of the errors involved for the three cases just cited no cognizance was taken of the symmetrical series impedances introduced by the transformers and machines, which are usually present in practice and whose effect would still further reduce the errors.

74. General Method for Zero-sequence Calculations.

In the foregoing every effort has been made to represent the circuits under consideration by means of equivalent circuits. This has the advantage that it tends toward the elimination of errors by the progressive inspectional checks of the computations in the successive simplification of the network. More complicated problems than those heretofore treated become quite involved when this is attempted. The following method is always applicable, regardless of the complexity of the problem.

Given any number of physically parallel conductors, a, b, c, . . . n, and designating the currents and voltages to ground. I_a, I_b, I_n, and E_a, E_b, . . . E_n, respectively,

$$\left.\begin{aligned} E_a &= Z_{aa}I_a + Z_{ab}I_b + \cdots Z_{an}I_n \\ E_b &= Z_{ab}I_a + Z_{bb}I_b + \cdots Z_{bn}I_n \\ &\ldots\ldots\ldots\ldots\ldots \\ E_n &= Z_{an}I_a + Z_{bn}I_b + \cdots Z_{nn}I_n \end{aligned}\right\} \tag{180}$$

These equations should preferably be set up on a zero-sequence basis, in which the self impedances Z_{aa}, Z_{bb}, . . . Z_{nn}, and the

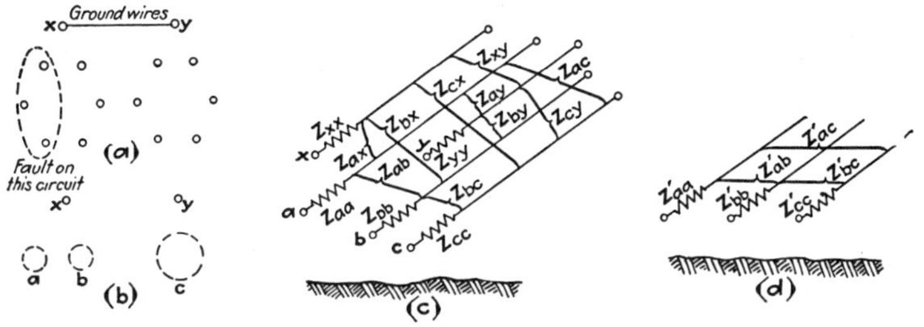

FIG. 92.—Equivalent zero-sequence network for four inductively coupled earth return circuits. (*a*) Physical configuration of actual conductors; (*b*) equivalent conductors; (*c*) equivalent network; (*d*) equivalent network when x and y are regarded as grounded.

mutual impedances Z_{ab}, . . . $Z_{(n-1)n}$ are three times the values determined by equations (169) and (170), or three times the values obtained from Fig. 75.

As a particular example consider the two twin-circuit lines of Fig. 92(*a*) which are equipped with two ground wires. To determine the zero-sequence network for a fault on the left-hand circuit of the left-hand tower, the conductor configuration may be replaced by the equivalent configuration shown in (*b*). It is well to retain the identity of the two ground wires because the great difference in their distances from the faulted circuit probably results in considerable difference in current through them. The twin circuit on the right-hand tower, on the other hand, can reasonably be replaced by a single equivalent circuit. A set of five simultaneous equations of the form shown by equation (180) may now be set up. These equations will involve the variables I_a, I_b, I_c, representing currents in the conductor groups; and I_x and I_y, representing currents in the ground wires

The physical conception of these constants is shown in Fig. 92(c) in which the self impedances are represented by series impedances in the conductors and the mutual impedances are indicated by the brackets connecting the respective conductors. For this particular case the potential of the ground wires is zero, so that the voltage equations for these two conductors are

$$\left.\begin{aligned} 0 &= Z_{ax}I_a + Z_{bx}I_b + Z_{cx}I_c + Z_{xx}I_x + Z_{xy}I_y \\ 0 &= Z_{ay}I_a + Z_{by}I_b + Z_{cy}I_c + Z_{xy}I_y + Z_{yy}I_y \end{aligned}\right\} \quad (181)$$

By solving these two simultaneous equations for I_x and I_y in terms of I_a, I_b, and I_c, and substituting in the remaining equations, three new voltage equations of the form of equations (180) will be obtained. These new equations, however, will have but three terms in each. The resultant network reduces to that shown in Fig. 92(d) in which the primed constants represent the new values. By this process the network has been replaced by an equivalent network without ground wires.

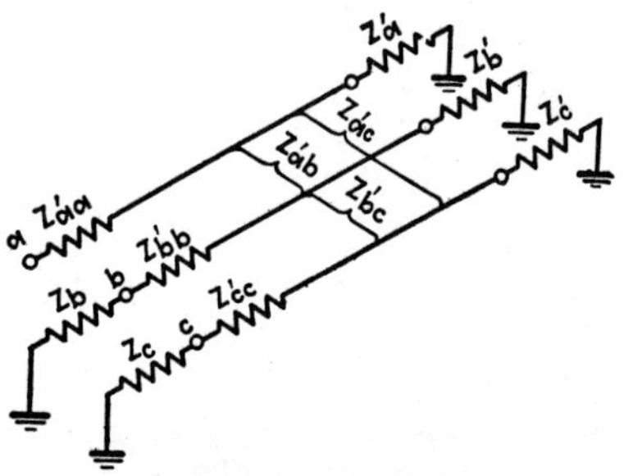

FIG. 93.—Inclusion of the equivalent circuit of a transmission line into the zero-sequence network.

To illustrate further how this circuit fits into the zero-sequence network, assume that circuit c is grounded through zero-sequence impedances Z_c and Z_c' at the near and far ends, respectively; that circuit b is grounded through the zero-sequence impedances Z_b and Z_b' at the near and far ends, respectively; that circuit a is grounded only at the far end through the zero-sequence impedance Z_a', and that the zero-sequence impedance of the network for a fault at the near end of circuit a is desired. The above impedances are inserted in the network as shown in Fig. 93. The resultant equations take the form

$$\left.\begin{aligned} E_a &= Z'_{aa}I_a + Z'_{ab}I_b + Z'_{ac}I_c + Z'_aI_a \\ 0 &= Z'_{ab}I_a + Z'_{bb}I_b + Z'_{bc}I_c + (Z_b + Z'_b)I_b \\ 0 &= Z'_{ac}I_a + Z'_{bc}I_b + Z'_{cc}I_c + (Z_c + Z'_c)I_c \end{aligned}\right\} \quad (182)$$

The answer sought is, of course, $\dfrac{E_a}{I_a}$ or the value of E_a for $I_a = 1$, which can be obtained most conveniently by solving the last two equations for I_b and I_c and inserting in the first.

Example. To illustrate the use of this method, let it be desired to determine the zero-sequence impedance of the 50-mile 66-kv. transmission system of Fig. 94(*a*), including two twin

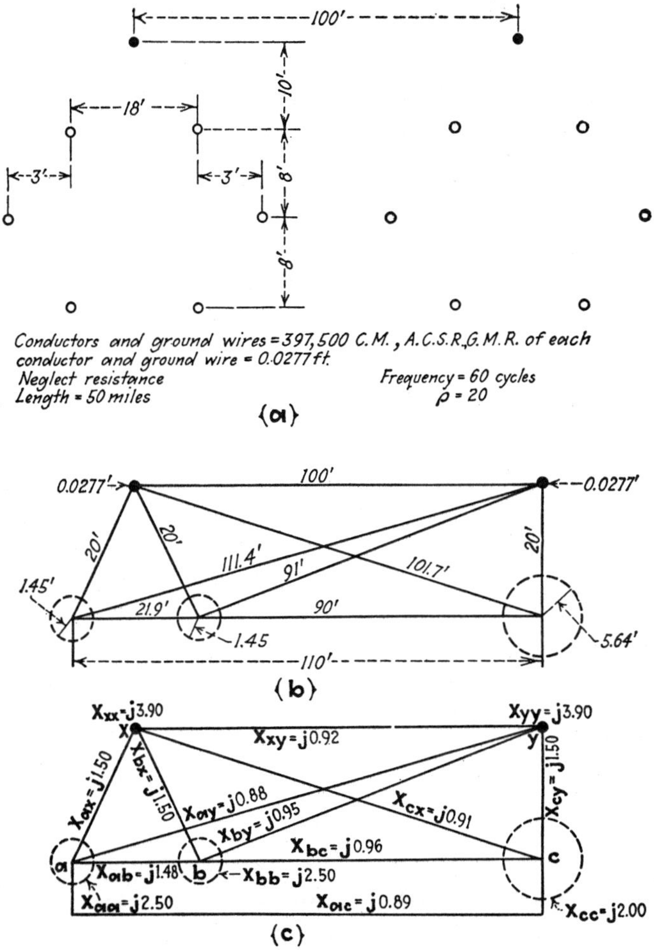

(*a*) *Circuit configuration and conductor data*
(*b*) *G.M.D. and G.M.R. data in feet*
(*c*) *Self and mutual reactances per phase per mile*

FIG. 94.—Transmission line circuit with ground wires used to illustrate zero-sequence impedance calculations.

circuits with solidly grounded transformers at each end. The fault will be assumed to occur on the extreme left-hand set of conductors. For this case, the zero-sequence current distribution between the three conductors of this circuit should not depart far from uniformity, so that the three conductors can be represented by a single equivalent conductor. Similarly, the second circuit from the left can be represented by an equivalent conductor. Because of the wide separation between the two towers supporting the twin circuit and assuming that the two circuits are grounded in a similar manner, the current distribution between the six conductors of the right-hand twin circuit should also be substantially uniform, so that the six conductors may be represented by a single equivalent conductor. The two ground wires cannot be combined, except as an approximation, into a single equivalent conductor because of the difference in current in the two conductors. The identity of the two ground wires will be retained to further illustrate the application of the method. With these simplifications the system may be represented by its equivalent conductors shown in Fig. 94(*b*). The *G.M.R.* and *G.M.D.* are obtained in the manner explained previously. For the wide separations it is sufficiently accurate to take as the mean separation the distance between the centers of gravity of the two systems under consideration. Probably the easiest way to determine the center of gravity is by the intersection of the medians of the triangle formed by the three conductors as apexes.

With the *G.M.D.* available, the self and mutual impedances to zero-sequence can be obtained from Fig. 82. These are shown in Fig. 94(*c*).

Since the ground wires are assumed to be grounded at each end through zero resistance their potential is zero and the following equations can be written by using the self and mutual impedances per mile from Fig. 94(*c*).

$$E_x = j[1.50I_a + 1.50I_b + 0.91I_c + 3.90I_x + 0.92I_y] = 0$$
$$E_y = j[0.88I_a + 0.95I_b + 1.50I_c + 0.92I_x + 3.90I_y] = 0$$

Solving for I_x and I_y

$$I_x = -[0.352I_a + 0.347I_b + 0.151I_c]$$
$$I_y = -[0.143I_a + 0.162I_b + 0.349I_c]$$

For circuits a, b, and c, the voltage equations are

$$E_a = j2.50I_a + j1.48I_b + j0.89I_c + j1.50I_x + j0.88I_y$$
$$E_b = j1.48I_a + j2.50I_b + j0.96I_c + j1.50I_x + j0.95I_y$$
$$E_c = j0.89I_a + j0.96I_b + j2.00I_c + j0.91I_x + j1.50I_y$$

Substituting the values of I_x and I_y in these equations, the results per mile are

$$E_a = j1.83I_a + j0.82I_b + j0.35I_c$$
$$E_b = j0.82I_a + j1.83I_b + j0.40I_c$$
$$E_c = j0.35I_a + j0.40I_b + j1.34I_c$$

The above impedances are for unit length of line. By multiplying by 50 they may be converted to those of the 50-mile line represented by the equivalent circuit shown between the dotted lines in Fig. 95(*b*) from which the ground-wire circuits have been eliminated.

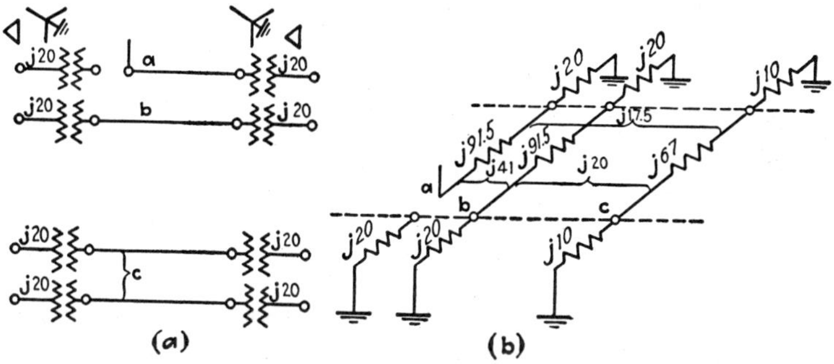

FIG. 95.—Zero-sequence equivalent network of circuit whose physical configuration is given in Fig. 94, without bussing on high-voltage sides of transformers.

Beyond this point the solution is determined by the system bussing arrangements and whether certain circuit-breakers are open or closed. For the system layout shown in Fig. 95(*a*), in which the high-voltage circuits are not bussed and the breaker in the *a* circuit is open, the transformer impedances may be included as shown in Fig. 95(*b*). The voltage equation for the *b* and *c* circuits may then be written

$$E_b = j41.0I_a + j131.5I_b + j20I_c = 0$$
$$E_c = j17.5I_a + j20I_b + j87I_c = 0$$

and solving for I_b and I_c

$$I_b = -0.292I_a$$
$$I_c = -0.134I_a$$

For the *a* conductor

$$E_a = j111.5I_a + j41.0I_b + j17.5I_c$$

and substituting the values of I_b and I_c

$$\frac{E_a}{I_a} = Z_0 = j97.2 \text{ ohms}$$

If the high-voltage lines had been bussed at both ends, as shown in Fig. 96(*a*), the equivalent circuit for solution would be changed to that shown in (*b*).

If the breaker in the *a* circuit had been closed, the equivalent circuit could have been obtained by merely closing the corresponding gap in the equivalent circuits. Of course, if this is the only condition to be analyzed, *i.e.*, with this breaker closed,

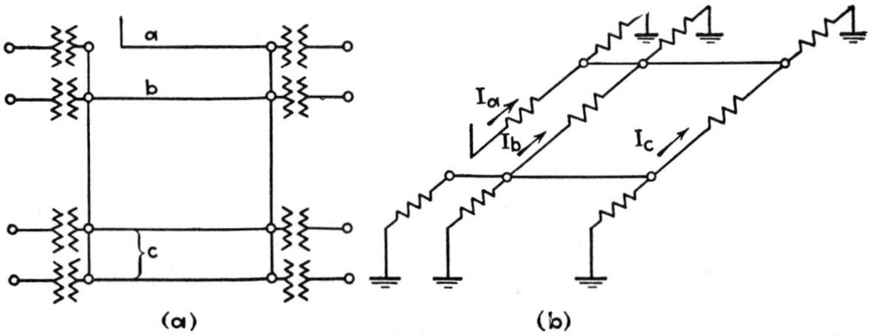

FIG. 96.—Zero-sequence equivalent network of Figs. 94 and 95 when the lines are bussed at both ends.

the equivalent circuit can be obtained easier by considering the *a* and *b* circuits as a unit, as was done for *c*, assuming equal current distribution between the six conductors.

75. Steel Ground Wires.

The zero-sequence impedance of circuits with steel ground wires is obtained in a manner similar to that indicated for circuits equipped with copper and aluminum ground wires, except that, whereas the resistance for the latter may usually be neglected, the resistance must be considered when dealing with steel ground wires. Data on the characteristics of steel conductors are given by Prof. H. B. Dwight.* The curves in Fig. 97 were calculated from these data and give the resistance and the *G.M.R.* of steel conductors of different grades as affected by current, for frequencies of 25 and 60 cycles.

Prof. Dwight's data are presented in the form of resistance and reactance (due to internal flux only) per mile. However, a knowledge of the equivalent *G.M.R.* fits into the present calculation more readily. These values can be obtained in the

* Resistance and Reactance of Commercial Steel Conductors, *Elec. Jour.*, p. 25, January, 1919.

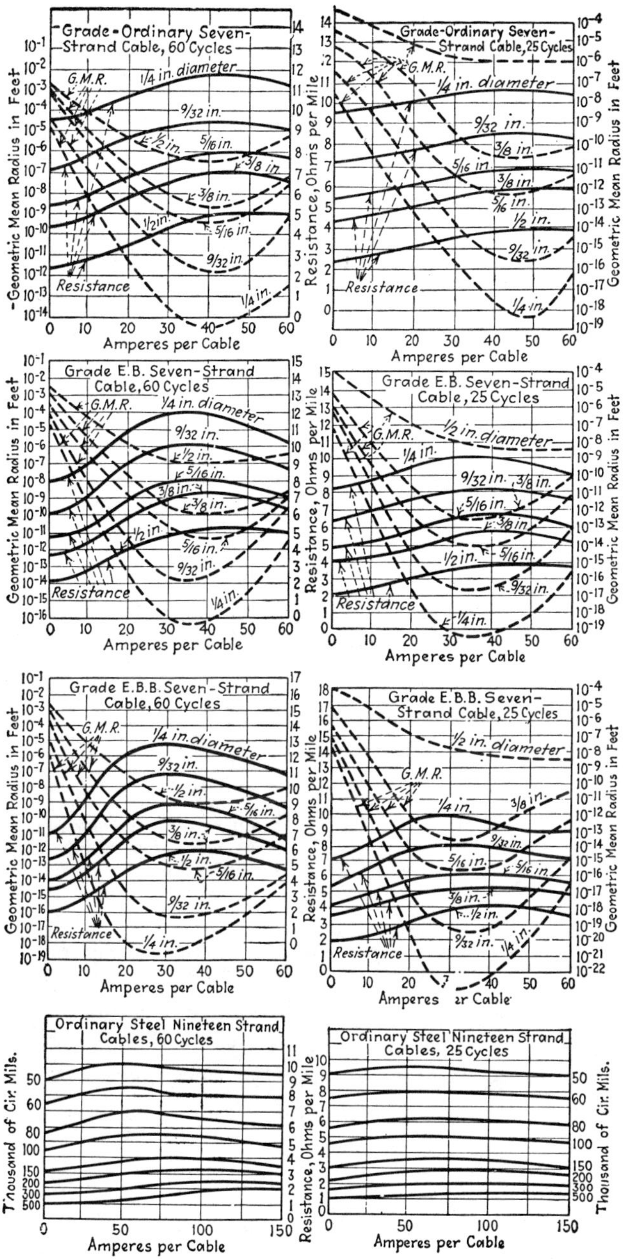

Fig. 97.—Electrical characteristics of steel ground wires.

following simple manner. Let x be the reactance in ohms per mile at 60 cycles due to the internal flux only, then

$$x = 0.2794 \log_{10} \frac{\text{radius}}{(G.M.R.)}$$

$$(G.M.R.) = \frac{\text{radius}}{\text{antilog}_{10} \dfrac{x}{0.2794}} \tag{183}$$

For 25 cycles 0.2794 should be changed to 0.1164.

As an example of the importance of the resistance of ground wires when steel is used, the same arrangement of conductors as given in Fig. 90 was chosen, except that the ground conductors

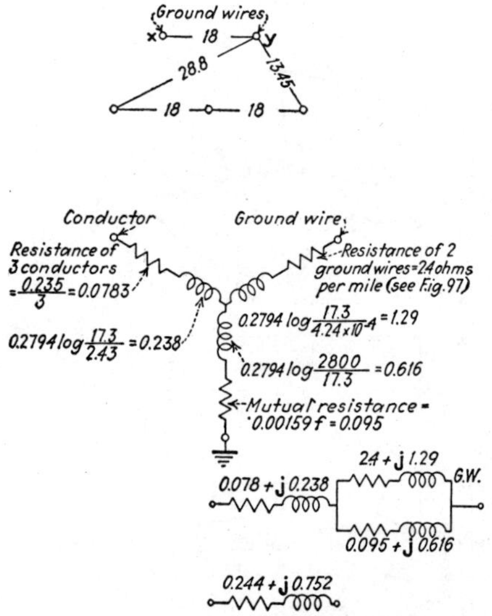

From Fig. 97 assuming ½ in. diameter Grade E.B.B. seven-strand cable and 60 amp. per wire at 60 cycles

$G.M.R. = 10^{-8}$ ft. $\rho = 100$

$G.M.R_{\text{ground wires}} = \sqrt{(d_{xy})(G.M.R.)_{\text{ground wires}}} = \sqrt{(18)(10^{-8})} = 4.24 \times 10^{-4}$ ft.

$G.M.D.$ of separation from Fig. 90 = 17.3 ft.

$G.M.R.$ of conductors from Fig. 90 = 2.43 ft.

Zero-sequence impedance = 3 (0.244 + 0.752) = 0.73 + 2.27 ohms per mile per phase. 14 per cent of total current flows through ground wire

FIG. 98.—Calculation of zero-sequence impedance of a single-circuit transmission line with two steel ground wires.

were assumed to consist of two ½-in. steel conductors. The details of calculation for the case are shown in Fig. 98. Neglecting the resistance components in this case gives a value of reactance equal to

$$j0.238 + \frac{(j1.29)(j0.616)}{j(1.29 + 0.616)} = j0.655$$

which is considerably smaller than the value ($j0.752$) given in Fig. 98. The use of steel conductors also increases the zero-sequence reactance over that obtained with aluminum or copper conductors as evidenced by the comparison with a similar layout using A.C.S.R. conductors (see Fig. 90) which gave a reactance of $j1.41$ ohms per mile per phase.

76. Copper-clad Ground Wires.

Copper-clad steel cables are sometimes used for ground wires. The increase in resistance* at 25 and 60 cycles may be taken as from 5 to 10 per cent over the direct-current resistance value for currents up to 200 amp.

The *G.M.R.* of round iron conductors of constant permeability μ assuming uniform current distribution is $a\epsilon^{-\frac{\mu}{4}}$. For $\mu = 1$, this reduces to the well-known figure 0.779. When alternating current flows through steel conductors, because of magnified skin-effect phenomena, most of the current flows near the surface. Nevertheless, it is always possible to express the result as an equivalent *G.M.R.* of this type, using an *effective* permeability. This effective permeability is not related directly to the actual permeability and is a function of current and frequency.

Some of the available test data* on copper-clad conductors express the results as a ratio of the internal inductance of the copper-clad conductor to the internal inductance of a copper conductor of the same diameter. It may be shown that this ratio is equal to the effective permeability defined above. The test results show a range for this value of from 4 to 20, but for most practical calculations a single value equal to 7 may be used. This gives a *G.M.R.* equal to $0.17a$.

77. Sequence Impedance Constants of Typical Circuits.

In order to develop a sense of proportion as to the order of magnitude of transmission-line impedances and the effect of adding ground wires, the constants of a number of typical transmission lines were calculated and tabulated in Table VIII. The two columns under the heading of positive-sequence impedance represent the ordinary impedance of a single and two paralleled circuits. The next general heading under zero-sequence impedance gives the accurate constants including resistance for the circuit conditions indicated at the head of the columns. The addition of ground wires may decrease the zero-sequence

* "Overhead System Reference Book," pp. 85 and 98; published by National Electric Light Association.

TABLE VIII.—CONSTANTS OF TYPICAL TRANSMISSION LINES

Configuration of conductors	Rated voltage in kilovolts	Number of ground wires	Positive-sequence impedance		Zero-sequence impedance			Zero-sequence reactance neglecting resistance in computations			Ratio: $\frac{X_0 \text{ of single circuit}}{X_1 \text{ of single circuit}}$	Ratio: $\frac{X_0 \text{ of double circuit}}{X_1 \text{ of double circuit}}$
			Single circuit	Double circuit	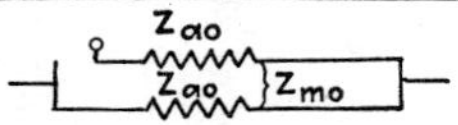 Z_{ao}	Z_{mo}	 Double circuit	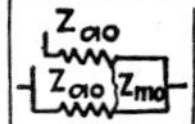 Z_{ao}	Z_{mo}	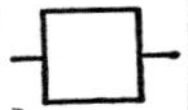 Double circuit		
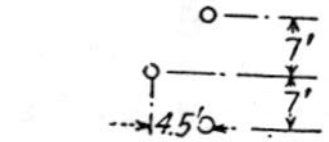 0 Copper $\rho = 20$	44	0	0.56 + j0.76	....	0.84 + j2.58	....	....	2.58	....	....	3.38	
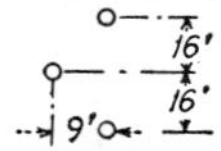 0000 Copper $\rho = 20$	110	0	0.28 + j0.81	....	0.56 + j2.3	....	....	2.34	....	....	2.89	
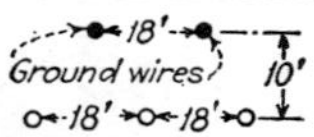 Conductors and ground wires 397,500 c.m. A.S.C.R.	110	0	0.24 + j0.81	....	0.52 + j2.85	....	....	2.85	....	....	3.52	
		1	0.24 + j0.81	....	0.46 + j1.76	....	....	1.76	....	....	2.17	
		2	0.24 + j0.81	....	0.41 + j1.43	....	....	1.43	....	....	1.77	
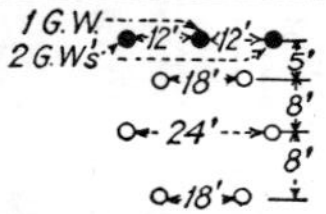 Conductors and ground wires 397,500 c.m. A.C.S.R. $\rho = 100$	66	0	0.24 + j0.72	0.12 + j0.350	0.52 + j2.75	0.29 + j1.77	0.40 + j2.26	2.75	1.77	2.26	3.82	6.45
		1	0.24 + j0.72	0.12 + j0.350	0.46 + j1.92	0.23 + j0.94	0.40 + j1.43	1.92	0.94	1.43	2.67	4.09
		2	0.24 + j0.72	0.12 + j0.350	0.42 + j1.57	0.17 + j0.62	0.29 + j1.09	1.57	0.62	1.09	2.18	3.12
Conductors 795,000 c.m. A.C.S.R.—two layers Ground wires 203,200 c.m. A.C.S.R.—one layer 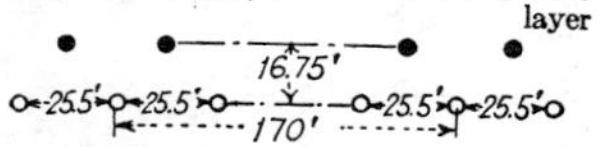	220	0	0.117 + j0.814	0.059 + j0.407	0.403 + j2.446	0.286 + j1.02	0.345 + j1.73	2.45	1.02	1.73	3.00	4.25
		4	0.117 + j0.814	0.059 + j0.407	0.28 + j1.42*	0.110 + j0.212*	0.195 + j0.81	1.42	0.233	0.826	1.74	2.00

* Four ground wires.

reactance as much as 50 per cent. As shown in the next set of columns the inclusion of the resistances of conductors has a negligible effect upon the value of zero-sequence reactance obtained. A convenient measure of zero-sequence reactance that is sometimes used is to express the value in terms of the positive-sequence reactance. These ratios are tabulated in the last two columns, from which it may be seen that 3.5 represents an approximate value to express this ratio for single circuits without ground wires, but for circuits with ground wires this value is much smaller, decreasing to 2.7 or 1.7 depending upon the effectiveness of the ground-wire system. For double circuits this ratio varies over a still greater range being 6.5 or 4.2 for lines without ground wires and 4.1, 3.1, or 2.0 for circuits with ground wires.

Problems

1. Assume the transmission system of Fig. 90 except that the ground wire is of 000 copper and that the value of ρ equals 10 meter-ohms. Compute the zero-sequence impedance of the circuit and the division of the return current between the ground wires and the earth.

2. Assume that the transmission system of Fig. 91 is equipped with grade EBB seven-strand steel cables and that the current is 50 amp. in each cable, assuming a wire of ½ inch diameter. Determine the zero-sequence impedance and division of current between ground wires and earth.

3. A power system with a transmission line as in Fig. 89 but without ground wires is paralleled at 100-ft. separation for 1 mile by a telephone circuit of two wires located 20 ft. above the ground. (*a*) Determine the voltage induced in the ground-return circuit of the telephone line due to electromagnetic induction resulting from the flow of 4,000 amp. through the parallel at a frequency of 60 cycles. (*b*) What reduction in voltage will be accomplished by the installation of a 4/0 copper ground wire on the power circuit as in Fig. 89 provided that the short-circuit current is not increased by the reduction of zero-sequence impedance? (*c*) If a 4/0 copper ground wire (shield wire) is located in a position 1 ft. above and midway between the telephone wires, which are located 10 in. apart horizontally, what per cent reduction will be accomplished if the connections to ground are of negligible impedance?

4. For the power system of Fig. 95 determine the impedance to zero-sequence as measured at the point *a* if the adjacent circuit-breaker is closed. Also determine the current division for zero-sequence.

5. For the case of Fig. 95, if the transformer terminals are bussed as in Fig. 96, determine the zero-sequence impedance for a fault at *a* with the adjacent circuit-breaker open.

CHAPTER IX

CONSTANTS OF LONG TRANSMISSION LINES

In the preceding chapters consideration was given to the characteristics of short transmission and distribution lines in which the effects of charging currents and the distributed nature of the constants could be neglected. The present chapter will discuss how these factors may be included in the characteristics of longer lines.

78. Positive- and Negative-sequence Characteristics.

For longer lines in which the capacitance is not negligible the positive-sequence characteristics may be specified by the ABC constants, which are defined by the equations

$$\begin{aligned} E_s &= AE_r + BI_r \\ I_s &= CE_r + AI_r \end{aligned} \tag{184}$$

where E_s and E_r are the voltages to neutral at the sending and receiving ends, respectively, and I_s and I_r are the corresponding line currents. These coefficients are the result of the hyperbolic solution of transmission lines taking into consideration the distributed character of the constants.*

$$\left.\begin{aligned} A &= \cosh \sqrt{ZY} = \left[1 + \frac{YZ}{2} + \frac{Y^2Z^2}{24} + \frac{Y^3Z^3}{720} + \cdots\right] \\ B &= \sqrt{\frac{Z}{Y}} \sinh \sqrt{ZY} = Z\left[1 + \frac{YZ}{6} + \frac{Y^2Z^2}{120} + \frac{Y^3Z^3}{5040} + \cdots\right] \\ C &= \sqrt{\frac{Y}{Z}} \sinh \sqrt{ZY} = Y\left[1 + \frac{YZ}{6} + \frac{Y^2Z^2}{120} + \frac{Y^3Z^3}{5040} + \cdots\right] \end{aligned}\right\} \tag{185}$$

* Various methods for calculating these constants are given by William Nesbit, on p. 57, "Electrical Characteristics of Transmission Circuits," 3d ed., Westinghouse Technical Night School Press, East Pittsburgh, Pa. These constants have been calculated and tabulated for different conductors, spacings, and lengths of line in the same book.

in which

$Z = R + jX$ = total line impedance = $(r + jx)l$.

$Y = G + jB$ = total line susceptance = $(g + jb)l$.

g = conductance to neutral in mhos per unit length of line (usually neglected).

b = capacitive susceptance to neutral in mhos per unit length of line.

The value of b may be obtained by taking the reciprocal of the capacitive reactances given in tables in the Appendix.

In the absence of tables, b may be obtained by means of the following formula:

$$b = \frac{14.64 \times 10^{-6}}{\log_{10} \frac{D}{a}} \text{ mhos per mile per phase at 60 cycles} \qquad (186)$$

in which D is the equivalent separation or $G.M.D.$ of the conductors* and a is the radius of the conductors.

Equivalent π. For many purposes, such as setting up the equivalent of the transmission line on calculating boards, it is desirable to know the equivalent π of the transmission line. This is a circuit of the form indicated in Fig. 99 having a series impedance and a shunt susceptance at each end. It can be easily shown that

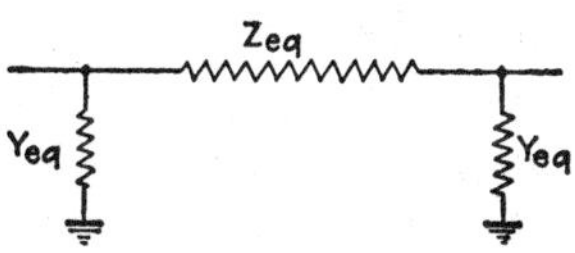

FIG. 99.—Equivalent circuit of a transmission line.

$$\left.\begin{aligned} Z_{equiv.} &= B = Z\left[1 + \frac{YZ}{6} + \frac{Y^2Z^2}{120} + \frac{Y^3Z^3}{5040} + \cdots\right] \\ Y_{equiv.} &= \frac{A - 1}{B} = Y\left[\frac{1}{2} - \frac{YZ}{24} + \frac{Y^2Z^2}{240} - \cdots\right] \end{aligned}\right\} \qquad (187)$$

For short lines $Y_{equiv.}$ may be neglected entirely; for longer lines the first term, namely $\frac{Y}{2}$ should be used; and only for the very long lines is it necessary to use more terms of the series.

Example. An idea of the relative importance of the terms may be obtained by calculating $Z_{equiv.}$ and $Y_{equiv.}$ for a 300-mile, 795,000 A.C.S.R. conductor line with 25-ft. spacing, 60 cycles.

* D is numerically equal to the cube root of the product of the three distances between conductors. It should not be confused with D used with the ABC constants.

$$
\begin{aligned}
Z_{equiv.} &= Z[\ 1.0 \\
&\qquad -0.0631 + j0.0093 \\
&\qquad +0.0012 - j0.0004 \\
&\qquad +0.0000 + j0.0000 + \cdots\ ] \\
&= Z[\ 0.9381 + j0.0089] \\
Y_{equiv.} &= Y[\ 0.5 \\
&\qquad +0.0158 - j0.0024 \\
&\qquad +0.0006 - j0.0002 + \cdots\ ] \\
&= Y[\ 0.5164 - j0.0026]
\end{aligned}
$$

It will be observed that for ordinary accuracy in short-circuit studies it is necessary, in calculating $Z_{equiv.}$ and $Y_{equiv.}$ for a 300-mile line, to consider only two terms and since Z and Y both vary linearly with the length of line it is necessary to consider only the first term for lines below 100 miles. In economic studies involving the determination of losses it may be necessary to carry the computation one term further.

79. Zero-sequence Susceptance.

In the calculation of the capacitive susceptance the surface of the earth may be considered an equipotential plane. The charge distribution on the surface will be such as to produce a field in the air equal to that produced by the charge on the conductor and an opposite charge on its image, at an equal distance below the earth's surface. Under these conditions the susceptance is

$$
b = \frac{0.2440 f 10^{-6}}{\log_{10} \dfrac{2h}{a}} \text{ mhos per mile} \tag{188}
$$

When the susceptance of three parallel conductors is desired, the $G.M.D.$ may be determined and inserted in formula (188). The derivation of these $G.M.D.$ for the ground return case follows:

Referring to Fig. 100(a), for unit positive charge (in absolute units) on the cylindrical conductor a and a unit negative charge upon its image, the potential of a above ground is

$$
V_a = 2 \log_\epsilon \frac{D_{aa}}{d_{aa}}
$$

and the potential of the point b is

$$
V_b = 2 \log_\epsilon \frac{D_{ab}}{d_{ab}}
$$

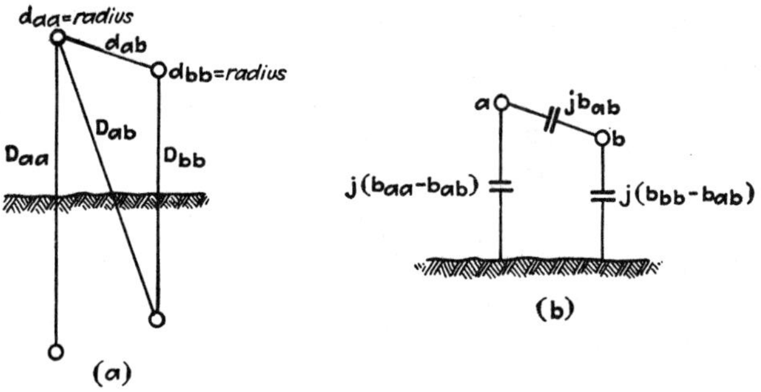

$$b_{aa} = \frac{0.244f10^{-6}\log_{10}\frac{D_{bb}}{d_{bb}}}{\Delta} \qquad b_{ab} = \frac{0.244f10^{-6}\log_{10}\frac{D_{ab}}{d_{ab}}}{\Delta}$$

$$b_{bb} = \frac{0.244f10^{-6}\log_{10}\frac{D_{aa}}{d_{aa}}}{\Delta} \qquad \Delta = \log_{10}\frac{D_{aa}}{d_{aa}}\log_{10}\frac{D_{bb}}{d_{bb}} - \left(\log_{10}\frac{D_{ab}}{d_{ab}}\right)^2$$

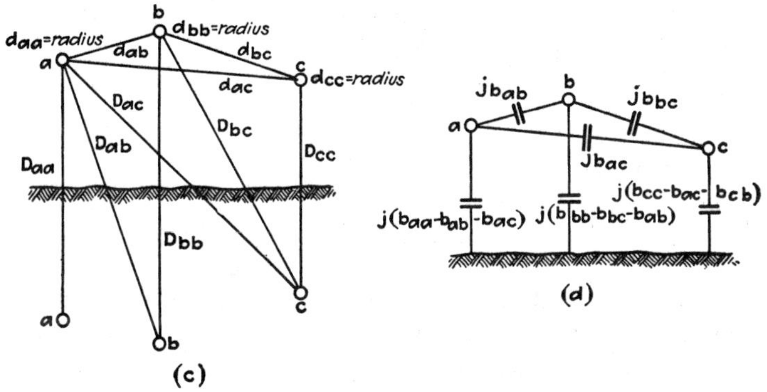

$$b_{aa} = \frac{0.244f10^{-6}}{\Delta}\left[\left(\log_{10}\frac{D_{bb}}{d_{bb}}\right)\left(\log_{10}\frac{D_{cc}}{d_{cc}}\right) - \left(\log_{10}\frac{D_{bc}}{d_{bc}}\right)^2\right]$$

$$b_{bb} = \frac{0.244f10^{-6}}{\Delta}\left[\left(\log_{10}\frac{D_{cc}}{d_{cc}}\right)\left(\log_{10}\frac{D_{aa}}{d_{aa}}\right) - \left(\log_{10}\frac{D_{ac}}{d_{ac}}\right)^2\right]$$

$$b_{cc} = \frac{0.244f10^{-6}}{\Delta}\left[\left(\log_{10}\frac{D_{aa}}{d_{aa}}\right)\left(\log_{10}\frac{D_{bb}}{d_{bb}}\right) - \left(\log_{10}\frac{D_{ab}}{d_{ab}}\right)^2\right]$$

$$b_{ab} = \frac{0.244f10^{-6}}{\Delta}\left[\left(\log_{10}\frac{D_{bc}}{d_{bc}}\right)\left(\log_{10}\frac{D_{ac}}{d_{ac}}\right) - \left(\log_{10}\frac{D_{ab}}{d_{ab}}\right)\left(\log_{10}\frac{D_{cc}}{d_{cc}}\right)\right]$$

$$b_{bc} = \frac{0.244f10^{-6}}{\Delta}\left[\left(\log_{10}\frac{D_{ab}}{d_{ab}}\right)\left(\log_{10}\frac{D_{ac}}{d_{ac}}\right) - \left(\log_{10}\frac{D_{bc}}{d_{bc}}\right)\left(\log_{10}\frac{D_{aa}}{d_{aa}}\right)\right]$$

$$b_{ac} = \frac{0.244f10^{-6}}{\Delta}\left[\left(\log_{10}\frac{D_{ab}}{d_{ab}}\right)\left(\log_{10}\frac{D_{bc}}{d_{bc}}\right) - \left(\log_{10}\frac{D_{ac}}{d_{ac}}\right)\left(\log_{10}\frac{D_{bb}}{d_{bb}}\right)\right]$$

$$\Delta = \left(\log_{10}\frac{D_{aa}}{d_{aa}}\log_{10}\frac{D_{bb}}{d_{bb}}\log_{10}\frac{D_{cc}}{d_{cc}}\right) + 2\left(\log_{10}\frac{D_{ab}}{d_{ab}}\log_{10}\frac{D_{bc}}{d_{bc}}\log_{10}\frac{D_{ac}}{d_{ac}}\right) - \left(\log_{10}\frac{D_{bb}}{d_{bb}}\right)\left(\log_{10}\frac{D_{ac}}{d_{ac}}\right)^2 - \left(\log_{10}\frac{D_{cc}}{d_{cc}}\right)\left(\log_{10}\frac{D_{ab}}{d_{ab}}\right)^2 - \left(\log_{10}\frac{D_{aa}}{d_{aa}}\right)\left(\log_{10}\frac{D_{bc}}{d_{bc}}\right)^2$$

FIG. 100.—Equivalent susceptance networks for two and three parallel circuits. Constants are in mhos per mile. For zero-sequence in mhos per phase per mile, divide the above values by 3.

Applying these same relations to the system of cylindrical conductors in Fig. 101, for unit charge on the three parallel conductors (one-third of a unit charge on each conductor), the potentials of the three conductors are

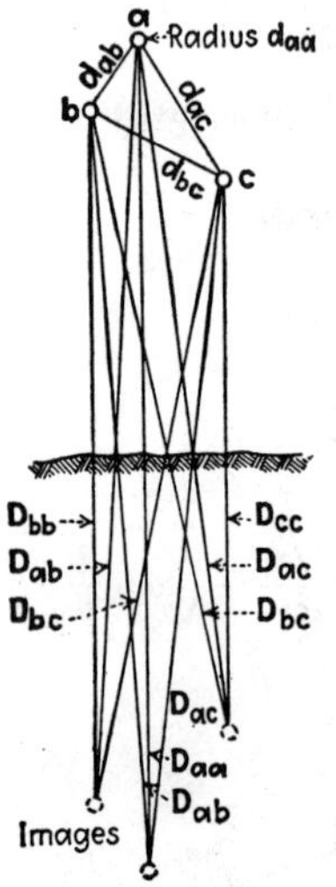

FIG. 101.—Notation for determination of zero-sequence susceptance.

$$V_a = \frac{2}{3}\log_\epsilon\frac{D_{aa}}{d_{aa}} + \frac{2}{3}\log_\epsilon\frac{D_{ab}}{d_{ab}} + \frac{2}{3}\log_\epsilon\frac{D_{ac}}{d_{ac}}$$

$$V_b = \frac{2}{3}\log_\epsilon\frac{D_{ab}}{d_{ab}} + \frac{2}{3}\log_\epsilon\frac{D_{bb}}{d_{bb}} + \frac{2}{3}\log_\epsilon\frac{D_{bc}}{d_{bc}}$$

$$V_c = \frac{2}{3}\log_\epsilon\frac{D_{ac}}{d_{ac}} + \frac{2}{3}\log_\epsilon\frac{D_{bc}}{d_{bc}} + \frac{2}{3}\log_\epsilon\frac{D_{cc}}{d_{cc}}$$

The distribution of charge between the three conductors in practical problems will not be far from uniform, but, to average out any difference, the potential of the three parallel conductors shall be taken as the average of the three potentials V_a, V_b, and V_c.

$$V = \frac{V_a + V_b + V_c}{3}$$

$$= \frac{2}{9}\log_\epsilon\frac{D_{aa}D_{bb}D_{cc}D_{ab}{}^2D_{bc}{}^2D_{ac}{}^2}{d_{aa}d_{bb}d_{cc}d_{ab}{}^2d_{bc}{}^2d_{ac}{}^2}$$

$$= 2\log_\epsilon\frac{\sqrt[9]{D_{aa}D_{bb}D_{cc}D_{ab}{}^2D_{bc}{}^2D_{ac}{}^2}}{\sqrt[9]{d_{aa}d_{bb}d_{cc}d_{ab}{}^2d_{bc}{}^2d_{ac}{}^2}} \quad (189)$$

The denominator in this expression is the **G.M.R. for susceptance** and is the same as the *G.M.R.* used for inductive reactance calculations, except that it does not contain the factor 0.779 which took the internal flux into consideration. For identical conductors $d_{aa} = d_{bb} = d_{cc} = a$, and the denominator may therefore be expressed as

$$G.M.R. = \sqrt[3]{aD^2} \quad (190)$$

The numerator, the **G.M.D.**, represents the ninth root of the product of the nine distances between the conductors and their images. For most work it is found sufficiently accurate to use as the *G.M.D.* twice the arithmetic mean of conductor heights h_1, h_2, and h_3

$$G.M.D. = 2\frac{h_1 + h_2 + h_3}{3} \quad (191)$$

Expressing the susceptance in practical units

$$b = \frac{0.244f10^{-6}}{\log_{10}\dfrac{G.M.D.}{G.M.R}} \text{ mhos per mile} \quad (192)$$

or expressed in mhos per phase for the zero-sequence

$$b_0 = \frac{0.0813f \times 10^{-6}}{\log_{10} \frac{G.M.D.}{G.M.R.}} \text{ mhos per mile per phase} \tag{193}$$

Example. For one of the circuits shown in Fig. 84 of Chap. VIII,

$$G.M.R. = \sqrt[3]{\frac{0.403}{12}(10.5)^2} = 1.55 \text{ ft.}$$

$$G.M.D. = \tfrac{2}{3}(45 + 53 + 61) = 106 \text{ ft.}$$

NOTE. Using the correct $G.M.D.$ gave a figure of 105.6 ft.

$$b_0 = \frac{0.0813 \times 60 \times 10^{-6}}{\log_{10} \frac{106}{1.55}} = 2.66 \times 10^{-6} \text{ mho per mile per phase}$$

80. Mutual Susceptance between Two Parallel Circuits.

It is sometimes desirable in making calculations on parallel circuits to express the charging-current characteristics in terms of self and mutual susceptance as in the equations

$$I_a = jb_{aa}E_a - jb_{ab}E_b \tag{194}$$

$$I_b = jb_{bb}E_b - jb_{ab}E_a \tag{195}$$

In these equations the coefficients b_{aa} and b_{bb} determine the charging current flowing in their respective circuits due to the potential on itself, whereas the coefficient b_{ab} determines the charging current flowing in one circuit when the other circuit alone is energized. These coefficients may be determined as follows:

Referring to Fig. 100(a), for charge q_a (in absolute units) on the cylindrical conductor a, the potential of a above ground is

$$V_a = 2q_a \log_\epsilon \frac{D_{aa}}{d_{aa}}$$

And the potential of point b above ground is

$$V_b = 2q_a \log_\epsilon \frac{D_{ab}}{d_{ab}}$$

in which d_{aa} is the radius of conductor a.

For a charge q_a on a and q_b on b

$$\left.\begin{aligned} V_a &= 2q_a \log_\epsilon \frac{D_{aa}}{d_{aa}} + 2q_b \log_\epsilon \frac{D_{ab}}{d_{ab}} \\ V_b &= 2q_b \log_\epsilon \frac{D_{bb}}{d_{bb}} + 2q_a \log_\epsilon \frac{D_{ab}}{d_{ab}} \end{aligned}\right\} \tag{196}$$

Solving for q_a and q_b in terms of V_a and V_b

$$q_a = \frac{2 \log_\epsilon \dfrac{D_{bb}}{d_{bb}}}{G} V_a - \frac{2 \log_\epsilon \dfrac{D_{ab}}{d_{ab}}}{G} V_b$$

$$q_b = \frac{2 \log_\epsilon \dfrac{D_{aa}}{d_{aa}}}{G} V_b - \frac{2 \log_\epsilon \dfrac{D_{ab}}{d_{ab}}}{G} V_a$$

in which

$$G = 4 \log_\epsilon \frac{D_{aa}}{d_{aa}} \log_\epsilon \frac{D_{bb}}{d_{bb}} - 4\left(\log_\epsilon \frac{D_{ab}}{d_{ab}}\right)^2$$

Converting to susceptances in practical units

$$b_{aa} = \frac{0.244 f 10^{-6} \log_{10} \dfrac{D_{bb}}{d_{bb}}}{\Delta} \text{ mhos per mile} \qquad (197)$$

$$b_{bb} = \frac{0.244 f 10^{-6} \log_{10} \dfrac{D_{aa}}{d_{aa}}}{\Delta} \text{ mhos per mile} \qquad (198)$$

$$b_{ab} = \frac{0.244 f 10^{-6} \log_{10} \dfrac{D_{ab}}{d_{ab}}}{\Delta} \text{ mhos per mile} \qquad (199)$$

in which

$$\Delta = \log_{10} \frac{D_{aa}}{d_{aa}} \log_{10} \frac{D_{bb}}{d_{bb}} - \left(\log_{10} \frac{D_{ab}}{d_{ab}}\right)^2$$

For multiconductor circuits the *G.M.D.* may be used with sufficient accuracy. Zero-sequence quantities are one-third of the above, for example,

$$b_{ab0} = \frac{0.0813 f 10^{-6} \log_{10} \dfrac{D_{ab}}{d_{ab}}}{\Delta} \text{ mhos per mile per phase} \qquad (200)$$

The equivalent network for the circuit as shown in Fig. 100 (*a*) is given in (*b*). Its derivation may be obtained from equations (194) and (195) as follows. The current flowing in b with one end grounded is equal to jb_{ab} for unit voltage on a, so that the susceptance connecting a and b must be equal to b_{ab}. The current flowing into a with b grounded is equal to jb_{aa} for unit voltage on a. Since a susceptance b_{ab} is already connected between a and b, the susceptance $(b_{aa} - b_{ab})$ must be connected between a and ground to permit current jb_{aa} to flow in a. Similar considerations prove that the susceptance between b and ground must be equal to $(b_{bb} - b_{ab})$.

The equivalent circuit for susceptance may readily be extended to take care of additional conductors or circuits. The value of the susceptances are determined by setting up a group of equations similar to (196). Thus the case of three conductors is shown in Fig. 100(c), and the corresponding equivalent circuit is given in (d). The accompanying equations define the susceptances for (d) and also the coefficients of the equations of the form of (194) and (195).

Example. As an example of such calculation, consider the twin-circuit line whose configuration is shown in Fig. 84. From previous calculations,

$$\begin{aligned} d_{aa} &= 1.55 \text{ ft.} = d_{bb} \\ d_{ab} &= 21.9 \text{ ft.} \\ D_{aa} &= 106 \text{ ft.} = D_{bb} \end{aligned}$$

Taking the ninth root of the product of the nine possible distances between one group of conductors and the images of the other group of conductors, $D_{ab} = 107.7$ ft.

NOTE. In general for most practical cases it is sufficiently accurate to take $D_{ab} = D_{aa}$.

$$\begin{aligned} \Delta &= \left(\log_{10}\frac{106}{1.55}\right)^2 - \left(\log_{10}\frac{107.7}{21.9}\right)^2 \\ &= (1.835)^2 - (0.692)^2 = 2.88 \\ b_{ab0} &= \frac{0.0813 \times 60 \times 10^{-6}}{2.88}\log_{10}\frac{107.7}{21.9} \\ &= 1.17 \times 10^{-6} \text{ mho per mile per phase} \end{aligned}$$

81. Susceptance Calculations for Circuits with Ground Wires.

For positive- and negative-sequence susceptance, the presence of the ground wires may ordinarily be neglected because the ground wire is usually located at such a position that the resulting dielectric field is very low. The effect of ground wires may be taken into account in the manner described below in connection with zero-sequence.

For zero-sequence, the presence of the ground wire should usually be taken into account. This may be done by setting up the equivalent circuit for zero-sequence susceptance as discussed in the preceding section in connection with Fig. 100. Thus, if conductor c of Fig. 100(c) is the ground wire which is taken as short-circuited to ground, then the equivalent circuit of Fig. 100(d) reduces to three terms, namely, the two self susceptances $(b_{aa} - b_{ab})$ and $(b_{bb} - b_{ab})$ and the mutual susceptance (b_{ab}).

82. Zero-sequence Equivalent Network for a Single Transmission Circuit.

For short lines the charging current may be neglected and the characteristics determined by the series resistance and reactance alone; but for longer lines it is necessary to take the distributed capacitance into consideration just as for the positive- and negative-sequence. The characteristics may be expressed either in terms of the ABC constants as in equations (185), or in terms of the equivalent series impedances and shunt admittances, as in equation (187) and Fig. 99.

Example. For one of the single circuits shown in Fig. 84 the zero-sequence constants for a 100-mile line are from Sec. 67 and 79:

$$
\begin{aligned}
Z_0 &= 100(0.521 + j2.76) = 52.1 + j276 \text{ ohms} \\
Y_0 &= 100(j2.66 \times 10^{-6}) = j2.66 \times 10^{-4} \text{ mho} \\
Y_0 Z_0 &= -0.0734 + j0.01386 \\
(Y_0 Z_0)^2 &= 0.0052 - j0.0020 \\
Z_{equiv.} &= Z_0\left[1 + \frac{-0.0734 + j0.01386}{6}\right. \\
&\qquad \left. + \frac{0.0052 - j0.0020}{120} + \cdots\right] \\
&= Z_0[1 + (-0.0122 + j0.00231) \\
&\qquad + (0.0000 + j0.0000) + \cdots] \\
&= (52.1 + j274)(0.9878 + j0.00231) \\
&= 50.8 + j271 \text{ ohms} \\
Y_{equiv.} &= Y_0\left[0.5 - \frac{-0.0734 + j0.01386}{24}\right. \\
&\qquad \left. + \frac{0.0052 - j0.0020}{240} + \cdots\right] \\
&= Y_0[0.5 - (-0.0031 + j0.0006) \\
&\qquad + (0.0000 - j0.0000) + \cdots] \\
&= (j2.66 \times 10^{-4})(0.5031 - j0.0006) \\
&= (0.0016 + j1.34)10^{-4} \text{ mho}
\end{aligned}
$$

The above indicates that for all lines below 100 miles it would be permissible for the degree of accuracy normally expected in short-circuit studies to disregard all but the first term, thus reducing the equivalent circuit to a series reactance equal to the total reactance and a shunt branch at each end equal to half the total susceptance. In certain cases the susceptance branches may be neglected entirely. The line may be so short that for all conditions the charging current will be negligibly small, or the fault may occur at such a point that the shunt susceptance is directly in parallel with the line impedance, in which case the impedance of the shunt susceptance, even for a

100-mile line, would be so large in comparison with the series impedance of the line as to by-pass an extremely small current.

83. Parallel Circuits with Distributed Inductive and Capacitive Coupling.*

For some cases it will be necessary to include also the effects of self and mutual capacitances as well as mutual inductances. The type of circuit under consideration is shown schematically in Fig. 102, which shows only a finite number of branches. The cases which usually occur in practice involve similar lines, and

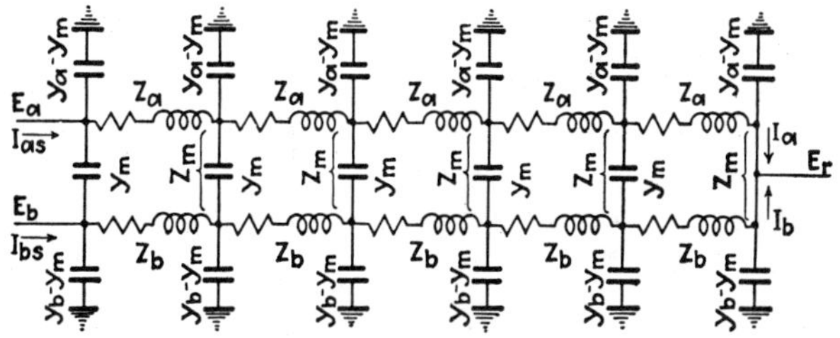

FIG. 102.—Schematic diagram of two parallel lines with distributed self and mutual constants.

for this case the equivalent network reduces to that shown in Fig. 103. A discussion of the derivation of this simplification follows:

In Fig. 102 let e_a and e_b be the voltage to ground at any point on lines a and b, respectively, i_a, i_b the current at any point in lines a and b, respectively, and

z_a = self impedance of line a, per unit length of line.

z_b = self impedance of line b, per unit length of line.

z_m = mutual impedance between line a and line b, per unit length of line.

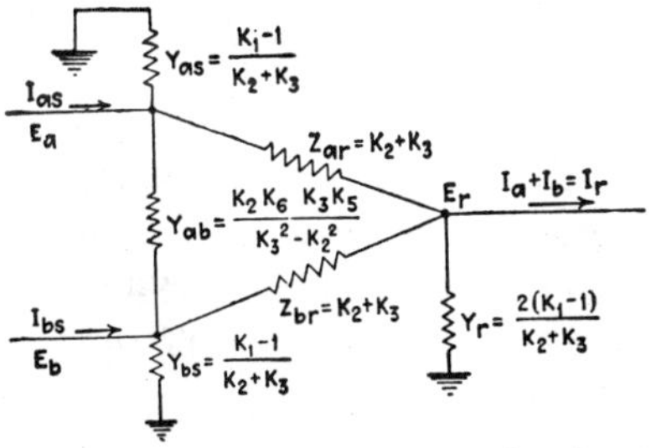

FIG. 103.—Zero-sequence equivalent network of two parallel transmission lines including distributed inductive and capacitive couplings.

*This section deals with a problem of infrequent application; its review, therefore, may well be postponed until the practical need arises. The problem and the method of attack employed are of interest from the theoretical point of view.

y_a = self admittance of line a, per unit length of line.
y_b = self admittance of line b, per unit length of line.
y_m = mutual admittance between line a and line b, per unit length of line.

The conductance components of the three admittances y_a, y_b, and y_m may usually be neglected. The susceptance components are defined in accordance with equations (196), (197), and (198).

The voltage to ground of any point, at a distance x from the receiving end, is equal to the receiver voltage E_r plus the self impedance drop due to i_a and the mutual impedance drop due to i_b. Expressed analytically:

$$e_a = E_r + \int_0^x i_a z_a dx + \int_0^x i_b z_m dx \tag{201}$$

The voltage of line b can be expressed similarly:

$$e_b = E_r + \int_0^x i_b z_b dx + \int_0^x i_a z_m dx \tag{202}$$

The current at any point in line a is equal to the load current I_a plus the charging current to ground and the charging current to the other conductor.

$$i_a = I_a + \int_0^x e_a y_a dx - \int_0^x e_b y_m dx \tag{203}$$

Similarly for the current in b:

$$i_b = I_b + \int_0^x e_b y_b dx - \int_0^x e_a y_m dx \tag{204}$$

These equations may be solved by a method of successive approximation that is frequently found of value in the practical solution of equations. Arguing from the short line in which the current is constant throughout the entire length but the voltage of which varies linearly, the first logical approximation is to assume

$$i_{a1} = I_a \tag{205}$$
$$i_{b1} = I_b \tag{206}$$

Substituting in (201) and (202)

$$e_{a1} = E_r + I_a z_a x + I_b z_m x \tag{207}$$
$$e_{b1} = E_r + I_b z_b x + I_a z_m x \tag{208}$$

and substituting (207) and (208) in (203) and (204), the second approximation is obtained.

$$i_{a2} = I_a + E_r y_a x + I_a y_a z_a \frac{x^2}{2} + I_b y_a z_m \frac{x^2}{2} - E_r y_m x - I_b y_m z_b \frac{x^2}{2} - I_a y_m z_m \frac{x^2}{2} \quad (209)$$

$$i_{b2} = I_b + E_r y_b x + I_b y_b z_b \frac{x^2}{2} + I_a y_b z_m \frac{x^2}{2} - E_r y_m x - I_a y_m z_a \frac{x^2}{2} - I_b y_m z_m \frac{x^2}{2} \quad (210)$$

and calling Z_a, Z_b, Z_m, Y_a, Y_b, and Y_m, the corresponding constants for the whole length of the line,

$$i_{a2} = I_{as} = [Y_a - Y_m]E_r + \left[1 + \frac{Y_a Z_a}{2} - \frac{Y_m Z_m}{2}\right] I_a + \left[\frac{Y_a Z_m}{2} - \frac{Y_m Z_b}{2}\right] I_b \quad (211)$$

$$i_{b2} = I_{bs} = [Y_b - Y_m]E_r + \left[\frac{Y_b Z_m}{2} - \frac{Y_m Z_a}{2}\right] I_a + \left[1 + \frac{Y_b Z_b}{2} - \frac{Y_m Z_m}{2}\right] I_b \quad (212)$$

Substituting (209) and (210) in turn in (201) and (202), integrating, and inserting the constants for the whole line there is obtained

$$e_{a2} = E_a = \left[1 + \frac{Y_a Z_a}{2} - \frac{Y_m Z_a}{2} + \frac{Y_b Z_m}{2} - \frac{Y_m Z_m}{2}\right] E_r + Z_a\left[1 + \frac{Y_a Z_a}{6} - \frac{Y_m Z_m}{3} + \frac{Y_b Z_m^2}{6 Z_a}\right] I_a + Z_m\left[1 + \frac{Y_a Z_a}{6} + \frac{Y_b Z_b}{6} - \frac{Y_m Z_m}{6} - \frac{Y_m Z_a Z_b}{6 Z_m}\right] I_b \quad (213)$$

$$e_{b2} = E_b = \left[1 + \frac{Y_b Z_b}{2} - \frac{Y_m Z_b}{2} + \frac{Y_a Z_m}{2} - \frac{Y_m Z_m}{2}\right] E_r + Z_b\left[1 + \frac{Y_b Z_b}{6} - \frac{Y_m Z_m}{3} + \frac{Y_a Z_m^2}{6 Z_b}\right] I_b + Z_m\left[1 + \frac{Y_a Z_a}{6} + \frac{Y_b Z_b}{6} - \frac{Y_m Z_m}{6} - \frac{Y_m Z_a Z_b}{6 Z_m}\right] I_a \quad (214)$$

These series may be expanded by the method indicated to any degree of accuracy desired. It is probable, however, that the expansion has already been extended sufficiently for all practical applications. The solution in the form given, however, does not lend itself well for coordination with other parts of the network. The desirable form is an equivalent network which represents the actual system, just as the equivalent π represents the simple transmission line. The equivalent circuit may then be set up on a calculating board or solved by the usual analytical processes.

It is known that any three-terminal network can be represented accurately by means of three impedances (or admittances) connected between terminals and an admittance or impedance connected to ground at each terminal.* Such a network is shown in Fig. 103. The value of the network constants can be obtained by applying unit voltage to one terminal with the two other terminals grounded. From a knowledge of the currents flowing in the three terminals, the impedance of each of the three branches connected to the energized terminal can be determined at once. This process repeated at the other terminals enables one to determine all of the impedances. This general method will be applied to our present problem for the particular case in which the lines are similar.

For this particular case $Z_a = Z_b = Z_L$ and $Y_a = Y_b = Y_L$, and equations (211) to (214) may be rewritten in the following form

$$E_a = K_1E_r + K_2I_a + K_3I_b \tag{215}$$
$$E_b = K_1E_r + K_3I_a + K_2I_b \tag{216}$$
$$I_{as} = K_4E_r + K_5I_a + K_6I_b \tag{217}$$
$$I_{bs} = K_4E_r + K_6I_a + K_5I_b \tag{218}$$

in which

$$\left.\begin{aligned}
K_1 &= 1 + \frac{Y_LZ_L}{2} - \frac{Y_mZ_L}{2} + \frac{Y_LZ_m}{2} - \frac{Y_mZ_m}{2} + \cdots \\
K_2 &= Z_L\left[1 + \frac{Y_LZ_L}{6} - \frac{Y_mZ_m}{3} + \frac{Y_LZ_m^2}{6Z_L} + \cdots\right] \\
K_3 &= Z_m\left[1 + \frac{Y_LZ_L}{3} - \frac{Y_mZ_m}{6} - \frac{Y_mZ_L^2}{6Z_m} + \cdots\right] \\
K_4 &= Y_L - Y_m + \cdots \\
K_5 &= 1 + \frac{Y_LZ_L}{2} - \frac{Y_mZ_m}{2} + \cdots \\
K_6 &= \frac{Y_LZ_m}{2} - \frac{Y_mZ_L}{2} + \cdots
\end{aligned}\right\} \tag{219}$$

* A more general discussion of the method of setting up equivalent networks is given in Chap. XII.

To determine Z_{ar}, Y_{as}, and Y_{ab} of Fig. 103, let

$$E_a = 1$$
$$E_b = E_r = 0$$

Substituting these values in (215) to (218), there is obtained

$$\left.\begin{aligned} 1 &= K_2 I_a + K_3 I_b \\ 0 &= K_3 I_a + K_2 I_b \\ I_{as} &= K_5 I_a + K_6 I_b \\ I_{bs} &= K_6 I_a + K_5 I_b \end{aligned}\right\} \quad (220)$$

Remembering that $I_a + I_b = I_r$, these equations may be solved with the following results

$$\left.\begin{aligned} I_r &= \frac{1}{K_2 + K_3} \\ I_{as} &= \frac{K_3 K_6 - K_2 K_5}{K_3^2 - K_2^2} \\ I_{bs} &= \frac{K_3 K_5 - K_2 K_6}{K_3^2 - K_2^2} \end{aligned}\right\} \quad (221)$$

In terms of the equivalent constants, it may be seen from Fig. 103 for the same conditions

$$\left.\begin{aligned} I_r &= \frac{1}{Z_{ar}} \\ I_{as} &= \frac{1}{Z_{ar}} + Y_{ab} + Y_{as} \\ I_{bs} &= -Y_{ab} \end{aligned}\right\} \quad (222)$$

Equating equations (221) and (222) and solving for the unknown equivalent constants

$$Z_{ar} = K_2 + K_3 \quad (223)$$

$$Y_{ab} = \frac{K_2 K_6 - K_3 K_5}{K_3^2 - K_2^2} \quad (224)$$

$$Y_{as} = \frac{K_5 + K_6 - 1}{K_2 + K_3} \quad (225)$$

By symmetry

$$Z_{br} = Z_{ar} \quad (226)$$
$$Y_{bs} = Y_{as} \quad (227)$$

In a similar manner, with the assumptions that

$$E_r = 1$$

and

$$E_a = E_b = 0$$

it may be shown that

$$Y_r = \frac{2(K_1 - 1)}{K_2 + K_3} \tag{228}$$

It may be observed from equations (219), that

$$K_5 + K_6 - 1 = K_1 - 1 \tag{229}$$

It therefore follows from equations (225), (227), and (229), that

$$Y_{bs} = Y_{as} = \frac{K_1 - 1}{K_2 + K_3} \tag{230}$$

The equivalent network thus obtained for the coupled circuits may be inserted in the complete system network and the currents at the terminals I_{as}, I_{bs}, and I_r, determined. During the simplification process the identity of the currents I_a and I_b in the line at the bussing point was lost. If these currents are desired, they may be obtained by solving equations (217) and (218), remembering that $I_a + I_b = I_r$, with the result that

$$I_a = \frac{I_r}{2} + \frac{I_{as} - I_{bs}}{2(K_5 - K_6)} \tag{231}$$

$$I_b = \frac{I_r}{2} - \frac{I_{as} - I_{bs}}{2(K_5 - K_6)} \tag{232}$$

The expansions expressed by equations (219) have been carried out to include the first-order terms only. This should be sufficiently accurate for all practical problems, since bussing points are usually employed on the longer lines.

Problems

1. Determine the positive-sequence constants per mile, z_1 and y_1, for the 60-cycle transposed transmission line shown in Fig. 89 neglecting the ground wire. Assuming that the transmission line is 200 miles in length, determine the A_1, B_1, and C_1 constants.

2. Find the equivalent circuit in the π and T forms for positive-sequence for the system described in Prob. 1.

3. For the case of Prob. 1, determine the zero-sequence constants per mile, z_0 and y_0, and also the A_0, B_0, and C_0 constants, assuming the average height above ground is 40 ft., $\rho = 100$ meter-ohms.

4. Find the equivalent circuit in the π and T forms for zero-sequence for the system described in Prob. 3.

5. Determine the zero-sequence susceptance per mile for the case of Prob. 1 taking into account the presence of the ground wire.

6. Assume that two identical transmission lines on separate rights-of-way be parallel at each end. Neglect the inductive coupling between circuits, and assume equivalent T networks for each line are as follows:

Positive- and negative-sequence constants

$$Z_{equiv.1} = 20 + j200$$

and

$$Y_{equiv.1} = +j0.001$$

Zero-sequence constants

$$Z_{equiv.0} = 50 + j600$$

and

$$Y_{equiv.0} = +j0.0005$$

If a constant positive-sequence voltage of 69 kv. from line-to-neutral is maintained at the sending end, find the phase *b* voltage-to-ground at the receiver for a fault on phase *a* at the same point, assuming the load is disconnected from the transmission line.

7. With the same system and fault as described in Prob. 6, assume that the breaker in the faulty line is opened at the sending end leaving the faulty line connected to the receiver bus. Determine the value of the voltage-to-ground on phase *b* at the sending end of the faulted line.

8. Consider a section of the line illustrated in Fig. 91 and assume that the left-hand circuit has impressed upon it 30-kv. zero-sequence. Determine the value of the voltage impressed on the right-hand circuit as the result of electric induction.

CHAPTER X

CONSTANTS OF CABLES

Practically all of the present-day commercial cables used for power transmission may be classified in one of the following groups:

1. Single-conductor.
2. Belted three-conductor:
   a. Round conductor.
   b. Sector conductor.
3. Shielded three-conductor (type H):
   a. Round conductor.
   b. Sector conductor.

The characteristics, both electrical and thermal, of the first two groups for normal operation are discussed in a very complete manner by D. M. Simmons in an article* from which much of the material presented here has been obtained. This volume is concerned only with the electrical characteristics which are required for the application of the method of symmetrical components. The properties of the single-conductor and belted three-conductor cables will be treated together, and later the properties of type H cable will be discussed.

An important consideration that requires repetition in connection with cable systems is the principle that regardless of the complexity of mutual inductive relations between component parts of individual phases, provided only that the three phases are symmetrical, the three sequences do not react upon each other. Three-conductor and type H cables by the nature of their construction inherently satisfy this condition; single-conductor cables may or may not. Unsymmetrical spacing and change in permeability resulting from different phase currents when certain devices to eliminate sheath currents are used may introduce dissymmetry. In general, the effect has a negligible

* Calculation of the Electrical Problems of Transmission by Underground Cables, *Elec. Jour.*, p. 366, August, 1925. Subsequently a revision of this article appeared in the *Elec. Jour.*, May to November, 1932.

influence upon short-circuit currents. When single-conductor armored cables are used, the effect of current upon the permeability of the steel may also affect symmetry.

84. Positive- and Negative-sequence Resistance.

The effective resistance to alternating current is larger than the resistance to direct current. The resistance to direct current may be taken as the resistance of a solid rod of the same cross section and the same length, increased by 2 per cent to take into account the effect of spiraling the strands. In three-conductor cables the twisting of the three conductors about each other produces an additional increase of about 2 per cent.

Skin Effect and Proximity Effect. The interior filaments of a conductor are surrounded by a larger number of magnetic lines of force than the filaments nearer the surface. For alternating current this results in an unequal distribution of current, a forcing of the current toward the surface of the conductor, called skin effect. This effect results in a larger resistance for alternating current than for direct current, the ratio of the two being called the skin-effect ratio. For small conductors this effect is negligible, but for larger conductors even at 60 cycles it may be quite pronounced. Table IX, taken from Simmons' article,

TABLE IX.—DIMENSIONS AND 60-CYCLE SKIN-EFFECT RATIO OF STRANDED COPPER CONDUCTORS AT 65°C.

Conductor size		Skin-effect ratio
Circular mils	Diameter, inches	
3,000,000	1.998	1.439
2,500,000	1.825	1.336
2,000,000	1.631	1.239
1,500,000	1.412	1.145
1,000,000	1.152	1.068
800,000	1.031	1.046
600,000	0.893	1.026
500,000	0.814	1.018
400,000	0.728	1.012
300,000	0.630	1.006

shows the magnitude of this ratio for a selected number of cables at 65°C. The ratio varies slightly with temperature.

The alternating magnetic flux in a conductor due to the current in a neighboring conductor gives rise to circulating currents, the effect of which is reflected in an apparent increase in resistance of the conductor. This effect is known as proximity effect. Only for the very largest sizes of cables is this effect important.

Sheath Currents of Single-conductor Cables. The flow of alternating current in the conductors of single-conductor cables induces alternating voltages in the cable sheaths. When the cable sheaths are solidly bonded, the voltages give rise to sheath currents and therefore additional losses. These losses may most conveniently be represented by correspondingly increasing the value of resistance assigned to the conductor and then ignoring the actual sheath currents. The increase in resistance for a typical case may be found as follows.

Assume three single-conductor cables arranged at the apexes of a triangle and let

x_m = mutual reactance in ohms per mile per phase between conductors and sheaths equal to the voltage per mile induced in a sheath by unit balanced three-phase currents in the conductors.
r_s = sheath resistance per phase in ohms per mile.
x_s = sheath reactance per phase in ohms per mile.
r_c = conductor resistance per phase in ohms per mile.

The induced sheath voltage per mile of cable for unit conductor current is then x_m, and the sheath current resulting therefrom is

$$\frac{x_m}{\sqrt{x_s^2 + r_s^2}}$$

The I^2R loss in watts or the increase in conductor resistance is

$$\Delta r = \frac{x_m^2 r_s}{x_s^2 + r_s^2} \tag{233}$$

The determination of the mutual reactance x_m is very similar to the determination of the ordinary self reactance of three aerial conductors, the only difference being that for aerial conductors the flux is integrated within the conductor and between the surface of the conductor and the center of the two other conductors, whereas for this case the flux is integrated between the sheath and the centers of the two conductors. This change in the integration limits results in the expression

$$x_m = 0.2794 \log_{10} \frac{2S}{r_4 + r_5} \text{ in ohms per phase per mile at 60 cycles} \tag{234}$$

in which

S = spacing between conductors in inches.
r_4 = inner radius of lead sheath in inches.
r_5 = outer radius of lead sheath in inches.

The quantity $\frac{r_4 + r_5}{2}$ is the mean radius or approximately the *G.M.R.* of the sheath.

The self reactance of the sheath x_s is also determined by the flux between the lead sheath and the centers of the conductors of the two other phases, since the disposition of the flux outside the sheath is the same whether unit current flows in the conductor or the sheath. Hence, the limits of integration are the same and it follows therefore that, to quite a high degree of accuracy,

$$x_s = x_m \tag{235}$$

The resistance per phase of the sheath is given by the formula

$$r_s = \frac{7{,}936s}{r_5^2 - r_4^2} \times 10^{-6}$$
$$= \frac{7{,}936s}{(r_5 + r_4)(r_5 - r_4)} \times 10^{-6} \text{ ohms per phase per mile}$$

in which s = resistivity of the lead sheath in microhm per centimeter cubed; equal at 50°C. to 25.2.

Inserting this value of s

$$r_s = \frac{0.2000}{(r_5 + r_4)(r_5 - r_4)} \text{ ohms per phase per mile} \tag{236}$$

Since $x_s = x_m$ the increment in resistance due to sheath currents is

$$\Delta r = \frac{x_m^2 r_s}{x_m^2 + r_s^2} \tag{237}$$

in which x_m and r_s are determined from (234) and (236), respectively.

Thus the resistance to positive- or negative-sequence is

$$r = r_c + \frac{x_m^2 r_s}{x_m^2 + r_s^2} \tag{238}$$

These considerations assumed a triangular disposition of the three individual cables. For other arrangements the geometric mean separation may be used with results sufficiently accurate for most practical purposes.

Example. Let it be desired to determine the resistance at 60 cycles of three 1,000,000-cir. mils single-conductor cables having a conductor insulation of $\frac{36}{64}$ in., and $\frac{1}{8}$ in. of lead sheath, run individually in fiber conduits 4.125 in. between centers, all in the same horizontal plane. The geometric mean separation is

$$S = \sqrt[3]{(4.125)^2(8.25)} = 5.20 \text{ in.}$$

From conductor tables (see Table IX), it may be found that the outside radius of the conductor is equal to 0.576 in. and, adding the insulation thickness,

$$r_4 = 1.138 \text{ in.}$$
$$r_5 = 1.263 \text{ in.}$$

From equation (234)

$$x_m = 0.279 \log_{10} \frac{2(5.20)}{1.138 + 1.263}$$
$$= 0.178 \text{ ohm per mile}$$

From equation (236)

$$r_s = \frac{0.200}{(2.401)(0.125)}$$
$$= 0.667 \text{ ohm per mile}$$

Substituting in equation (237)

$$\Delta r = \frac{(0.178)^2(0.667)}{(0.178)^2 + (0.667)^2}$$
$$= 0.044 \text{ ohm per mile}$$

From wire tables it may be found that the direct-current resistance at 65°C. is 0.0655, but due to skin effect (see Table IX) this should be increased to (1.068)(0.0655) or 0.070 ohm per mile. The total resistance including the sheath loss is then

$$r = 0.070 + 0.044 = 0.114 \text{ ohm per mile per phase}$$

The sheath loss can conveniently be eliminated by interposing insulation in the sheath, to prevent a complete sheath circuit being formed. This may give rise to quite large induced voltages in the sheaths which introduce a hazard to life or difficulties due to electrolysis. Some cables are broken up at several points to limit the induced voltage per section. Other devices are also utilized to prevent the flow of the sheath currents without producing dangerously high voltages. The reader is referred to cable specialists for more detailed information. Papers by Halperin and Miller,* and Arnold† will be found useful for this

* HALPERIN and MILLER, Reduction of Sheath Losses in Single-conductor Cables, *Trans. A.I.E.E.*, vol. 48, p. 399, April, 1929.

† ARNOLD, A. H. M., The Theory of Sheath Losses in Single-conductor Lead-covered Cables, *Jour. I.E.E.*, vol. 67, p. 69, 1929.

purpose. Such schemes do not invalidate the use of the simplified system of symmetrical components as long as the constants of all the phases are symmetrical. Caution must be exercised in those schemes involving reactances for limiting the current in which the permeability is altered by the current strength. The reactances in the different phases may not be equal in such cases.

Sheath loss for three-conductor cables is negligibly small except for the very large cables. The external magnetic field is almost completely annulled by the close proximity of the three conductors, but for the larger sizes local fields produce eddy currents in that portion of the sheath adjacent to the conductor.

85. Zero-sequence Resistance.

The zero-sequence resistance of the conductors alone (neglecting the sheath and ground circuits) is equal to three times the resistance of the three conductors in parallel. In this case the total currents in the individual conductors are in phase, so that the adjacent conductors tend to produce still greater distortion (proximity effect) of the current distribution. However, according to unpublished data of F. Wollaston, the net effect of the proximity effect is very small. This is probably due to the high contact resistance arising from either surface films or imperfect mechanical contact which prevents current flow between adjacent layers. Within individual layers the stranding also prevents the flow of current by a continuous transposition of the individual strands. The zero-sequence resistance of the conductors, per phase, is then equal to the direct-current resistance of one conductor multiplied by the skin-effect ratio of one conductor. Currents are produced in the sheath both for single-conductor and three-conductor cables when zero-sequence currents flow. Sheath losses and the resistance of the earth-return path, which are involved in the total zero-sequence resistance, will be discussed in connection with the zero-sequence reactance.

86. Positive- and Negative-sequence Reactance.

The positive- and negative-sequence reactance of **three-conductor cables** is determined in a manner analogous to that for aerial conductors. Assuming uniform current distribution throughout the section of the conductor, the formula becomes

TABLE X.—CHARACTERISTICS OF ROUND THREE-CONDUCTOR CABLES (60 CYCLES)

Circular mils or A.W.G. (B & S)	Diameter, inches	Resistance,* ohms per mile	*G.M.R.*, one conductor	Positive-negative sequence		Zero-sequence				Sheath		Positive-negative sequence		Zero-sequence				Sheath	
				Series reactance, ohms per mile	Shunt capacitive† reactance, ohms per mile	*G.M.R.*, three-conductors	Series reactance,‡ ohms per mile	Series resistance,§ ohms per mile	Shunt capacitive† reactance, ohms per mile	Thickness, inches	Resistance, ohms per mile 50°C.	Series reactance, ohms per mile	Shunt capacitive† reactance, ohms per mile	*G.M.R.*, three conductors	Series reactance,‡ ohms per mile	Series resistance,§ ohms per mile	Shunt capacitive† reactance, ohms per mile	Thickness, inches	Resistance, ohms per mile 50°C.
				$\frac{4}{64}$-in. conductor insulation, $\frac{2}{64}$-in. belt insulation							(1.0 kv.)	$\frac{5}{64}$-in. conductor insulation, $\frac{3}{64}$-in. belt insulation							(1.0 kv.)
6	0.184	2.50	0.067	0.177	6,530	0.185	0.307	8.38	12,360	$\frac{8}{64}$	1.96								
4	.232	1.58	.084	.167	5,700	.220	.282	6.62	10,850	$\frac{8}{64}$	1.68								
3	.260	1.25	.094	.162	5,300	.241	.271	5.99	10,100	$\frac{8}{64}$	1.58								
2	.292	0.987	.106	.157	4,950	.263	.264	5.40	9,500	$\frac{8}{64}$	1.47								
1	.332	.787	.126	.152	4,550	.297	.247	4.90	8,800	$\frac{8}{64}$	1.37								
0	.373	.618	.141	.150	4,250	.327	.239	4.43	8,250	$\frac{8}{64}$	1.27								
00	.418	.494	.159	.146	3,950	.361	.230	4.03	7,750	$\frac{8}{64}$	1.18								
000	.470	.391	.178	.143	3,700	.398	.222	3.66	7,200	$\frac{8}{64}$	1.09								
0000	.528	.310	.200	.140	3,350	.439	.217	3.31	6,620	$\frac{8}{64}$	1.00								
250,000	.575	.266	.221	.....	.....	.....	.....	....	......	..	....	0.144	3,800	0.490	0.223	2.53	7,560	$\frac{8}{64}$	0.755
300,000	.630	.220	.242	.....	.....	.....	.....	....	......	..	....	.141	3,600	.531	.217	2.43	7,120	$\frac{8}{64}$	.737
350,000	.681	.190	.262	.....	.....	.....	.....	....	......	..	....	.140	3,450	.568	.213	2.28	6,800	$\frac{9}{64}$	.698
400,000	.728	.166	.280	.....	.....	.....	.....	....	......	..	....	.138	3,250	.601	.210	2.16	6,510	$\frac{9}{64}$	.664
450,000	.772	.148	.297	.....	.....	.....	.....	....	......	..	....	.137	3,150	.634	.204	2.06	6,220	$\frac{9}{64}$	.636
500,000	.814	.134	.313	.....	.....	.....	.....	....	......	..	....	.136	3,000	.665	.201	1.97	6,000	$\frac{9}{64}$	.611

* Alternating-current resistance based upon 100 per cent conductivity at 65°C. including 2 per cent allowance for spiral of strands and 2 per cent allowance for spiral of conductors.

† For specific inductive capacity = 3.5.

‡ Reactance based upon all return current in the sheath; none in the ground.

§ Resistance based upon all return current in the sheath; none in the ground.

TABLE X.—CHARACTERISTICS OF ROUND THREE-CONDUCTOR CABLES (60 CYCLES).—(*Continued*)

Circular mils or A.W.G. (B & S)	Diameter, inches	Resistance,* ohms per mile	*G.M.R.*, one conductor	Positive-negative sequence		Zero-sequence				Sheath		Positive-negative sequence		Zero-sequence				Sheath	
				Series reactance, ohms per mile	Shunt capacitive† reactance, ohms per mile	*G.M.R.*, three conductors	Series reactance,‡ ohms per mile	Series resistance,§ ohms per mile	Shunt capacitive† reactance, ohms per mile	Thickness, inches	Resistance, ohms per mile 50°C.	Series reactance, ohms per mile	Shunt capacitive† reactance, ohms per mile	*G.M.R.*, three conductors	Series reactance,‡ ohms per mile	Series resistance,§ ohms per mile	Shunt capacitive† reactance, ohms per mile	Thickness, inches	Resistance, ohms per mile 50°C.
				$\frac{8}{64}$-in. conductor insulation, $\frac{3}{64}$-in. belt insulation (2.0 to 3.0 kv. incl.)								$\frac{7}{64}$-in. conductor insulation, $\frac{4}{64}$-in. belt insulation (4.0 to 6.0 kv. incl.)							
6	0.184	2.50	0.067	0.200	8,350	0.208	0.326	7.21	14,950	$\frac{8}{64}$	1.57	0.210	9,150	0.220	0.339	6.79	16,470	$\frac{8}{64}$	1.43
4	.232	1.58	.084	.186	7,250	.245	.301	5.87	13,350	$\frac{8}{64}$	1.43	.195	8,000	.257	.315	5.51	14,800	$\frac{8}{64}$	1.31
3	.260	1.25	.094	.180	6,800	.266	.291	5.33	12,600	$\frac{8}{64}$	1.36	.188	7,500	.278	.304	5.00	14,000	$\frac{8}{64}$	1.25
2	.292	0.987	.106	.174	6,400	.289	.282	4.83	12,000	$\frac{8}{64}$	1.28	.182	6,950	.302	.293	4.53	13,100	$\frac{8}{64}$	1.18
1	.332	.787	.126	.168	5,850	.323	.264	4.39	11,120	$\frac{8}{64}$	1.20	.176	6,450	.336	.276	4.15	12,400	$\frac{8}{64}$	1.12
0	.373	.618	.141	.163	5,450	.353	.256	3.98	10,360	$\frac{8}{64}$	1.12	.171	6,000	.367	.266	3.77	11,650	$\frac{8}{64}$	1.05
00	.418	.494	.159	.160	5,050	.388	.246	3.64	9,700	$\frac{8}{64}$	1.05	.166	5,650	.401	.256	3.46	10,900	$\frac{8}{64}$	0.99
000	.470	.391	.178	.155	4,750	.426	.237	3.33	9,100	$\frac{8}{64}$	0.98	.162	5,250	.439	.248	3.15	10,200	$\frac{8}{64}$	.92
0000	.528	.310	.200	.152	4,400	.467	.231	3.04	8,550	$\frac{8}{64}$	.91	.157	4,900	.481	.239	2.89	9,550	$\frac{8}{64}$	.86
250,000	.575	.266	.221	.149	4,150	.504	.223	2.85	8,150	$\frac{8}{64}$	.86	.154	4,650	.518	.235	2.43	9,130	$\frac{9}{64}$	.72
300,000	.630	.220	.242	.145	3,950	.544	.221	2.35	7,700	$\frac{9}{64}$	.71	.150	4,400	.558	.229	2.26	8,650	$\frac{9}{64}$	.68
350,000	.681	.190	.262	.143	3,750	.582	.216	2.23	7,330	$\frac{9}{64}$	.68	.149	4,200	.596	.223	2.11	8,300	$\frac{9}{64}$	.64
400,000	.728	.166	.280	.142	3,600	.616	.212	2.09	7,000	$\frac{9}{64}$	.64	.147	4,000	.630	.220	2.03	7,960	$\frac{9}{64}$	.62
450,000	.772	.148	.297	.141	3,450	.648	.209	2.01	6,800	$\frac{9}{64}$	.62	.145	3,900	.663	.217	1.92	7,700	$\frac{9}{64}$	.59
500,000	.814	.134	.313	.140	3,350	.678	.206	1.90	6,500	$\frac{9}{64}$	.59	.143	3,750	.693	.213	1.84	7,450	$\frac{9}{64}$	.57

* Alternating-current resistance based upon 100 per cent conductivity at 65°C. including 2 per cent allowance for spiral of strands and 2 per cent allowance for spiral of conductors.

† For specific inductive capacity = 3.5.

‡ Reactance based upon all return current in the sheath; none in the ground.

§ Resistance based upon all return current in the sheath; none in the ground.

TABLE X.—CHARACTERISTICS OF ROUND THREE-CONDUCTOR CABLES (60 CYCLES).—(*Continued*)

Circular mils or A.W.G. (B & S)	Diameter, inches	Resistance,* ohms per mile	*G.M.R.*, one conductor	Positive-negative sequence		Zero-sequence				Sheath		Positive-negative sequence		Zero-sequence				Sheath	
				Series reactance, ohms per mile	Shunt capacitive† reactance, ohms per mile	*G.M.R.*, three conductors	Series reactance,‡ ohms per mile	Series resistance,§ ohms per mile	Shunt capacitive† reactance, ohms per mile	Thickness, inches	Resistance, ohms per mile 50°C.	Series reactance, ohms per mile	Shunt capacitive† reactance, ohms per mile	*G.M.R.*, three conductors	Series reactance,‡ ohms per mile	Series resistance,§ ohms per mile	Shunt capacitive† reactance, ohms per mile	Thickness, inches	Resistance, ohms per mile 50°C.
				$\frac{8}{64}$-in. conductor insulation, $\frac{4}{64}$-in. belt insulation						(7.0 to 7.5 kv. incl.)		$\frac{11}{64}$-in. conductor insulation, $\frac{5}{64}$-in. belt insulation						(15.0 kv.)	
6	0.184	2.50	0.067	0.219	9,800	0.233	0.340	6.55	16,950	$\frac{8}{64}$	1.35	0.242	11,500	0.264	0.359	5.89	18,900	$\frac{8}{64}$	1.13
4	.232	1.58	.084	.203	8,650	.270	.317	5.30	15,350	$\frac{8}{64}$	1.24	.224	10,200	.303	.335	4.73	17,300	$\frac{8}{64}$	1.05
3	.260	1.25	.094	.195	8,050	.291	.306	4.82	14,500	$\frac{8}{64}$	1.19	.216	9,600	.325	.323	4.28	16,500	$\frac{8}{64}$	1.01
2	.292	0.987	.106	.189	7,500	.315	.295	4.38	13,700	$\frac{8}{64}$	1.13	.209	9,050	.350	.313	3.90	15,650	$\frac{8}{64}$	0.97
1	.332	.737	.126	.182	6,950	.349	.278	3.97	12,850	$\frac{8}{64}$	1.06	.200	8,400	.385	.297	3.55	14,750	$\frac{8}{64}$	.92
0	.373	.618	.141	.176	6,450	.380	.269	3.65	12,200	$\frac{8}{64}$	1.01	.194	7,800	.417	.284	3.26	13,940	$\frac{8}{64}$	.88
00	.418	.494	.159	.171	6,000	.414	.260	3.34	11,450	$\frac{8}{64}$	0.95	.188	7,250	.451	.275	2.98	13,200	$\frac{8}{64}$	.83
000	.470	.391	.178	.166	5,650	.452	.251	3.06	10,750	$\frac{8}{64}$	.89	.181	6,750	.490	.268	2.46	12,480	$\frac{9}{64}$	.69
0000	.528	.310	.200	.161	5,250	.495	.245	2.50	10,200	$\frac{9}{64}$	.73	.175	6,300	.533	.259	2.26	11,800	$\frac{9}{64}$	.65
250,000	.575	.266	.221	.158	5,000	.532	.238	2.37	9,550	$\frac{9}{64}$	.70	.171	6,000	.572	.250	2.16	11,250	$\frac{9}{64}$	.63
300,000	.630	.220	.242	.154	4,750	.572	.232	2.20	9,100	$\frac{9}{64}$	.66	.167	5,650	.612	.244	1.99	10,600	$\frac{9}{64}$	.59
350,000	.681	.190	.262	.151	4,500	.610	.226	2.05	8,700	$\frac{9}{64}$	.62	.164	5,400	.650	.238	1.90	10,200	$\frac{9}{64}$	.57
400,000	.728	.166	.280	.149	4,300	.644	.222	1.97	8,400	$\frac{9}{64}$	.60	.161	5,200	.684	.234	1.82	9,800	$\frac{9}{64}$	.55
450,000	.772	.148	.297	.148	4,150	.677	.219	1.89	8,100	$\frac{9}{64}$	.58	.160	5,000	.717	.230	1.74	9,550	$\frac{9}{64}$	.53
500,000	.814	.134	.313	.146	4,000	.708	.215	1.81	7,850	$\frac{9}{64}$	.56	.157	4,850	.748	.224	1.66	9,250	$\frac{9}{64}$	.51

* Alternating-current resistance based upon 100 per cent conductivity at 65°C. including 2 per cent allowance for spiral of strands and 2 per cent allowance for spiral of conductors.

† For specific inductive capacity = 3.5.

‡ Reactance based upon all return current in the sheath; none in the ground.

§ Resistance based upon all return current in the sheath; none in the ground.

$$x = 0.2794 \log_{10} \frac{S}{0.779a} \text{ ohms per mile per phase for 60 cycles} \tag{239}$$

in which a = outside radius of the conductors and 0.779 = factor to convert to *G.M.R.** For cables it is convenient to express S, a, and the *G.M.R.* in inches.

The departure from uniform distribution, while important in its effect upon the resistance, is of little consequence in its effect upon the reactance. In Table X are evaluated the reactances for the most common range of cable sizes met in practice. These reactances were calculated by equation (239) for a frequency of 60 cycles.

The positive- and negative-sequence reactances for **single-conductor cables,** when sheath currents are present, may be determined from the following expression:

$$x = 0.2794 \log_{10} \frac{S}{0.779a} - \frac{x_m^3}{x_m^2 + r_s^2} \text{ ohms per mile per phase} \tag{240}$$

The last term represents the correction for the presence of sheath currents. The negative sign arises from the fact that the current in the sheath is in a direction opposite to that in the conductor, thus tending to limit the flux to the region between the conductor and the sheath. The last term corresponds to equation (238) with x_m substituted for r_s and is derived by considering the current in the sheath and the component of voltage which it induces in the conductor in quadrature to the conductor current.

For sector-shaped conductors no accurate data are available, but Simmons may be quoted as authority for the statement that the reactance is from 5 to 10 per cent lower than for round conductors of the same area and same thickness of insulation.

87. Zero-sequence Reactance.

Cable sheaths commonly have definite connections to ground at several points and considerable leakage to ground throughout the length of the cable ducts. Hence it is obvious that the

* For most underground cables the number of strands is so large that the factor 0.779, based on solid conductors, may be used without appreciable error. For cables with a small number of strands the factors from Table VI, Sec. 61, may be used or the *G.M.R.* may be used directly in inches from Table X.

method of making these connections and their resistances will have important effects upon the zero-sequence path. Calculations of the zero-sequence impedance can be made practically for two cases as follows:

Case 1. Return current in sheath and earth in parallel.

Case 2. Return current in sheath alone.

The actual value will lie between the limits given by these cases, depending upon the length, the grounding, and leakage. In Case 1 it is assumed that the ground connections and length of cable are such that maximum earth current obtains; in other words, the calculations are based upon the distribution of current between sheath and earth which will occur in the middle of a long section of cable with well-grounded sheath. In Case 2 the return current may be confined to the sheath because of the relatively high resistance of ground connections and the small effect of leakage in short lengths of cables in well-dried cable ducts. The method of analysis to be employed will make use of an equivalent circuit from which the zero-sequence impedance can be obtained for both cases.

FIG. 104.—Cable in a buried duct.

For the analysis of the zero-sequence reactance of cables, consider the conductor shown in Fig. 104, completely buried in the ground with a non-magnetic non-conducting material insulating the conductor from ground. The radius of the conductor will be denoted by a and the radius of the intervening space by b. Since the conductors are usually located near the surface of the earth, one would expect considerable dissymmetry in the current distribution in the earth. Carson* has shown, however, that for the cases met in practice the effect of the changed current distribution is not very great, increasing the earth-return part of the impedance on the order of 5 to 10 per cent over the value that would be obtained if the earth extended indefinitely above as well as below the conductor. In view of the fact that cables are usually supplied with lead sheaths which are then in parallel with the earth circuit whose impedance is much higher than the resistance of the sheaths, it follows that it is unnecessary to determine the earth-return circuit precisely. Hence, the above correction can in general be neglected.

* CARSON, JOHN R., Ground Return Impedance: Underground Wire with Earth Return, *Bell System Tech. Jour.*, p. 94, January, 1929.

Carson's results are surprising, in that one would at first expect that the impedance of the conductor when near the surface of the earth would be much larger than if completely surrounded by earth of infinite extent in all directions. However, Rudenberg[(137)] has arrived at this same result by a different method of reasoning. Rudenberg has further shown that the impedance of such a symmetrical circuit due to the flux in the earth is, for commercial frequencies and normal ground resistance, and for unit length, equal to

$$\pi^2 f + j2\omega \log_\epsilon \left(\frac{0.178\sqrt{\frac{\rho}{f}}}{b} \right) \text{ in absolute units} \tag{241}$$

The reactance due to the flux in the air and in the conductor is

$$j2\omega\left(\log_\epsilon \frac{b}{a} + \frac{1}{2}\right)$$

or

$$j2\omega \log_\epsilon \frac{b}{0.779a} \tag{242}$$

Adding the conductor resistance to expressions (241) and (242), the total self impedance of the conductor and ground return is

$$z_g = r_{conductor} + \pi^2 f + j2\omega \log_\epsilon \frac{0.178\sqrt{\frac{\rho}{f}}}{0.779a}$$

in absolute units.

$$= r_{conductor} + 0.00159f + j0.004657f \log_{10} \frac{2{,}160\sqrt{\frac{\rho}{f}}}{(G.M.R.)_{conductor}} \text{ in ohms per mile} \tag{243}$$

This expression is identical with that* which determines the impedance of aerial wires with ground return. The latter is practically independent of the height of conductor above the ground, and this expression is independent of the radius of air section in which the conductor lies. Equation (243) can in turn be reduced to the form of equation (169) giving

* See equation (163), Sec. 64.

$$z_g = r_{conductor} + 0.00159f + j0.004657f \log_{10} \frac{D_e}{G.M.R.} \text{ ohms per mile} \quad (244)$$

in which $D_e = 2{,}160\sqrt{\frac{\rho}{f}}$ [Eq. (166)]

In a similar manner it may be shown that the mutual impedance between two conductors in the cylindrical air tunnel can be reduced to the form of equation (170).

$$z_{gm} = 0.00159f + j0.004657f \log_{10} \frac{D_e}{d_{ab}} \text{ ohms per mile} \quad (245)$$

The expressions (244) and (245), changed to per phase values, are represented in curve form in Figs. 82(a), (b) and (c), in which D_e, $G.M.R.$, and d_{ab} are expressed in feet, and ρ in meter-ohms.

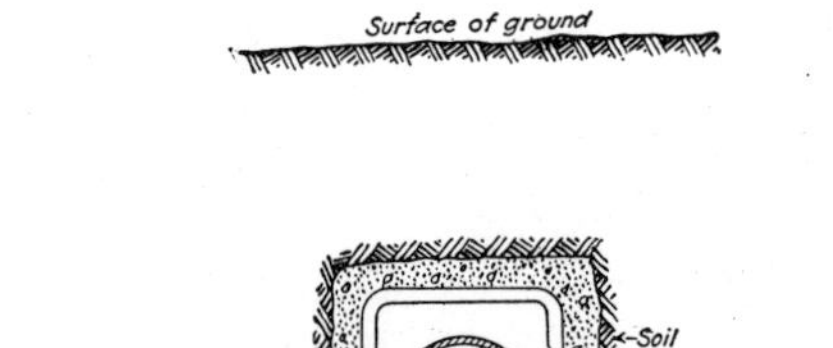

Fig. 105.—Calculation of earth or sheath return impedance of buried cable. For sheath or ground return circuits use these values directly and for zero-sequence per phase multiply by 3. Results are in ohms per mile.

With these relations established, the zero-sequence impedance of cables can be determined in a manner analogous to that used for the case of the transmission line with ground wires.*

The geometric configuration of the conductors of a **three-conductor cable** in an underground duct is shown in Fig. 105(a). The distance from the surface is immaterial. The equivalent circuit representing the impedance between conductors, sheath, and ground is indicated in Fig. 105(b). The mutual impedance between the sheath and conductors considered as a unit is given by equation (245), in which the geometric mean spacing is

* For a discussion of the equivalent circuits used, refer to Sec. 73.

approximately $\frac{r_4 + r_5}{2}$. This determines the star branch next to ground. The self impedance between conductors and ground is given by equation (244). This self impedance is equal to the sum of the star branches in the conductor and ground legs, so that the impedance associated with the conductor leg is the difference between the self and mutual impedances, or

$$r_{conductors} + 0.00159f + j0.004657f \log_{10} \frac{D_e}{(G.M.R.)_{conductors}} - 0.00159f - j0.004657f \log_{10} \frac{2D_e}{r_4 + r_5} = r_{conductors} + j0.004657f \log_{10} \frac{r_4 + r_5}{2(G.M.R)_{conductors}} \quad (246)$$

Similar considerations apply for the impedance of the sheath branch, but in this case the self and mutual impedances are equal, except for the sheath resistance.

With the sheath grounded at both ends, the impedance between conductor and ground is obtained by solving the circuit shown in Fig. 105(c). The distribution of current between sheath and ground can also be determined from this circuit.

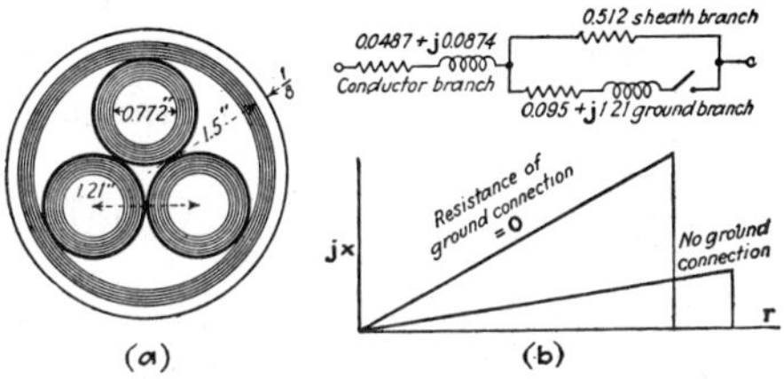

FIG. 106.—Calculation of the earth and sheath return impedances of a special 450,000-cir. mil three-conductor cable. The resistance of each conductor at 65°C. is 0.146 ohm per mile. For 60 cycles and damp earth, D_e = 2,800 ft. For sheath or earth return use the foregoing values directly. For zero-sequence multiply by 3 to obtain results in ohms per mile per phase.

If the sheath is grounded at one end only, the voltage between sheath and ground per ampere per mile is equal to the mutual impedance.

The zero-sequence impedance is obtained by multiplying the above quantities by 3.

Example 1. As an illustration of such a case, let it be desired to obtain the impedance of a single three-conductor cable, whose characteristics are shown by the section of cable in Fig. 106, when buried in damp earth: D_e = 2,800 ft.; resistance per conductor = 0.146 ohm per mile, 60 cycles.

$$G.M.R. \text{ of three conductors} = \sqrt[3]{(0.779)(0.386)(1.210)^2} = 0.761 \text{ in.}$$

$$G.M.\text{ spacing, conductor to sheath} = \frac{r_4 + r_5}{2} = \frac{1.5 + 1.625}{2} = 1.5625\text{ in.}$$

$$\text{Mutual reactance} = 0.004657 \times 60 \log_{10} \frac{2{,}800 \times 12}{1.5625} = 1.21$$

Reactance of conductor branch =

$$0.004657 \times 60 \log_{10} \frac{1.5625}{0.761} = 0.0874$$

$$\text{Resistance of conductor branch} = \frac{0.146}{3} = 0.0487$$

$$\text{Resistance of sheath branch} = \frac{0.20}{3.125 \times 0.125} = 0.512$$

Mutual resistance $= 0.00159 \times 60 = 0.095$ ohm.

The equivalent network is shown in Fig. 106(*b*).

Paralleling the sheath and ground branches and adding the impedance of the conductor branch,

$$\frac{0.512(0.095 + j1.210)}{0.512 + (0.095 + j1.210)} + (0.0487 + j0.0874) = 0.474 + j0.260$$

The zero-sequence impedance is then

$$3(0.474 + j0.260) = 1.42 + j0.78 \text{ ohms per mile per phase}$$

The absolute value of this impedance is 1.62 ohms per mile per phase

Comparing this value of zero-sequence impedance with the positive- and negative-sequence impedances per phase

$$(0.146 + j0.169),$$

it will be observed that for this particular case the ratio of zero-sequence resistance to positive-sequence resistance is 10 and the ratio of zero-sequence reactance to positive-sequence reactance is 4.6.

If the cable is not grounded or the ground connections are of such high resistance that no appreciable current flows through the ground, then the zero-sequence impedance is the sum of the impedances in the conductor and sheath branches, namely,

$$(0.0487 + j0.0874) + 0.512 = 0.561 + j0.0874$$

or in terms of ohms per mile per phase, $(1.68 + j0.262)$. The absolute value of this impedance is 1.70 ohms, which is practically equal to that for the ground connections of zero resistance.

The impedance diagram in which the two conditions are represented by closing and opening the switch, and the vector representation of these impedances, are shown in Fig. 106(b).

Example 2. To illustrate further these calculations, determination of the zero-sequence impedance per phase of the four three-

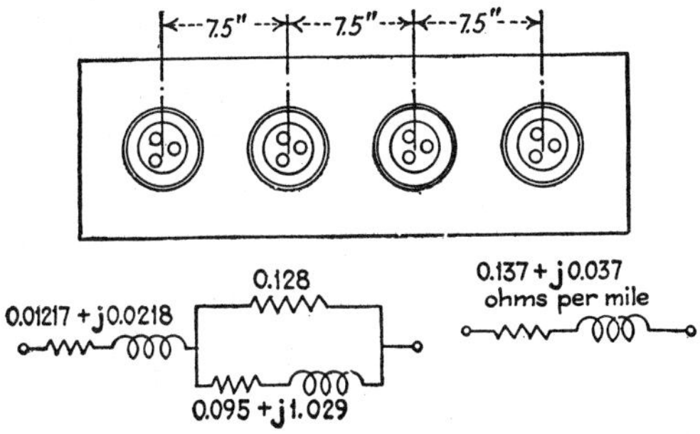

G.M.R. of three conductors in one cable $= \sqrt[3]{(0.779)(0.386)(1.210)^2} =$ 0.761 in.

G.M.R. of conductors of four circuits $= \sqrt[16]{(0.761)^4(7.5)^6(15)^4(22.5)^2} =$ 5.775 in.

G.M.D. spacing three conductors to surrounding sheath $= \dfrac{r_4 + r_5}{2} = \dfrac{1.5 + 1.625}{2} = 1.563$ in.

G.M.D. spacing of conductors and sheaths $= \sqrt[16]{(1.563)^4(7.5)^6(15)^4(22.5)^2} =$ 6.913 in.

$$\text{Mutual reactance} = 0.004657 \times 60 \log_{10} \frac{2{,}800 \times 12}{6.913} = 1.029 \text{ ohms}$$

$$\text{Reactance of conductor branch} = 0.004657 \times 60 \log_{10} \frac{6.913}{5.775} = 0.0218 \text{ ohm}$$

$$\text{Resistance of conductor branch} = \frac{0.146}{3 \times 4} = 0.01217 \text{ ohm}$$

$$\text{Resistance of sheath branch} = \frac{0.200}{4 \times 3.125 \times 0.125} = 0.128 \text{ ohm}$$

Mutual resistance $= 0.00159 \times 60 = 0.095$ ohm

Zero-sequence impedance (no ground connection) $= 3(0.01217 + j0.0218 + 0.128) = 0.421 + j0.066$

Zero-sequence impedance (with ground connection) $= 3(0.137 + j0.37) = 0.411 + j0.111$

FIG. 107.—Calculation of zero-sequence impedance of four three-conductor cables in parallel. A description of the individual cable is given in Fig. 106.

conductor cables in the conduits shown in Fig. 107 will be made. The calculations in the figure are self-explanatory. The results show that if the ground connections have zero resistance the zero-sequence impedance per phase is $0.411 + j0.111$ ohm per mile, the absolute value of which is 0.425. If the connection resistance is very high the cable impedance approaches the value corresponding to the condition in which no current flows through the ground. The cable impedance for this condition is

$$0.421 + j0.066,$$

the absolute value of which is 0.426. Here again the absolute value is practically unchanged by the nature of the ground connection.

From Fig. 105 it may be seen that the zero-sequence impedance for lead-covered cables with no ground return circuit (sheath only) is

$$z_o = \frac{0.60}{(r_5 + r_4)(r_5 - r_4)(\text{number of sheaths})} + 3(r_{\textit{paralleled conductors}}) + j0.014f \log_{10} \frac{(G.M.D.)_{\textit{separation}}}{(G.M.R.)_{\textit{conductors}}} \quad (247)$$

This formula is usually sufficiently accurate for practical purposes because the nature of the ground return circuit for cable systems is usually quite indefinite, being mixed up with water-pipe circuits, and the difficulty of assuring a ground connection which is low relative to the sheath resistance.

In Table X are given the zero-sequence series reactances for the commonly used three-conductor cables assuming no earth-return currents. The sheath resistances are also given which will be of assistance in calculating the zero-sequence impedances, for example, assuming no ground return circuit the zero-sequence resistance of a single three-conductor cable is equal to the resistance of one conductor plus three times the given sheath resistance. The *G.M.R.* of the individual conductors and of the group will also be of assistance in these calculations.

The zero-sequence impedance of solidly bonded **single-conductor cables** can be obtained in a manner analogous to that described for the four three-conductor cables of Example 2. Where special devices are employed to reduce sheath losses, the zero-sequence impedance is largely dependent upon the degree to which this is accomplished. When the sheaths are completely open-circuited, the return current must flow through the ground. Impedance devices have been used which represent a mean between solidly bonding and open-circuiting. Other devices are arranged to insert high impedance in the sheath circuit for the flow of positive-sequence current and low impedances for the flow of zero-sequence current.

88. Shunt Capacitive Reactance.

In Fig. 108 are given curves for the determination of the positive-, negative-, and zero-sequence shunt capacitive reactance for single-conductor and belted three-conductor cables. The results in each case are given in terms of ohms to neutral per phase. This was done for the two-fold purpose of providing a convenient scale and to present the data in the most usable form for the most common application of symmetrical components: short-circuit studies. These particular curves were calculated for 60 cycles and for a uniform specific inductive capacity of 3.5. For other values of these constants the capacitive reactance can be calculated by the formula indicated in the caption under the figure. The data* from which these curves were prepared are based upon semigraphical and analytical calculations of geometric factors of cables which indicate an accuracy to within 1 per cent.

For **three-conductor cables** the positive- and negative-sequence values differ from the zero-sequence values, but for **single-conductor cables** these values are all equal. The capacitance of sectored cables is larger than for the equivalent round conductor cables, so that X_c is smaller. The correction factor to be applied to the constants for round conductor cables of the same conductor section and insulation thickness to obtain the corresponding constants for the sectored belted cable is given by the bottom curve in Fig. 108.

The foregoing premised a uniform insulating material, which may not always be true. Further, a value of $k = 3.5$ was assumed. Most paper-insulated cables run about 10 per cent less than this value. The usual range is given in Table XI.

Table XI.—Values of Specific Inductive Capacity

Material	Value of k
Impregnated paper	3.0 to 4.0
Varnished cambric	4.0 to 6.0
Rubber	4.0 to 9.0

In Table X are given the positive-, negative-, and zero-sequence shunt capacitive reactances for the commonly used three-conductor cables.

* Calculation of the Electrical Problems of Transmission by Underground Cables, *Elec. Jour.*, p. 366, August, 1925.

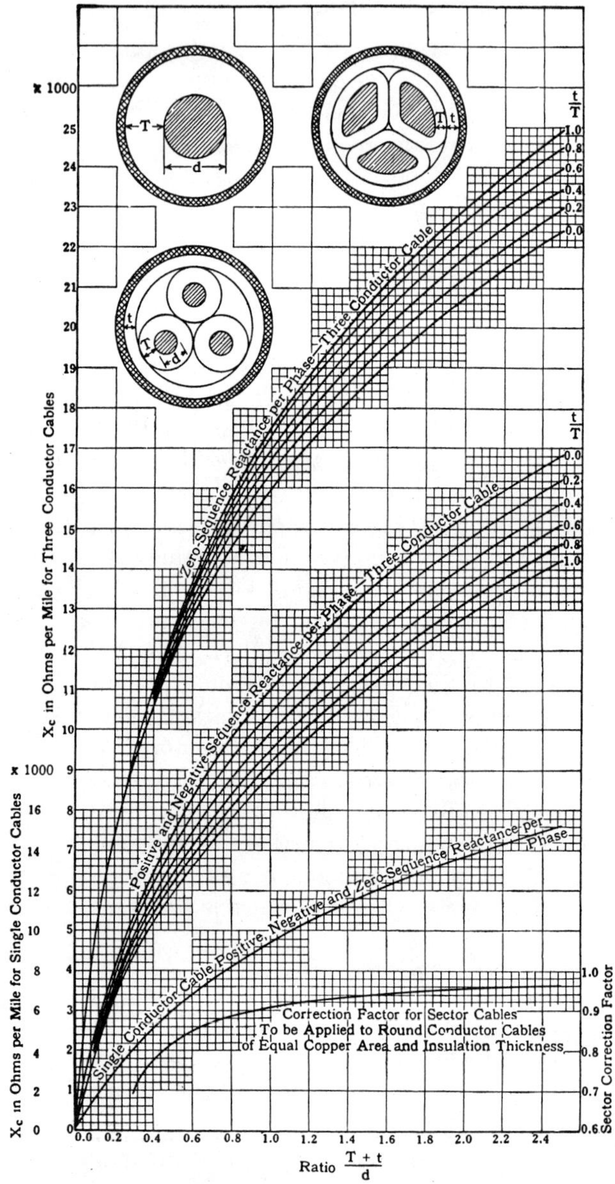

FIG. 108*a*.—Shunt capacitive reactance to neutral of single- and three-conductor cables in ohms per mile for 60 cycles.

$$X_c' = \frac{3.5}{k}\frac{60}{f}\frac{1}{\text{length in miles}}X_c$$

89. Shielded or Type H Cable.

In type H cable, copper tape is wound around all or a portion of the conductor insulation. A typical construction of this cable is shown in Fig. 109. The purpose of the tape is to control the electrostatic stress, reduce corona formation, and decrease

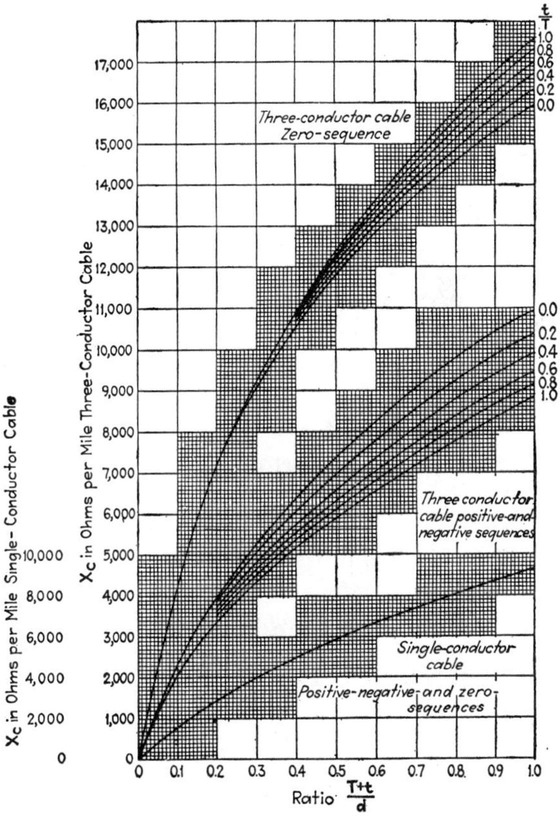

Fig. 108b.—Shunt capacitive reactance to neutral of single- and three-conductor cables in ohms per mile for 60 cycles.

$$X_c' = \frac{3.5}{k} \frac{60}{f} \frac{1}{\text{length in miles}} X_c$$

the thermal resistance. The tape is made very thin in order to minimize the circulating current under normal operating conditions and thus minimize the loss.

The effect of these circulating currents upon the positive- and negative-sequence reactance and resistance can be calculated by the method discussed for determining the effect of sheath currents for three single-conductor lead-covered cables. For short-circuit calculations these currents may be neglected. The zero-sequence impedance is likewise affected to only a very small extent by the circulating currents in the shields. Calculations of the impedance between the paralleled conductors with the paralleled sheath and shields as return for a three-conductor 450,000-cir. mil type H cable show that only 1 per cent of the total current flows through the shields. This effect may therefore be neglected and the impedance calculated as though the shields were not present.

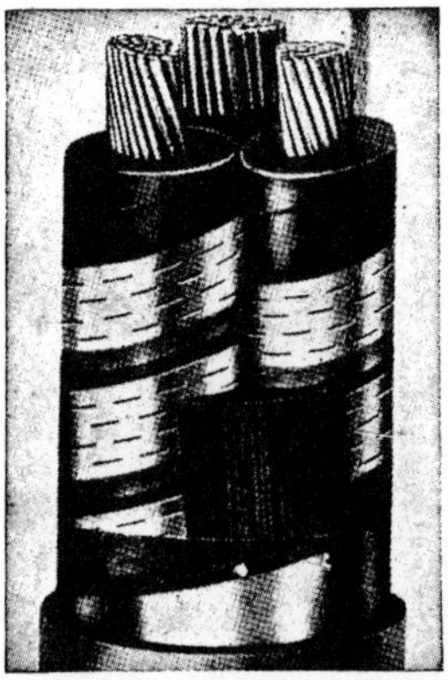

FIG. 109.—Three-conductor type H cable.

In practice the shields and sheath are connected together at frequent points along the cable so that the shunt capacitive reactance may be calculated on the basis of three single-conductor cables in which the shields represent the sheaths. The shunt capacitive reactances per phase will be the same for the positive-, negative-, and zero-sequences. In the absence of constants for sectored cables, the shunt capacitive reactance of a single-conductor cable having the same insulation thickness T and the same mean conductor insulation periphery is a very close approximation. If the mean periphery of the conductor insulation is designated P, then the diameter of the round conductor having the same insulation thickness is

$$d = \frac{P}{\pi} - T$$

and

$$\frac{T}{d} = \frac{1}{\frac{P}{\pi T} - 1} \tag{248}$$

The shunt capacitive reactance for the positive-, negative-, and zero-sequences can then be determined from the curve for single-conductor cable in Fig. 108. The same data are replotted in Fig. 110 directly as a function of $\frac{T}{P}$.

90. Submarine Cable.

For cables lying on the sea floor, at considerable depth below the surface, the return current path includes not only the sheath and earth but also the sea water. Using Rudenberg's assumption of radial symmetry of current flow in both the sea water and the earth, the current distribution will not differ from that for the current flow in either the sea water or the earth alone. The voltage drops in both mediums must be equal at all points of their surface contacts, from which it follows that their current densities at the contact surfaces, and hence the division of current between the two mediums, must vary inversely as their respective resistivities. Since from Table VII the resistivity of sea water is approximately one-hundredth that of damp earth, practically all the current flows through the sea water. Thus the impedance may be determined by using the value of ρ corresponding to sea water in the formulas previously outlined.

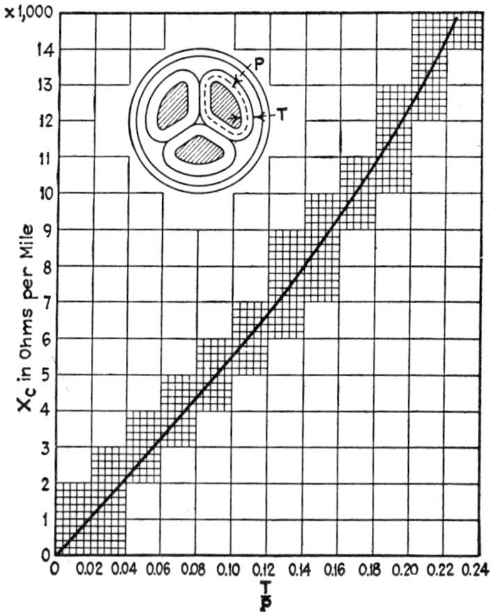

FIG. 110.—Positive-, negative-, and zero-sequence shunt capacitive reactance to neutral for type H cables in ohms per mile for 60 cycles and $k = 3.5$.

91. Constants of Typical Cables.

In Table XII are shown the constants, calculated by the methods outlined, of a number of typical cables. These values should be useful as an approximate check upon calculations, as they enable one to orient his sense of values. The effect of the elimination of sheath currents is reflected principally in the values of the positive-sequence resistance and is important only when the spacing between the single-conductor cables is large.

TABLE XII.—CONSTANTS OF TYPICAL CABLES
(In ohms per phase per mile)

Cable	Frequency	Voltage	Positive- and negative-sequence					Zero-sequence, sheath return only, solidly bonded sheaths		
			A-c. resistance†		Series reactance		Shunt capacitive reactance	Resistance†	Series reactance	Shunt capacitive reactance
			No sheath current	Include sheath current	No sheath current	Include sheath current				
Single conductor; 1,000,000 cir. mils; $\frac{36}{64}$-in. insulation; $\frac{1}{8}$-in. sheath; three cables spaced 4.125 in. horizontally	60	26,000	0.070	0.114	0.297	0.285	5,800	0.737	0.119	5,800
Single conductor; 500,000 cir. mils; 670 mils insulation; $\frac{9}{64}$-in. sheath; three cables spaced 6, 6, 8.49 in.	60	138,000	0.134	0.209	0.386	0.358	8,300	0.752	0.158	8,300
Single conductors; 250,000 cir. mils; $\frac{3}{32}$-in. insulation; $\frac{7}{64}$-in. sheath; three sheaths in contact and 0000 copper neutral wire	60	210	0.264	0.239	0.181	0.180	2,400	1.45*	0.69*	2,400
Three conductor; type *H* round; 500,000 cir. mils; $\frac{14}{64}$-in. insulation	60	15,000	0.134	.....	0.168	.....	3,600	1.74	0.220	3,600
Three conductor; belted round; 500,000 cir. mils; $\frac{8}{64}$-in. conductor insulation; $\frac{4}{64}$-in. belt insulation	60	7,500	0.134	.....	0.146	.....	4,000	1.82	0.215	7,850

* Neutral wire included in sheath circuit.
† Conductor at 65°C., sheath at 50°C.

Problems

1. Assume three single-conductor cables of 4/0 stranded copper conductors with $\frac{12}{32}$-in. insulation and a $\frac{1}{8}$-in. lead sheath. The three cables composing the circuit are arranged to give an equilateral spacing of 8 in. between centers. Determine the positive-, negative- and zero-sequence impedances, assuming the cable sheaths are insulated from ground.

2. What will be the per cent rise in voltage at the open end of a 30-mile section of the three-phase cable described in Fig. 106 when energized by balanced voltages? Use a method of calculation similar to that discussed in Sec. 78.

3. Determine the zero-sequence impedance of the three-conductor cable of Fig. 106 if the sheath is paralleled by a 4/0 stranded copper conductor located 8 in. from the center of the cable, assuming no return current in the earth.

4. A communication circuit located in a lead sheath cable parallels a power cable for 1 mile at a separation of 100 ft. Determine the voltage induced in the telephone wires for the case of a conductor-to-ground fault causing 1,000 amp. to flow in the conductor through the parallel. Assume that the power and telephone sheaths are grounded through connections of zero resistance.

5. Assume the three-conductor cable as shown in Fig. 106 with the sheath solidly grounded. The voltage impressed on this cable has a third harmonic of 180 cycles which constitutes a zero-sequence. Find the lengths of cable which will give open-circuit and short-circuit resonances to the third harmonic voltage. Assume $f = 60$ and $\rho = 200$.

6. Assume a three-phase circuit consisting of three single-conductor cables with sheath return and a source of negligible impedance. Determine the analytical expression for the current for a line-to-sheath fault by the method of symmetrical components and compare the result with that by the single-phase method. Show that the analytical expressions obtained by these two methods are identical.

CHAPTER XI

POWER SYSTEM VOLTAGES AND CURRENTS UNDER FAULT CONDITIONS*

It has been rather common practice to base the requirements of system apparatus on normal load conditions and on balanced three-phase short-circuits. The effects of unbalanced faults have been either ignored or considered to a limited extent only. However, as shown in this chapter, two systems of the same voltage, which, for particular fault locations have identical values of three-phase short-circuit current, may have radically different voltages and currents for *unbalanced* faults. Thus, the relations between the sequence impedances have an important bearing on many applications and may result in higher currents than for three-phase short-circuits or higher voltages than for ordinary operation. High currents are important because of mechanical stresses in rotating machines, transformers, reactors, circuit-breakers, and bus structures. High voltages are important in apparatus or line insulation and in lightning arresters. Combinations of voltage and current are important in circuit-breaker and relay applications. The relation between the sequence impedances or the range of voltages and currents under unbalanced fault is important in system problems such as grounding, stability, and inductive coordination with communication circuits.

The value of voltages and currents that obtain under abnormal conditions on power systems thus is of great importance in varied applications. In previous chapters the methods of analysis have been described and boundary conditions for certain limiting cases have been defined. This chapter gives a systematic presentation of the ranges of voltages and currents that may occur on a power system under fault conditions.

92. Assumptions.

In this study, the system will be assumed to be symmetrical up to the point of fault, or stated otherwise, the system will be

* Based on a paper of the same title by R. D. Evans and S. H. Wright.[58]

assumed symmetrical except for the fault itself. The types of faults which will be considered are as follows:

1. Three-phase (3ϕ):
   Either line-to-line-to-line (*L-L-L*), or
   Three lines-to-ground (3*L-G*).
2. Single line-to-ground (*L-G*).
3. Double line-to-ground (2*L-G*).
4. Line-to-line (*L-L*).

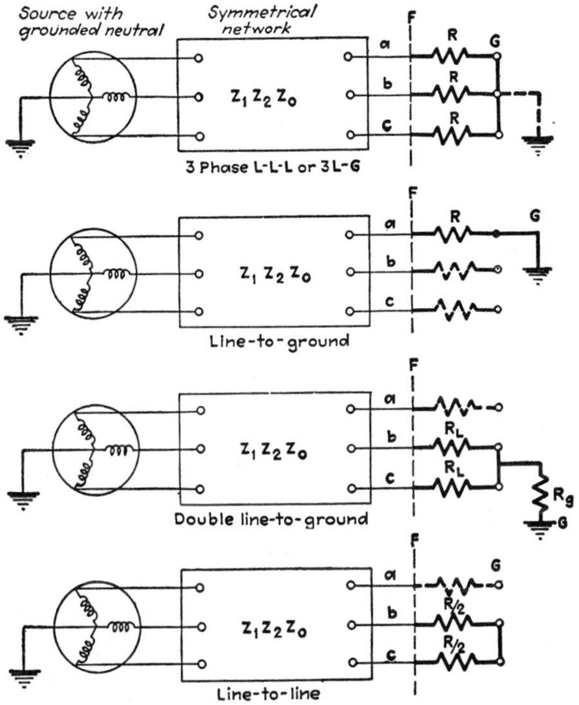

FIG. 111.—Types of faults on three-phase systems.

These types are illustrated schematically in Fig. 111. It will be noted that fault resistance as indicated by the diagram will be taken into account. For the sake of simplicity, the different types of faults are assumed to occur on particular phases, as indicated. This, of course, introduces no limitation on the analysis since the system is assumed to be symmetrical up to the point of fault. Fault currents, and the voltages between conductors and from conductors to ground at the point of fault, can be determined by the methods discussed previously. It

TABLE XIII.—FAULT CURRENTS*

Type of fault	Vector expression, effect of fault resistance included†	Magnitude when fault resistance equals $R_0 = R_1 = R_2 = 0$
3ϕ	$I_a = \dfrac{E_g}{Z_1 + R}$	$I_a = I_b = I_c = \dfrac{E_g}{X_1}$
L-L	$I_b = \dfrac{-j\sqrt{3}E_g}{Z_1 + Z_2 + R}$ $I_c = -I_b$	$I_b = I_c = \dfrac{\sqrt{3}E_g}{X_1 + X_2}$
L-G	$I_a = \dfrac{3E_g}{Z_0 + Z_1 + Z_2 + 3R}$	$I_a = \dfrac{3E_g}{X_0 + X_1 + X_2}$
$2L$-G‡	$I_b = \dfrac{-\sqrt{3}E_g}{2\Delta_v}[\sqrt{3}(Z_2 + R_L) + j(2Z_0 + Z_2 + 3R_L + 6R_g)]$ $I_c = \dfrac{-\sqrt{3}E_g}{2\Delta_v}[\sqrt{3}(Z_2 + R_L) - j(2Z_0 + Z_2 + 3R_L + 6R_g)]$ $I_g = I_b + I_c = 3I_0$ $= \dfrac{-3E_g}{\Delta_v}(Z_2 + R_L)$ $\Delta_v = (Z_1 + R_L)(Z_2 + R_L) + (Z_1 + Z_2 + 2R_L)(Z_0 + R_L + 3R_g)$	$I_b = I_c = \dfrac{\sqrt{3}E_g}{\Delta_M}\sqrt{X_0^2 + X_0X_2 + X_2^2}$ $I_g = \dfrac{3E_g}{\Delta_M}X_2$ $\Delta_M = X_1X_2 + X_0(X_1 + X_2)$

* See Fig. 111 for definition of impedances.

† If fault has reactance as well as resistance, replace the latter by the fault impedance.

‡ It may readily be shown that $I_b = I_c$ for all values of R_g when $R_L = R_1 = R_2 = 0$.

will be assumed that all the generated e.m.fs. may be reduced to a single positive-sequence set of generated voltage* E_g, and that the equivalent circuits may be represented by the series impedances Z_1, Z_2, and Z_0 for the positive-, negative-, and zero-sequence networks, respectively.

93. Formulas for Line Currents and Voltages between Line Conductors and to Ground.

The method of symmetrical components has been applied in the development of formulas for the short-circuit currents in the different phases, and the voltages between the line conductors, and between the line conductors and ground at the point of fault. The more important of these formulas, which include the effect of fault resistance, are summarized in Table XIII for currents and Table XIV for voltages.

94. Basis for Current and Voltage Curves.

The general formulas for currents and voltages referred to in the preceding section are complicated to such an extent that it is difficult to visualize readily the range of voltages and currents that may obtain. This is due to the fact that there are several types of faults to be considered, and that for each type of fault there is a wide range of values which the impedance of the system elements may have. Further complication arises from the fact that in machines there are two values of X_1 which apply during transient conditions.

Range of the Sequence Impedances. In general, positive- and negative-sequence impedances of a system are of the same order of magnitude, whereas the zero-sequence impedance may vary through a very wide range from being very small to very large in comparison with the positive-sequence impedance. Also, in general, the positive-sequence resistance R_1 and the negative-sequence resistance R_2 are small in comparison with the positive- and negative-sequence reactances. Consequently, the effect of these two resistances on the magnitude of the fault voltages and currents is relatively small. Arc resistance may be very low, particularly during the interval following flashover, and, of course, may be zero, as in the case of a metallic connection.

* The method of obtaining a single equivalent generated e.m.f. is discussed in Sec. 113, Chap. XII.

TABLE XIV.—FAULT VOLTAGES*

Type of fault	Vector expression, effect of fault resistance included†	Magnitude when fault resistance equals $R_0 = R_1 = R_2 = 0$
3ϕ	$E_a = E_g \dfrac{R}{Z + R}$	$E_a = 0$
L-L	$E_a = E_g \dfrac{2Z_2 + R}{Z_1 + Z_2 + R}$	$E_a = E_g \dfrac{2X_2}{X_1 + X_2}$
	$E_b = -E_g \dfrac{\frac{1}{2}R + j\frac{\sqrt{3}}{2}R + Z_2}{Z_1 + Z_2 + R}$	$E_b = E_c = E_g \dfrac{X_2}{X_1 + X_2}$
	$E_c = -E_g \dfrac{\frac{1}{2}R - j\frac{\sqrt{3}}{2}R + Z_2}{Z_1 + Z_2 + R}$	$E_C = E_g \dfrac{3X_2}{X_1 + X_2}$
L-G	$E_a = E_g \dfrac{3R}{Z_0 + Z_1 + Z_2 + 3R}$	$E_a = 0$
	$E_b = \dfrac{-\sqrt{3}E_g}{2}\left[\dfrac{\sqrt{3}(Z_0 + R) + j(Z_0 + 2Z_2 + 3R)}{Z_0 + Z_1 + Z_2 + 3R}\right]$	$E_b = E_c = \sqrt{3}E_g \dfrac{\sqrt{X_0^2 + X_0X_2 + X_2^2}}{X_0 + X_1 + X_2}$
	$E_c = \dfrac{-\sqrt{3}E_g}{2}\left[\dfrac{\sqrt{3}(Z_0 + R) - j(Z_0 + 2Z_2 + 3R)}{Z_0 + Z_1 + Z_2 + 3R}\right]$	$E_A = \sqrt{3}E_g \dfrac{X_0 + 2X_2}{X_0 + X_1 + X_2}$
2L-G‡	$E_a = \dfrac{3E_g}{\Delta_v}(Z_2 + R_L)(Z_0 + R_L + 2R_g)$	$E_a = \dfrac{3E_g}{\Delta_M}X_0X_2$
	$E_b = \dfrac{-\sqrt{3}E_g}{2\Delta_v}[\sqrt{3}(Z_2 + R_L)(R_L + 2R_g) + jR_L(2Z_0 + Z_2 + 3R_L + 6R_g)]$	$E_b = E_c = 0$
	$E_c = \dfrac{-\sqrt{3}E_g}{2\Delta_v}[\sqrt{3}(Z_2 + R_L)(R_L + 2R_g) - jR_L(2Z_0 + Z_2 + 3R_L + 6R_g)]$	$E_C = E_a$
	$\Delta_v = (Z_1 + R_L)(Z_2 + R_L) + (Z_1 + Z_2 + 2R_L)(Z_0 + R_L + 3R_g)$	$\Delta_M = X_1X_2 + X_0(X_1 + X_2)$

* See Fig. 111 for definition of impedances.

† If fault has reactance as well as resistance, replace the latter by the fault impedance.

‡ It may readily be shown that $E_b = E_c$ for all values of R_g, when $R_L = R_1 = R_2 = 0$.

In view of this, and the complication resulting from the consideration of the effect of arc resistance, it has been deemed advisable to consider only the case of zero resistance in the fault. In the case of zero-sequence, however, the resistance R_0 may be small or may be very large as a result of a neutral grounding resistor. For this reason it is necessary to arrange the curves to cover a wide range of zero-sequence resistance and reactance.

As a result of the study of the range of values for impedances, the number of variables has been so reduced that sets of curves may be plotted with values of zero-sequence impedance as abscissas, and voltages or currents as ordinates. Static apparatus, as previously pointed out, has identical values of positive- and negative-sequence impedances. This also represents a kind of mean value for rotating machines, especially for the reactances effective at the instant of short-circuit. This suggests a set of curves plotted for equal values of positive- and negative-sequence reactances, supplemented by additional sets of curves to indicate the probable maximum and minimum values of the ratio of negative- to positive-sequence impedance, taking into account the characteristics of all types of synchronous machines.

Ratio of Sequence Impedances Used in Curves. The basis for selecting the values of these ratios will now be considered. The principal machine reactances that apply to transient conditions are zero-sequence reactance X_0, the negative-sequence reactance X_2, and the positive-sequence reactance X_1; X_1 may be either the subtransient or transient reactance, depending, respectively, on whether the initial high decrement component of current is considered or neglected. The ratio of X_2 to X_1 for commercial machines will normally lie within the range of 0.5 to 1.5, though it is quite possible with special machines to somewhat exceed this range. The higher ratio of X_2 to X_1 is obtained in machines without dampers, and for this class of machines the average ratio may be taken as 1.5. Machines with dampers or their equivalent, such as solid-rotor turbine generators, always have X_2 less than the transient reactance (X_1), and the ratio of X_2 to X_1 for machines of this class may be as low as 0.5. It may be well to point out that all machine reactances vary to some extent with saturation; also, that the value of the negative-sequence reactance of a machine varies somewhat with external reactance because of distortion resulting

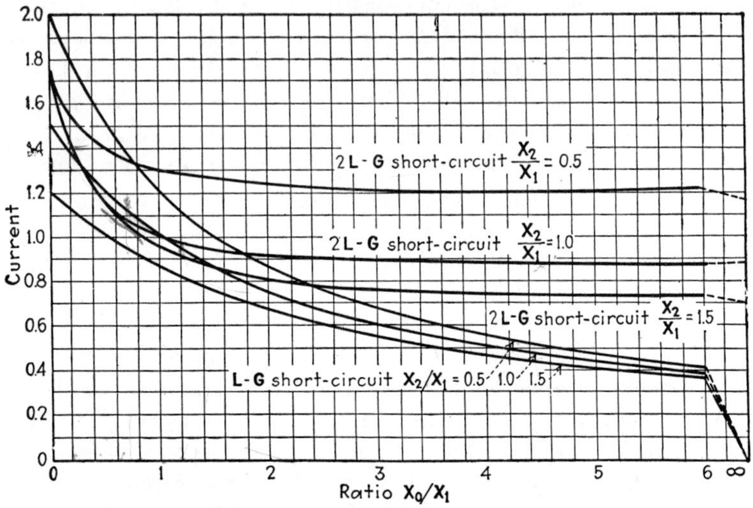

FIG. 112.—Curves of fault currents *vs.* system reactances for single and double line-to-ground faults. Each curve is labeled to indicate the type of fault and the ratio of X_2/X_1. All currents are expressed as a ratio to the three-phase short-circuit current. For these curves, all resistances are assumed equal to zero.

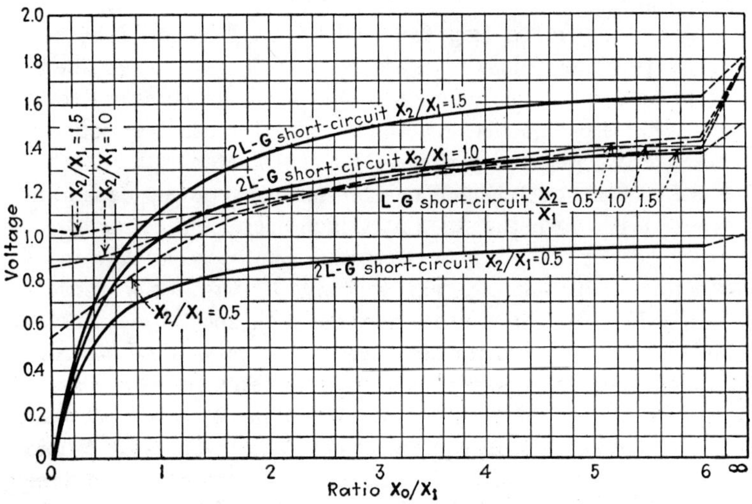

FIG. 113.—Curves of fault voltages *vs.* system reactances for single and double line-to-ground faults. Each curve is labeled to indicate the type of fault and the ratio of X_2/X_1. The voltages are from line-to-ground and are expressed as a ratio to the normal line-to-neutral voltages. For these curves, all resistances are assumed equal to zero.

from the inequality of direct and quadrature axis subtransient reactances. However, except perhaps for special machines or cases where high accuracy is required for a wide range of external circuit reactance, it is permissible to use a single value of any one machine reactance. These considerations have led to the selection of sets of curves for three ratios of X_2 to X_1, namely, 0.5, 1.0, and 1.5.

Basis for Comparing Systems. The final consideration in plotting these curves for unbalanced conditions was to find a basis suitable for comparing all systems. For this purpose the following reference quantities have been taken: *for line-to-ground voltages, the normal line-to-neutral voltage; for line-to-line voltages, the normal line-to-line voltage; for currents, the three-phase short-circuit current; and for impedances, the positive-sequence reactance.* Thus all values are plotted not in units such as volts, amperes, and ohms, but simply as ratios or decimal fractions of the reference quantities. All voltages are computed for the point F (not G) of the diagrams in Fig. 111.

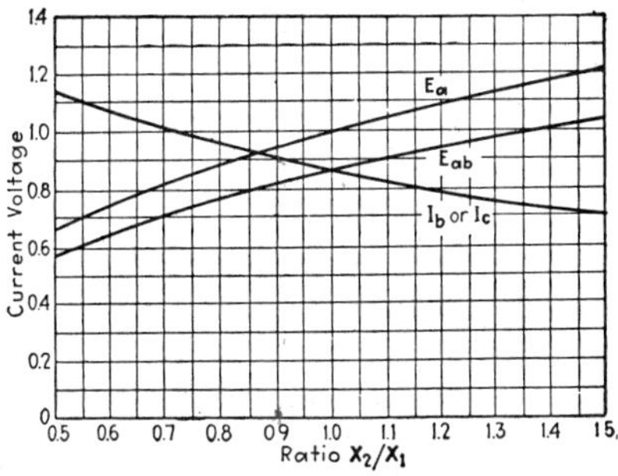

FIG. 114.—Curves of fault voltages and currents *vs.* system reactances for line-to-line faults. Line-to-ground and line-to-line voltages are expressed as ratios to their respective normal values. Current is expressed as a ratio to the three-phase short-circuit current. All resistances are assumed equal to zero.

95. Fault Current and Voltage Curves.

Curves prepared in accordance with the preceding discussion are shown in Figs. 112 to 116 inclusive. Figures 112 and 113 show the ranges of line currents and line-to-ground voltages respectively for single and double line-to-ground faults, on systems for which the three sequence reactances only require consideration; *i.e.*, fault resistances, R_0, R_1, and R_2 are taken equal to zero. Figure 114 shows the range of line-to-ground voltages, line-to-line voltages, and line currents that obtain for line-to-line short-circuits. Figures 115 and 116 show the range of currents and voltages, respectively, that obtain for single and double line-to-ground faults on systems for which the zero-sequence resistance requires consideration; the curves are plotted with the ratio $\frac{R_0}{X_1}$ as abscissa, with families of curves to

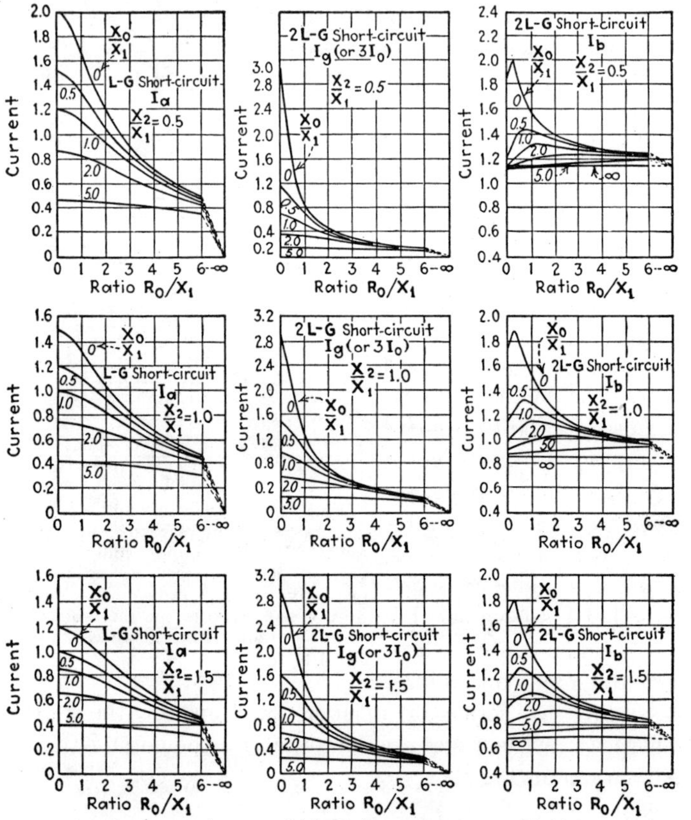

(a) Line current Single line-to-ground fault (b) Ground current Double line-to-ground fault (c) Line current Double line-to-ground fault

FIG. 115.—Curves of fault currents *vs.* system impedances. The legend with each group of curves indicates the type of fault, the current plotted, and the ratio of X_2/X_1. The individual curves in each group are for various values of the ratio of X_0/X_1. All currents are expressed as a ratio to the three-phase short-circuit current.

cover the range of the ratio of $\frac{X_0}{X_1}$, and groups of curves to cover the three ratios of $\frac{X_2}{X_1}$ of 0.5, 1.0, and 1.5.

96. Discussion of Curves.

The ranges of line currents, ground currents, line-to-ground voltages, and line-to-line voltages that exist on a system under unbalanced fault conditions as shown by the curves have been summarized in Table XV. One of the most important facts brought out by these curves is that the *line-to-ground* voltage may rise to *twice* the normal line-to-neutral voltage, *i.e.*, to a value

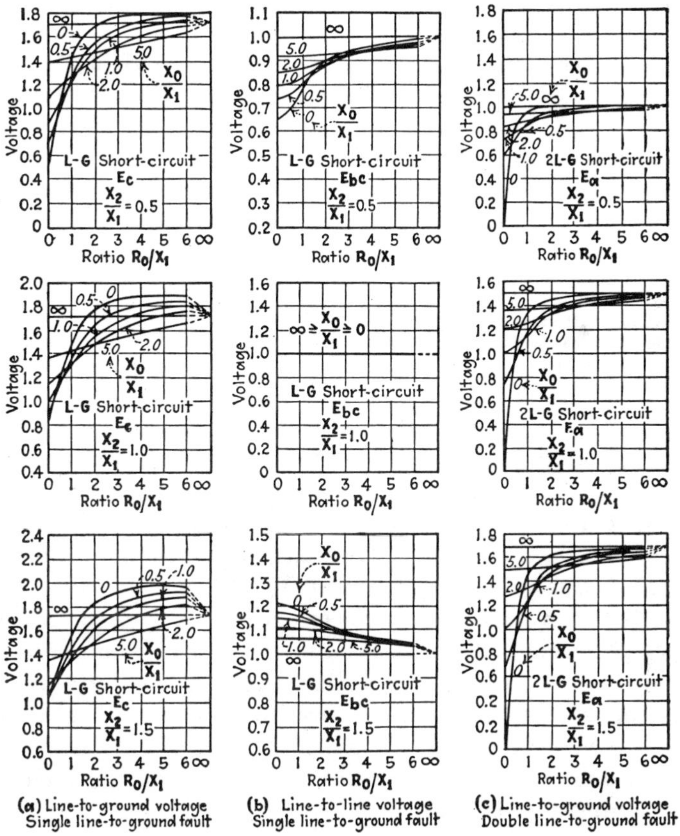

FIG. 116.—Curves of fault voltages *vs.* system impedances. The legend with each group of curves indicates the type of fault, the voltage plotted, and the ratio of X_2/X_1. The individual curves in each group are for the various values of the ratio of X_0/X_1. Line-to-ground and line-to-line voltages are expressed as ratios to their respective normal values.

somewhat greater than the normal *line-to-line* voltage—a fact that is not generally appreciated. It is well to note that the ratios of line current and ground current to the three-phase short-circuit current may be as great as 2.0 and 3.0, respectively; also, that the ground current of a double line-to-ground fault

TABLE XV.—RANGE OF FAULT CURRENTS AND VOLTAGES

Type of fault	Current ratio*		Voltage ratio	
	Line	Ground	*L-G*†	*L-L*‡
L-L	0.69–1.15	....	0.67–1.20	0.58–1.04
L-G	0.00–2.00	0.00–2.00	0.58–1.97	0.67–1.20
2*L-G*	0.69–2.02	0.00–3.00	0.00–1.80§	0.00–1.04§

Based on

$$\frac{R_1}{X_1} = \frac{R_2}{X_1} = 0; \quad 0 \leq \frac{R_0}{X_1} \leq \infty$$

and

$$\frac{X_2}{X_1} = 0.5 \text{ to } 1.5; \quad 0 \leq \frac{X_0}{X_1} \leq \infty$$

* Reference is the three-phase short-circuit current.

† Reference is the normal line-to-neutral voltage.

‡ Reference is the normal line-to-line voltage.

§ These voltages are identical; only the line-to-ground voltage curves have been plotted.

may be 50 per cent greater than the maximum value of a single line-to-ground fault.

Examination of the curves shows that the maximum values of fault currents are obtained generally with low values of X_0 and X_2, and, correspondingly, that the maximum values of voltages occur with high values of X_0 and X_2. However, this is not the case for zero-sequence resistance R_0, as the maximum values of currents and voltages are obtained with intermediate values of R_0. The effect of adding zero-sequence resistance to a system having no resistance is, first, to increase the line current on a double line-to-ground fault; subsequent increase of R_0 reduces line currents as well as ground currents and increases line-to-ground voltages.

97. Systems Giving Maximum Voltage or Current Ratios.

The power-system layouts which give the maximum values of voltage or current ratios are accurately defined by the sequence impedance ratios used in plotting the curves. However, it may be well to describe the limiting conditions in other terms. The simplest illustration is that of several generators connected to a bus; a low ratio of X_0 to X_1 is obtained for one machine or for several machines in parallel, provided that in every case the machine neutral is solidly grounded and provided that all generators have two-thirds pitch windings. For such a case the zero-sequence reactance may be as low as 1 per cent, and the ratio of X_0 to X_1 may readily be less than 0.2. A high ratio of X_0 to X_1 may, of course, be obtained by introducing a neutral reactor. Another method is to ground the neutral of only one of several machines operating in parallel; for such a case the machines should have a full pitch winding and therefore a relatively high value of X_0. Considering a fault on the bus, the effective value of the positive-sequence reactance is $\frac{X_1}{n}$, where X_1 is the reactance per machine and n the number of machines; similarly, the effective value of the negative-sequence reactance is $\frac{X_2}{n}$; and, finally, the effective value of the zero-sequence reactance is, of course, $\frac{X_0}{1}$, since only one machine has its neutral grounded. Thus, it is clear that a relatively high ratio of X_0 to X_1 may be obtained if there are several machines connected to a bus and only one machine has its neutral grounded. Incidentally, under these conditions the single line-to-ground fault current may become several times the three-phase short-circuit current of a single machine, which may produce unusually high stresses in the machine whose neutral is grounded.

Another illustration of a case where the ratio $\frac{X_0}{X_1}$ is low, is that of a fault on the grounded-star side of delta-star step-up transformers supplied by waterwheel type generators. The zero-sequence reactance is merely that of the transformers, while the positive-sequence reactance is the transformer reactance plus the reactance of the waterwheel generators, which is normally several times that of the transformers. For faults on the transmission line remote from the grounded star-delta transformers, the zero-sequence reactance may become relatively large and the ratio of X_0 to X_1 may be considerably greater than unity.

A high ratio of X_2 to X_1 for synchronous machines alone occurs for machines without dampers, while a low ratio occurs for machines with dampers (considering transient reactance). For system faults this ratio has correspondingly high or low values, depending upon the type of machine, when the fault occurs at machine terminals; this ratio approaches unity for faults remote from machine buses, *i.e.*, with external reactance in series with the machines.

98. Effect of Method of Grounding. Zero-sequence Impedance Ratio.

Examination of the curves of Figs. 115 and 116 shows that the ranges of voltages and currents under unbalanced fault conditions are controlled principally by the ratio of Z_0 to X_1; *i.e.*, by the method of system grounding and the values of the grounding impedances.

The common classification of systems with respect to grounding is:

1. Direct-grounded.
2. Grounded through impedance:
   a. Resistor.
   b. Reactor.
3. Free neutral.*

Systems are usually said to be direct-grounded if the neutrals of one or more generators or transformers are connected directly to ground. Systems grounded through impedances may approach direct-grounded systems on one hand and free neutral systems on the other, depending upon the magnitude of the grounding impedance. In fact, the zero-sequence impedance of a direct-grounded system may actually be higher than that of a system of similar capacity and extent, which is grounded through a neutral impedance of low value. Obviously, a classification of systems into grounded or ungrounded without further restrictions is unsatisfactory because it takes into account only the impedance between the ground and the neutral points of apparatus, instead of the total effective zero-sequence impedance which includes line and apparatus impedance as well as that of grounding resistors and reactors. A more accurate classification is suggested by the curves of Figs. 115 and 116. These curves cover the range of system grounding from a direct-grounded system to a free neutral system with negligible capacitance, and they suggest that the ratio of Z_0 to X_1 be used as a criterion,

* For the purpose of this classification, system grounding through neutral reactance which resonates with line capacitance, such as the Petersen coil, Bauch transformer, or Jonas dissonance coil, may be viewed as free neutral systems, since the object of these reactance devices is to nullify line capacitance. In this discussion no consideration is given to the phenomena of arcing grounds.

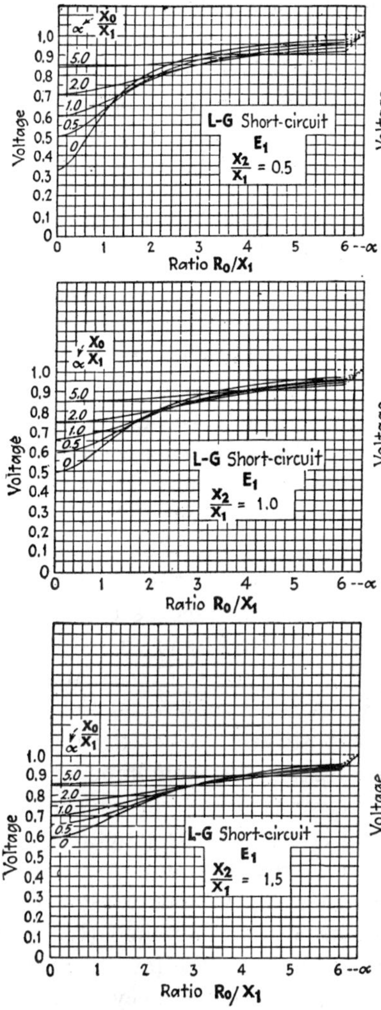

FIG. 117.—Curves of positive-sequence voltage at the fault *vs.* system impedances for single line-to-ground fault.

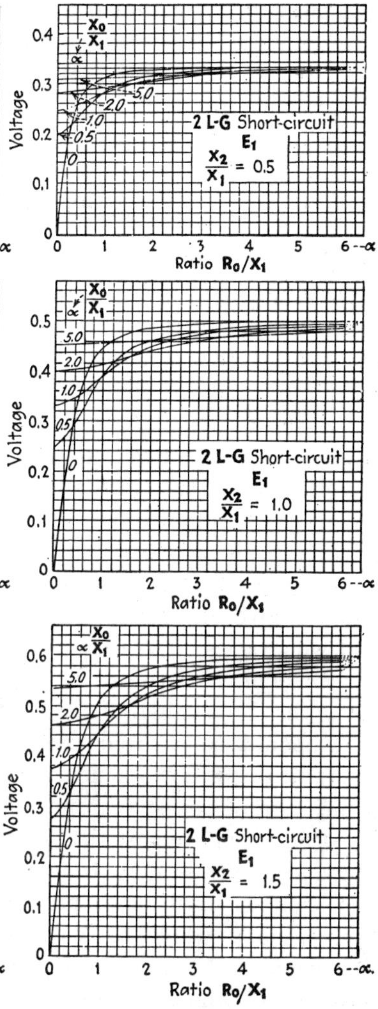

FIG. 118.—Curves of positive-sequence voltages at the fault *vs.* system impedances for double line-to-ground fault.

The legend with each group of Figs. 117 and 118 indicates the ratio of X_2/X_1. The individual curves in each group are for various values of the ratio of X_0/X_1 Voltages are expressed as ratios to the normal line-to-neutral voltage.

or a measure of the *degree of grounding*, low values of this ratio corresponding to a *solidly grounded* system and high values to a free neutral system. Hence, from the standpoint of system performance, the criterion is not whether there is an impedance between one or more neutral points and ground, but whether the ratio of X_0 or R_0 to X_1 is high or low. The ratio of Z_0 to X_1 is the **zero-sequence impedance ratio.**

99. Positive-sequence Voltage at the Fault.

Different methods of grounding a system and different values of the zero-sequence impedance ratio have rather complicated effects on the positive-sequence voltage used in stability analysis. Since any type of unbalanced fault may be replaced by a three-phase short-circuit through a symmetrical three-phase impedance, it has been suggested that the positive-sequence component of voltage E_1 at the fault is a measure of the stability of a system under fault conditions.[(28),(35)] Accordingly, Figs. 117 and 118 have been plotted to show the range of E_1 for single and double line-to-ground faults, respectively. These curves have been plotted with the ratios of E_1 to normal voltage as ordinates, and with ratios of R_0 to X_1 as abscissas; families of curves cover the range of the ratios of X_0 to X_1, while groups of these families cover the three values of the ratio of X_2 to X_1 of 0.5, 1.0, and 1.5.

It should be realized that the positive-sequence voltage at the fault is, for given generator voltages and system impedances, a measure of the synchronizing power that may be exchanged between machines past the point of fault. The drop in positive-sequence voltage may cause an underexcited and overloaded synchronous machine to pull out of step, without causing the system as a whole to lose synchronism. Consequently, the reduction of the positive-sequence voltage at the fault is also a measure of the effect of grounding on the tendency of these machines or of the system as a whole to pull out of step.

Problems

1. A power system has the following impedances as measured from the 110-kv. line:

$$\left.\begin{aligned} Z_1 &= + j20 \text{ per cent} \\ Z_2 &= + j20 \text{ per cent} \\ Z_0 &= 40 + j10 \text{ per cent} \end{aligned}\right\} \text{based on 200,000 kva.}$$

Find the largest fault current and the highest line-to-ground voltage for (*a*) single line-to-ground fault; (*b*) double line-to-ground fault; (*c*) line-to-line fault; (*d*) three-phase fault. Make use of the curves of Figs. 114, 115, and 116.

2. Plot a curve for the neutral voltage of power systems for a line-to-ground fault assuming that the zero-sequence resistance is all in a grounding resistor and all the zero-sequence reactance is in a delta-star transformer. Use data from curves of Chap. XI assuming $X_2 = X_1$ and $X_0 = 0.5X_1$.

3. Circuit-breaker duty is sometimes measured in kilovolt-amperes obtained by multiplying the r.m.s. value of the current through a breaker pole before opening and the r.m.s. value of the voltage that appears across the breaker after opening. Assume that the system of Prob. 1 is subjected to a zero-resistance three-phase fault to ground. Assuming that the poles open individually in the order of least duty expressed in kilovolt-amperes and that the high frequency and wave distortion may be ignored, find the duty on each pole.

4. Assume the same conditions as in Prob. 3 but without resistance, and determine the circuit-breaker duty opening one pole at a time as before. The above system impedance ratios are those which obtain at the sending end of a transmission line supplied by a grounded star-delta transformer. Explain why one pole of a breaker on such a system may open appreciably ahead of the two others.

5. A generator and a step-up transformer connected delta on the low-voltage side and grounded-star on the transmission-line side has impedances of $Z_1 = +j40$ per cent, $Z_2 = +j35$ per cent, $Z_0 = +j10$ per cent. The transmission line has the following impedances for a 10-mile section: $Z_1 = +j5$ per cent, $Z_2 = +j5$ per cent, $Z_0 = +j15$ per cent. Assume that it is desired to limit the line-to-ground voltage to 120 per cent of normal for a line-to-ground fault. What is the length of line that can be protected by grounding the system at the step-up transformer station? In this calculation neglect the receiver-end characteristics.

CHAPTER XII

SIMULTANEOUS FAULTS

Experience has shown that a certain number of the faults on power systems occur concurrently at different points in the circuits. These "simultaneous" faults may be encountered as the result of a ground at one location raising the voltage on a sound phase so that flashover occurs at a second point. They may also occur on separate circuits which are physically close together, as, for example, in the case of two circuits on the same structure for which the flashovers are produced by the same lightning disturbance.

Simultaneous faults are important even though relatively infrequent because relay systems giving satisfactory operation for a fault at a single location may fail to isolate simultaneous faults. A scheme for obtaining correct relay operation under these conditions has been proposed by R. M. Smith.* He showed the applicability of this scheme to a particular case by calculations of fault currents and voltages using the methods developed by the authors and presented in this chapter in an amplified form. G. Oberdorfer[174] published the first analysis of this problem in 1930 and this was followed by a more comprehensive treatment of the same subject by Edith Clarke[70] in 1931.

This chapter will present in some detail a method for calculating simultaneous faults at two locations, after which the method of attack for more than two simultaneous faults will be outlined. The discussion reviews the special case of a fault at a single location with several sources of e.m.f. differing in magnitude or phase position.

100. Simplest Equivalent Network.

Before taking up the solution of simultaneous faults, it is desirable to discuss briefly certain methods of representing and simplifying networks. The simplest general form of network

* Suggestions for Avoiding Faulty Action of Ground Relay, *Elec. World*, vol. 95, p. 1092, May 31, 1930.

for a particular problem is determined by the number of terminals whose identity must be retained. For the calculation of simultaneous faults these terminals are of three types, namely, (1) the fault locations, (2) the e.m.f. sources whose magnitude and phase relation are not identical, and (3) the neutral of the system, required for cases involving shunt load. Minimum number of branches in the simplest general network is obtained when branches connect all terminals without additional star points or parallel branches. Thus 10 branches are, in general, required to represent a network involving two separate sources

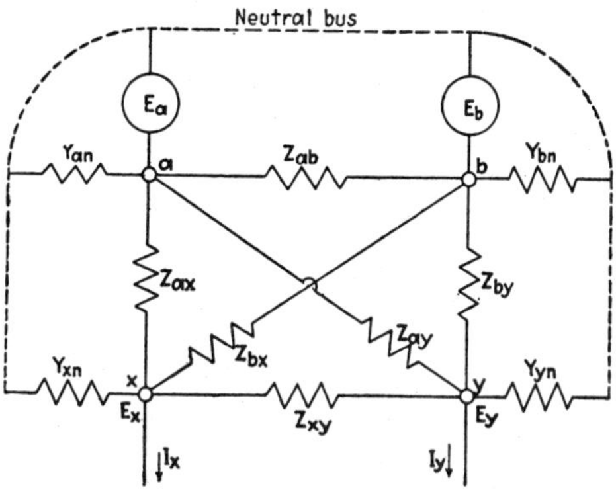

(a)

$$I_x = \frac{E_a}{Z_{ax}} + \frac{E_b}{Z_{bx}} + \frac{E_y}{Z_{xy}} - E_x\left(\frac{1}{Z_{ax}} + \frac{1}{Z_{bx}} + \frac{1}{Z_{xy}} + Y_{xn}\right)$$

$$I_y = \frac{E_a}{Z_{ay}} + \frac{E_b}{Z_{by}} + \frac{E_x}{Z_{xy}} - E_y\left(\frac{1}{Z_{ay}} + \frac{1}{Z_{by}} + \frac{1}{Z_{xy}} + Y_{yn}\right)$$

(b)

FIG. 119.—Simplest equivalent network for a system with four line terminals and one neutral terminal. (a) Diagram of connections; (b) equations for terminal currents expressed in terms of terminal voltages and network constants.

of e.m.f., two fault locations, and shunt loads from line-to-neutral. Such a network is illustrated in Fig. 119.

101. Determination of the Branches of the Form Z_{ax}.

The series impedance branches of the equivalent network for any specified number of terminals will be designated by the impedance symbol Z with two subscripts to denote the ends of the branch; thus Z_{ax} represents the impedance of the branch between the points a and x. The shunt loads may be represented by the admittance symbol Y with similar subscripts, such as Y_{an}.

Determination of Branch Impedances by the Calculating Board. The various branches of the equivalent network may conveniently be determined by means of the alternating-current calculating board. Consider the network of Fig. 120(a). If the identity of terminals a, b, x, and y is to be retained, the simplest equivalent network is that shown in (*c*). The branches may be obtained by applying voltage to one terminal and short-circuiting the remaining terminals, then measuring the current flowing at each terminal as illustrated in (*b*). The vector ratio of the source voltage to the current in any terminal gives

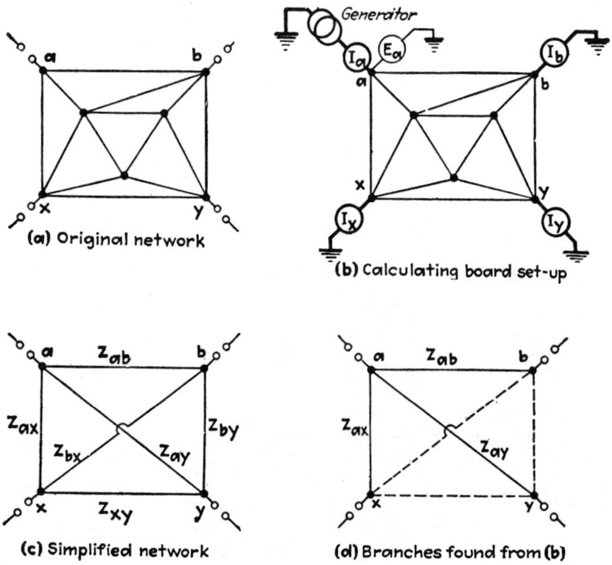

FIG. 120.—Method of simplifying network particularly suited to the alternating-current calculating board.

the impedance of the branch between that terminal and the source, thus

$$\frac{E_a}{I_x} = Z_{ax} \tag{249}$$

From the measurements indicated in (*b*), the network branches shown by full lines in (*d*) are obtained. This procedure must be repeated using other terminals as the source until all of the branches of the equivalent network are obtained. It will be noted that checks will be obtained on the branch impedances.

The network of Fig. 120(*a*) showed only series branches between the terminals *a*, *b*, *x*, and *y*. If shunt loads were taken from this network, a maximum of four additional branches in the equivalent circuit from the terminals to neutral would be required. These shunt branches may be obtained by measuring the current in the source as well as in the remaining terminals.

The value of the branch admittance is determined by the expression

$$Y_{an} = \frac{I_a - (I_b + I_x + I_y)}{E_a} \tag{250}$$

The admittance of the remaining branches may be obtained by using other terminals as the sources.

Determination of Branch Impedances by Analytical Methods. The simplest equivalent network may also be obtained analytically in a similar manner or by means of network reduction methods. The latter procedure will be illustrated subsequently in connection with the determination of the positive-sequence impedance constants.

102. Circuit Equations for Simplest Equivalent Network.

Probably the principal advantage of reducing the network to its simplest equivalent form is the ease with which the basic equations relating voltages and currents at the terminals may be obtained since they may be written from inspection. For present purposes it is convenient to assume the positive direction of current flow as outward at every terminal and to express these currents in terms of the voltages of the several terminals and the network constants. Thus for the network of Fig. 119, the currents drawn from the points x and y may be written as follows:

$$\left.\begin{aligned} I_x &= \frac{E_a - E_x}{Z_{ax}} + \frac{E_b - E_x}{Z_{bx}} + \frac{E_y - E_x}{Z_{xy}} - E_x Y_{xn} \\ I_y &= \frac{E_a - E_y}{Z_{ay}} + \frac{E_b - E_y}{Z_{by}} + \frac{E_x - E_y}{Z_{xy}} - E_y Y_{yn} \end{aligned}\right\} \tag{251}$$

These equations may be rewritten as follows:

$$\left.\begin{aligned} I_x &= \frac{E_a}{Z_{ax}} + \frac{E_b}{Z_{bx}} + \frac{E_y}{Z_{xy}} - E_x\left(Y_{xn} + \frac{1}{Z_{ax}} + \frac{1}{Z_{bx}} + \frac{1}{Z_{xy}}\right) \\ I_y &= \frac{E_a}{Z_{ay}} + \frac{E_b}{Z_{by}} + \frac{E_x}{Z_{xy}} - E_y\left(Y_{yn} + \frac{1}{Z_{ay}} + \frac{1}{Z_{by}} + \frac{1}{Z_{xy}}\right) \end{aligned}\right\} \tag{252}$$

Because of the symmetry of these equations it is readily apparent that similar expressions may be obtained for additional terminals.

103. Impedance-drop Constants of the Form D_{xx} and D_{xy}.

For certain calculations it is desirable to carry the network analysis a step farther and to define the set of constants known as impedance-drop constants* of the form D_{xx} and D_{xy}. Thus, for a source at a and for current drawn from the network at the points x and y, the voltages at these points may be written as

$$\left.\begin{aligned} E_x &= E_a - D_{xx}I_x - D_{xy}I_y \\ E_y &= E_a - D_{xy}I_x - D_{yy}I_y \end{aligned}\right\} \quad (253)$$

The determination of the drop constants may be made by means of the alternating-current calculating board as illustrated

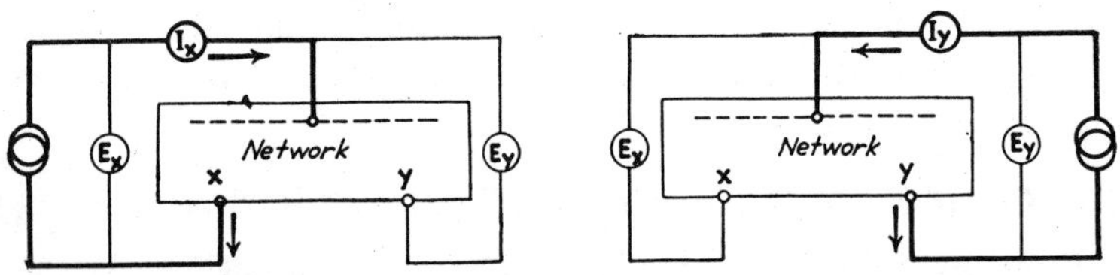

$$D_{xx} = \frac{E_x}{I_x} \qquad D_{yy} = \frac{E_y}{I_y}$$

$$D_{xy} = \frac{E_y}{I_x} \qquad D_{xy} = \frac{E_x}{I_y}$$

(a) (b)

FIG. 121.—Determination of impedance-drop constants by means of the alternating-current calculating board.

in Fig. 121. Impress voltage at x and measure the current I_x and the voltages E_x and E_y. The drop constants are

$$D_{xx} = \frac{E_x}{I_x} \quad (254)$$

and

$$D_{xy} = \frac{E_y}{I_x} \quad (255)$$

By applying voltage at y and measuring the voltages E_y and E_x and the current I_y, the impedance-drop constants obtained are

$$D_{yy} = \frac{E_y}{I_y} \quad (256)$$

$$D_{xy} = \frac{E_x}{I_y} \quad (257)$$

The drop constant D_{xx} is merely the impedance of the network between the point x and neutral, while the drop constant D_{xy}

* Note that the symbols Z_{xx} and Z_{xy} are not infrequently used instead of D_{xx} and D_{xy}. The notation here used has been chosen to avoid confusion with the symbol for the impedance of the branch between the terminals denoted by subscripts as discussed previously.

is the mutual impedance drop from the neutral to the point y due to the flow of unit current into the network at the neutral and out at the point x.

104. Discussion of Network Restraints.

For the solution of any short-circuit problem it is convenient to make use of the restraints imposed by the impedance constants of the different sequence networks and also by the type of fault. For the case of two simultaneous faults it becomes necessary to determine the value of 12 unknown quantities, namely, the three sequence currents and the three sequence voltages at the two fault locations. For this purpose 12 independent equations must be set up. Since the system is assumed to be balanced, the three sequence networks will be independent of each other and therefore it will be possible to obtain:

1. From a consideration of the positive-sequence network impedances, two equations relating the positive-sequence voltages and currents at each fault location.
2. From a consideration of the negative-sequence network impedances, two equations relating the negative-sequence voltages and currents at each fault location.
3. From a consideration of the zero-sequence network impedances, two equations relating the zero-sequence voltages and currents at each fault location.

In this manner six of the requisite equations may be obtained.

At each point of fault, three imaginary leads may be brought out and the particular conditions for the fault imposed upon them. Regardless of the characteristics of the fault, the restraints imposed will always give three equations relating sequence voltages and currents at that point, and since two faults are involved six additional equations will thus be obtained. It follows then that this procedure produces a sufficient number of equations to permit solving of the unknown quantities.

105. Two Simultaneous Faults with Two Sources of E.M.F.

The solution of the case of two simultaneous faults with two sources of e.m.f. will now be undertaken. The solution will be illustrated by application to a particular case. For this purpose consider the network of Fig. 122(a) which is identical with that of Fig. 23, Chap. IV, with the point x corresponding to the point F. The two faults are assumed to occur at x at an intermediate point on one line, and at the point y on the right-hand bus. Let it be assumed further that the generated voltages of the machines A and B are identical, each being equal to E_{a1}, and that the generated voltages of the machines C and D are identical, each being equal to E_{c1}. The shunt loads will be neglected for this particular illustration so that it is necessary to consider only four terminal points, the voltages of which are E_{a1}, E_{c1}, E_x, and E_y. The sequence components of the currents and the

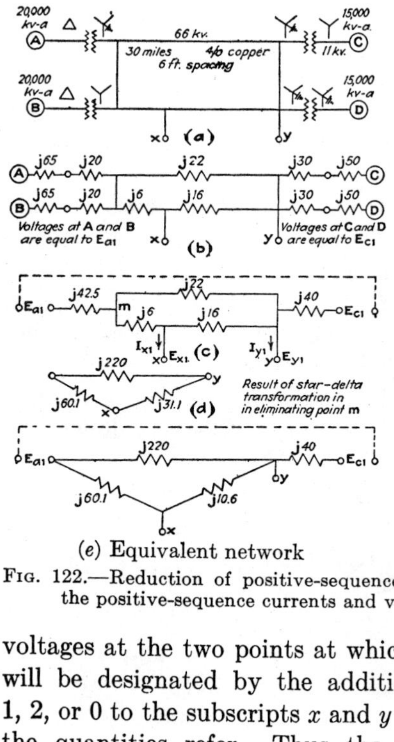

(e) Equivalent network

(f) General equations

$$I_{x1} = \frac{E_{a1}}{Z_{ax1}} + \frac{E_{c1}}{Z_{cx1}} + \frac{E_{y1}}{Z_{xy1}} - E_{x1}\left(\frac{1}{Z_{ax1}} + \frac{1}{Z_{cx1}} + \frac{1}{Z_{xy1}}\right)$$

$$I_{y1} = \frac{E_{a1}}{Z_{ay1}} + \frac{E_{c1}}{Z_{cy1}} + \frac{E_{x1}}{Z_{xy1}} - E_{y1}\left(\frac{1}{Z_{ay1}} + \frac{1}{Z_{cy1}} + \frac{1}{Z_{xy1}}\right)$$

(g) Equations for the specific network

$$I_{x1} = -j0.0166E_{a1} - j0.0943E_{y1} + j0.111E_{x1}$$

$$I_{y1} = -j0.0045E_{a1} - j0.0250E_{c1} - j0.0943E_{x1} + j0.124E_{y1}$$

FIG. 122.—Reduction of positive-sequence network to two equations relating the positive-sequence currents and voltages at the two fault points.

voltages at the two points at which faults are assumed to occur will be designated by the addition of the subscript numbers 1, 2, or 0 to the subscripts x and y to indicate the point to which the quantities refer. Thus the positive-sequence component of fault current at the point x will be designated as I_{x1}. Similarly the sequence impedance of the various branch impedances is indicated by the addition of the numerals 1, 2, or 0, for example, Z_{ax1}.

106. Restraints Imposed by the Sequence Network Constants.

It was pointed out in the previous section that six of the 12 equations required for the solution of the simultaneous faults are determined from the restraints imposed by the impedance of the sequence networks. The determination of these sequence-impedance constants will now be described.

Positive-sequence Network Constants. The positive-sequence impedance diagram of the assumed system is shown in Fig. 122(*b*). The network may be simplified as illustrated in (*c*), (*d*), and (*e*), eliminating the point *m* by means of a star-delta transformation.* It will be noted that Fig. 122(*e*) is in the form of the simplest equivalent network of the type illustrated in Fig. 120, but in this particular case there are no shunt branches and the impedance branches Z_{ac} and Z_{cx} are infinite.

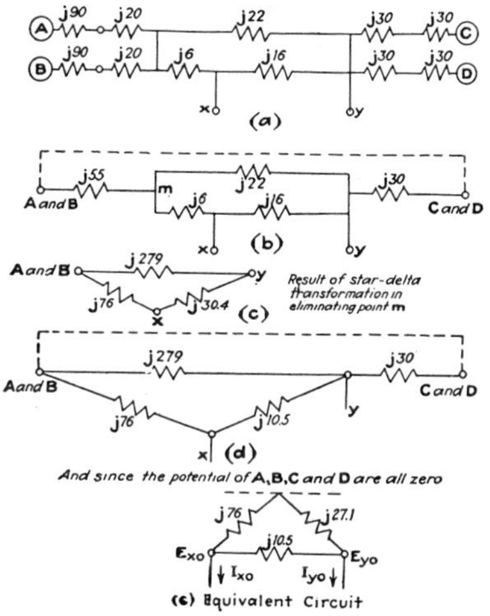

(*f*) Drop constants

$$D_{xx2} = \frac{(j76)(j10.5 + j27.1)}{j(76 + 10.5 + j27.1)} = j25.2$$

$$D_{xy2} = \frac{j76(j27.1)}{j(76 + 10.5 + j27.1)} = j18.1$$

$$D_{yy2} = \frac{(j27.1)(j10.5 + j76)}{j(76 + 10.5 + j27.1)} = j20.6$$

(*g*) Equations relating terminal voltages and currents

$$E_{x2} = -j25.2I_{x2} - j18.1I_{y2}$$
$$E_{y2} = -j18.1I_{x2} - j20.6I_{y2}$$

FIG. 123.—Reduction of negative-sequence network to two equations relating the negative-sequence currents and voltages at the two fault points.

The expression for voltages and currents at the various terminals may now be written by comparison with equations (252) or by inspection with the results given in item (*f*) of Fig. 122.

The drop constants of the positive-sequence network are not used in the calculation of two simultaneous faults with *two sources of e.m.f.;* these constants are, however, used for the calculation with a single source as described in Sec. 113.

Negative-sequence Network Constants. The negative-sequence network is handled in a manner similar to that of the positive-sequence except that the impedance-drop constants are used for all cases regardless of the number of e.m.f. sources involved. The negative-sequence network has the same number

* This formula is given in the Appendix.

of branches as the positive-sequence network but in general includes no sources of generated e.m.f. The negative-sequence network of the system shown in Fig. 122(a) is represented by the impedance diagram of Fig. 123(a). By means of successive steps of network simplification, the network is reduced to the form shown in Fig. 123(e). The impedance-drop constants D_{xx2}, D_{yy2}, and D_{xy2} may readily be computed from (e)

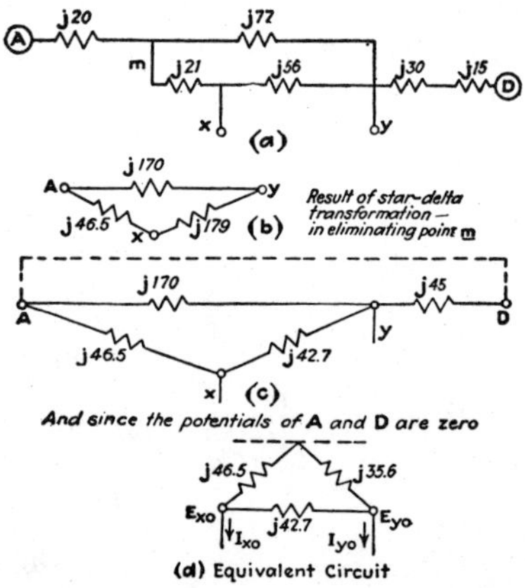

(e) Drop constants

$$D_{xx0} = \frac{(j46.5)(j42.7 + j35.6)}{j(46.5 + 42.7 + 35.6)} = j29.2$$

$$D_{xy0} = \frac{(j46.5)(j35.6)}{j(46.5 + 42.7 + 35.6)} = j13.3$$

$$D_{yy0} = \frac{(j35.6)(j42.7 + j46.5)}{j(46.5 + 42.7 + j35.6)} = j25.4$$

(f) Equations relating terminal voltages and currents

$$E_{x0} = -j29.2I_{x0} - j13.3I_{y0}$$
$$E_{y0} = -j25.4I_{y0} - j13.3I_{x0}$$

Fig. 124.—Reduction of zero-sequence network to two equations relating the zero-sequence currents and voltages at the two fault points.

with the results shown in (f). The equations for the terminal voltages E_{x2} and E_{y2} in terms of the currents I_{x2} and I_{y2} and the network constants are

$$\left.\begin{aligned} E_{x2} &= -D_{xx2}I_{x2} - D_{xy2}I_{y2} \\ E_{y2} &= -D_{xy2}I_{x2} - D_{yy2}I_{y2} \end{aligned}\right\} \quad (258)$$

whose numerical values are given in Fig. 123(g). In general, the real and imaginary parts of D_{xx2}, D_{xy2}, and D_{yy2} will be positive.

Zero-sequence Network Constants. The zero-sequence network is handled in a manner similar to that of the negative-sequence network. The steps are shown in Fig. 124, resulting finally in the simplest equivalent network shown in (d). The zero-sequence voltages E_{x0} and E_{y0} may be expressed in terms of the sequence currents I_{x0} and I_{y0} and the network constants by means of the equations

$$\left.\begin{aligned} E_{x0} &= -D_{xx0}I_{x0} - D_{xy0}I_{y0} \\ E_{y0} &= -D_{yy0}I_{y0} - D_{xy0}I_{x0} \end{aligned}\right\} \quad (259)$$

The impedance-drop constants D_{xx0}, D_{yy0}, and D_{xy0} may be obtained readily from the network of Fig. 124(d). The numerical

Voltages and currents are expressed in terms of star components.

Line-to-ground faults

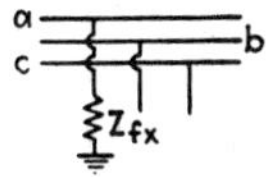

$$I_{x2} = I_{x1} \qquad I_{x0} = I_{x1}$$
$$E_{x1} = -E_{x2} - E_{x0} + 3Z_{fx}I_{x1}$$

Phase a

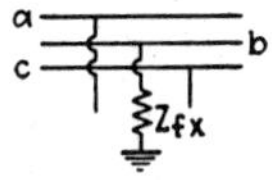

$$I_{x2} = aI_{x1} \qquad I_{x0} = a^2I_{x1}$$
$$E_{x1} = -a^2E_{x2} - aE_{x0} + 3Z_{fx}I_{x1}$$

Phase b

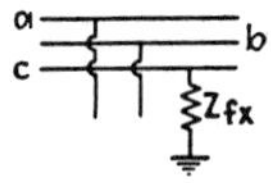

$$I_{x2} = a^2I_{x1} \qquad I_{x0} = aI_{x1}$$
$$E_{x1} = -aE_{x2} - a^2E_{x0} + 3Z_{fx}I_{x1}$$

Phase c

Line-to-line faults

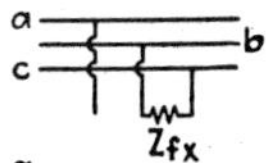

$$I_{x2} = -I_{x1} \qquad I_{x0} = 0$$
$$E_{x1} = E_{x2} + Z_{fx}I_{x1}$$

Phases b and c

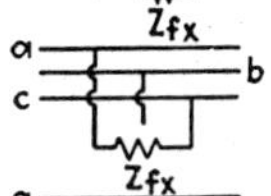

$$I_{x2} = -aI_{x1} \qquad I_{x0} = 0$$
$$E_{x1} = a^2E_{x2} + Z_{fx}I_{x1}$$

Phases a and c

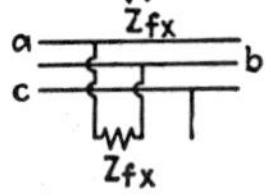

$$I_{x2} = -a^2I_{x1} \qquad I_{x0} = 0$$
$$E_{x1} = aE_{x2} + Z_{fx}I_{x1}$$

Phases a and b

Double line-to-ground faults

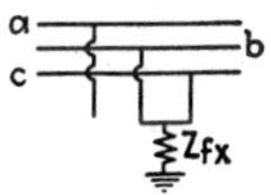

$$I_{x0} + I_{x2} = -I_{x1}$$
$$E_{x1} = E_{x2} \qquad E_{x2} = E_{x0} - 3Z_{fx}I_{x0}$$

Phases b and c

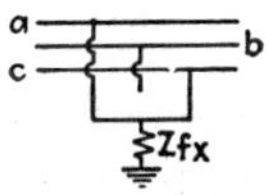

$$I_{x0} + aI_{x2} = -a^2I_{x1}$$
$$E_{x1} = a^2E_{x2} \qquad E_{x2} = a^2(E_{x0} - 3Z_{fx})I_{x0}$$

Phases a and c

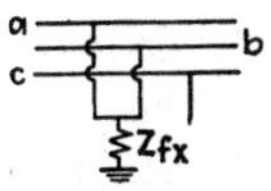

$$I_{x0} + a^2I_{x2} = -aI_{x1}$$
$$E_{x1} = aE_{x2} \qquad E_{x2} = a(E_{x0} - 3Z_{fx}I_{x0})$$

Phases a and b

Three-phase faults

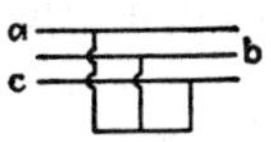

$$I_{x0} = 0$$
$$E_{x1} = 0 \qquad E_{x2} = 0$$

Not to ground

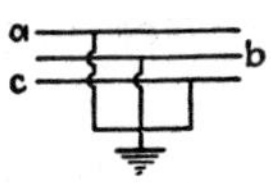

$$E_{x1} = 0 \qquad E_{x2} = 0 \qquad E_{x0} = 0$$

To ground

FIG. 125.—Equations relating sequence quantities at point x for different kinds of faults.

value of these constants substituted in equations gives the equations of item (*f*).

107. Restraints Imposed by the Type of Fault.

The restraints imposed upon the sequence quantities at the point of fault vary with the type of fault and the particular phase or phases involved. In Fig. 125 a table is given showing the relations between the sequence quantities for the most important types of faults. These relations also include the effect of fault impedances. The derivation of these relations is illustrated in Chap. III, Sec. 16 to 19 inclusive. It will be observed that three equations are obtained for each type of fault.

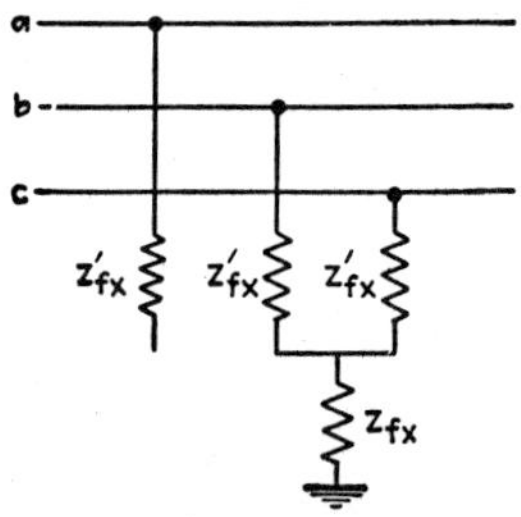

FIG. 126.—Device for including equal impedances in the conductor paths in addition to the common impedance Z_{fx}.

With the exception of the double line-to-ground fault, the fault impedance could just as well have been included as a part of the positive-, negative-, and zero-sequence networks. For example, the single line-to-ground fault could have been applied through the impedance Z_{fx} inserted in all three phases, but, since only one phase carries current, the presence of the two other impedances is immaterial. Similarly, for a line-to-line fault, the fault could have been applied through the impedance $\frac{Z_{fx}}{2}$ in series with each phase. This device can be used to advantage to include the effect of series resistance in two of the faulted phases for a double line-to-ground fault. The diagram for such a case is illustrated in Fig. 126 which shows an impedance Z'_{fx} in each leg of the imaginary leads coming from the line. This device is limited, of course, to equal impedances in the two legs.

Similar equations can be obtained for faults applied at the point *y*.

108. Solution by Combination of Restraints Due to Network Constants and Types of Faults.

Having obtained the two equations relating the negative-sequence currents and voltages from equations (258) and two relating the zero-sequence currents and voltages from equations

(259) and the six relations from the terminal restraints, it is possible to eliminate the eight negative- and zero-sequence unknown quantities, obtaining therefrom two equations relating the positive-sequence currents and voltages at the fault points. This has been done with the results as tabulated in Fig. 127 for the most important combinations of faults at the two points.

In order to illustrate how these relations are obtained, the principal steps for a typical case will be given. Assume a single line-to-ground fault at x on phase a and a single line-to-ground fault at y on phase b. In the table of Fig. 125 it is shown that

$$I_{x2} = I_{x1} \tag{260}$$
$$I_{x0} = I_{x1} \tag{261}$$
$$E_{x1} = -E_{x2} - E_{x0} + 3Z_{fx}I_{x1} \tag{262}$$

and also that

$$I_{y2} = aI_{y1} \tag{263}$$
$$I_{y0} = a^2I_{y1} \tag{264}$$
$$E_{y1} = -a^2E_{y2} - aE_{y0} + 3Z_{fy}I_{y1} \tag{265}$$

Substituting equations (258) and (259) in (262) and (265), there results

$$E_{x1} = D_{xx2}I_{x2} + D_{xy2}I_{y2} + D_{xx0}I_{x0} + D_{xy0}I_{y0} + 3Z_{fx}I_{x1}$$

and

$$E_{y1} = a^2D_{yy2}I_{y2} + a^2D_{xy2}I_{x2} + aD_{yy0}I_{y0} + aD_{xy0}I_{x0} + 3Z_{fy}I_{y1}$$

In turn, substituting the relations from (260), (261), (263), and (264) and collecting terms

$$E_{x1} = (D_{xx2} + D_{xx0} + 3Z_{fx})I_{x1} + (aD_{xy2} + a^2D_{xy0})I_{y1} \tag{266}$$
$$E_{y1} = (a^2D_{xy2} + aD_{xy0})I_{x1} + (D_{yy2} + D_{yy0} + 3Z_{fy})I_{y1} \tag{267}$$

These equations check with those tabulated for the second case in Fig. 127.

It will be observed that having *given the type of faults and the phase involved and also the self and mutual drop constants from the negative- and zero-sequence networks, it is merely necessary to substitute the values of* E_{x1} *and* E_{y1}, *obtained from Fig.* 127, *into the equations for* I_{x1} *and* I_{y1}, *obtained from the positive-sequence network.* The solution of the two resulting equations gives the values of I_{x1} and I_{y1}, from which the other unknown quantities at the point of fault are derived.

109. Simultaneous Faults at More than Two Locations.

The method just described can be extended to the calculation of simultaneous faults at any number of points. For three simultaneous faults, 18 quantities comprising the six sequence currents and voltages at each of the three locations are unknown. From the positive-sequence network it is possible to obtain three

Single line-to-ground faults at both points

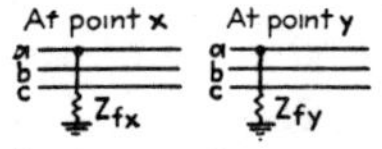

$$E_{x1} = (D_{xx2} + D_{xx0} + 3Z_{fx})I_{x1} + (D_{xy2} + D_{xy0})I_{y1}$$
$$E_{y1} = (D_{xy2} + D_{xy0})I_{x1} + (D_{yy2} + D_{yy0} + 3Z_{fy})I_{y1}$$

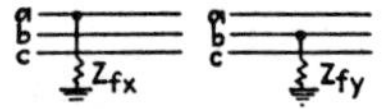

$$E_{x1} = (D_{xx2} + D_{xx0} + 3Z_{fx})I_{x1} + (aD_{xy2} + a^2D_{xy0})I_{y1}$$
$$E_{y1} = (a^2D_{xy2} + aD_{xy0})I_{x1} + (D_{yy2} + D_{yy0} + 3Z_{fy})I_{y1}$$

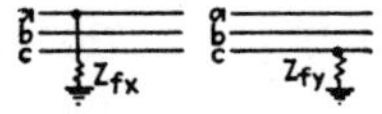

$$E_{x1} = (D_{xx2} + D_{xx0} + 3Z_{fx})I_{x1} + (a^2D_{xy2} + aD_{xy0})I_{y1}$$
$$E_{y1} = (aD_{xy2} + a^2D_{xy0})I_{x1} + (D_{yy2} + D_{yy0} + 3Z_{fy})I_{y1}$$

Double line-to-ground faults at both points.

$$E_{x1} = -(D_{xx2}I_{x2} + D_{xy2}I_{y2}) \qquad E_{y1} = -a^2(D_{xy2}I_{x2} + D_{yy2}I_{y2})$$

in which

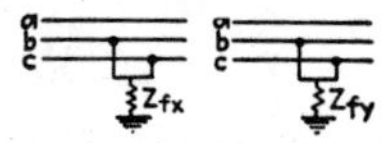

$$I_{x2} = \frac{1}{KN - LM}\{[LD_{xy0} - N(D_{xx0} + 3Z_{fx})]I_{x1} + [L(D_{yy0} + 3Z_{fy}) - ND_{xy0}]I_{y1}\}$$
$$I_{y2} = \frac{1}{KN - LM}\{[M(D_{xx0} + 3Z_{fx}) - KD_{xy0}]I_{x1} + [MD_{xy0} - K(D_{yy0} + 3Z_{fy})]I_{y1}\}$$
$$K = D_{xx2} + D_{xx0} + 3Z_{fx} \quad L = D_{xy2} + D_{xy0} \quad M = L$$
$$N = D_{yy2} + D_{yy0} + 3Z_{fy}$$

$$E_{x1} = -(D_{xx2}I_{x2} + D_{xy2}I_{y2}) \qquad E_{y1} = -a(D_{yy2}I_{y2} + D_{xy2}I_{x2})$$

in which

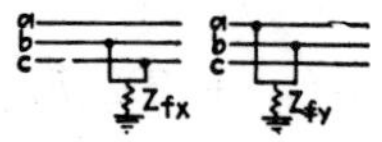

$$I_{x2} = \frac{1}{KN - LM}\{[aLD_{xy0} - N(D_{xx0} + 3Z_{fx})]I_{x1} + [a^2L(D_{yy0} + 3Z_{fy}) - aND_{xy0}]I_{y1}\}$$
$$I_{y2} = \frac{1}{KN - LM}\{[M(D_{xx0} + 3Z_{fx}) - aKD_{xy0}]I_{x1} + [aMD_{xy0} - a^2K(D_{yy0} + 3Z_{fy}]I_{y1}\}$$
$$K = D_{xx2} + D_{xx0} + 3Z_{fx} \quad L = D_{xy2} + a^2D_{xy0} \quad M = D_{xy2} + aD_{xy0}$$
$$N = D_{yy2} + D_{yy0} + 3Z_{fy}$$

$$E_{x1} = -(D_{xx2}I_{x2} + D_{xy2}I_{y2}) \qquad E_{y1} = -a^2(D_{yy2}I_{y2} + D_{xy2}I_{x2})$$

in which

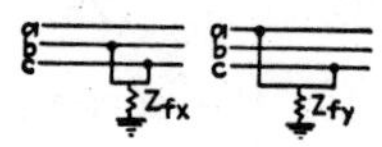

$$I_{x2} = \frac{1}{KN - LM}\{[a^2LD_{xy0} - N(D_{xx0} + 3Z_{fx})]I_{x1} + [aL(D_{yy0} + 3Z_{fy}) - a^2ND_{xy0}]I_{y1}\}$$
$$I_{y2} = \frac{1}{KN - LM}\{[M(D_{xx0} + 3Z_{fx}) - a^2KD_{xy0}]I_{x1} + [a^2MD_{xy0} - aK(D_{yy0} + 3Z_{fy})I_{y1}\}$$
$$K = D_{xx2} + D_{xx0} + 3Z_{fx} \quad L = D_{xy2} + aD_{xy0} \quad M = D_{xy2} + a^2D_{xy0}$$
$$N = D_{yy2} + D_{yy0} + 3Z_{fy}$$

FIG. 127.—For descriptive legend see opposite page.

equations of the type developed in Fig. 122, which express the currents I_{x1}, I_{y1}, and I_{z1} in terms of E_{x1}, E_{y1}, and E_{z1}, the constants of the positive-sequence network, and the generated e.m.fs. From the negative-sequence network the three following equations can be derived.

$$\left.\begin{aligned} E_{x2} &= -D_{xx2}I_{x2} - D_{xy2}I_{y2} - D_{xz2}I_{z2} \\ E_{y2} &= -D_{xy2}I_{x2} - D_{yy2}I_{y2} - D_{yz2}I_{z2} \\ E_{z2} &= -D_{xz2}I_{x2} - D_{yz2}I_{y2} - D_{zz2}I_{z2} \end{aligned}\right\} \quad (268)$$

Similar equations relating the zero-sequence constants can also be set up. In this manner nine of the requisite equations are determined.

Line-to-line faults at both points

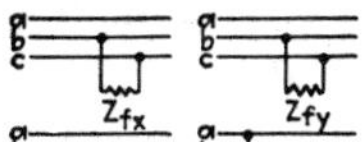

$$E_{x1} = (D_{xx2} + Z_{fx})I_{x1} + D_{xy2}I_{y1}$$
$$E_{y1} = D_{xy2}I_{x1} + (D_{yy2} + Z_{fy})I_{y1}$$

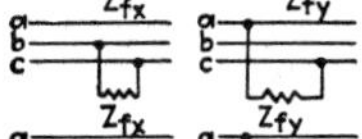

$$E_{x1} = (D_{xx2} + Z_{fx})I_{x1} + aD_{xy2}I_{y1}$$
$$E_{y1} = a^2D_{xy2}I_{x1} + (D_{yy2} + Z_{fy})I_{y1}$$

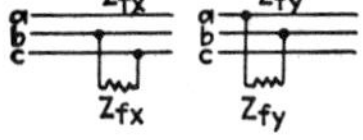

$$E_{x1} = (D_{xx2} + Z_{fx})I_{x1} + a^2D_{xy2}I_{y1}$$
$$E_{y1} = aD_{xy2}I_{x1} + (D_{yy2} + Z_{fy})I_{y1}$$

Line-to-line fault at x and single line-to-ground fault at y

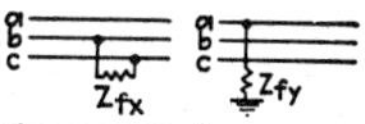

$$E_{x1} = (D_{xx2} + Z_{fx})I_{x1} - D_{xy2}I_{y1}$$
$$E_{y1} = -D_{xy2}I_{x1} + (D_{yy2} + D_{yy0} + 3Z_{fy})I_{y1}$$

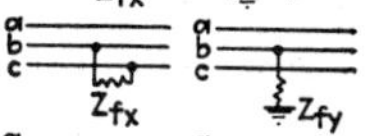

$$E_{x1} = (D_{xx2} + Z_{fx})I_{x1} - aD_{xy2}I_{y1}$$
$$E_{y1} = -a^2D_{xy2}I_{x1} + (D_{yy2} + D_{yy0} + 3Z_{fy})I_{y1}$$

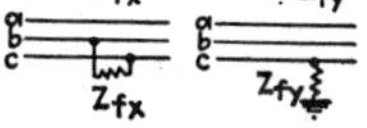

$$E_{x1} = (D_{xx2} + Z_{fx})I_{x1} - a^2D_{xy2}I_{y1}$$
$$E_{y1} = -aD_{xy2}I_{x1} + (D_{yy2} + D_{yy0} + 3Z_{fy})I_{y1}$$

Double line-to-ground fault at x and single line-to-ground fault at y

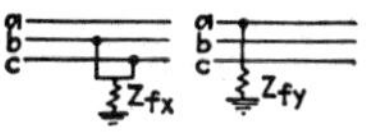

$$E_{x1} = \frac{1}{D_{xx2} + D_{xx0} + 3Z_{fx}}\{D_{xx2}(D_{xx0} + 3Z_{fx})I_{x1} - [D_{xy0}D_{xx2} + D_{xy2}(D_{xx0} + 3Z_{fx})]I_{y1}\}$$
$$E_{y1} = -\frac{D_{xy0}D_{xx2} + D_{xy2}(D_{xx0} + 3Z_{fx})}{D_{xx2} + D_{xx0} + 3Z_{fx}}I_{x1} + \left[(D_{yy2} + D_{yy0} + 3Z_{fy}) - \frac{(D_{xy2} - D_{xy0})^2}{D_{xx2} + D_{xx0} + 3Z_{fx}}\right]I_{y1}$$

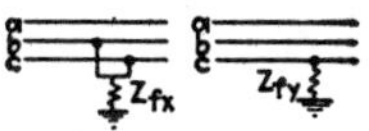

$$E_{x1} = \frac{1}{D_{xx2} + D_{xx0} + 3Z_{fx}}\{D_{xx2}(D_{xx0} + 3Z_{fx})I_{x1} - [a^2D_{xy0}D_{xx2} + aD_{xy2}(D_{xx0} + 3Z_{fx})]I_{y1}\}$$
$$E_{y1} = -\frac{aD_{xy0}D_{xx2} + a^2D_{xy2}(D_{xx0} + 3Z_{fx})}{D_{xx2} + D_{xx0} + 3Z_{fx}}I_{x1} + \left[D_{yy2} + D_{yy0} + 3Z_{fy} - \frac{D^2_{xy2} + D_{xy2}D_{xy0} + D^2_{xy0}}{D_{xx2} + D_{xx0} + 3Z_{fx}}\right]I_{y1}$$

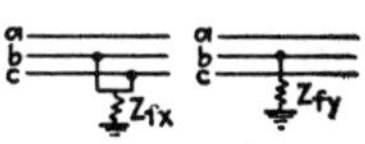

$$E_{x1} = \frac{1}{D_{xx2} + D_{xx0} + 3Z_{fx}}\{D_{xx2}(D_{xx0} + 3Z_{fx})I_{x1} - [aD_{xx2}D_{xy0} + a^2D_{xy2}(D_{xx0} + 3Z_{fx})]I_{y1}\}$$
$$E_{y1} = -\frac{a^2D_{xy0}D_{xx2} + aD_{xy2}(D_{xx0} + 3Z_{fx})}{D_{xx2} + D_{xx0} + 3Z_{fx}}I_{x1} + \left[D_{yy2} + D_{yy0} + 3Z_{fy} - \frac{D^2_{xy2} + D_{xy2}D_{xy0} + D^2_{xy0}}{D_{xx2} + D_{xx0} + 3Z_{fx}}\right]I_{y1}$$

FIG. 127 (*Continued*).—Equations relating positive-sequence quantities at fault locations for simultaneous faults at two points in terms of negative- and zero-sequence constants. Each pair of equations can be represented in the form:
$E_{x1} = kI_{x1} + mI_{y1}$; $E_{y1} = nI_{x1} + lI_{y1}$

At each of the locations, the terminal restraints are imposed, which, depending upon the character and phases of the faults, provide three additional equations of the type shown in Fig. 125 for each location. These restraints supply the nine equations still required.

110. Simultaneous Faults for the Case of a Single Source of E.M.F.

In the calculation of system short-circuit currents a simplifying assumption frequently used is that all of the generated e.m.fs. are identical in magnitude and phase. In this case the positive-sequence network can be handled much like the negative- and zero-sequence networks. Neglecting for the moment the generated e.m.fs. the self and mutual drop constants for the positive-sequence

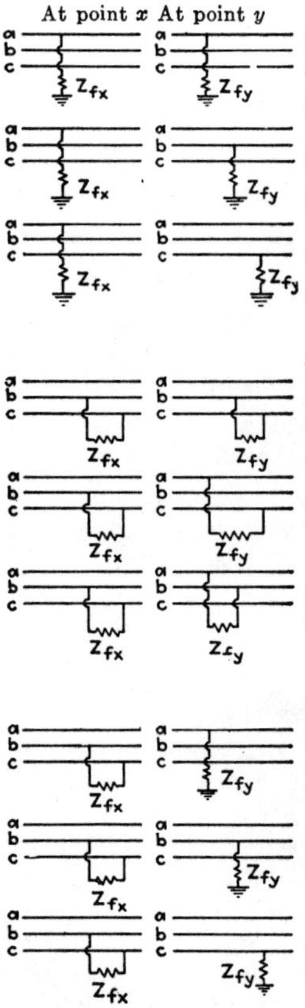

Single line-to-ground faults at both points

$$E_g = (D_{xx1} + D_{xx2} + D_{xx0} + 3Z_{fx})I_{x1} + (D_{xy1} + D_{xy2} + D_{xy0})I_{y1}$$
$$E_g = (D_{xy1} + D_{xy2} + D_{xy0})I_{x1} + (D_{yy1} + D_{yy2} + D_{yy0} + 3Z_{fy})I_{y1}$$

$$E_g = (D_{xx1} + D_{xx2} + D_{xx0} + 3Z_{fx})I_{x1} + (D_{xy1} + aD_{xy2} + a^2D_{xy0})I_{y1}$$
$$E_g = (D_{xy1} + a^2D_{xy2} + aD_{xy0})I_{x1} + (D_{yy1} + D_{yy2} + D_{yy0} + 3Z_{fy})I_{y1}$$

$$E_g = (D_{xx1} + D_{xx2} + D_{xx0} + 3Z_{fx})I_{x1} + (D_{xy1} + a^2D_{xy2} + aD_{xy0})I_{y1}$$
$$E_g = (D_{xy1} + aD_{xy2} + a^2D_{xy0})I_{x1} + (D_{yy1} + D_{yy2} + D_{yy0} + 3Z_{fx})I_{y1}$$

Line-to-line faults at both points

$$E_g = (D_{xx1} + D_{xx2} + Z_{fx})I_{x1} + (D_{xy1} + D_{xy2})I_{y1}$$
$$E_g = (D_{xy1} + D_{xy2})I_{x1} + (D_{yy1} + D_{yy2} + Z_{fy})I_{y1}$$

$$E_g = (D_{xx1} + D_{xx2} + Z_{fx})I_{x1} + (D_{xy1} + aD_{xy2})I_{y1}$$
$$E_g = (D_{xy1} + a^2D_{xy2})I_{x1} + (D_{yy1} + D_{yy2} + Z_{fy})I_{y1}$$

$$E_g = (D_{xx1} + D_{xx2} + Z_{fx})I_{x1} + (D_{xy1} + a^2D_{xy2})I_{y1}$$
$$E_g = (D_{xy1} + aD_{xy2})I_{x1} + (D_{yy1} + D_{yy2} + Z_{fy})I_{y1}$$

Line-to-line fault at x and single line-to-ground fault at y

$$E_g = (D_{xx1} + D_{xx2} + Z_{fx})I_{x1} + (D_{xy1} - D_{xy2})I_{y1}$$
$$E_g = +(D_{xy1} - D_{xy2})I_{x1} + (D_{yy1} + D_{yy2} + D_{yy0} + 3Z_{fy})I_{y1}$$

$$E_g = (D_{xx1} + D_{xx2} + Z_{fx})I_{x1} + (D_{xy1} - aD_{xy2})I_{y1}$$
$$E_g = +(D_{xy1} - a^2D_{xy2})I_{x1} + (D_{yy1} + D_{yy2} + D_{yy0} + 3Z_{fy})I_{y1}$$

$$E_g = (D_{xx1} + D_{xx2} + Z_{fx})I_{x1} + (D_{xy1} - a^2D_{xy2})I_{y1}$$
$$E_g = +(D_{xy1} - aD_{xy2})I_{x1} + (D_{yy1} + D_{yy2} + D_{yy0} + 3Z_{fy})I_{y1}$$

FIG. 128.—Equations relating positive-sequence quantities at fault locations for two simultaneous faults. All generated voltages equal.

network can be determined as indicated in Fig. 123 or 124. The voltage equations for the positive-sequence network can then be written in terms of these impedances and the generated e.m.f. E_g as follows:

$$\left.\begin{aligned} E_g &= E_{x1} + D_{xx1}I_{x1} + D_{xy1}I_{y1} \\ E_g &= E_{y1} + D_{xy1}I_{x1} + D_{yy1}I_{y2} \end{aligned}\right\} \quad (269)$$

The values of E_{x1} and E_{y1} are also defined in terms of the restraints due to the type of fault as given in Fig. 127 which relations are in the form

$$\left.\begin{aligned} E_{x1} &= kI_{x1} + mI_{y1} \\ E_{y1} &= nI_{x1} + lI_{y1} \end{aligned}\right\} \quad (270)$$

After substituting E_{x1} and E_{y1} from equations (270) in equations (269), the value of I_{x1} and I_{y1} may be expressed in terms of the generated voltage E_g and the system constants, with the following result:

$$\begin{aligned} E_g &= (k + D_{xx1})I_{x1} + (m + D_{xy1})I_{y1} \\ E_g &= (n + D_{xy1})I_{x1} + (l + D_{yy1})I_{y1} \end{aligned}$$

from which

$$I_{x1} = \frac{(l + D_{yy1}) - (m + D_{xy1})}{(k + D_{xx1})(l + D_{yy1}) - (m + D_{xy1})(n + D_{xy1})} E_g \quad (271)$$

$$I_{y1} = \frac{(k + D_{xx1}) - (n + D_{xy1})}{(k + D_{xx1})(l + D_{yy1}) - (m + D_{xy1})(n + D_{xy1})} E_g \quad (272)$$

The coefficients of equations (271) and (272), when developed in terms of the positive-, negative- and zero-sequence quantities, result in quite symmetrical forms as may be seen from the table of Fig. 128, which has been prepared for the cases which result in the more simple expressions.

111. Current and Voltage Distribution for Simultaneous Faults.

Having obtained I_{x1} and I_{y1} at the fault points, the other sequence currents at these points may be obtained through the use of the relations shown in Fig. 125 and the sequence voltages by use of equations (258) and (259). The current distribution within the individual networks can be obtained either analytically by working back through the networks in a reverse manner, as in obtaining the self and mutual impedances for the entire network, or by means of a calculating board by applying the calculated voltages to the fault points.

112. Equivalent Circuits Representing Fault Conditions.

The solution of the equations from Fig. 127 in conjunction with those from the positive-sequence network will, in general, offer little difficulty. It is of interest, however, to provide a complete solution utilizing the calculating board.

The solution of the equations from Fig. 127 may be expressed in the form of equations (270). For the special case in which $m = n$, considerable simplification results. This condition holds for only the first case of each group in Fig. 127. The positive-sequence network for these special cases may be represented by the equivalent circuit shown in Fig. 129, in which the mutual impedance is the common branch and the fork branches are equal to $(k - m)$ and $(l - n)$ for the x and y points, respectively. That this represents the equivalent of equations (270) may readily be verified as follows:

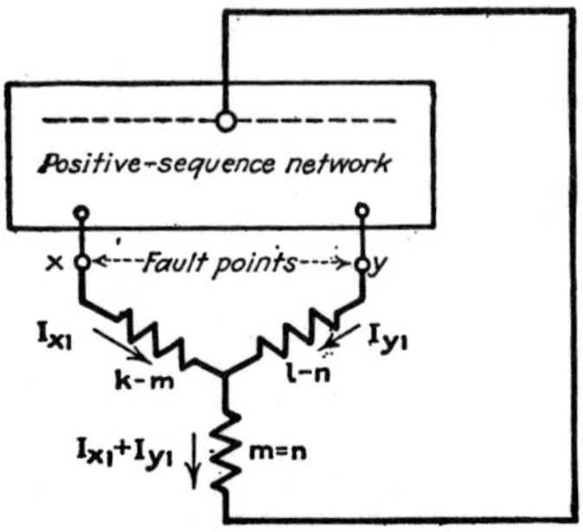

FIG. 129.—Equivalent positive-sequence diagram for faults at x and y for the special case in which $m = n$.

$$\left.\begin{aligned} E_{x1} &= m(I_{x1} + I_{y1}) + (k - m)I_{x1} = kI_{x1} + mI_{y1} \\ E_{y1} &= n(I_{x1} + I_{y1}) + (l - n)I_{y1} = nI_{x1} + lI_{y1} \end{aligned}\right\} \quad (273)$$

The cases in which $m \neq n$, are more involved, and for their representation use will be made of a device[70] described by Miss Edith Clarke. Equations (270) may be written:

$$\left.\begin{aligned} E_{x1} &= (k-n)I_{x1} + \frac{m+n}{2}(I_{x1} + I_{y1}) + \frac{m-n}{2}(I_{y1} - I_{x1}) \\ E_{y1} &= (l-m)I_{y1} + \frac{m+n}{2}(I_{x1} + I_{y1}) + \frac{m-n}{2}(I_{y1} - I_{x1}) \end{aligned}\right\} \quad (274)$$

Figure 130 shows the equivalent circuit of the positive-sequence network which will represent this condition. The impedance $\frac{m - n}{2}$ is paralleled by a source of potential which must be varied so that the current flowing through the impedance $\frac{m - n}{2}$ is equal to $(I_{y1} - I_{x1})$. It may develop that the impedance $\frac{m - n}{2}$ involves a negative resistance in which case the impedance $\frac{n - m}{2}$ may be

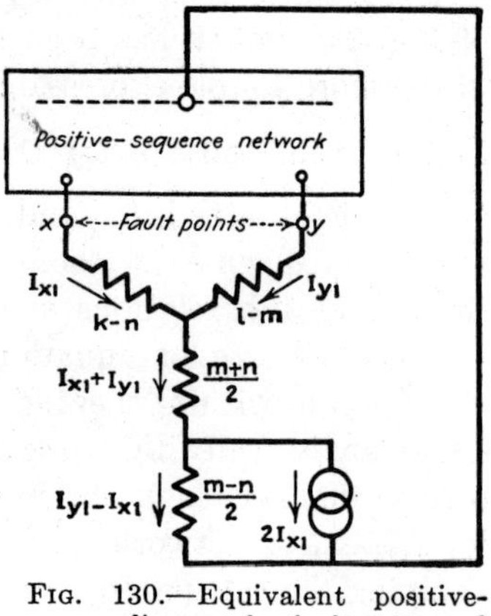

FIG. 130.—Equivalent positive-sequence diagram for faults at x and y for the general case. The impedance $\frac{n - m}{2}$ may be used in place of $\frac{m - n}{2}$; when this is done, the current through the shunt source of potential should be $2I_{y1}$ instead of $2I_{x1}$.

used instead and the current through it adjusted to $(I_{x1} - I_{y1})$. If $\frac{m-n}{2}$ is used, the shunt voltage may be varied until the current flowing through it is $2I_{x1}$ (in both magnitude and phase) and, when $\frac{n-m}{2}$ is used, until the current is $2I_{y1}$.

Special Cases. Two special cases, due to Miss Edith Clarke,[70] are worthy of consideration. The solution for the currents and voltages at the fault points by these methods are about as complicated as for the ordinary analytical methods. Their principal value is in providing an equivalent circuit for stability studies.

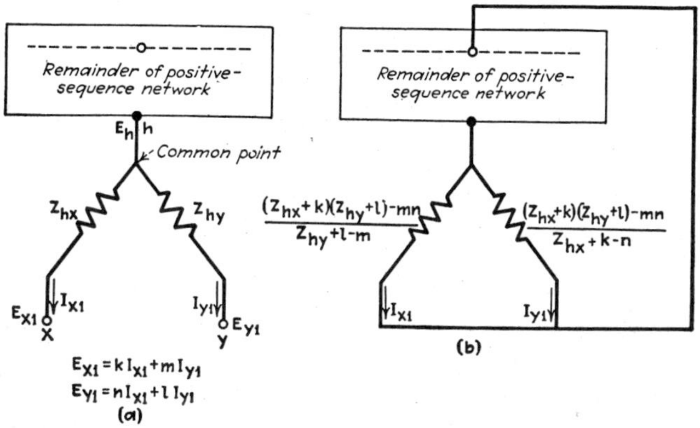

FIG. 131.—Equivalent positive-sequence network for two simultaneous faults on two unloaded feeders radiating from a common point. (*a*) Actual circuit; (*b*) equivalent circuit.

a. Equivalent Circuit for Simultaneous Faults on Unloaded Feeders Radiating from a Common Point. In Fig. 131(*a*) let the faults occur at the points x and y on the feeders radiating from the common point h somewhere in the positive-sequence network. The positive-sequence impedances of the branches between h and the fault points are Z_{hx} and Z_{hy}. The positive-sequence voltages and currents at x and y are defined in terms of the negative- and zero-sequence network constants by equations (270). The remainder of the positive-sequence network is indicated by the box.

Now, from the figure,

$$\left.\begin{aligned} E_h &= E_{x1} + Z_{hx}I_{x1} \\ E_h &= E_{y1} + Z_{hy}I_{y1} \end{aligned}\right\} \quad (275)$$

Substituting (270)

$$E_h = (k + Z_{hx})I_{x1} + mI_{y1} \tag{276}$$
$$E_h = nI_{x1} + (l + Z_{hy})I_{y1} \tag{277}$$

Solving

$$I_{x1} = \frac{Z_{hy} + l - m}{(Z_{hx} + k)(Z_{hy} + l) - mn} E_h \tag{278}$$

$$I_{y1} = \frac{Z_{hx} + k - n}{(Z_{hx} + k)(Z_{hy} + l) - mn} E_h \tag{279}$$

The reciprocals of these coefficients are the equivalent impedances to be placed in the x and y branches and are shown in Fig. 131(b). This circuit may be simplified a little further by paralleling the two impedances, in which case the single equivalent impedance reduces to

$$\frac{(Z_{hx} + k)(Z_{hy} + l) - mn}{Z_{hx} + k + Z_{hy} + l - m - n} \tag{280}$$

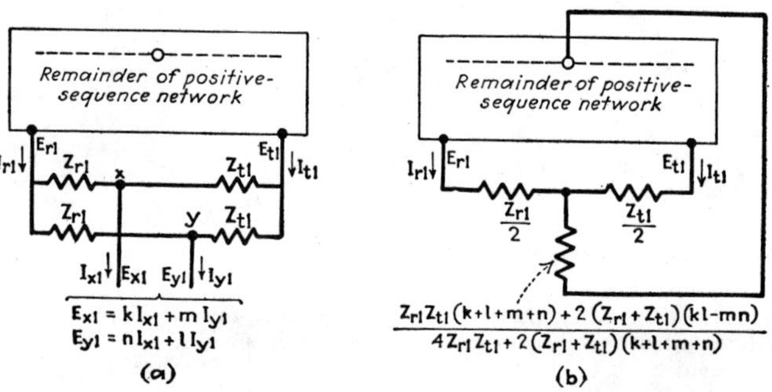

FIG. 132.—Equivalent positive-sequence network for two simultaneous faults on two transmission lines of equal impedances bussed at both ends. (a) Actual circuit; (b) equivalent circuit.

b. Equivalent Circuit for Simultaneous Faults on Two Transmission Lines of Equal Impedances Bussed at Both Ends. Another important case that may arise is that of a double simultaneous fault at a given point along the line of a double-circuit transmission line. In this case the lines will be bussed at points on either side of the faults and the positive-sequence impedance between the fault points and the buses will be equal for both lines. The equivalent circuit and explanation of the circuit constants are sufficiently developed in Fig. 132 to be self-explanatory.

113. Single Fault for the Case of Several Sources of E.M.F.

It has been pointed out in Chap. II that synchronous machines are represented individually in the positive-sequence network by an impedance and a source of e.m.f. which in general will be different in magnitude and phase position for each machine. For short-circuit calculation, as pointed out in Chap. IV, Sec. 24, it is usually permissible to assume that these e.m.fs. are identical in magnitude and phase position. The more general case where this is not permissible will now be considered.

Network solutions with several separate sources of e.m.f. may readily be obtained on the alternating-current calculating board, since independent adjustment of magnitude and phase position of the e.m.f. sources is provided.

The case of a fault at a single location on a system with several sources of e.m.f. which are not identical in magnitude and phase position may be considered as a special case of the problem involving simultaneous faults. Using this method, the solution may be obtained by following the steps outlined in the previous section for two simultaneous faults by merely ignoring the currents at points other than the first fault location. Thus equations (252), (258), and (259) reduce to

$$I_{x1} = \frac{E_a}{Z_{ax}} + \frac{E_b}{Z_{bx}} + \cdots - E_{x1}\left(Y_x + \frac{1}{Z_{ax}} + \frac{1}{Z_{bx}} \cdots\right) \qquad \text{[Eq. (252}a\text{)]}$$

$$E_{x2} = -D_{xx2}I_{x2} \text{ [Eq. (258}a\text{)]}$$

$$E_{x0} = -D_{xx0}I_{x0} \text{ [Eq. (259}a\text{)]}$$

for a fault at a single location. These, together with the proper set of equations from Fig. 127 for the type of fault, provide enough equations to solve for all the unknown voltages and currents.

An alternative method* may be used which is based on the reduction of the positive-sequence network with its several sources of e.m.f. to a single equivalent impedance and a single equivalent source of e.m.f. Considering the simplest equivalent general network for two sources and one fault point and using the notation of Fig. 119, the total fault current for a single three-phase zero impedance short-circuit at the point x is

$$I_x = \frac{E_a}{Z_{ax}} + \frac{E_b}{Z_{bx}} \tag{281}$$

* This is based upon a more general method described in Appendix I of bibliography item 28.

The relation that the same fault current is to be obtained from the equivalent network gives the equation

$$I_x = \frac{E_{equiv.}}{Z_{equiv.}} \tag{282}$$

If the equivalent impedance is assumed to have the same value as exists when the e.m.fs. are identical then

$$\frac{1}{Z_{equiv.}} = \frac{1}{Z_{ax}} + \frac{1}{Z_{bx}} \tag{283}$$

The e.m.f. of the equivalent source is then given by

$$\frac{E_{equiv.}}{Z_{equiv.}} = \frac{E_a}{Z_{ax}} + \frac{E_b}{Z_{bx}} \tag{284}$$

or

$$E_{equiv.} = \frac{Z_{bx}E_a + Z_{ax}E_b}{Z_{ax} + Z_{bx}} \tag{285}$$

The method using equivalent impedance and equivalent voltage sources has been outlined for the case of two sources, but it is readily apparent that additional sources can be taken care of by adding suitable terms to the right-hand members of equations (283) and (284).

Problems

1. Two grounded neutral generators A and B, operating at no load, are paralleled through a transmission circuit. The reactances on the base of the rating of machine A are as follows:

Machine A	Machine B	Transmission Line
$x_1 = 0.30$	$x_1 = 0.15$	$x_1 = 0.25$
$x_2 = 0.30$	$x_2 = 0.15$	$x_2 = 0.25$
$x_0 = 0$	$x_0 = 0$	$x_0 = 0.75$

What are the fault currents for a simultaneous line-to-ground fault on phase a at the terminal of each machine? Neglect all resistance and the capacitance of the transmission line.

2. Determine the fault currents for short-circuits between phases b and c on the terminals of machine A and between phases a and b on the terminals of machine B of Prob. 1.

3. A 5,000-kva., 2,200-volt, 80 per cent power-factor generator is connected to a transmission line whose reactance is 0.2 ohm per phase. Resistance loads of 4.0 ohms are taken off across phases b and c at the terminals, 5.0 ohms across phases b and c at the end of the line, and 5.0 ohms across phases a and c also at the end of the line. If x_1 of the machine is 1.0 ohm and x_2 is 0.5 ohm per phase and the positive-sequence internal voltage is 1,500 volts to neutral, what are the voltages between the phases at the machine terminals and at the end of the line?

CHAPTER XIII

DETERMINATION OF SEQUENCE QUANTITIES FROM PHASE QUANTITIES

Frequently it is desired to determine the value of sequence quantities when only the ordinary single-phase instrument readings are available. For three-phase systems involving zero-sequence, six unknown quantities are required to define the three vectors. These unknown quantities may be expressed in terms of three amplitudes and three phase angles with respect to an arbitrary reference, or in terms of six amplitudes, such as the three line-to-line voltages and the three line-to-neutral voltages. For three-phase systems in which zero-sequence does not appear, three amplitudes are sufficient to define the vectors with respect to each other though an additional relation is required to relate them to a reference vector. In this chapter will be presented several methods using analytical expressions, graphical constructions, or charts by means of which the sequence quantities may be determined from the phase quantities. The developments will follow along two general lines, those for the general case and those for the special case in which the three vectors form a closed triangle, *i.e.*, the case in which the zero-sequence quantity is equal to zero.

114. Determination of Vectors from Amplitudes.

Before undertaking the determination of the sequence quantities, methods will be described for obtaining the vectors from the amplitudes of the phase quantities. When zero-sequence is absent the line-to-neutral voltages may be plotted to form a closed triangle so that their phase angles with respect to each other are readily determined. In case the zero-sequence quantities are present, it is necessary to obtain six relations such as given by the six amplitudes of the three line-to-line voltages and the three line-to-neutral voltages. The phase relation of the line-to-neutral voltage vectors may be obtained by first plotting the line-to-line voltages to form a closed triangle as indicated in Fig. 133. The line-to-neutral voltages may then be plotted from the corresponding apexes of the triangle, and the common intercept O will define the neutral point of the system.

In case the zero-sequence quantity is absent and only the amplitudes of the line-to-line voltages are available, the line-to-neutral voltages may be obtained as indicated in Fig. 134. Form the triangle from the voltages E_A, E_B, and E_C. From any two apexes draw two lines, such as ce and ad, to bisect the opposite sides. Their intercept at O' gives the neutral point of the system. Thus the voltage vectors E_a, E_b, and E_c are the corresponding line-to-neutral voltages.

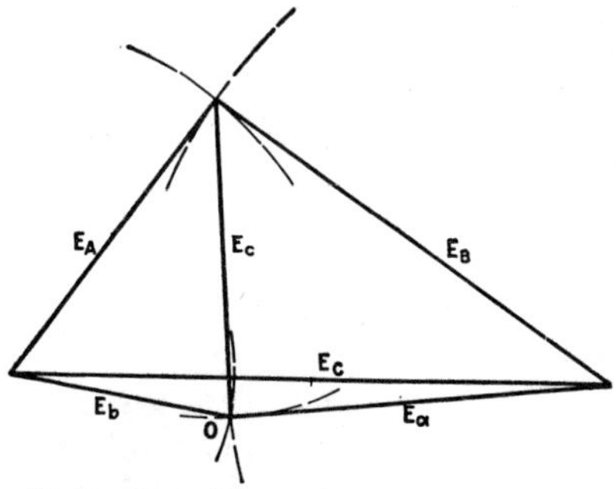

FIG. 133.—Determination of line-to-neutral vectors from the amplitudes of E_a, E_b, E_c, E_A, E_B and E_C.

The validity of this theorem can be proved by showing that the vector sum of E_a, E_b, and E_c vanishes in the construction of Fig. 134, that is

$$E_a + E_b + E_c = 0 \quad (286)$$

The construction of Fig. 134 gives

$$E_a = O'a = \tfrac{2}{3}(dc + ca)$$
$$= \tfrac{2}{3}(\tfrac{1}{2}E_A + E_B)$$
$$= \tfrac{1}{3}E_A + \tfrac{2}{3}E_B \quad (287)$$

Similarly,

$$E_b = \tfrac{1}{3}E_B + \tfrac{2}{3}E_C \quad (287a)$$

and

$$E_c = \tfrac{1}{3}E_C + \tfrac{2}{3}E_A \quad (287b)$$

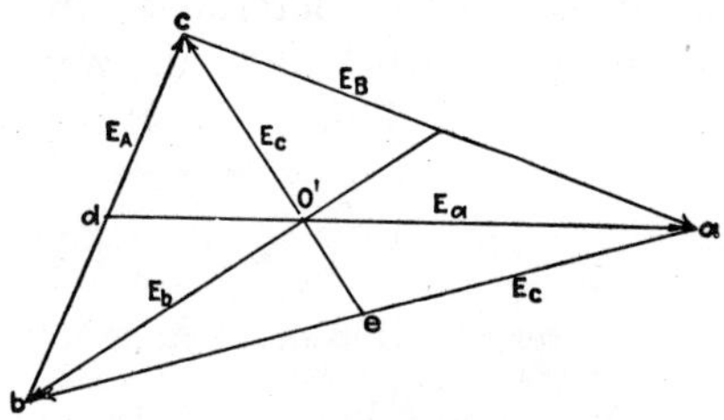

FIG. 134.—Construction showing that the intersection of the medians is the neutral point of the system of three vectors comprising the sides of the triangle.

Adding equations (287) gives equation (286) and proves the construction. It shows that E_a, E_b, and E_c are the particular set of star voltages which produce the delta voltages E_A, E_B, and E_C and have no zero-sequence component.

115. Analytical Expressions.

The sequence quantities expressed as vectors can be obtained from the vector value of the phase quantities with the aid of the fundamental equations (27), (28), and (29) given on next page.

$$E_{a0} = \tfrac{1}{3}(E_a + E_b + E_c)$$
$$E_{a1} = \tfrac{1}{3}(E_a + aE_b + a^2E_c)$$
$$E_{a2} = \tfrac{1}{3}(E_a + a^2E_b + aE_c)$$

When the three vectors form a closed triangle, *i.e.*, when $E_{a0} = 0$, the expressions for the positive- and negative-sequences become, after substituting $E_a = -E_b - E_c$ in equations (28) and (29),

$$E_{a1} = \tfrac{1}{3}[(a-1)E_b + (a^2-1)E_c] \quad (288)$$

$$= -\frac{1}{\sqrt{3}}(E_b e^{-j30°} + E_c e^{j30°}) \quad (289)$$

or

$$= -\frac{e^{-j30°}}{\sqrt{3}}(E_b + E_c e^{j60°}) \quad (290)$$

And

$$E_{a2} = \tfrac{1}{3}[(a^2-1)E_b + (a-1)E_c] \quad (291)$$

$$= -\frac{1}{\sqrt{3}}(E_b e^{j30°} + E_c e^{-j30°}) \quad (292)$$

$$= -\frac{e^{j30°}}{\sqrt{3}}(E_b + E_c e^{-j60°}) \quad (293)$$

For this special case other methods are available which do not require a knowledge of the phase values in vector form. Kennelly,[(37)] Sah,[(38)] and Norman[(40)] have developed expressions which give the positive- and negative-sequence components in terms of the amplitudes of the three unbalanced vectors forming a closed triangle. These are

$$\bar{E}_{a1}^2 = \tfrac{1}{2}(\bar{A}_m^2 + \bar{A}_s^2) \quad (294)$$

and

$$\bar{E}_{a2}^2 = \tfrac{1}{2}(\bar{A}_m^2 - \bar{A}_s^2) \quad (295)$$

in which

$$\bar{A}_m^2 = \tfrac{1}{3}(\bar{E}_a^2 + \bar{E}_b^2 + \bar{E}_c^2) \quad (296)$$

$$\bar{A}_s^2 = \sqrt{\tfrac{1}{3}(\bar{E}_a + \bar{E}_b + \bar{E}_c)(\bar{E}_a + \bar{E}_b - \bar{E}_c)(\bar{E}_a - \bar{E}_b + \bar{E}_c)(-\bar{E}_a + \bar{E}_b + E_c)} \quad (297)$$

It is interesting to note, as Dr. Kennelly has observed, that "$\bar{A}_m^2$ is the mean of the squares of the three unbalanced sides and $\bar{A}_s^2$ is the square of the equilateral side of an unbalanced system having the same area as the unbalanced triangle."[(37)] Evans and Heumann have developed expressions (not previously published) that reduce to the above but in addition enable one to obtain the phase relations of the sequence components. They have simplified the expressions by letting $\bar{E}_a = 1$, $\dfrac{\bar{E}_b}{\bar{E}_a} = x$, and $\dfrac{\bar{E}_c}{\bar{E}_a} = y$. The analytical expressions for the various quantities are given in Fig. 135.

116. Graphical Constructions.

The methods for determining the sequence components of three-phase systems by graphical construction may conveniently be divided into two groups: (*a*) methods for the determination

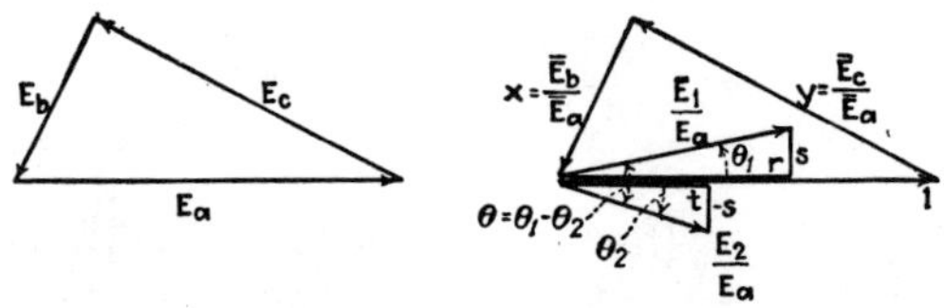

$$r = \frac{1}{2} + \frac{1}{2\sqrt{3}}\sqrt{(1+x+y)(1+x-y)(1-x+y)(-1+x+y)}$$

$$t = \frac{1}{2} - \frac{1}{2\sqrt{3}}\sqrt{(1+x+y)(1+x-y)(1-x+y)(-1+x+y)}$$

$$s = \frac{y^2 - x^2}{2\sqrt{3}}$$

$$\theta_1 = \tan^{-1}\frac{s}{r}$$

$$\theta_2 = \tan^{-1} - \frac{s}{t}$$

$$\bar{E}_{a1} = \sqrt{r^2 + s^2}\,\bar{E}_a$$

$$\bar{E}_{a2} = \sqrt{t^2 + s^2}\,\bar{E}_a$$

FIG. 135.—Analytical expressions for positive- and negative-sequence components of a set of vectors without a zero-sequence component. (*Evans & Heumann.*)

and subsequent elimination of the zero-sequence component, and (*b*) methods for the determination of the positive- and negative-sequence components from vectors not containing a zero-sequence component.

a. Determination and Elimination of Zero-sequence Components. One method for determining the zero-sequence component is shown in Fig. 136. It is based on the fundamental equation (27) and consists in taking one-third of the vector sum of the three phase vectors. The method is simple and gives directly the quantity desired.

Another method, shown in Fig. 137(*a*), requires the determination of the neutral point of the system with zero-sequence components eliminated. As described in Sec. 114, the neutral point of this system of vectors is found at the point O', the intersection of two median lines. The vector connecting the points O and O' gives directly the zero-sequence voltage in its proper phase position with respect to the line-to-neutral vectors E_a, E_b, and E_c.

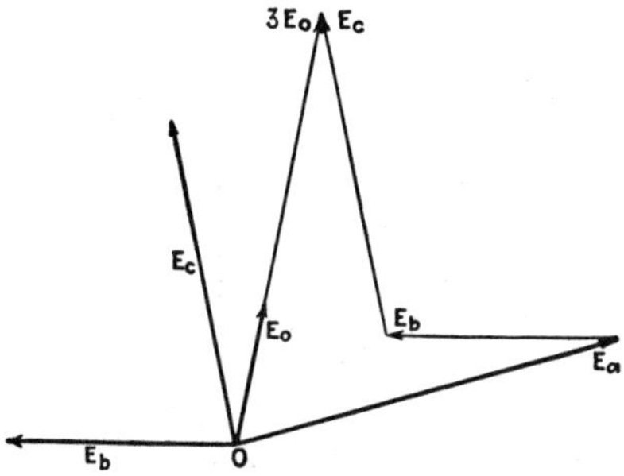

FIG. 136.—Determination of the zero-sequence component by taking one-third of the sum of the phase quantities.

Either of these methods provides the zero-sequence component from which the three new vectors indicated in Fig. 137(*b*) as $(E_a - E_0)$, $(E_b - E_0)$, and $(E_c - E_0)$ can be obtained and which include only the positive- and negative-sequence components.

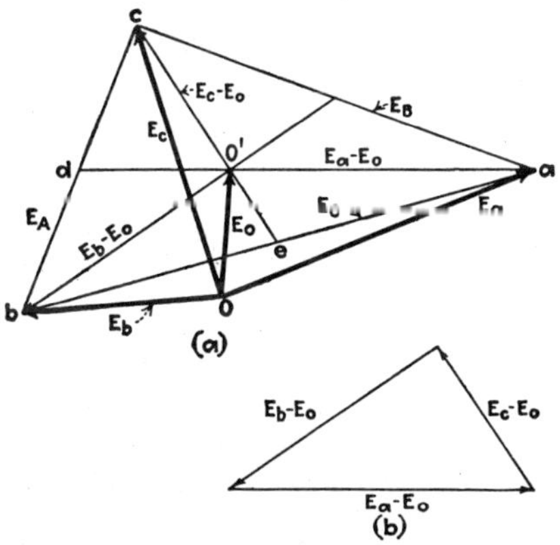

FIG. 137.—Determination of zero-sequence component from intersection of medians. (*a*) Determination of E_o; (*b*) closed triangle after eliminating E_o.

b. Determination of Positive- and Negative-sequence Components. Referring to the fundamental equations (28) and (29), these may be rewritten as

$$E_{a1} = \tfrac{1}{3}(E_a + E_b e^{j120°} + E_c e^{j240°}) \qquad (298)$$

$$E_{a2} = \tfrac{1}{3}(E_a + E_b e^{j240°} + E_c e^{j120°}) \qquad (299)$$

From the standpoint of graphical constructions, these signify that to determine E_{a1}, E_b must be rotated through 120 deg. and E_c through 240 deg., both in a counterclockwise direction and the resultant vectors added together with E_a. The resultant vector is equal to $3E_{a1}$. Similar relations hold for E_{a2}. These constructions are shown in Fig. 138.

After rotating E_b and E_c, one-third of the sum of the three vectors may be obtained conveniently by constructing a triangle whose apexes are the termini of the three vectors whose sum is desired. The intersection of the median lines of the triangle so formed locates the terminus of the vector E_{a1}. While this construction is not shown, it may be verified by trial in Fig.

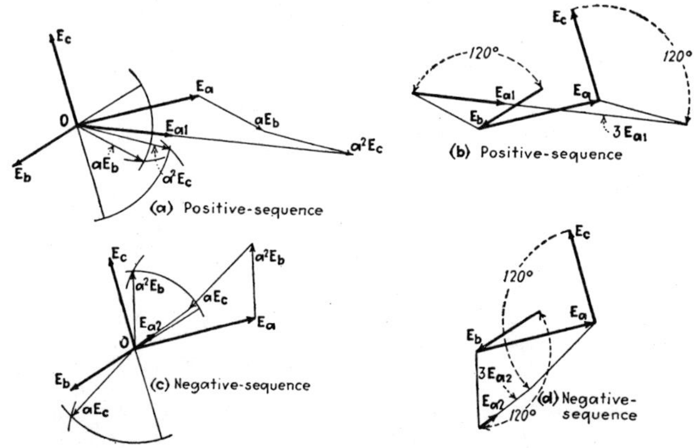

FIG. 138.—Graphical construction for positive- and negative-sequence components.

138(*a*). A similar construction is applicable to the negative-sequence system. This construction, applied to positive- or negative-sequence, corresponds to that described in Fig. 137 for zero-sequence.

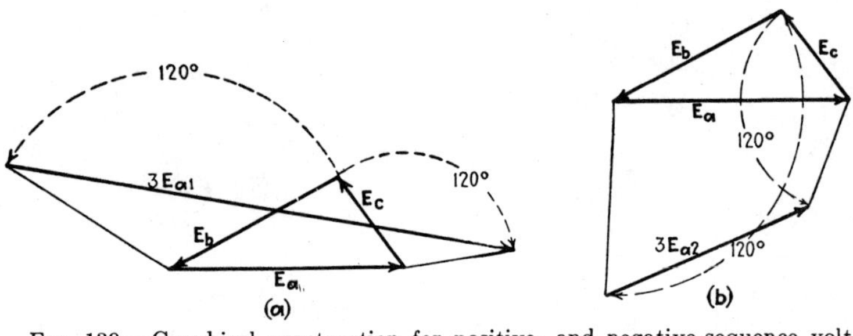

FIG. 139.—Graphical construction for positive- and negative-sequence voltages when the voltages form a closed triangle ($E_0 = 0$). Vectors rotated 120 deg (*a*) Positive-sequence, (*b*) negative-sequence.

For three-phase three-wire systems or more generally when the zero-sequence components are equal to zero, a number of interesting constructions are available. Most of these are based on the fact that for this case the vectors may be drawn as the sides of a closed triangle. One of these methods is given in

Fig. 139, which, in fact, is the same as that shown in Fig. 138 except that the construction is on the triangle. The method is self-explanatory and appears as a very direct application of the

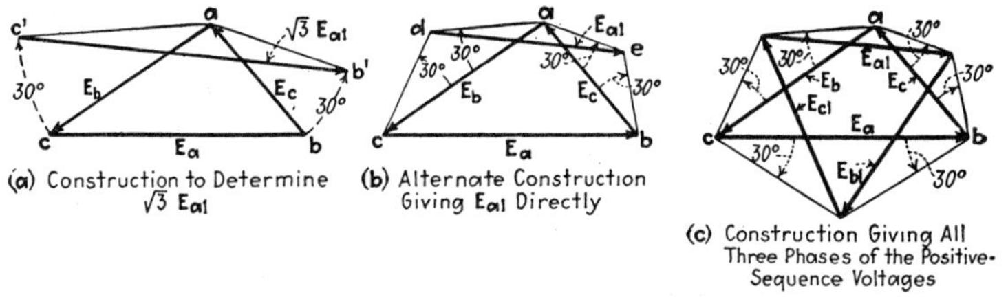

FIG. 140.—Graphical construction for positive-sequence voltages when voltages form a closed triangle ($E_{ao} = 0$). Vectors rotated 30 deg. (*Genkin.*)

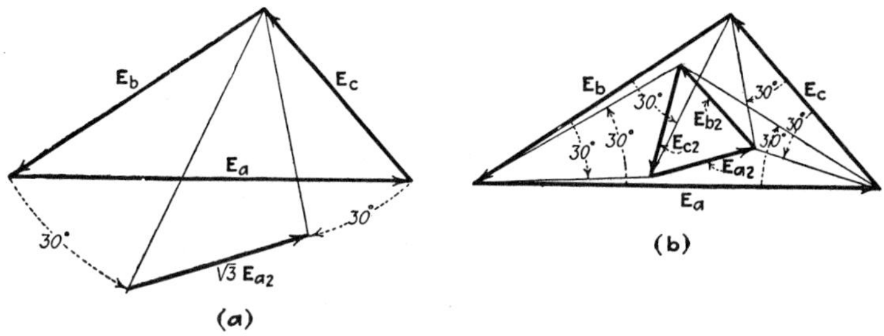

FIG. 141.—Graphical construction for negative-sequence voltages when voltages form a closed triangle ($E_{ao} = 0$). Vectors rotated 30 deg. (*Genkin.*)

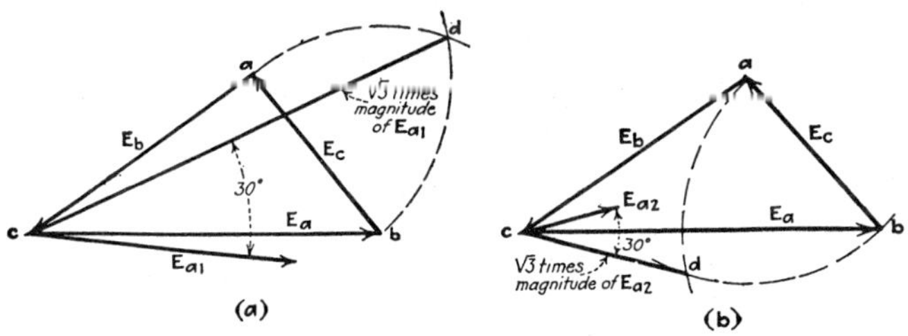

FIG. 142.—Graphical construction for positive- and negative-sequence voltages when voltages form a closed triangle ($E_o = 0$). Vectors rotated 60 deg. (*a*) Construction for E_{a1}; (*b*) construction for E_{a2}.

fundamental theory. The particular construction illustrated in Fig. 139 is included in the German Electrical Standards.*

In Fig. 140(*a*) is shown a construction by Genkin,[134] based on equation (289). Vectors E_b and E_c are each rotated outward from the triangle through an angle of 30 deg. from their common apex as a center and the terminals connected together. The

* Regeln für die Bewertung und Prüfung von elektrischen Maschinen, §16.

resultant vector gives the correct phase position of E_{a1}, but its amplitude is equal to $\sqrt{3}\bar{E}_{a1}$. Similar procedures apply to the other apexes and provide the positive-sequence components of voltage in the remaining phases. For the negative-sequence the vectors must be rotated inward, with the lines connecting the terminals as before. This gives the vector of the proper

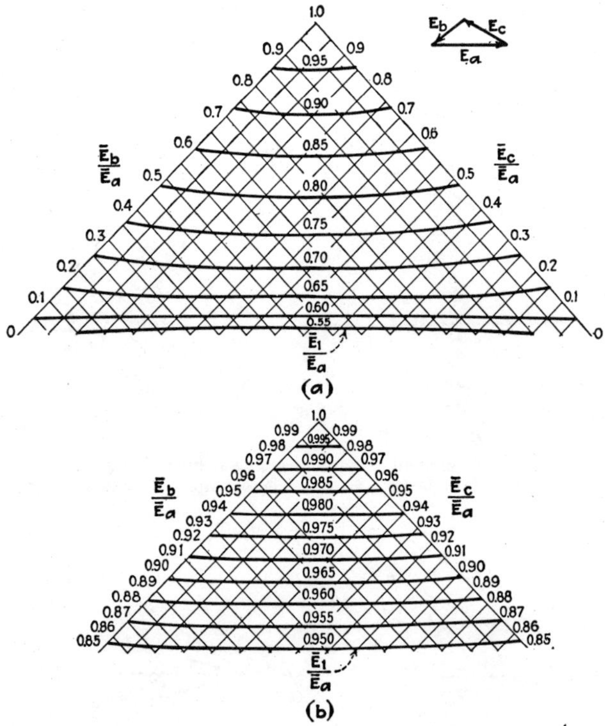

FIG. 143.—Positive-sequence component of systems of closed triangles.

phase position but whose magnitude is $\sqrt{3}\bar{E}_{a2}$. This construction is shown in Fig. 141(a).

A modification of this method has also been given by Genkin whereby the result gives not only the correct phase position but also the correct magnitude. This modification is shown in Fig. 140(b) in which isosceles triangles are constructed upon the sides of the triangle by drawing lines making angles of 30 deg. with the sides. The apexes of these triangles are connected together giving the desired vector. The method lends itself to the determination of the positive- and negative-sequence vectors in each of the three phases as illustrated in Figs. 140(c) and 141(b), respectively.

The preceding two methods may be briefly described as based on (1) the rotation of two vectors through 120 deg., and (2) on the rotation of two vectors through 30 deg. A third method is also available which can be similarly described as being based on the rotation of but one vector through 60 deg. This method

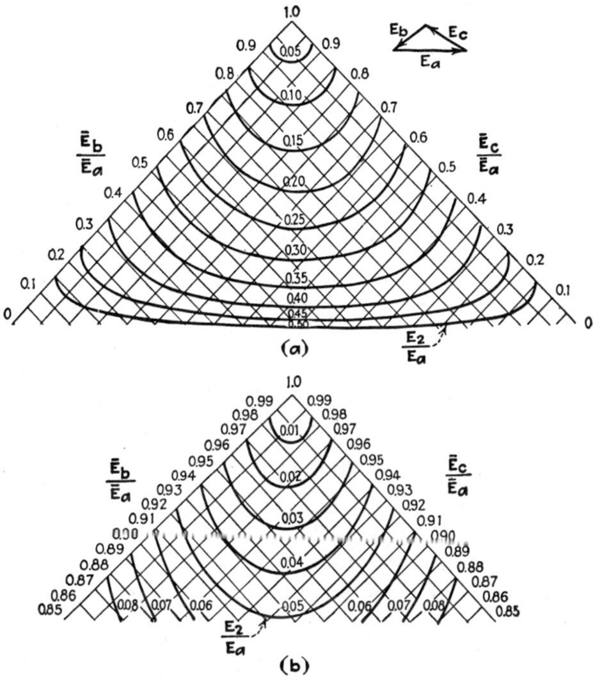

FIG. 144.—Negative-sequence component of systems of closed triangles.

is shown in Fig. 142 and is based on equations (290) and (293). Given the three vectors formed by the closed triangle abc, constructing an equilateral triangle externally on any side such as ab (which is equivalent to rotating that side through 60 deg.), and connecting the apex d of that triangle to the apex c, opposite the side used as a base for the equilateral triangle, gives a line whose length is equal to $\sqrt{3}$ times the positive-sequence component. Similarly, constructing the triangle internally and connecting the apexes, a line is obtained whose length is equal to $\sqrt{3}$ times the negative-sequence component as shown in (b). The positions of these lines do not give the correct phase position of the components. For the positive-sequence it is necessary to rotate the vector backward 30 deg. as shown in (*a*), and for the negative-sequence component forward through 30 deg. as shown in (*b*).

Of the three types of construction described, that of Fig. 139 is the most fundamental and direct but involves an operation upon two quantities through inconvenient angles; that of Figs. 140 and 141 also involves operation upon two quantities, but the graphical construction is somewhat easier; but the construction of Fig. 142 is the simplest when only the magnitude of the sequence components is desired, as it involves operation on only one quantity by means of a very simple construction.

117. Charts.

The positive- and negative-sequence components of voltages or currents on three-phase three-wire systems or more generally where $E_{a0} = 0$ may conveniently be determined by the charts of Figs. 143 and 144 for positive- and negative-sequence, respectively. These charts are based on the vectors forming a closed triangle as indicated in the figures. The ratios of the two shorter sides to the longest side are used as coordinates with the point so located as to determine the magnitude of the ratio of positive-sequence voltage to the voltage of the longest side which was chosen as a reference. Figures 143(*b*) and 144(*b*) have been drawn to a larger scale in order to give better accuracy for the ordinary small unbalances encountered in practice. The negative-sequence components are determined in a similar manner. If only the amplitude of three voltage vectors is known, then it is not possible to determine whether the triangle should be plotted as shown above the reference line E_a or below it. Consequently, it is not possible to determine whether the larger component is the positive- or the negative-sequence component. This, in general, can be determined from some other information. However, in the usual problem, the positive-sequence quantity will predominate.

118. Unbalance Factor.

The ratio of the negative- to positive-sequence amplitudes is commonly called the unbalance factor and is included in the German Electrical Standards. This factor was proposed by

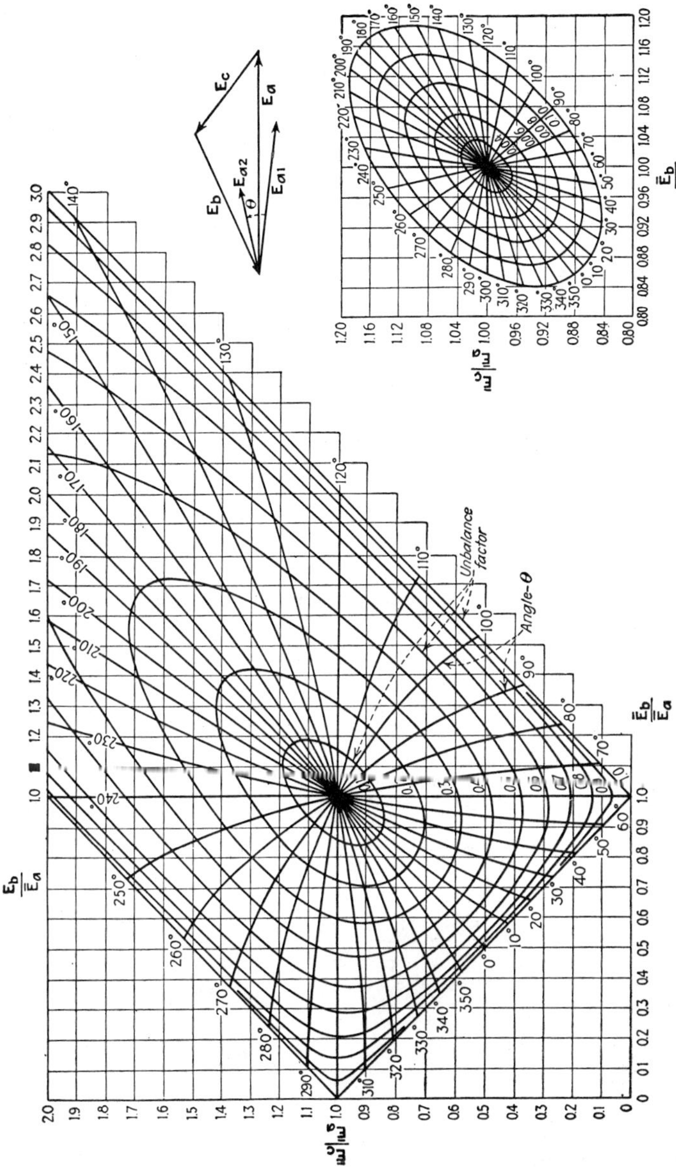

FIG. 145.—Determination of unbalance factor.

Evans and Pierce for measuring the unbalance in voltages and currents on three-phase three-wire systems. They also devised means of measuring voltage and current unbalance factors.*

The unbalance factor may conveniently be determined by means of **charts,** such as those given in Fig. 145. In this figure the ratios $\frac{\bar{E}_b}{\bar{E}_a}$ and $\frac{\bar{E}_c}{\bar{E}_a}$ are used as coordinates and the point thus determined gives the unbalance factor by interpolation of a family of curves. The angle plotted in the diagram gives the angle by which the vector E_{a2} leads the vector E_{a1}. The smaller chart of Fig. 145 gives the unbalance factor on a larger scale and is convenient for estimating the amount of unbalance on systems that are approximately balanced.

The first work along this line was published by R. Dubusc[(143)] in 1927 and amplified by von G. Hauffe[(161)] in 1929 and by the authors in this work.

Problems

1. The line-to-line voltages of a system measure $E_A = 75$, $E_B = 65$, and $E_C = 40$. Determine the positive- and negative-sequence components in the three phases by graphical construction.

2. Using the values of phase voltages from Prob. 1 determine the positive- and negative-sequence components, the unbalance factor, and the angle between the components by means of the charts in Figs. 143 to 145.

3. Given the voltages:

$$E_a = 25 - j2.0$$
$$E_b = -13 - j2.0$$
$$E_c = 0 + j15$$

Determine the three sequence components graphically.

4. Prove the relations of Fig. 135

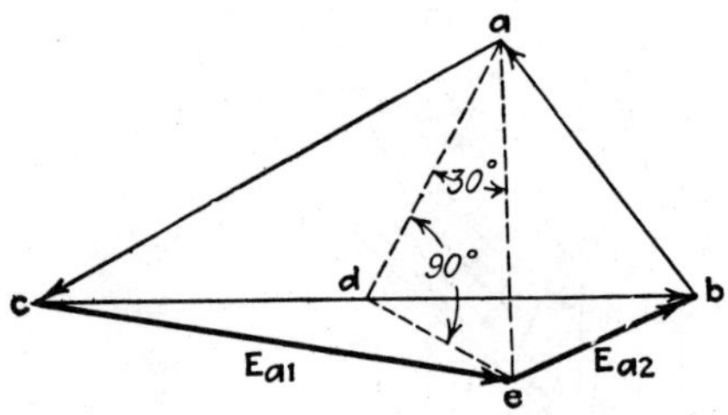

5. Dubusc in his report to the International Congress on Large Electrical Networks in 1927 has given the construction shown in the accompanying figure to obtain the positive- and negative-sequence components. Given the voltage triangle *abc*, draw the median *ad*. On this median construct the 30-deg. right triangle *adc*. The positive- and negative-sequence voltage components relative to the *a* phase are then given by *ce* and *eb*, respectively. Prove this relation.

* U. S. Patents 1,535,589 and 1,567,582.

CHAPTER XIV

MEASUREMENT OF SEQUENCE VOLTAGES AND CURRENTS

Sequence components may be measured by means of instruments or meters operating on well-known principles requiring only special external networks or in some cases special windings. In all the previous discussions, sequence quantities have been treated as component vectors of the *single-phase* vector quantities measured in the different phases. For the proper understanding of the present discussion, it is preferable to consider the sequence quantities as having an existence as real and definite as the single-phase quantities; for example, positive-sequence voltage is just as real as a line-to-line voltage and will be measured in a similar manner by a meter element and an external network.

Sequence quantities that can be measured include for each sequence the same quantities that can be measured in a single-phase circuit, namely, voltage, current, real power, reactive volt-amperes, power factor, volt-amperes, etc. In addition, the sequence quantities may be combined in various ways for measurement. The sequence quantities may be used to operate indicating, integrating, and recording instruments and meters. In addition, they may be used to operate relays of the various kinds, such as protective relays and voltage regulators.

The first proposal for measuring sequence quantities, such as voltage or current, was the result of a discussion between Fortescue, Chubb, and Slepian. They contemplated the use of rotating machines to eliminate one sequence to permit the measurement of another sequence quantity. The first measurements, however, were made by R. D. Evans who proposed static networks to segregate a desired sequence quantity in the presence of several sequence quantities. A large number of different networks have been worked out by Evans, Allcutt, Fortescue, and others. The first device actuated by a sequence component, embodied in a commercial form, was the negative-sequence overcurrent relay. Subsequently, voltage regulators were built to be operated by positive-sequence voltage. At the present time, other devices responding to sequence components are being used for a number of purposes and are being considered for many other applications.

On a three-phase four-wire system there may exist the three sequence voltages. Consequently, in order to measure all of these sequence quantities on a commercial system, it is necessary to have available positive-, negative-, and zero-sequence voltmeters. On a three-phase three-wire system the zero-sequence voltage is indeterminate and the corresponding voltmeter is unnecessary and inapplicable. Similar relations hold for the sequence currents, and positive-, negative-, and zero-sequence ammeters are necessary. In general, positive- and negative-sequence voltmeters are similar but are different from the zero-sequence voltmeters which are simpler. The sequence ammeters may likewise be classified as positive- and negative-sequence ammeters and zero-sequence ammeters.

In this chapter on sequence measuring devices, the various types of sequence voltmeters and ammeters will be studied. In the following chapter the various combinations of sequence voltages and currents in wattmetric devices will be considered.

119. Measurement of Zero-sequence Voltage.

The zero-sequence voltage, as shown previously, is determined from the line-to-neutral voltages by means of equation (27):

$$E_0 = \tfrac{1}{3}(E_a + E_b + E_c)$$

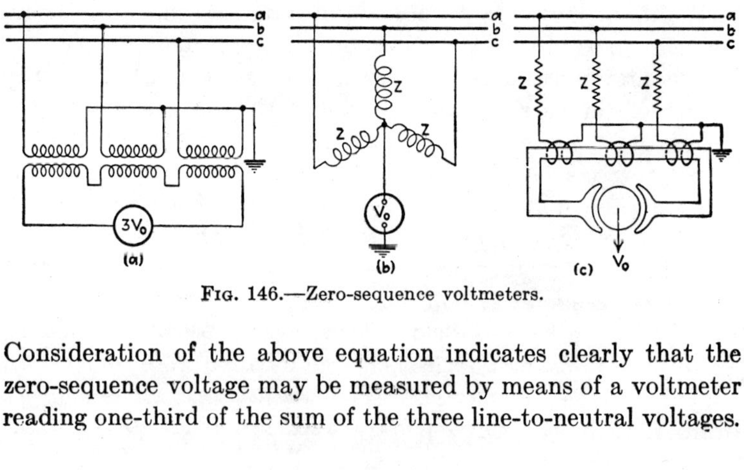

FIG. 146.—Zero-sequence voltmeters.

Consideration of the above equation indicates clearly that the zero-sequence voltage may be measured by means of a voltmeter reading one-third of the sum of the three line-to-neutral voltages.

Several alternative methods for measuring zero-sequence voltage are shown in Fig. 146. The form most commonly used is shown in Fig. 146(*a*); it makes use of three potential transformers and a standard voltmeter calibrated to read one-third of the voltage actually impressed across it. Figure 146(*b*) requires three accurately balanced impedances with the voltmeter connected between the star point and neutral or ground. Calibration must take into account the three parallel impedances $\frac{Z}{3}$, in series with the meter across the calibrating voltage. The special winding arrangement of (*c*) may also be used. For calibration the three terminals connected to phases *a*, *b*, and *c* should be connected together to form one side of the circuit upon which the calibrating voltage is impressed.

120. Measurement of Positive- and Negative-sequence Voltages.

The positive-sequence voltage as shown previously is determined from the line-to-neutral voltages by means of equation (28):

$$E_1 = \tfrac{1}{3}(E_a + aE_b + a^2E_c)$$

The negative-sequence voltage is determined in a similar manner by equation (29):

$$E_2 = \tfrac{1}{3}(E_a + a^2E_b + aE_c)$$

Equation (28) defines the conditions which must be satisfied by the network for the meter to measure positive-sequence voltage. Thus, the meter must respond with equal magnitude to voltages from phases *a*, *b*, and *c*, individually, but the currents through the meter must have relative phase angles of 0, 120, and 240 deg., respectively, corresponding to 1, *a*, and a^2, respectively. The phase shift of zero degrees may be obtained by using a very large resistance external to the meter winding. Phase shifts of 120 and 240 deg. by external impedances alone would involve negative resistance. This difficulty is overcome by reversing the connections of windings *b* and *c* with respect to winding *a* as indicated in Fig. 147. The effective value of the meter current is indicated by

$$I_M = \frac{E_a}{Z_a} - \frac{E_b}{Z_b} - \frac{E_c}{Z_c} \qquad (300)$$

With this connection the power-factor angles of the impedances in series with windings a, b, and c, from comparison of equations (28) and (300), should be 0, +60, and −60 deg., respectively, for the positive-sequence voltmeter.

For measuring the negative-sequence voltage, it may be shown in a similar manner, from equations (29) and (300), that the impedances Z_a, Z_b, and Z_c should have the values given in connection with Fig. 147(b).

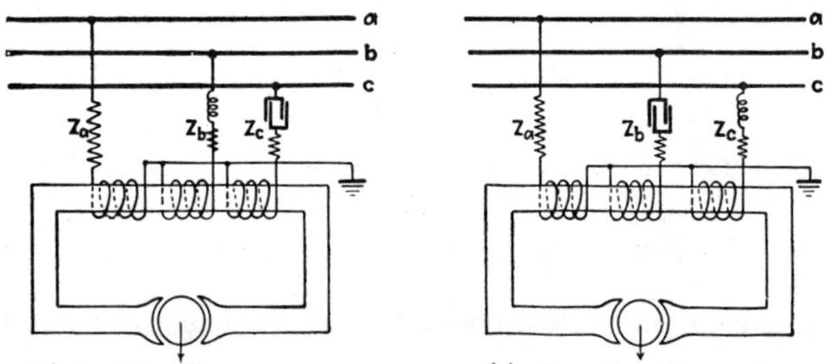

General Relations

$$Z_b = Z_a e^{+j60°} \qquad Z_b = Z_a e^{-j60°}$$
$$Z_c = Z_a e^{-j60°} \qquad Z_c = Z_a e^{+j60°}$$

Typical Values

$$Z_a = R \qquad Z_a = R$$
$$Z_b = R(0.5 + j0.866) \qquad Z_b = R(0.5 - j0.866)$$
$$Z_c = R(0.5 - j0.866) \qquad Z_c = R(0.5 + j0.866)$$

Phase rotation $a\ b\ c$
for positive-sequence voltage

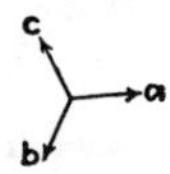

FIG. 147.—Positive- and negative-sequence voltmeters.

121. Common Form of Positive- and Negative-sequence Voltmeters Applicable in Absence of Zero-sequence Voltage.

In general the voltage vectors of a three-phase system possess three degrees of freedom, and therefore the network branches must possess a corresponding number, which in the networks of Fig. 147 were obtained by the three impedances shown. However, if the zero-sequence voltage were eliminated, the meter and network would require only two degrees of freedom. The elimination of zero-sequence voltage may, of course, be obtained by using line-to-line voltages instead of line-to-neutral voltages. Special windings in the voltmeter are unnecessary as the instrument may be made to read directly by appropriate connection and adjustment of the external network. Thus positive-sequence voltage may be obtained directly by means of the simplified network due to C. T. Allcutt, illustrated in Fig. 148. Since this network is the one most commonly used, an analysis of its operation will be given. The basic equations of this type of network may be derived from an application of Kirchhoff's laws giving the equations

$$\left.\begin{aligned} E_a - E_c &= Z_B I_B + M(I_A + I_B) \\ E_c - E_b &= Z_A I_A + M(I_A + I_B) \end{aligned}\right\} \quad (301)$$

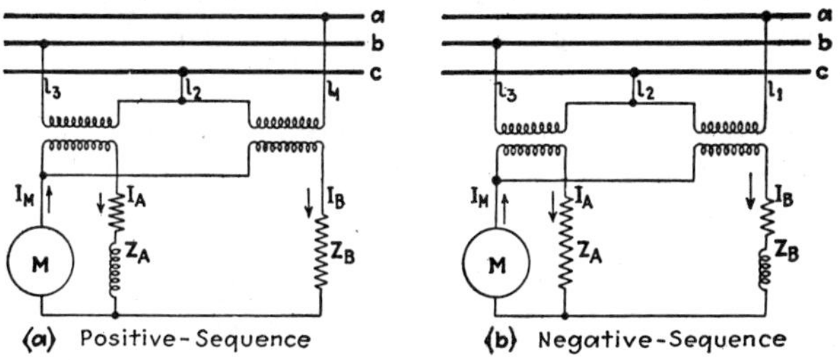

General Relations

$Z_A = -a^2 Z_B$ $\qquad$ $Z_B = -a^2 Z_A$

Typical Values

$$\begin{aligned} Z_B &= R \\ Z_A &= Re^{j60°} \\ &= R(0.50 + j0.866) \end{aligned} \qquad \begin{aligned} Z_A &= R \\ Z_B &= Re^{j60°} \\ &= R(0.50 + j0.866) \end{aligned}$$

Phase rotation $a\ b\ c$
for positive-sequence voltage

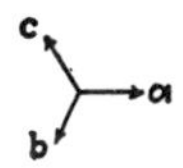

FIG. 110. Positive and negative-sequence voltmeters. Applicable in absence of zero-sequence voltage. The most commonly used connections.

Multiplying the first equation given above by Z_A and the second by Z_B and adding them, and remembering that $I_A + I_B = I_M$, there is obtained:

$$I_M = \frac{Z_A(E_a - E_c) + Z_B(E_c - E_b)}{Z_A Z_B + MZ_A + MZ_B} \quad (302)$$

Expressing E_a, E_b, and E_c in terms of the sequence voltages, the meter current I_M becomes

$$I_M = \frac{(1 - a)(Z_A + aZ_B)}{Z_A Z_B + M(Z_A + Z_B)} E_1 + \frac{(1 - a^2)(Z_A + a^2 Z_B)}{Z_A Z_B + M(Z_A + Z_B)} E_2 \quad (303)$$

The condition for the meter to measure only positive-sequence voltage is that the coefficient of E_2 in equation (303) must equal zero. Hence, the necessary condition defining the positive-sequence segregating network is

$$Z_A = -a^2 Z_B = Z_B\, e^{+j60°} = Z_B(0.5 + j0.866) \quad (304)$$

For the negative-sequence segregating network a similar expression may be obtained by setting the coefficient of E_1 in equation (303) equal to zero which gives:

$$Z_B = -a^2 Z_A = Z_A\, e^{+j60^\circ} = Z_A(0.5 + j0.866) \qquad (305)$$

The network most commonly used with the positive- or negative-sequence voltmeter consists of two branches, one a resistor and the other an inductive impedance with a power-factor angle of +60 deg. as indicated in the tables accompanying Fig. 148.

The quantity measured by the positive-sequence voltmeter of Fig. 148 is obtained by substituting equation (304) in equation (303) which gives

$$I_{M1} = \frac{3E_1}{Z_B + M(1 - a)} = \frac{3E_1}{Z_B + \sqrt{3}Me^{-j30^\circ}} \qquad (306)$$

$$= \frac{3E_1 e^{+j60^\circ}}{Z_A + \sqrt{3}Me^{+j30^\circ}} \qquad (306a)$$

The corresponding expression for the negative-sequence voltmeter is given by

$$I_{M2} = \frac{3E_2}{Z_B + M(1 - a^2)} = \frac{3E_2}{Z_B + \sqrt{3}Me^{+j30^\circ}} \qquad (307)$$

$$= \frac{3E_2 e^{-j60^\circ}}{Z_A + \sqrt{3}Me^{-j30^\circ}} \qquad (307a)$$

Vector Diagram and Range of Impedance Values. Equation (306) may be used as a basis for a vector diagram for the positive-sequence network of Fig. 148. In this diagram, shown in Fig. 149, resistance is plotted as the abscissa and reactance as the ordinate, plotted upward if inductive and downward if condensive. If the meter impedance M is zero, the current through the meter will lag behind the positive-sequence voltage of phase a by the power-factor angle of the impedance Z_B, and if Z_B is a pure resistance the current will be in phase with that voltage. If the meter impedance is not zero, then the equivalent impedance of the meter and network may be obtained by adding to the impedance Z_B the impedance M multiplied by $\sqrt{3}$ and shifted through 30 deg. in a clockwise direction as indicated in the diagram. The current through the meter will be equal to three times the positive-sequence voltage divided by the effective impedance OD and the phase of the current through the meter element will be shifted back of the positive-sequence voltage of phase a by the angle that OD makes with the X-axis as indicated by the position of I_M. It is important that the transformer connections shown in Fig. 148 be followed carefully for the above conditions to be fulfilled.

According to equation (304), the only condition which must be satisfied, so that the meter reading be proportional to the positive-sequence voltage, is that the power-factor angle of Z_A lead that of Z_B by 60 deg. By plotting this relation, the diagram of Fig. 149 may be used to visualize the phase-angle range of the impedances Z_A and Z_B. The limitation of avoiding negative resistance in both quantities fixes the range of the impedance angles of these quantities to the values

$$Z_A = R(e^{-j30°} \text{ to } e^{+j90°})$$
$$Z_B = R(e^{-j90°} \text{ to } e^{+j30°})$$

Furthermore, it is not feasible to reach the extreme range of either $e^{+j90°}$ or $e^{-j90°}$ because of the fact that all circuits have some resistance. The most important case is that in which Z_B is pure resistance and Z_A is made up of a resistance and inductance, the combination having a power-factor angle of +60 deg.

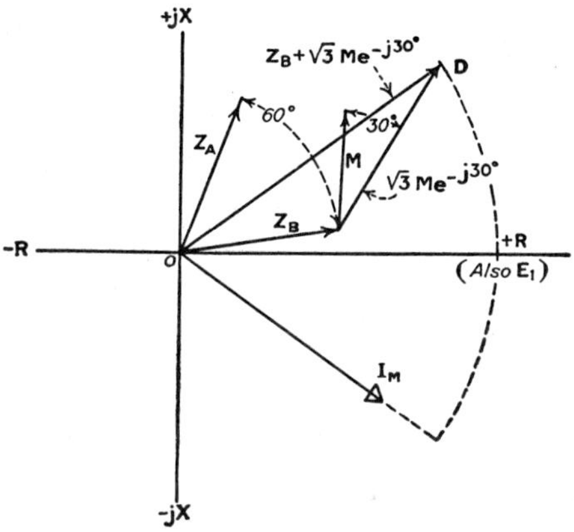

FIG. 149.—Diagram for studying positive-sequence segregating networks of the type shown in Fig. 148.

Phase angle of meter current = negative of phase angle of OD.

$$\text{Magnitude of meter current} = \frac{3E_1}{\text{impedance } (OD)}$$

A negative-sequence diagram similar to that of Fig. 149 for the positive-sequence can be derived from equation (307). It is to be noted, however, that the phase-angle shift for the negative-sequence will be different from that for positive-sequence.

Conversion of Sequence Networks—Interchange of Leads or Network Impedances. A positive-sequence voltmeter may be converted into a negative-sequence voltmeter by the mere **interchange of** two **leads** between the network and the system whose voltage is being measured. Thus the positive-sequence voltmeter of Fig. 148(*a*) becomes a negative-sequence voltmeter by tapping network leads l_2 and l_3 to phases *b* and *c*, respectively. Also, a positive-sequence network can be converted into a negative-sequence network by the interchange of the phase-shifting impedances, for example, by interchanging the impedances of branches Z_A and Z_B of the network of Fig. 148. This interchangeability of the positive- and negative-sequence networks is a result which is borne out by analysis and is to be expected since the mere interchange of two leads on a three-phase system converts a positive-sequence voltage to a negative-sequence voltage and *vice versa*.

In connection with Fig. 148(*a*), it should be pointed out that if leads l_1, l_2, and l_3 are tapped to phases *abc*, *bca*, and *cab*, respectively, the voltage applied to the meter will be identical in magnitude, but the phase angle will successively be shifted by 120 deg. in the same sense as the positive-sequence voltages. In the case of the negative-sequence voltmeter of Fig. 148(*b*), the tapping of leads l_1, l_2, and l_3 to phases *abc*, *bca*, and *cab*, respectively, cause the meter to read identical in magnitude, but the phase angle of the voltages applied to the meter will successively be shifted by 120 deg. in the same direction as the negative-sequence voltage considering phase rotation *abc* for positive-sequence voltage.

122. Second Form of Positive- and Negative-sequence Voltmeters Applicable in Absence of Zero-sequence.

The positive- and negative-sequence networks that have just been described are characterized by the fact that the impedance angles of the two impedances differ by 60 deg. and one transformer connection is reversed. Another form of network, which may be used when the zero-sequence component of voltage is zero, is characterized by the fact that no reversal in transformer connection is required but the impedance angles of the two impedances must differ by 120 deg. A network of the latter class is indicated in Fig. 150. The network may be made with the impedances

$$Z_B = R(0.5 - j0.866)$$
$$Z_C = R(0.5 + j0.866)$$

for which one branch is a condensive impedance and the other branch is an inductive impedance. If these conditions are fulfilled, and the transformer connections of Fig. 150 followed, the current through the meter will have the same value relative to E_1 as that in Fig. 148.

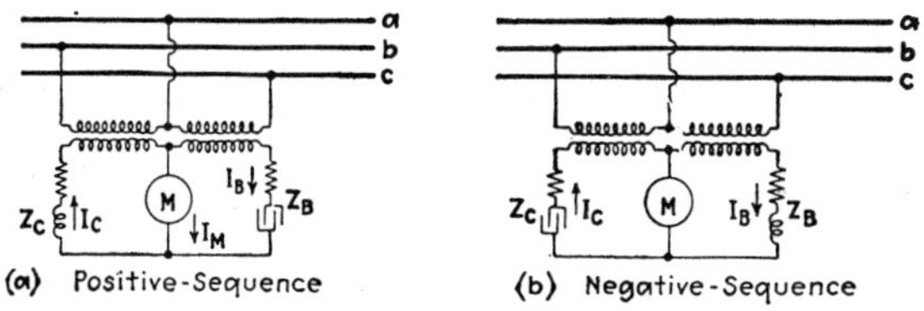

General Relation

$$Z_C = aZ_B \qquad\qquad Z_B = aZ_C$$

Typical Values

$$Z_B = R(0.500 - j0.866) \qquad Z_B = R(0.500 + j0.866)$$
$$Z_C = R(0.500 + j0.866) \qquad Z_C = R(0.500 - j0.866)$$

Phase rotation $a\ b\ c$
for positive-sequence voltage

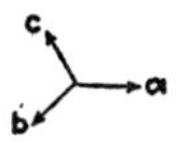

FIG. 150.—Positive- and negative-sequence voltmeters. Segregating network with impedance angles differing by 120° instead of 60° as in Fig. 148.

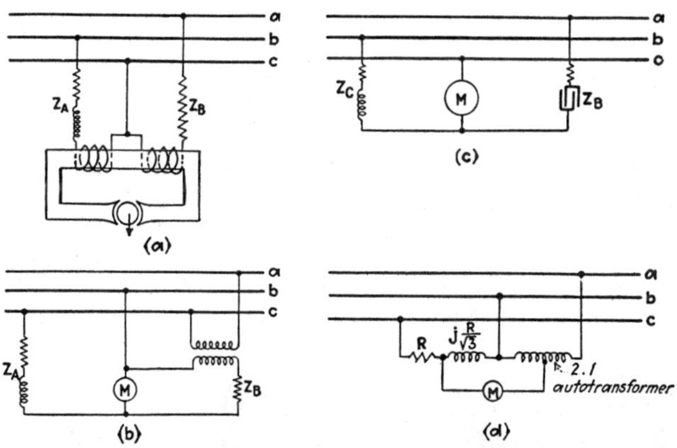

FIG. 151.—Connections for eliminating transformers in positive-sequence voltmeters. (*a*) Special windings with network constants similar to Fig. 148; (*b*) Fig. 148 with one transformer omitted; (*c*) Fig. 148 with both transformers omitted. (*d*) Negative-sequence voltmeters may be obtained by merely interchanging leads or network impedances.

123. Reduction and Elimination of Voltage Transformers for Sequence Networks.

In some cases it will be desirable to reduce the number of potential transformers required with the positive- or negative-sequence voltmeters or even to eliminate them completely. Various schemes for this purpose are outlined in Fig. 151. The connection of (*a*) with two windings on the meter obviously avoids the necessity for potential transformers, since the connection of the meter winding may be made in such a manner as to give the same result as with the transformer connections of Fig. 148. In case the use of a voltmeter with special windings is not desired, one of the potential transformers of Fig. 148 may be omitted as indicated in Fig. 151(*b*). If there is no objection to the use of a condenser as an element in the network, then the transformers of Fig. 150 may be eliminated and the connection reduced to that of Fig. 151(*c*). A fourth method involving only one autotransformer is shown in Fig. 151(*d*).

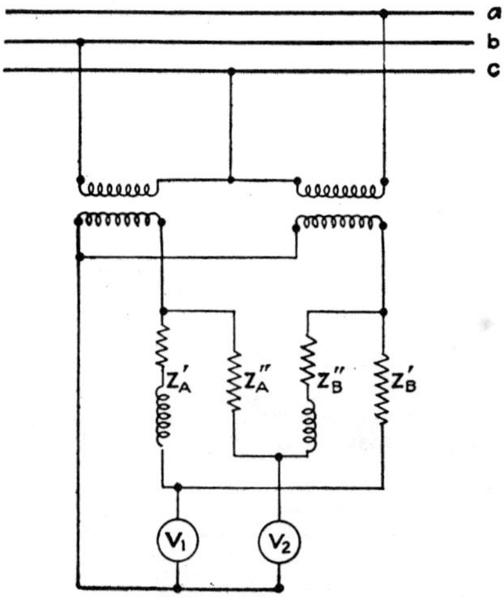

General Relations

$$Z'_A = -a^2 Z'_B$$
$$Z''_B = -a^2 Z''_A$$

Typical Values

$$Z''_A = Z'_B = R$$
$$Z'_A = Z''_B = R(0.500 + j0.866)$$

FIG. 152.—Positive- and negative-sequence voltmeters for simultaneous measurement in meters V_1 and V_2, respectively.

124. Simultaneous Measurement of Sequence Voltages.

Sometimes it is desired to measure simultaneously both the positive- and negative-sequence voltages. The simplest network for this purpose is a combination of (*a*) and (*b*) of Fig. 148, as indicated in Fig. 152. In case it is desired to measure all three sequence voltages simultaneously, the connection of Fig. 146(*a*) may be used for the zero-sequence with auxiliary potential transformers to give the equivalent of Fig. 152 to obtain the positive- and negative-sequence voltages. In case special meter windings can be used, it is possible to superpose on the connection of Fig. 146(*c*) for zero-sequence, Fig. 151(*a*) for positive-sequence and a similar one for negative-sequence, with the branch impedances Z_A and Z_B interchanged.

125. Measurement of Zero-sequence Current.

The zero-sequence current, as shown previously, is determined from the line currents by means of the expression

$$I_0 = \tfrac{1}{3}(I_a + I_b + I_c) \tag{308}$$

Consideration of the above equation indicates clearly that the zero-sequence current may be measured by means of an ammeter reading either (1) one-third of the sum of the line currents or (2) the neutral or ground current. Several alternative methods for measuring zero-sequence current are shown in Fig. 153. In

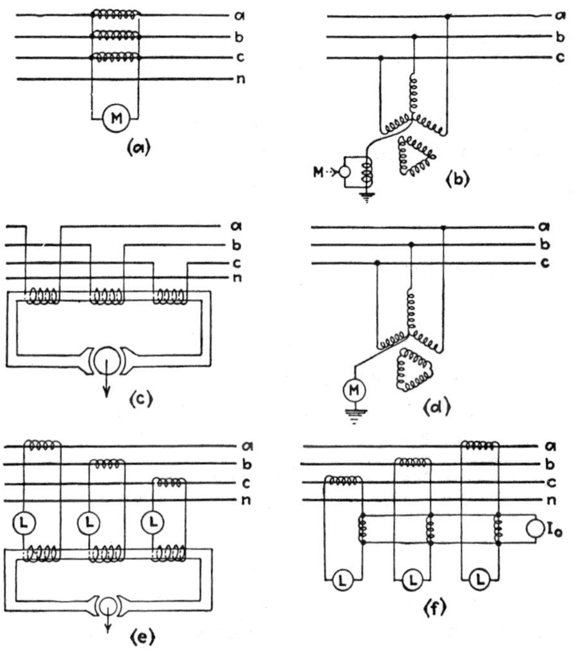

FIG. 153.—Zero-sequence ammeters.

the schemes of (*a*), (*c*), (*e*), and (*f*) the sum of the three line currents are used, while in the schemes of (*b*) and (*d*) the neutral or ground current is used. The most common connections make use of current transformers but direct measuring devices are shown in (*c*) and (*d*). In connection with the schemes of (*b*) and (*d*), it should perhaps be pointed out that it is essential that all the zero-sequence current flows through the neutral wire or ground connection in which the meter is installed, or that this current bears a known proportionality to the total current. The scheme used in Fig. 153(a) requires the bussing of all current transformers. This restriction may be avoided by the use of separate ammeter windings as in (e), or by the addition of separate current transformers as in (f). In the latter case it may be noted that if the current transformers have a ratio of 3 to 1, the ammeter will read zero-sequence current directly. With the schemes of (e) or (f) the ordinary current transformer loads, indicated by L in the figures, may be connected in delta, for example, without interfering with the measurement of zero-sequence current.

126. Measurement of Positive- and Negative-sequence Currents.

The positive-sequence current, as shown previously, may be determined from the line currents by means of the following expression

$$I_1 = \tfrac{1}{3}(I_a + aI_b + a^2I_c) \tag{309}$$

The negative-sequence current is determined in a similar manner by means of

$$I_2 = \tfrac{1}{3}(I_a + a^2I_b + aI_c) \tag{310}$$

From equation (309) it will be seen that the meter must respond with equal magnitudes to currents from phases a, b, and c, individually, but the currents through the meter must have relative phase angles of 0, 120, and 240 deg., respectively, corresponding to 1, a, and a^2, respectively. Figure 154 shows a meter and network suitable for measuring positive-sequence current according to equation (309). This meter connection corresponds closely to Fig. 147 for measuring positive-sequence voltage. The response of the meter of Fig. 154 is given by

$$I_{M1} = \frac{S_a}{Z_a + S_a}I_a - \frac{S_b}{Z_b + S_b}I_b - \frac{S_c}{Z_c + S_c}I_c \tag{311}$$

where Z_a, Z_b, and Z_c represent the impedance in the path through the meter windings of phases a, b, and c, respectively; and S_a, S_b, and S_c represent the shunt impedances in parallel with Z_a, Z_b, and Z_c, respectively. In this analysis the mutual impedance between windings will be ignored because it is assumed to be small in comparison with the series impedances

Z_a, Z_b, and Z_c. Subsequently, other analyses will be given which definitely take into account the impedance of the meter windings. It will be noted that equation (311) is in the form

$$I_{M1} = K_a I_a - K_b I_b - K_c I_c \tag{312}$$

where K_a, K_b, and K_c are the coefficients of I_a, I_b, and I_c in equation (311), respectively. By comparison with equation

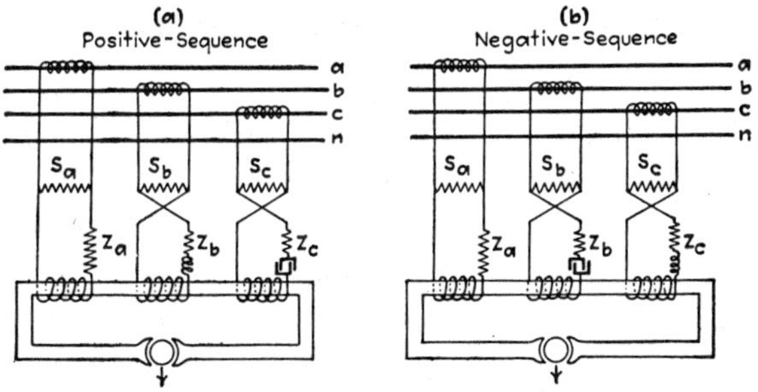

General Relation

$$\frac{S_b}{Z_b + S_b} = \frac{S_a}{Z_a + S_a} e^{+j60°} \qquad \frac{S_b}{Z_b + S_b} = \frac{S_a}{Z_a + S_a} e^{-j60°}$$

$$\frac{S_c}{Z_c + S_c} = \frac{S_a}{Z_a + S_a} e^{-j60°} \qquad \frac{S_c}{Z_c + S_c} = \frac{S_a}{Z_a + S_a} e^{+j60°}$$

Typical Values

$$\frac{S_a}{Z_a + S_a} = Ke^{j0°} \qquad \frac{S_a}{Z_a + S_a} = Ke^{j0°}$$

$$\frac{S_b}{Z_b + S_b} = Ke^{j60°} \qquad \frac{S_b}{Z_b + S_b} = Ke^{-j60°}$$

$$\frac{S_c}{Z_c + S_c} = Ke^{-j60°} \qquad \frac{S_c}{Z_c + S_c} = Ke^{+j60°}$$

Phase rotation $a\ b\ c$
for positive-sequence voltage

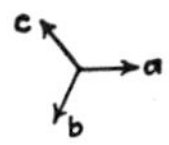

FIG. 154.—Positive- and negative-sequence ammeters.

(309) it will be seen that the meter of Fig. 154 will measure the positive-sequence current when

$$\left.\begin{aligned} K_a &= K \\ K_b &= -aK = Ke^{-j60°} \\ K_c &= -a^2K = Ke^{+j60°} \end{aligned}\right\} \tag{313}$$

in which K is numeric. Hence positive-sequence current is measured by the meter of Fig. 154 if (1) the meter responds equally to phases a, b, and c, individually, and (2) the current of winding a is in phase with the line current of phase a, the current of winding b lags the current of phase b by 60 deg., and the current of winding c leads the line current of phase c by 60 deg.

For measuring negative-sequence current it may be shown in a similar manner from equations (310) and (312) that the positive-sequence network may be used, provided the impedance branches Z_b and S_b are interchanged with those of branches Z_c and S_c, or provided that the line currents from phases b and c are interchanged.

127. Common Form of Positive- and Negative-sequence Ammeters Applicable in Absence of Zero-sequence Current.

The positive- and negative-sequence ammeters of Fig. 154 have been considered, in spite of the fact that they are not the most practical schemes, because they use in simple form the most general relations involved in the segregating networks. Means for simplifying the meter and network will now be considered. In general, the current vectors of a three-phase system possess three degrees of freedom and therefore the network branches must possess a corresponding number. However, if the zero-sequence current were eliminated, the network would require only two degrees of freedom. It is possible to eliminate the zero-sequence current before it enters the network and thus to require the connection of the network to only two phases of the system. The elimination of zero-sequence current is a subject of considerable importance in sequence measurement and for this reason it will be discussed in a subsequent section. For the purposes of the present discussion, it will be assumed that the zero-sequence current is non-existent as is the case with the three-phase three-wire system. Another method of simplifying the meter and network is to eliminate the special meter windings by obtaining the same result through a connection of the external impedances. These considerations lead to the type of network illustrated in Figs. 155 and 157.

The simplest type of positive- or negative-sequence network applicable where zero-sequence currents do not exist was proposed by C. T. Allcutt and is illustrated in Fig. 155. This form of network is frequently considered for commercial devices and for that reason an analysis of its operation will be given, taking into account the impedance of the meter element. By using the notation of Fig. 155, the expression for the meter current may be written

$$I_M = -\frac{Z_c}{Z_b + Z_c + M} I_c - \frac{Z_b}{Z_b + Z_c + M} I_b \tag{314}$$

In terms of sequence currents, I_M becomes

$$I_M = -\frac{(aZ_c + a^2Z_b)I_1 + (a^2Z_c + aZ_b)I_2}{Z_b + Z_c + M} \tag{315}$$

The condition for the meter to measure only positive-sequence current is that the coefficient of I_2 be zero, which gives

$$Z_c = -a^2Z_b = Z_b e^{+j60^\circ} \tag{316}$$

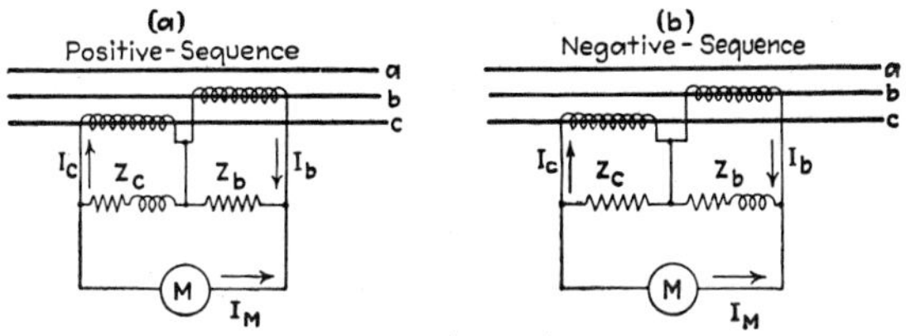

General Relation

$Z_c = -a^2Z_b$ $\qquad$ $Z_b = -a^2Z$

Typical Values

$Z_b = R$ $\qquad$ $Z_c = R$

$Z_c = R(0.500 + j0.866)$ $\qquad$ $Z_b = R(0.500 + j0.866)$

Phase rotation $a\ b\ c$
for positive-sequence voltage

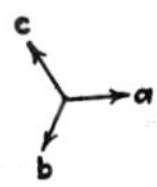

FIG. 155.—Positive- and negative-sequence ammeters applicable when zero-sequence current is not present.

In a similar manner the characteristics of the network for measuring negative-sequence current may be derived from equation (315) by making the coefficient of I_1 equal zero which gives

$$Z_b = -a^2Z_c = Z_c e^{+j60^\circ} \tag{317}$$

In the table associated with Fig. 155 are given not only the general relations which must be satisfied for the meter to measure positive- or negative-sequence current but also the typical values which are commonly used.

The quantity measured by the meter element of the network of Fig. 155 for positive- and negative-sequence connections is obtained by substituting in equation (315) the necessary relations for the segregation of sequence quantities indicated by equations (316) and (317), which give for the positive-sequence current I_{M1} and the negative-sequence current I_{M2} the following:

$$I_{M1} = \frac{1}{1 + \frac{M}{\sqrt{3}Z_b}\epsilon^{-j30^\circ}} = \frac{1}{1 + \frac{\overline{M}}{\sqrt{3}\,\overline{Z}_b}\epsilon^{j(\theta_M - \theta_{Z_b} - 30^\circ)}} \tag{318}$$

$$I_{M2} = \frac{1}{1 + \frac{M}{\sqrt{3}Z_b}\epsilon^{+j30^\circ}} = \frac{1}{1 + \frac{\overline{M}}{\sqrt{3}\,\overline{Z}_b}\epsilon^{j(\theta_M - \theta_{Z_b} + 30^\circ)}} \tag{319}$$

in which θ_M and θ_{Z_b} are the impedance angles of the impedances M and Z_b, respectively.

Vector Diagram. Equation (318) for positive-sequence current may be plotted graphically in the manner illustrated by Fig. 156. In this case the solution is obtained by plotting $\frac{M}{Z_b}$ passing through the point $(1 + j0)$. Rotate $\frac{M}{Z_b}$ clockwise 30 deg. and divide by $\sqrt{3}$. The reciprocal of the distance OS then gives the factor by which I_{a1} must be multiplied to obtain the amplitude of the meter current. The angle AOD equal to the angle AOS gives the angle of shift of I_{M1} with respect to I_{a1}, the positive-sequence current in phase a. A similar diagram may be made for negative-sequence, but it is to be noted particularly that the phase shifts are different.

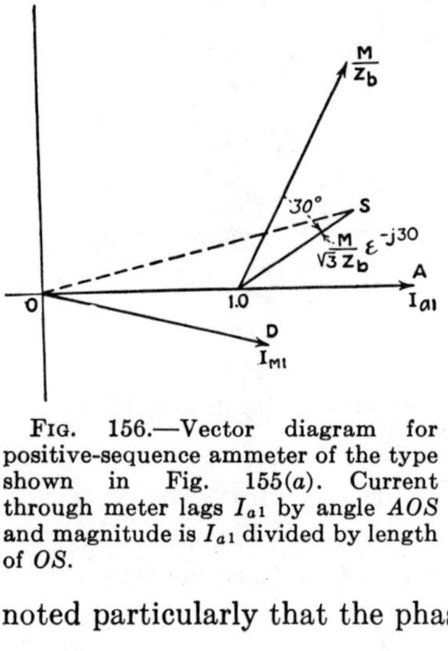

FIG. 156.—Vector diagram for positive-sequence ammeter of the type shown in Fig. 155(*a*). Current through meter lags I_{a1} by angle AOS and magnitude is I_{a1} divided by length of OS.

128. Bridge Type of Positive- and Negative-sequence Ammeters.

The bridge type of network for measuring positive- and negative-sequence currents is due to C. L. Fortescue. This type of network, as illustrated in Fig. 157, is applicable to three-phase three-wire systems, and, with special connections for eliminating zero-sequence current as discussed later, it may be used on three-phase four-wire systems. This type of network was developed originally for the simultaneous measurement of both the positive-sequence current (current in branch M_1) and the negative-sequence current (current in branch M_2). In the analysis of this type of network, refer to Fig. 157 for the notation. A necessary condition for the simultaneous measurement of both positive- and negative-sequence currents is that the meter impedances be equal, so that letting

$$\left.\begin{aligned} M_1 &= M_2 = M \\ \Delta_m &= R + Z + 2M \end{aligned}\right\} \quad (320)$$

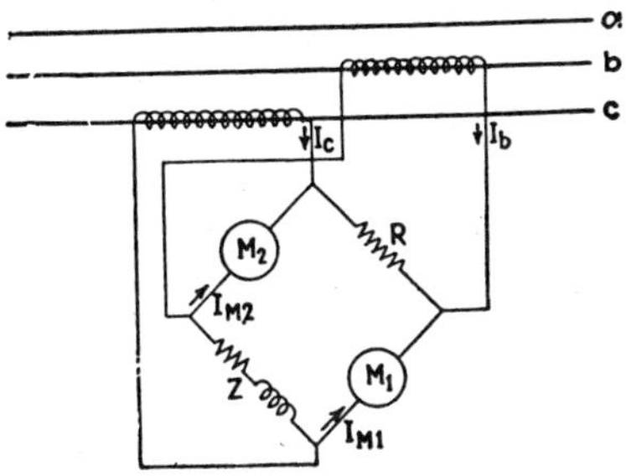

General Relation

$Z + M = -a^2(R + M) = e^{j60}(R + M)$

FIG. 157.—Bridge-type positive- and negative-sequence ammeters. The bridge-type network as shown is applicable only when zero-sequence current is not present.

The meter current I_{M1} for positive-sequence and I_{M2} for negative-sequence may be written from inspection as follows:

$$\left.\begin{aligned} I_{M1} &= -\frac{R + M}{\Delta_m} I_b - \frac{Z + M}{\Delta_m} I_c \\ I_{M2} &= -\frac{Z + M}{\Delta_m} I_b - \frac{R + M}{\Delta_m} I_c \end{aligned}\right\} \quad (321)$$

Expressing I_b and I_c in terms of sequence currents, equations (321) become

$$\left.\begin{aligned} I_{M1} &= -\frac{I_1}{\Delta_m}[a^2(R + M) + a(Z + M)] - \frac{I_2}{\Delta_m}[a(R + M) + a^2(Z + M)] \\ I_{M2} &= -\frac{I_1}{\Delta_m}[a^2(Z + M) + a(R + M)] - \frac{I_2}{\Delta_m}[a(Z + M) + a^2(R + M)] \end{aligned}\right\} \quad (322)$$

Examination of equations (322) shows that if

$$(Z + M) = -a^2(R + M),$$

$$\left.\begin{aligned} I_{M1} &= +\frac{I_1(R + M)}{\Delta_m}(1 - a^2) = I_1 \\ I_{M2} &= \frac{I_2(R + M)}{\Delta_m}(1 - a^2) = I_2 \end{aligned}\right\} \quad (323)$$

Thus it will be seen that the currents in branches M_1 and M_2 of the bridge-type network of Fig. 157 give in correct magnitude and phase position the positive- and negative-sequence currents, respectively, of phase a.

It is interesting to observe that if M_2 of Fig. 157 be made equal to zero, the network reduces to that of Fig. 155(a).

Vector Diagram. The bridge-type network lends itself particularly well to exposition by means of vector diagrams as given in Fig. 158. The line currents of Fig. 157 divide

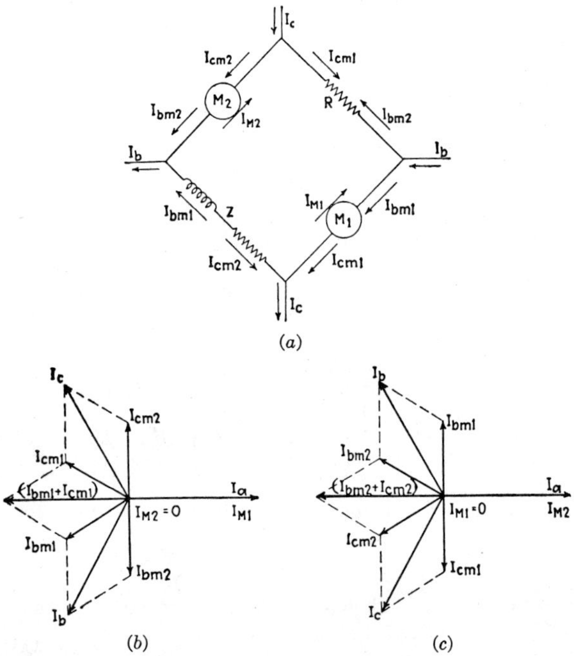

FIG. 158.—Vector analysis for positive- and negative-sequence bridge-type ammeters of Fig. 157. (a) Current division of network of Fig. 157. (b) Meter currents for line currents of positive-sequence. (c) Meter currents for line currents of negative-sequence.

into components in the network branches M_1, M_2, R, and Z as indicated in Fig. 158(a). Thus the current I_b divides into the components I_{bm1} and I_{bm2} flowing through the branches Z and M_1, and R and M_2, respectively. Similarly the current I_c divides into the components I_{cm1} and I_{cm2} flowing through the branches R and M_1, and Z and M_2, respectively. Figure 158(b) and (c) give vector diagrams for the branch currents corresponding to line currents I_a, I_b, and I_c, of positive-sequence and negative-sequence, respectively. It will be noted in both diagrams that the components I_{bm1} and I_{bm2} are equal in magnitude. but they lag by 30 deg. and lead by 30 deg. the resultant current I_b. Also, in both diagrams the components I_{cm1} and I_{cm2} are equal in magnitude but lead by 30 deg. and lag by 30 deg. the resultant current I_c. This division results from the fact that in the network of Fig. 157 the branches

$$(Z + M) = (R + M)e^{j60^\circ}.$$

In the case of Fig. 158(*b*) for positive-sequence line currents, the total current in the branch M_1 is (I_{m1}) which is equal to $-(I_{bm1} + I_{cm1})$, assuming the positive sense as indicated by the arrows in Fig. 158. It will be observed that the current I_{m1} is equal in magnitude and in phase with the positive-sequence current of phase *a*. The current in branch M_2 is zero since the components I_{bm2} and I_{cm2} exactly cancel each other.

In the case of Fig. 158(*c*) for negative-sequence line currents, the analysis is similar to that given in connection with Fig. 158(*b*) for positive-sequence. In the present case $I_{m1} = 0$ and I_{m2} is identical in magnitude and in phase with the negative-sequence current of phase *a*.

In connection with Fig. 158(*b*) and (*c*), it should be noted that the positive- and negative-sequence currents, although plotted as though they were equal and in phase, may actually have any magnitude or phase position. By superposition it can be shown that, regardless of the unbalance in line currents of a three-phase three-wire system, the meter branches M_1 and M_2 will carry only positive- and negative-sequence currents, respectively. The type of analysis just used can conveniently be applied to check the correctness of any assumed network. The mathematical analysis of the different schemes has been given principally because it is more general and more readily permits defining the type of network which must be used to obtain a segregating network.

129. Elimination of Zero-sequence Current.

The elimination of zero-sequence current is a matter of importance not only in sequence measurements but also for relaying[60] and other applications. Consequently, the various connections for this purpose will be considered by themselves and then a few typical combinations with positive- (or negative-) sequence ammeters will be described. In Fig. 154 the zero-sequence current was prevented from actuating the moving element by the method of connecting the windings into the network, the element thus being rendered unresponsive to the zero-sequence current.

The zero-sequence current may, of course, be eliminated from load circuits supplied from current transformers if the secondaries are connected in delta as shown in Fig. 159(*a*). This connection of current transformers for eliminating zero-sequence current corresponds to the line-to-line connection of

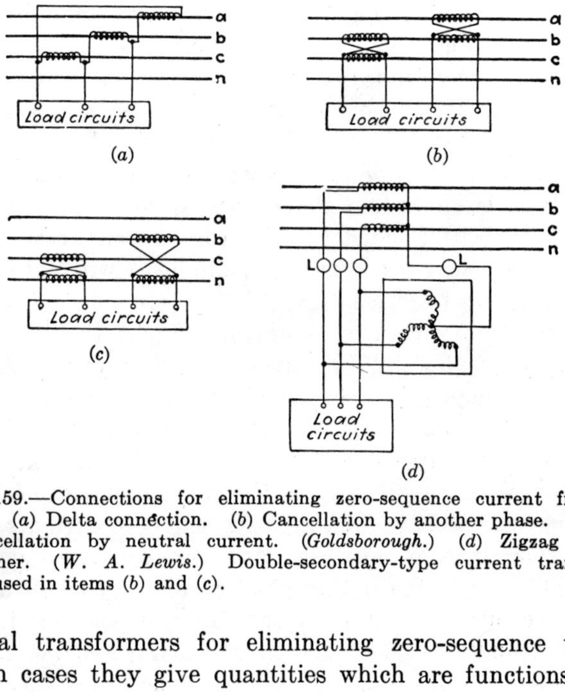

FIG. 159.—Connections for eliminating zero-sequence current from load circuits. (*a*) Delta connection. (*b*) Cancellation by another phase. (*Genkin.*) (*c*) Cancellation by neutral current. (*Goldsborough.*) (*d*) Zigzag shunting transformer. (*W. A. Lewis.*) Double-secondary-type current transformers may be used in items (*b*) and (*c*).

potential transformers for eliminating zero-sequence voltage. In both cases they give quantities which are functions of the positive- and negative-sequence components in the circuit to which the instrument transformers are connected. For certain instrument connections line currents are required, and for these cases the cross connection from another phase, as illustrated in Fig. 159(*b*), will of course accomplish the elimination of zero-sequence current since the line currents have equal values of zero-sequence current. Where the total neutral current (or a known proportion of it) is available, the zero-sequence current from the different phases may be eliminated by the connection of Fig. 159(*c*). It is to be noted for this case that the current transformer for the neutral conductor should have a ratio so as to give only one-third of the current of the corresponding line conductors; this difference in ratio is required because the neutral current contains three times the zero-sequence current that flows in an individual phase conductor. Another type of transformer connection is illustrated in Fig. 159(*d*) and operates on the principle of providing no path for zero-sequence current through the load circuits but including in shunt therewith a low-impedance path for zero-sequence current and a high impedance path for positive- or negative-sequence currents.

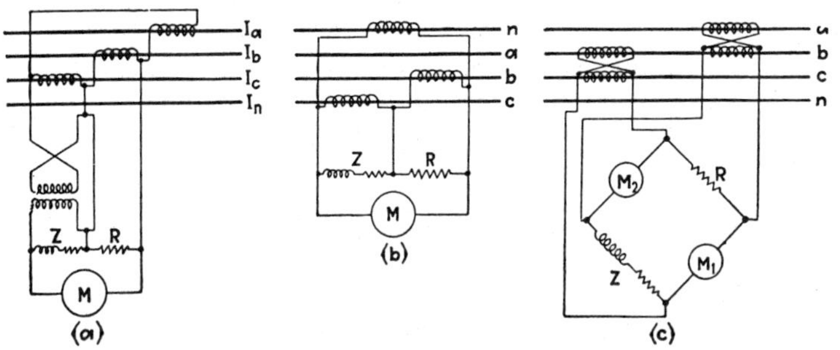

FIG. 160.—Typical positive- and negative-sequence ammeters for four-wire service. (*a*) Negative-sequence ammeter. Zero-sequence current eliminated by delta connection. (*b*) Positive-sequence ammeter. Zero-sequence current eliminated by means of return current. (*c*) Positive-sequence ammeter. Zero-sequence current eliminated by cross connection of two phases.

The current transformer is of the zigzag type with three separate cores and two equal windings on each. It should be noted that the main current transformers are connected in star which is the most frequently occurring case. For this connection it is permissible to introduce load circuits both before and after the shunt connections added to effect the elimination of zero-sequence components from phase currents.

The use of the connections of Fig. 159 will now be considered for several typical sequence ammeters. For example, the negative-sequence network of Fig. 155(*b*) can be adapted for four-wire service by using current transformer connections of Fig. 159(*a*) or (*d*) and an additional current transformer for reversing the polarity of one current as shown in Fig. 160(*a*). The positive-sequence ammeter of Fig. 155 may also be operated from the current transformer connections of Fig. 159(*b*) or (*c*), a variation of the latter being illustrated in Fig. 160(*b*). The positive- and negative-sequence bridge-type ammeters of Fig. 157 may also be operated from the current transformer connections of Fig. 159(*b*) or (*c*), the former being illustrated in Fig. 160(*c*).

130. Simultaneous Measurements of Sequence Currents.

In case it is desired to measure simultaneously both the positive- and negative-sequence currents, the bridge-type network of Fig. 157 may be used for three-wire circuits and the network of Fig. 160(*c*) for four-wire circuits, as this connection requires the minimum number of current transformers and network branches. If it is desired to measure the zero-sequence current in addition, it is merely necessary to add to some positive- and negative-sequence bridge network the connections necessary to measure zero-sequence current. This may be done by means of the connections of Fig. 153(*f*) or by inserting an ammeter in the star connection of Fig. 159(*d*).

131. Frequency and Transient Errors.

The **departure of frequency from normal** will always cause an error in the response of measuring devices for both positive- and negative-sequence quantities but may not cause error in zero-sequence devices. Consider first the frequency errors for the different sequence voltmeters. There will be a small error in the response of the positive-sequence voltmeter to positive-sequence voltage of abnormal frequency. For negative-sequence voltage of normal frequency the components of current through the positive-sequence voltmeter just cancel, producing zero response. Variation from normal frequency in either direction will prevent cancellation, thus producing an error in response. If the zero-sequence voltage is eliminated from the positive-sequence voltmeter by means of instrument transformer connections, the zero-sequence voltage during departures from normal frequency cannot affect the meter reading. Similar remarks apply to negative-sequence. The zero-sequence voltmeter will not respond to positive- or negative-sequence voltages provided the zero-sequence voltage is obtained by instrument transformer connections.

The character of the frequency error can conveniently be analyzed by considering the effects in connection with a typical instrument, such as the positive-sequence voltmeter of Fig.

148(*a*). For the purpose of this study the meter impedance may be ignored and the network assumed to consist of

$$\begin{aligned} Z_B &= R \\ Z_A &= \frac{R}{2}\left(1 + j\frac{\sqrt{3}f}{60}\right) \end{aligned} \tag{324}$$

where f is the system frequency whose normal value is 60 cycles. The response of the positive-sequence voltmeter will be obtained by substituting the above impedance values in equation (303) which gives for the meter currents the following relations for positive-sequence voltage:

$$I_{M1} \propto \frac{\left(1 + \frac{f}{60}\right)E_1}{1 + j\frac{\sqrt{3}f}{60}} \tag{325}$$

and for negative-sequence voltage:

$$I_{M2} \propto \frac{\left(1 - \frac{f}{60}\right)E_2}{1 + j\frac{\sqrt{3}f}{60}} \tag{326}$$

Examination of equation (325) will show that a 5 per cent variation in the system frequency of the positive-sequence voltage will cause the current in the positive-sequence voltmeter to be in error by 1.1 per cent. There is also a shift in the phase of the voltage applied to the voltmeter which amounts to about one degree lag or lead for system frequencies of 105 and 95 per cent normal, respectively.

The positive-sequence voltmeter does not respond to negative-sequence voltage of normal frequency, 60 cycles in this case, since the components of the meter current just cancel each other. However, when the frequency is either 5 per cent above or below normal, the meter will measure about 5 per cent of the negative-sequence voltage on the system. The total error is of course the combination of the errors due to positive- and negative-sequence voltage responses. Thus on a normal system, the negative-sequence voltage is usually relatively small and the total error will be only of the order of 1 per cent even for a 5 per cent variation in frequency from normal. However, in the case of the negative-sequence voltmeter, the frequency errors may be of importance because of the fact that the negative-sequence voltage is normally zero. Thus a 5 per cent variation in frequency from normal will cause a normally high positive-sequence voltage to produce a response in the negative-sequence voltmeter corresponding to about 5 per cent of normal. It should be clearly understood, however, that variation from normal frequency is not a matter of much importance at the present time because power-system frequencies are closely regulated in order to provide time-keeping service.

Another type of error that may arise is due to **electrical transients** which include (1) the meter response to the direct-current components of transients in voltages or currents in the measured circuits, and (2) the lag in inductive branches of the segregating network in following changes in the measured circuit quantities.

The meter response to the direct-current component is of brief duration and for many purposes it may be ignored. This error may be minimized, if desired, by shunting the meter element by a low-resistance, high-inductance branch, thus by-passing the direct-current component but not the alternating-current component of the transient current. There is also the error due to the time lag of current building up in the inductive branch of a sequence ammeter network, but fortunately this error is very small and will reduce to about 2.7 per cent in 1 cycle and less than 0.1 per cent in 2 cycles on a 60-cycle system. Thus the error is entirely negligible for ordinary instruments but may require consideration in the interpretation of oscillographic records of sequence voltages during the first half cycle. There is also a similar error in connection with the sequence voltmeters, but this error is of much less magnitude and is entirely negligible.

The presence of **harmonics** on a system may constitute another possible source of error in the response of sequence instruments. This problem of course can be considered as merely one of an excessive variation in frequency from normal. Thus by considering equation (326), it can be shown that a 10 per cent fifth harmonic, which is of negative-sequence, will cause a positive-sequence meter response of about 5 per cent of the normal meter reading. Thus while it appears that the error in meter response may be quite large, the magnitudes of the harmonics fortunately are usually small so that the resulting error can be ignored. If desired, a low-resistance high-inductance branch may be placed in series with the meter element and thus reduce the effect of these harmonics to a negligible value.

In concluding this discussion of the errors of the sequence measuring devices, it should perhaps be pointed out that while there is a theoretical possibility of these errors, they have not, as a matter of fact, been of much practical importance so that it has not been necessary, except in a few isolated instances, to consider means for minimizing these errors.

132. Adjustment and Testing of Segregating Networks.

The adjustment of segregating networks is best made by determining accurately the impedances including the reactance and particularly the *alternating-current resistance* of the network elements. The procedure is most readily followed by considering a typical network, such as that of the positive-sequence voltmeter of Fig. 148(a). The value of the reactance of the branch Z_A should first be obtained and then its resistance adjusted to give the power-factor angle of 60 deg., and finally the resistance of the branch Z_B should be adjusted to have the same impedance as the branch Z_A. These data are sufficient to establish a segregating network. However, as a practical matter, the impedances of the various elements are approximately determined and the final adjustments are made on the combination. A given voltage is impressed between terminals l_1 and l_2 in parallel with l_3 and then between terminals l_3 and l_2 in parallel with l_1, and the network adjusted to give equal deflections of the voltmeter. The transformer leads connected to the c phase are then separated so that the same voltage as before can be applied to both transformers in parallel. The meter reading should then be $\sqrt{3}$ times the former deflection. The final check is to connect the voltmeter and network to a balanced three-phase supply. If the network is correctly adjusted, the meter will read the line-to-neutral voltage correctly for one phase rotation and zero voltage if two of the leads are interchanged.

For the adjustment of positive- or negative-sequence current networks a similar procedure may be followed. For the network of Fig. 155(a) the steps are almost identical with those given above for the positive-sequence voltmeter. In the case of the current networks, it may be necessary to disconnect any current transformer whose primary side is open-circuited. So long as the voltage across the transformer is small so that saturation does not occur this is unnecessary, but for larger voltages saturation will decrease the exciting impedance so that appreciable current is by-passed.

The effect of the impedance of the measuring element of the voltmeter or ammeter on the adjustment of the segregating network will now be considered. Examination of the equations for the various networks for the voltmeters show that they will respond only to voltages of the desired sequence without placing any restriction on the meter impedance. Thus the voltages impressed on the positive-sequence voltmeter elements of Figs. 147 to 155 inclusive, are affected in both magnitude and phase relation by the meter impedances, but they do not affect the kind of quantity to which the meter responds, which is positive-sequence voltage only. This statement also applies to the sequence ammeters except for the case of the bridge-type networks when used to measure both positive- and negative-sequence currents. For this case the impedance of one meter is a part of the segregating network for the other meter. Thus for a network of this type, such as Fig. 157, it is necessary for the two meter branches to be equal and that the meter impedances and the other branches bear the relation

$$Z + M = (R + M)e^{j60^\circ} \tag{327}$$

133. Application of Sequence Voltage and Current Devices.

The sequence voltage and current instruments and relays are applicable for a wide variety of uses. Several of these devices have been made available in commercial form by the Westinghouse Electric and Manufacturing Company who pioneered in this development. They possess the advantage of measuring directly the quantity used in the analysis of power systems under unbalanced conditions. Sequence instruments required for particular purposes will usually be less in number and never in excess of the required number of single-phase instruments to give the complete indication of system performance. In addition to the advantage in the smaller number of devices, the sequence instruments possess the very real advantage of giving the data in a more practical form, which is required for many applications. When single-phase instruments are used the data for the various uses can be obtained only by making calculations from the combined readings, as for example by the methods outlined in the previous chapter.

The application of the sequence instruments will first be reviewed in a general way and this will be followed by a more detailed description of three of the more common applications. Knowledge of system voltage is of importance for many applications, as, for example, the tendency of synchronous motors to pull out of step on the occurrence of a dip in the system voltage. For this purpose a positive-sequence voltmeter is particularly convenient. Hence a single meter of the sequence segregating type will give more satisfactory results than three instruments of the conventional single-phase type. Another illustration is the use of the negative-sequence ammeter to determine the amount of unbalance arising from the single-phase applications of certain loads, such as those due to electric furnaces. For this purpose a single negative-sequence ammeter will give more satisfactory results than three single-phase instruments.

A further advantage of the sequence measuring devices of the negative- and zero-sequence types arises from the fact that these instruments usually carry no current when the system is in normal condition but have a positive indication under abnormal conditions; consequently, they provide an ideal basis for relay operation since more sensitive and faster operation can usually be secured from relays which have to distinguish between a zero and a finite value instead of between two differences in magnitude.

134. Positive-sequence Graphic Voltmeter.

The positive-sequence voltage, as pointed out previously, is frequently required for analysis of power-system operation particularly in respect to the effect of faults on systems reducing the system voltage and tending to pull synchronous motors out of step. For this purpose the positive-sequence recording voltmeter is being used. The meter element itself is of the conventional type used for single-phase purposes. The difference, however, is in the external network which consists of the sequence segregating network of Fig. 148(*a*). The network, consisting of merely reactor and resistor elements, is extremely simple and will not occupy any considerable space. A typical network built by the Westinghouse Electric and Manufacturing Company is illustrated in Fig. 161. As a matter of interest a standard resistor for a recording wattmeter is shown alongside for comparison. In connection with the positive-sequence voltmeter it should perhaps be pointed out that two potential transformers are required when connected line-to-line, or three when connected line-to-neutral. To obtain equivalent data it is necessary to use the same number of instrument transformers with sequence-segregating or single-phase devices but the number of meters will be a minimum if sequence-segregating devices are used.

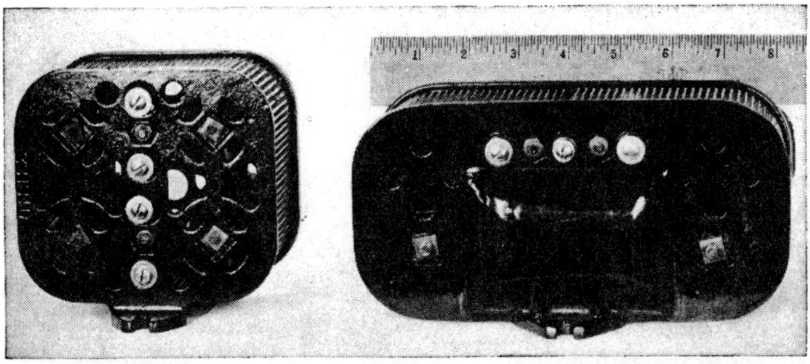

(a) (b)

FIG. 161.—Relative size of network and standard meter external resistor. (a) Standard resistor for recording wattmeter; (b) positive-sequence network for recording voltmeter.

135. Positive-sequence Voltage Regulator.

It is a matter of operating experience that the method of connecting the voltage-regulator operating coil to the machine terminals through a single potential transformer does not always result in the proper action of the voltage regulator due to the fact that the regulator responds to what is happening on that particular phase only. The voltage on the unloaded phases of three-phase machines supplying a single-phase railway load will be appreciably above the loaded phase. Similar conditions exist on a power system at times of unbalanced faults. Therefore, since the throwing of a single-phase load or a single-phase fault on to one phase may result, under certain conditions, in a rise rather than a drop in voltage on one of the other phases, it frequently happens that a regulator will cause the exciter voltage to build down when it should build up, if the potential transformer happens to be connected to one of the unloaded phases. An example of this rise in voltage is given in Fig. 162 which shows an oscillogram taken on a 120-kva., 1,100-volt, three-phase,

60-cycle generator. Full load at unity power factor was thrown on one phase and the current on the loaded phase, and the voltage between all three terminals recorded. The voltage *AB* actually rises about 6.5 per cent when the single-phase load is thrown on phase *CA*. In the oscillogram, voltage *AB* stays above normal for a period of about 15 cycles; its fall

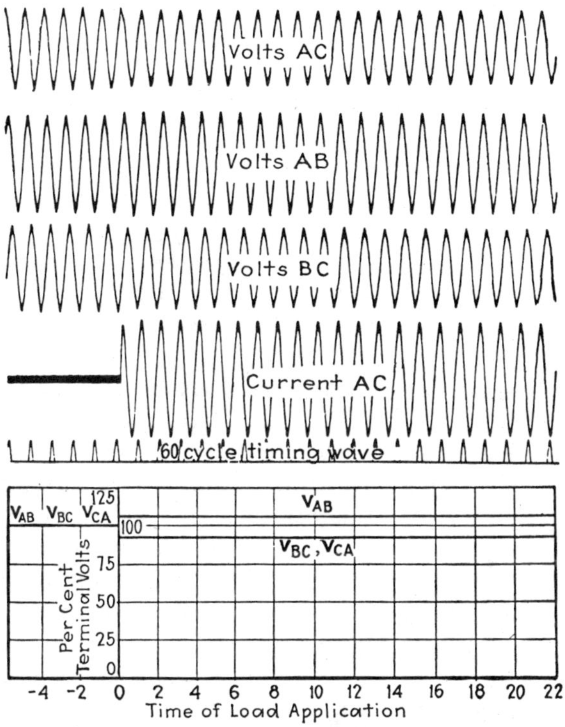

FIG. 162.—Oscillogram showing rise in voltage on phase *AB* when a resistance load is applied to phase *AC*. The curves at the bottom give the r.m.s. voltages corrected for drop in generator speed.

from the maximum value is due principally to the decrease in speed of the prime mover in supplying the additional load. In this case it is obvious that with a voltage regulator connected across terminals *A* and *B*, the exciter voltage would have been reduced whereas it was really required to build up since the load was demagnetizing the main machine. A positive-sequence network, however, supplying voltage to the regulator coil would have caused it to respond correctly. In addition the positive-sequence voltage regulator gives the same indication irrespective of the phase to which the unbalanced load or fault is applied, which is obviously much more satisfactory than the condition with the single-phase voltage connection. The need for the positive-sequence voltage regulator was recognized about the same time as quick-response excitation was introduced. It is obvious that if excitation is to be changed in a very short space of time, it is exceedingly important that the change be made in the desired direction. For applications where slow-response exciters are used, the positive-sequence voltage regulator is not ordinarily used since the circuit-breaker will generally isolate the fault before there has been much opportunity for the exciter voltage to change. Of course, if a system were supplying heavy single-phase loads on different phases, a positive-sequence voltage regulator would provide the best means of determining the proper mean value of machine excitation.

In connection with the application of the positive-sequence voltage regulator there is one special problem introduced owing to the fact that with the vibrating type of regulator it may be desirable alternately to increase and decrease the voltage applied to the regulator coil in order to introduce anti-hunting features. This was accomplished in the regulators operating on a single-phase voltage by short-circuiting a resistance in series with the regulator coil. This same principle can be introduced with the positive-sequence voltage network of the type of Fig. 148(*a*). As pointed out before, the meter branch of the network including the meter element and external resistance can be varied without affecting the kind of voltage impressed on the coil, *i.e.*, the adjustment of the external resistance will still leave positive-sequence voltage impressed on the coil.

136. Negative-sequence Overcurrent Relay.

The negative-sequence current network of Fig. 155(*b*) or 157 may be used in connection with a relay instead of an ammeter. Such a negative-sequence current relay may be used to provide protection against single-phase operation and phase reversal. Its principal application to date has been for protecting induction motors against single-phase operation, the excessive heating of which is discussed in Chap. XVII on Induction Motors. A commercial form of the negative-sequence relay which makes use of the bridge-type network is the type CQ built by the

Westinghouse Electric and Manufacturing Company. Sequence relays involve one feature which requires consideration, namely, the provision for changing tap settings. Such changes in taps ordinarily alter the meter impedance, and this, in turn, will affect the current flowing through the meter element since all of the current networks have shunt paths in parallel with the meter element. In case networks of the type of Fig. 155(*b*) are used, a change in the meter impedance affects the magnitude and phase angle of the current flowing through the operating coil of the relay, but it responds only to negative-sequence current. In the case of the negative-sequence bridge-type network of Fig. 157, a change in the meter impedance alters not only the response to negative-sequence current but also permits the flow of positive-sequence current through the relay element. Hence it is essential that the relay give constant impedance for the full range of tap settings. This is accomplished in the type CQ relay, a diagram of which is shown in Fig. 163, by introducing a dummy coil in branch M_1 equivalent in impedance to that of the relay element M_2. The connections of Fig. 163 were also selected so that, if desired, a relay element could be substituted for the dummy coil and thus permit measurement of the positive-sequence current in one branch and

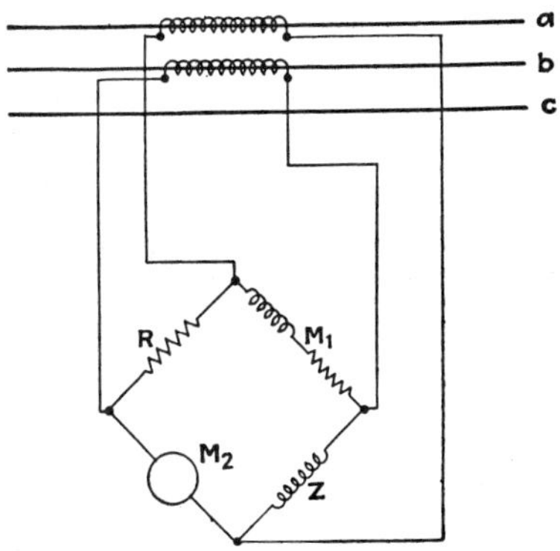

FIG. 163.—Negative-sequence relay.

FIG. 164.—Negative-sequence over-current relay.

negative-sequence current in another branch of the network. The constant impedance characteristic of the type CQ* relay is accomplished by shifting contacts which vary the amount of secondary load on the relay in such a manner as to secure constant impedance independent of relay setting. The physical appearance of the type CQ phase-balance and reverse phase-current relay is shown in Fig. 164.

137. Polyphase Sequence Networks and Instruments.

The sequence segregating voltage and current devices that have been described have been of the *single-phase* type, and it

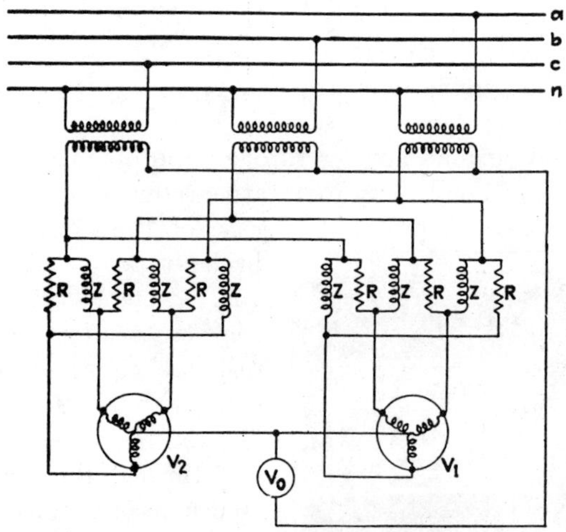

FIG. 165.—Typical polyphase sequence voltage segregating networks together with positive-, negative-, and zero-sequence voltmeters.

seems desirable to describe a few typical polyphase devices. The polyphase sequence segregating network can readily be made by combining three single-phase networks with the networks connected to their different phases in cyclic order. The polyphase relay may be of the induction-motor or synchronous type; in either event there would be a distributed polyphase winding. In general, the connections for three single-phase devices operating from different phases will require a few additional network elements in excess of the minimum required for polyphase

* Further information on the detailed characteristics of the negative-sequence overcurrent relay are given by J. V. Breisky.(21)

service; hence, several typical polyphase sequence segregating voltage and current devices will be described.

A typical connection for measuring polyphase positive- and negative-sequence voltages is shown in Fig. 165. In connection with this scheme it should perhaps be pointed out that zero-sequence current may flow through both the positive- and

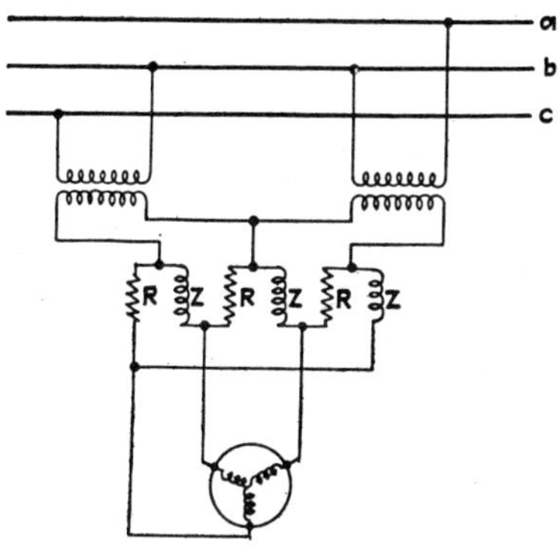

FIG. 166.—Simplified polyphase positive-sequence voltmeter.

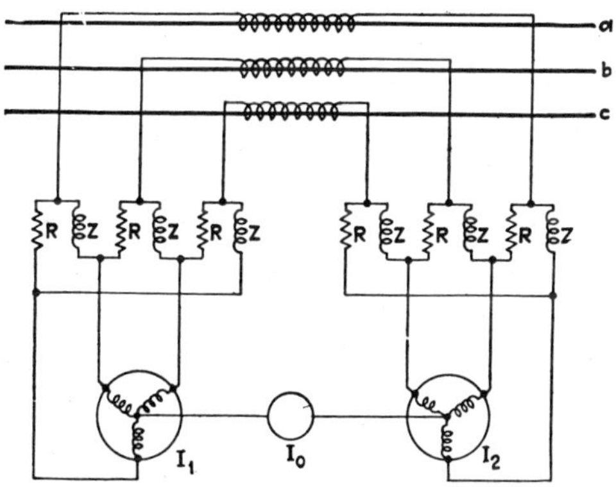

FIG. 167.—Typical polyphase sequence current segregating network together with positive-, negative-, and zero-sequence ammeters.

negative-sequence polyphase meters. However, since zero-sequence currents flow equally in each phase of the distributed polyphase winding there will be no tendency to operate the meter. The particular set of connections of Fig. 165 incidentally shows the meter for measuring zero-sequence voltage. If only the polyphase positive-sequence voltage is desired, and line-to-line potential transformers are available, then the connections of Fig. 166 may be used. In connection with the network of Figs. 165 and 166 it may be observed that the positions of Z and R may be interchanged, or the phase connections of the potential transformers may be reversed in order to cause a positive-sequence meter to measure negative-sequence voltage and *vice versa*.

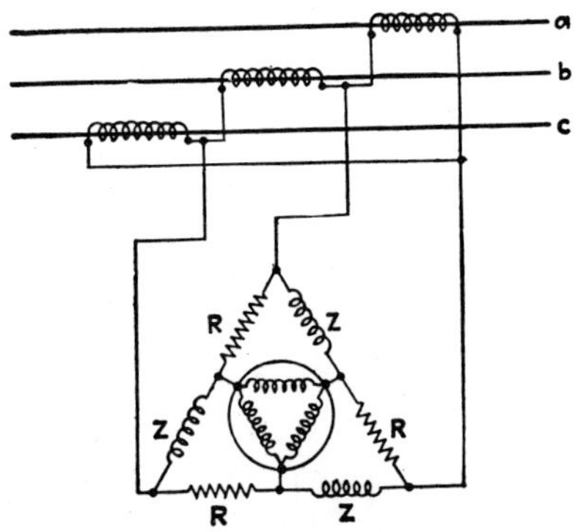

FIG. 168.—Simplified polyphase negative-sequence ammeter.

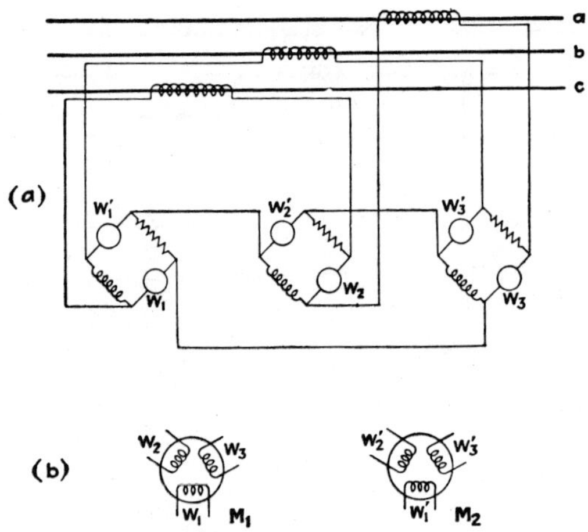

FIG. 169.—Polyphase sequence ammeter using bridge-type networks. (*a*) Schematic connection; (*b*) meter winding arrangement.

A typical set of polyphase sequence segregating ammeter connections is shown in Fig 167. Positive-, negative-, and zero-sequence currents are measured by the instruments marked I_1, I_2, and I_0, respectively. Here again zero-sequence current circulates through the polyphase meters but produces no resultant effect. Figure 168 shows a simplified form of negative-sequence ammeter. In this case the zero-sequence current is eliminated by means of the delta connection of current transformers. The bridge-type networks of Fig. 157 may be used as a basis of the polyphase sequence meters of Fig. 169. In this case six terminals are required for each polyphase meter winding as illustrated in (*b*); the connections for the individual phases are shown in the schematic diagram of (*a*).

Problems

1. A positive-sequence voltmeter whose meter impedance is 500 ohms and whose resistance branch of the segregating network is 2,000 ohms has 100 per cent voltage applied to the primary of each transformer singly. What are the meter readings for the two cases?

2. Prove that the connections of Fig. 151(*d*) measure positive-sequence voltage.

3. In a certain oscillograph test it was desired to measure the positive-sequence current. Only two current transformers and a positive-sequence bridge-type network were available, but since the system was grounded this connection did not eliminate the zero-sequence current. However, a set of three current transformers giving $3I_0$ for relay operation constituted part of the installation and, while these could not be disturbed for cross-connection purposes, a resistance was inserted in this circuit so that the drop across it could be used to buck down the zero-sequence drop in the bridge-type network. What should be the connection and the relation between the resistance of the oscillograph shunt and the resistance inserted in the relay circuit?

4. A certain polyphase device requires for its operation a positive-sequence set of currents of 5 amp. and only a single-phase source is available. The impedance of the device per phase is $0.02 + j0.01$ ohm. The different phase currents are obtained by connecting the three windings in series and shunting the individual windings by the impedances Z_a, Z_b, and Z_c. If Z_a be equal to 0.025 ohm, what must be the values of Z_b and Z_c? Advantage may be taken of the fact that since the windings are insulated from each other, the effect of a negative resistance may be obtained by reversing the terminals of a winding.

5. A positive-sequence voltmeter of the type shown in Fig. 148(*a*) has a meter element of $400 + j80$ ohms, and $R = 5{,}000$ ohms. By rewinding the element with three times the turns of finer wire, the element impedance is $3{,}600 + j720$ ohms. Calculate the value of R to obtain the same sensitivity

6. In Prob. 5, compare the currents in the segregating elements for 50 volts negative-sequence applied to both the original voltmeter and the rewound combination.

CHAPTER XV

THE MEASUREMENT OF POWER QUANTITIES ON POLYPHASE CIRCUITS

This chapter reviews the measurement of power quantities on single-phase and polyphase circuits, the latter including both balanced and unbalanced conditions. Conventional wattmeter, reactive volt-ampere meter, and power-factor meter connections for three-phase circuits are examined with respect to both balanced and unbalanced conditions. It is shown that a number of the conventional connections with two or more elements give incorrect indications with unbalanced currents and voltages though giving correct indications with balanced currents and voltages. The discussion continues with an analysis of the sequence power quantities including positive-, negative-, and zero-sequence wattmeters, reactive volt-ampere meters, and power-factor meters. The chapter concludes with a discussion of the flow of the various power quantities on unbalanced circuits with various kinds of loads.

138. Power Expressed in Terms of Voltage and Current Vectors.

Power in an alternating-current circuit is at any instant equal to the product of the instantaneous values of current and voltage. In Chap. II, equation (16), it was shown that the instantaneous value of a sinusoidally varying voltage is

$$e = \frac{\sqrt{2}}{2}(E\epsilon^{j\omega t} + \hat{E}\epsilon^{-j\omega t})$$

The corresponding expression for the instantaneous value of current is

$$i = \frac{\sqrt{2}}{2}(I\epsilon^{j\omega t} + \hat{I}\epsilon^{-j\omega t}) \tag{328}$$

The instantaneous value of power is then

$$ei = \tfrac{1}{2}(E\hat{I} + EI\epsilon^{j2\omega t}) + \tfrac{1}{2}(\hat{E}I + \hat{E}\hat{I}\epsilon^{-j2\omega t}) \tag{329}$$

The two parts on the right-hand side are conjugate to each other. $E\hat{I}$ and $\hat{E}I$ are conjugate vectors of equal magnitude independent of time, lying in opposite quadrants. The double-frequency rotational vectors $EI\epsilon^{j2\omega t}$ and $\hat{E}\hat{I}\epsilon^{-j2\omega t}$ are also conjugate to each other and have the same absolute values as $E\hat{I}$ and $\hat{E}I$. Thus it will be seen that the expression for instantaneous power consists of a constant term with a sinusoidal term of double frequency superposed. These relations are illustrated in Fig. 170.

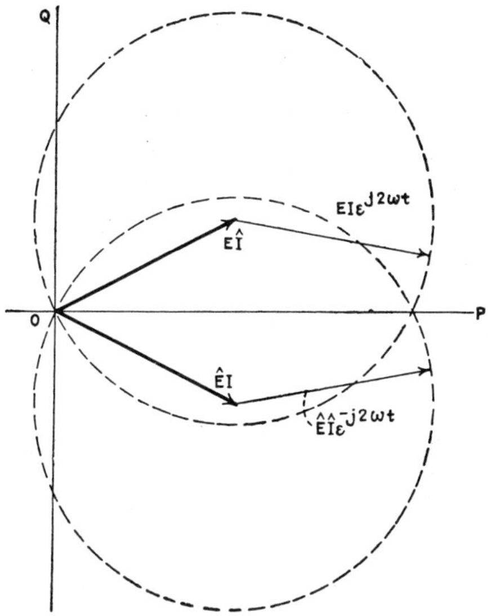

FIG. 170.—Power diagram showing double-frequency rotating vectors.

It can be seen that the average values of the double-frequency components are zero so that the mean value of power is

$$P = \tfrac{1}{2}(E\hat{I} + \hat{E}I) \tag{330}$$

Now if $E = \bar{E}\epsilon^{j\alpha}$ and $I = \bar{I}\epsilon^{j\beta}$, then

$$\begin{aligned} P &= \tfrac{1}{2}[\bar{E}\bar{I}\epsilon^{j(\alpha-\beta)} + \bar{E}\bar{I}\epsilon^{-j(\alpha-\beta)}] \\ &= \frac{\bar{E}\bar{I}}{2}[\{\cos(\alpha-\beta) + j\sin(\alpha-\beta)\} \\ &\qquad + \{\cos(\alpha-\beta) - j\sin(\alpha-\beta)\}] = \bar{E}\bar{I}\cos(\alpha-\beta) \end{aligned} \tag{331}$$

This is the familiar form of expression that is commonly used. It will be observed that the real parts of either $E\hat{I}$ or $\hat{E}I$ are equal to this value and from this point of view, having once defined P as the real part of $E\hat{I}$ or $\hat{E}I$, either expression can be used.

139. Reactive Volt-amperes Expressed in Terms of Voltage and Current Vectors.

For recurrent phenomena, the reactive volt-amperes consumed by a load are equal to the product of the voltage and the component of current lagging 90 deg. behind the voltage. Starting with the definition of reactive volt-amperes as equal to $\bar{E}\bar{I}\sin(\alpha - \beta)$, it is possible to show that the imaginary parts of the above expressions $E\hat{I}$ and $\hat{E}I$ are equal in magnitude to the reactive volt-amperes. Since $E\hat{I}$ and $\hat{E}I$ are conjugates of each other, their imaginary parts are the negative of each other. The choice of the expression to use is dependent upon the conventional sign associated with inductive or capacitive, lagging or leading reactive volt-amperes. The authors of the earlier papers presented before the American Institute of Electrical Engineers chose to assume inductive reactive volt-amperes as positive. The same convention was adopted by C. L. Fortescue in presenting for the first time a vector expression(5) for power quantities. Others have chosen the opposite convention. A search of the literature as to the use of the sign associated with inductive reactive volt-amperes was recently made by the authors. This search revealed a preponderance of the use of the positive sign for this purpose. Most of the references cited used reactive volt-amperes only incidentally in the development of some other main subject. Of even greater significance than mere numerical superiority is the fact that of those authors such as Kennelly, Jackson, Fortescue, and Slepian, whose subject matter dealt primarily with reactive volt-amperes, all preferred the use of the positive sign. In the absence of standardization, the authors have followed the conclusions resulting from this search in considering inductive reactive volt-amperes as positive. With the adoption of this convention real and reactive volt-amperes will have signs corresponding to $(R + jX)\bar{I}^2$.

The drop across an inductive reactance X is equal to jXI and using the expression $E\hat{I}$, Q, the reactive volt-amperes, is equal to $(jXI)(\hat{I})$ or $jX\bar{I}^2$. On the other hand, for $\hat{E}I$, Q is equal to $(-jX\hat{I})(I)$ or $-jX\bar{I}^2$. From the foregoing it is evident that for the sign of Q to agree with the convention adopted, $E\hat{I}$ must be chosen to express power and reactive volt-amperes, so that

$$P + jQ = E\hat{I} \tag{332}$$

140. Power Quantities on Single-phase Circuits.

The power quantities on single-phase circuits as used in this volume consist of the power quantity P; the reactive volt-ampere quantity Q, which is sometimes referred to as reactive power; and power factor PF. These power quantities have physical significance in accordance with common usage. Thus the power P indicates the rate of consuming energy in a circuit and is commonly measured in watts or kilowatts. The reactive volt-amperes Q consumed by a load is commonly measured in volt-amperes or kilovolt-amperes, usually abbreviated "vars" or "kilovars." Power factor is defined in terms of the power quantities used in this discussion as follows:

$$PF = \frac{P}{\sqrt{P^2 + Q^2}} \tag{333}$$

141. Power Quantities on Polyphase Circuits.

The amount of power which must be supplied to polyphase circuits is, of course, the sum of the power requirements for the several phases. Thus for a three-phase system with phases a, b, and c, the total power

$$P_T = P_a + P_b + P_c \tag{334}$$

The total reactive volt-amperes Q_T are, of course, equal to the sum of the reactive volt-ampere requirements of the several phases. Thus for the three-phase system

$$Q_T = Q_a + Q_b + Q_c \tag{335}$$

It follows, therefore, that

$$\begin{aligned} P_T + jQ_T &= (P_a + P_b + P_c) + j(Q_a + Q_b + Q_c) \\ &= (P_a + jQ_a) + (P_b + jQ_b) + (P_c + jQ_c) \\ &= E_a\hat{I}_a + E_b\hat{I}_b + E_c\hat{I}_c \end{aligned} \tag{336}$$

On a balanced three-phase system, power factor has the same significance as on a single-phase system. Similarly, total power factor under unbalanced conditions will be defined in terms of P_T and Q_T in a manner identical with that for the single-phase circuit. Thus the power factor for a three-phase system even under unbalanced conditions* is

$$PF_T = \frac{P_T}{\sqrt{P_T^2 + Q_T^2}} = \frac{(P_a + P_b + P_c)}{\sqrt{(P_a + P_b + P_c)^2 + (Q_a + Q_b + Q_c)^2}} \tag{337}$$

* *A.I.E.E. Standards*, 1922 ed., §3243.

These relations for total power, reactive volt-amperes, and power factor, in terms of the quantities for the individual phases, may best be clarified by reference to Fig. 171.

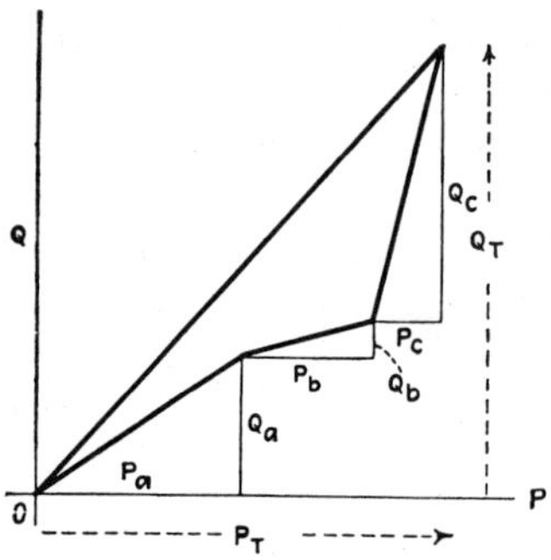

FIG. 171.—Power quantities on an unbalanced three-phase circuit.

142. Measuring Devices for Single-phase Power Quantities.

Measuring devices for single-phase power quantities include the wattmeter, the reactive volt-ampere meter, the single-phase power-factor meter, and are indicated schematically in Fig. 172 for conventional indicating instruments. The watt-meter shown in Fig. 172(*a*) has a rotating part which operates against a spring and a potential coil with external resistance to cause the current through the coil to be in phase with the voltage of the circuit to which it is connected. The reactive volt-ampere meter shown in Fig. 172(*b*) is quite similar but has a reactance in series with the voltage coil which causes the current through it to lag 90 deg. behind the voltage of the

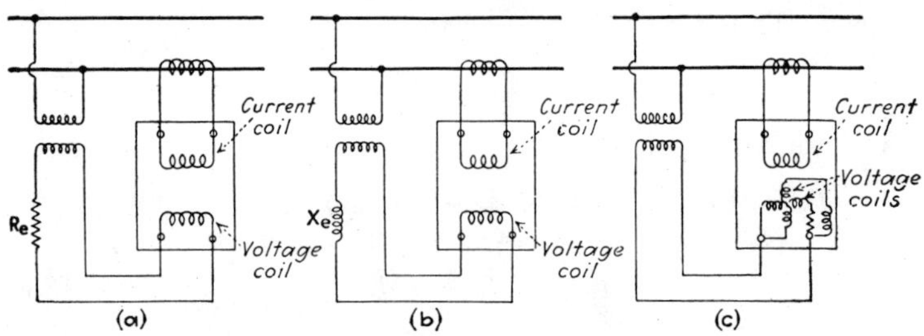

FIG. 172.—Electrodynamic instruments for indicating single-phase power quantities. (*a*) Wattmeter. (*b*) Reactive volt-ampere meter. (*c*) Power-factor meter (or reactive-factor meter when suitably calibrated).

circuit to which it is connected. A conventional form of the single-phase power-factor meter or reactive-factor meter is illustrated in (*c*). Such a meter may be calibrated to read power factor, reactive factor of the circuit, or the difference in phase angle between voltage and current. It may be pointed out that the power-factor and reactive-factor meter differ from the wattmeter and reactive volt-ampere meter of (*a*) and (*b*) in that the latter operate against a spring whereas the former take a position depending upon the relative phase of voltage and current. In general, the reactive-factor meter is used for determining the excitation of machines because its scale is not so crowded for the usual high power-factor operating range. Because of the similarity of the action of the reactive-factor and power-factor meters the subsequent discussion will be restricted to power-factor meters since that is the more commonly used factor.

Phase-angle Correction. In connection with the simple circuit shown in Fig. 172(*a*) and (*b*), it may be observed that the current through the voltage coil should have phase angles of exactly 0 deg. and 90 deg., respectively, from the phase of the voltage of the circuit to which the instrument is connected. This relation is approximately realized by the use of a non-inductive external resistor R_e and an external reactance coil X_e. However, due to the fact that the voltage coil will have some reactance, and the reactance coil X_e will have some resistance, this solution remains approximate.

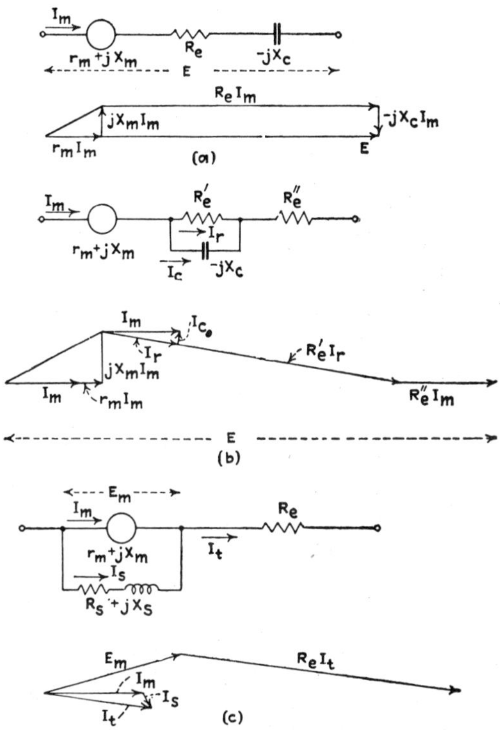

FIG. 173.—Methods for correcting phase angle in potential circuits of wattmeters.

Full control of the phase angle of the current in the voltage coil of wattmeters may be obtained in several ways as shown in Fig. 173. In this figure, (*a*) shows a series capacitor whose impedance is just equal to that of the reactance of the meter element. In this manner the effect of meter reactance is annulled.

The scheme shown in Fig. 173(*b*) utilizes a capacitor in shunt with all or a part of the external resistance. The vector diagram is self-explanatory in showing the shift of the various vector quantities which result in the meter current I_m being in phase with the voltage across the potential circuit. Condensers cannot conveniently be divided into small divisions. The method of Fig. 173(*b*) is therefore of particular value in that the small corrections in phase angle may be accomplished by adjusting the value of the external resistance shunted by the condenser. Figure 173(*c*) shows a third scheme which consists of shunting the meter element by a highly inductive shunt.

One method of correcting the phase angle of the potential coil of a reactive volt-ampere meter is shown in Fig. 174. This consists of shunting the meter element with a condenser. The vector diagram illustrates the relation between the various vector quantities.

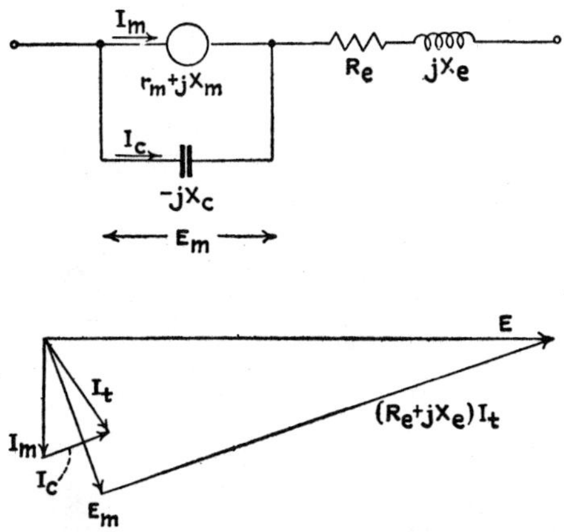

FIG. 174.—Method for correcting phase angle in potential circuits of reactive volt-ampere meters.

The phase-angle corrections, the possibility of which has been demonstrated in connection with Figs. 173 and 174, will be assumed in the subsequent discussion, and the shunting devices will not be shown in the main circuit connections since such procedure obscures the essential phenomena.

143. Total Power Quantities in Terms of Sequence Quantities.

It has been pointed out in Sec. 141 that the total power quantities on a three-phase system may be expressed in terms of the phase quantities as follows:

$$P_T + jQ_T = E_a\hat{I}_a + E_b\hat{I}_b + E_c\hat{I}_c \text{ [Eq. (336)]}$$

Each of the above three expressions may be expanded in terms of their sequence voltages and currents. For term $E_a\hat{I}_a$,

$$
\begin{aligned}
E_a\hat{I}_a &= (E_0 + E_1 + E_2)(\hat{I}_0 + \hat{I}_1 + \hat{I}_2) \\
&= E_0\hat{I}_0 + E_0\hat{I}_1 + E_0\hat{I}_2 \\
&\quad + E_1\hat{I}_0 + E_1\hat{I}_1 + E_1\hat{I}_2 \\
&\quad + E_2\hat{I}_0 + E_2\hat{I}_1 + E_2\hat{I}_2
\end{aligned}
$$

The two other terms $E_b\hat{I}_b$ and $E_c\hat{I}_c$ can be expanded in a similar manner. The results of these expansions are shown in tabular form in Table XVI, the tabulated values representing

TABLE XVI.—EXPANSION OF SINGLE-PHASE POWER EXPRESSION IN TERMS OF SEQUENCE COMPONENTS TO SHOW THAT TOTAL POWER IS EQUAL TO $3(E_1\hat{I}_1 + E_2\hat{I}_2 + E_0\hat{I}_0)$

Meter quantity	$E_1\hat{I}_1$	$E_1\hat{I}_2$	$E_1\hat{I}_0$	$E_2\hat{I}_1$	$E_2\hat{I}_2$	$E_2\hat{I}_0$	$E_0\hat{I}_1$	$E_0\hat{I}_2$	$E_0\hat{I}_0$
$E_a\hat{I}_a$	1	1	1	1	1	1	1	1	1
$E_b\hat{I}_b$	1	a	a^2	a^2	1	a	a	a^2	1
$E_c\hat{I}_c$	1	a^2	a	a	1	a^2	a^2	a	1
Total	3	0	0	0	3	0	0	0	3

the coefficients of the quantities given at the top of any column. Adding the coefficients in the respective columns, it will be noted that the coefficients of six of the products disappear leaving the total power quantity in terms of the sequence voltages and currents

$$P_T + jQ_T = 3(E_1\hat{I}_1 + E_2\hat{I}_2 + E_0\hat{I}_0) \tag{338}$$

Examination of the above equation suggests that the total power quantities are made up of the positive-, negative-, and zero-sequence components in each of the three phases. Sequence power quantities may therefore be defined in a manner similar to that for the quantities of single-phase circuits.

The positive-sequence power quantities per phase are

$$P_1 + jQ_1 = E_1\hat{I}_1 \tag{339}$$

The negative-sequence power quantities per phase are

$$P_2 + jQ_2 = E_2\hat{I}_2 \tag{340}$$

The zero-sequence power quantities per phase are

$$P_0 + jQ_0 = E_0\hat{I}_0 \tag{341}$$

Thus the total power quantities on the three phases may also be expressed as

$$P_T + jQ_T = 3[(P_1 + P_2 + P_0) + j(Q_1 + Q_2 + Q_0)] \tag{342}$$

144. Power Quantities on Unbalanced Three-phase Circuits Analyzed by Sequence Components.

There are a number of conventional wattmeter and reactive volt-ampere meter connections which are in use today. Some of these are correct for balanced or unbalanced conditions, while others give correct results under balanced conditions but incorrect results under unbalanced conditions. In examining these connections for measuring power quantities it is convenient to

TABLE XVII.—COEFFICIENTS OF SEQUENCE PRODUCTS OF VOLTAGE AND CONJUGATE OF CURRENT FOR VARIOUS PHASE VOLTAGE AND CURRENT COMBINATIONS

Item	Meter quantity	Sequence voltage-current products								
		$E_1\hat{I}_1$	$E_1\hat{I}_2$	$E_1\hat{I}_0$	$E_2\hat{I}_1$	$E_2\hat{I}_2$	$E_2\hat{I}_0$	$E_0\hat{I}_1$	$E_0\hat{I}_2$	$E_0\hat{I}_0$
1	$E_a\hat{I}_a$	1	1	1	1	1	1	1	1	1
2	$E_a\hat{I}_b$	a	a^2	1	a	a^2	1	a	a^2	1
3	$E_a\hat{I}_c$	a^2	a	1	a^2	a	1	a^2	a	1
4	$E_b\hat{I}_a$	a^2	a^2	a^2	a	a	a	1	1	1
5	$E_b\hat{I}_b$	1	a	a^2	a^2	1	a	a	a^2	1
6	$E_b\hat{I}_c$	a	1	a^2	1	a^2	a	a^2	a	1
7	$E_c\hat{I}_a$	a	a	a	a^2	a^2	a^2	1	1	1
8	$E_c\hat{I}_b$	a^2	1	a	1	a	a^2	a	a^2	1
9	$E_c\hat{I}_c$	1	a^2	a	a	1	a^2	a^2	a	1
10	$E_A\hat{I}_a$	$a-a^2$	$a-a^2$	$a-a^2$	a^2-a	a^2-a	a^2-a	0	0	0
11	$E_A\hat{I}_b$	a^2-1	$1-a$	$a-a^2$	$1-a^2$	$a-1$	a^2-a	0	0	0
12	$E_A\hat{I}_c$	$1-a$	a^2-1	$a-a^2$	$a-1$	$1-a^2$	a^2-a	0	0	0
13	$E_B\hat{I}_a$	$1-a$	$1-a$	$1-a$	$1-a^2$	$1-a^2$	$1-a^2$	0	0	0
14	$E_B\hat{I}_b$	$a-a^2$	a^2-1	$1-a$	$a-1$	a^2-a	$1-a^2$	0	0	0
15	$E_B\hat{I}_c$	a^2-1	$a-a^2$	$1-a$	a^2-a	$a-1$	$1-a^2$	0	0	0
16	$E_C\hat{I}_a$	a^2-1	a^2-1	a^2-1	$a-1$	$a-1$	$a-1$	0	0	0
17	$E_C\hat{I}_b$	$1-a$	$a-a^2$	a^2-1	a^2-a	$1-a^2$	$a-1$	0	0	0
18	$E_C\hat{I}_c$	$a-a^2$	$1-a$	a^2-1	$1-a^2$	a^2-a	$a-1$	0	0	0

make use of the sequence components. For this purpose the method, illustrated in Table XVI, has been extended in Table XVII to give the coefficients of the nine products of the sequence components for the 18 possible products of a line current and a line-to-neutral or a line-to-line voltage. This table may be used for determining either the real or reactive component of each of the nine sequence power products which may be measured by a meter element with the understanding that the current through the voltage coil must be in phase with the circuit voltage in order to measure the power component or to lag 90 deg. behind the circuit voltage in order to measure the reactive volt-ampere component.

It will be noted that for the three-phase four-wire system the total power quantities may include three sequence components, positive-, negative-, and zero-sequence; consequently, three meter elements are required to measure the total power. These measurements may be made either in terms of the three sequence components in line with equation (338) or in terms of the three phase components, equation (336), or with various other combinations. On the three-phase three-wire system, the total power quantities will include only two sequence components, namely, the positive- and negative-sequence components. Thus it is clear that the total power quantities may be measured by means of two wattmeter-type elements which may be either of the single-phase or of the sequence types. In case of symmetrical systems with balanced loads, only the positive-sequence power quantities exist and these may be measured by one meter of either the single-phase or positive-sequence type.

145. Analysis of Meter Connections for Power Quantities on Three-phase Circuits.

In this discussion the meter reading will be designated by the symbol $P_M + jQ_M$, in which P_M represents the value obtained when wattmeters are utilized to make the measurements and Q_M represents the value obtained when reactive volt-ampere meters are utilized to make the measurements. As stated previously, reactive volt-ampere meters differ from wattmeters only in that reactance instead of resistance is inserted in series in the voltage element.

Three-element Wattmeter for Four-wire Service. The conventional connection of wattmeters for measuring total power or reactive volt-amperes using three meter elements is indicated in Fig. 175. This meter measures power quantities in accordance with equation (336). It has previously been shown by means of

Table XVI that this meter connection measures three times the positive-, negative-, and zero-sequence quantities in accordance with equation (338). By adjusting the impedance of the potential circuit so that the current through the potential coil is in phase with the applied voltage the meter will measure the total power. Similarly, by adjusting the impedance in the potential coil circuit so that the current lags 90 deg. behind the potential applied to this circuit the meter will measure reactive volt-amperes.

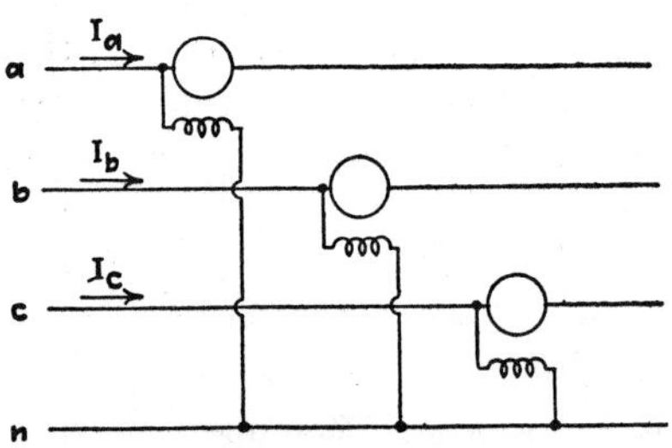

FIG. 175.—Conventional three-element wattmeter connection.

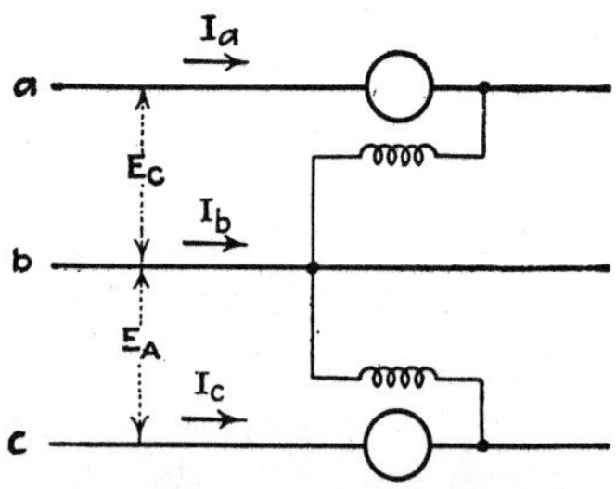

FIG. 176.—Conventional two-wattmeter method of measuring total power on a three-phase three-wire system.

Conventional Two-element Wattmeter for Three-wire Service. The conventional connection for measuring total power on three-phase three-wire systems is indicated in Fig. 176. It will be noted that the two-meter elements measure $E_C\hat{I}_a$ and $E_A\hat{I}_c$, respectively. Referring to Table XVII these quantities are represented by items (16) and (12) and when the polarities are adjusted so that the difference of these items is taken, the meter measures

$$P_M + jQ_M = 3(E_1\hat{I}_1 + E_2\hat{I}_2) - 3a^2E_1\hat{I}_0 - 3aE_2\hat{I}_0 \quad (343)$$

On three-phase three-wire systems, $\hat{I}_0$ is, of course, zero. Consequently the meter measures merely

$$P_M + jQ_M = 3(E_1\hat{I}_1 + E_2\hat{I}_2) \quad (344)$$

Thus the meter connections of Fig. 176 measure the total power quantities for all positive- and negative-sequence voltage and current combinations but do not respond correctly in case zero-sequence current is present.

On three-phase four-wire systems the zero-sequence current may be present and the last two terms of equation (343) may introduce quite appreciable errors. However, if the zero-sequence component is eliminated from the phase currents as explained in the discussion of Fig. 159(*c*), then the last two terms of equation (343) disappear and the meter reads the sum of the positive- and negative-sequence power. This is a closer approximation to the true total power in a four-wire circuit with grounded neutral than the connection of Fig. 176, as the zero-sequence power quantity is the product of E_0 and $\hat{I}_0$ which are usually small in themselves, whereas the term $E_1\hat{I}_0$ of equation (343) involves one large term.

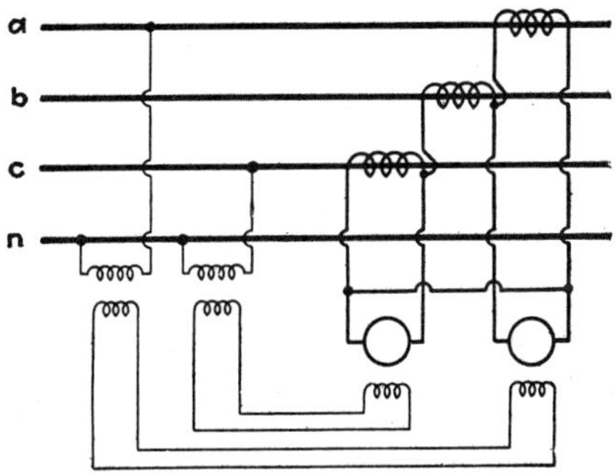

FIG. 177.—Wattmeter connection for measuring total power quantities with balanced voltages and unbalanced currents.

Special Wattmeter Connection for Balanced Voltages and Unbalanced Currents. It is of interest to consider the wattmeter connection of Fig. 177 which is sometimes recommended for measuring the total power quantities on a system with balanced voltages but with unbalanced currents. Examination of this connection with the aid of Table XVII shows that

$$\begin{aligned} P_M + jQ_M &= E_c(\hat{I}_c - \hat{I}_b) + E_a(\hat{I}_a - \hat{I}_c) \\ &= 3E_1\hat{I}_1 + 3E_2\hat{I}_2 - 3aE_0\hat{I}_1 - 3a^2E_0\hat{I}_2 \end{aligned} \tag{345}$$

Under balanced voltage conditions, E_2 and E_0 are both equal to zero and thus equation (345) reduces to

$$P_M + jQ_M = 3(E_1\hat{I}_1) \tag{346}$$

Referring to equation (338), it will be seen that when E_2 and E_0 are equal to zero the total power is $3E_1\hat{I}_1$. A single positive-sequence wattmeter will measure correctly the total power on a circuit with balanced voltages but unbalanced currents. Thus for balanced voltages a single positive-sequence wattmeter measures the same quantity as the two-element meter of Fig. 177.

On the other hand, a single wattmeter whose current element is connected in phase *a* and whose voltage element is connected between phase *a* and neutral will not measure the total power quantities. This may be verified from item (1) of Table XVII which states that the meter in this connection measures

$$(E_1\hat{I}_1 + E_1\hat{I}_2 + E_1\hat{I}_0)$$

which evidently is not the total power quantity.

Power-factor Meters for Polyphase Circuits. The ordinary indicating power-factor meters for three-phase circuits are built with a single-phase current winding and three voltage windings. These instruments do not give correct indication under unbalanced conditions though they are correct for balanced conditions. They are used generally for operating purposes as on switchboards or for measurements on balanced circuits. An indicating power-factor meter giving correct operation under unbalanced conditions is relatively complicated and for this reason is not used commercially. However, some of the recording power-factor meters are built with two elements, one measuring total watts and the other total reactive volt-amperes which are combined to give a correct indication of power factor under unbalanced conditions.

146. Cross Connections for Measuring Reactive Volt-amperes.

The measurement of total reactive volt-amperes of polyphase circuits, as has been pointed out in the previous section, may be made with the same connections as described for measuring power, with the exception that the potential coil circuit of each meter element must be altered so that the current through the coil lags the applied voltage by 90 deg. This method of measuring total reactive volt-amperes seems so straightforward that further discussion would perhaps seem unnecessary. However, practically all the schemes that have been used to date have been based on the use of the ordinary wattmeter with a line current and with the voltage of another phase in order to obtain the required 90-deg. displacement. These schemes generally give correct operation for balanced conditions but give the difference of the positive- and negative-sequence reactive volt-amperes under unbalanced conditions when zero-sequence current is not present, and a still less significant quantity when zero-sequence current is present. Most of these schemes were proposed before the method of symmetrical components came into general use and thus were proposed before the significance of the quantities measured by these meters could be determined. It is interesting, therefore, to study some of these proposals by the method of symmetrical components.

A common connection proposed for measuring reactive volt-amperes is shown in Fig. 178. The three meter elements measure

$$P_M + jQ_M = E_A\hat{I}_a + E_B\hat{I}_b + E_C\hat{I}_c \quad (347)$$

which from Table XVII may be shown to be equal to

$$P_M + jQ_M = j\sqrt{3}[3(E_1\hat{I}_1 - E_2\hat{I}_2)] \quad (348)$$

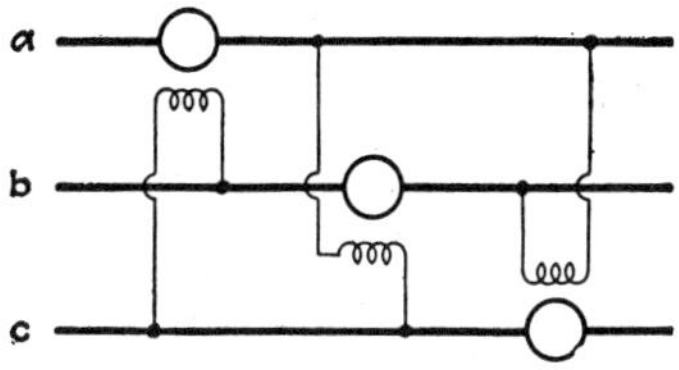

FIG. 178.—Cross connection scheme using wattmeters of the ordinary type, suitable for measuring reactive volt-amperes on systems with balanced voltages and unbalanced currents.

The presence of the j on the right-hand side indicates that wattmeters used with this cross connection measure quantities proportional to the imaginary terms within the brackets. Thus under balanced voltages the wattmeters of Fig. 178 measure the total reactive volt-amperes of the circuit. In spite of the use of three meter elements, this connection under unbalanced conditions fails to measure the true total reactive volt-amperes, namely, the imaginary component of $3(E_1\hat{I}_1 + E_2\hat{I}_2 + E_0\hat{I}_0)$. For balanced conditions, for which this connection measures reactive volt-amperes correctly, a single wattmeter energized, for example, by E_A and I_a might just as well have been used.

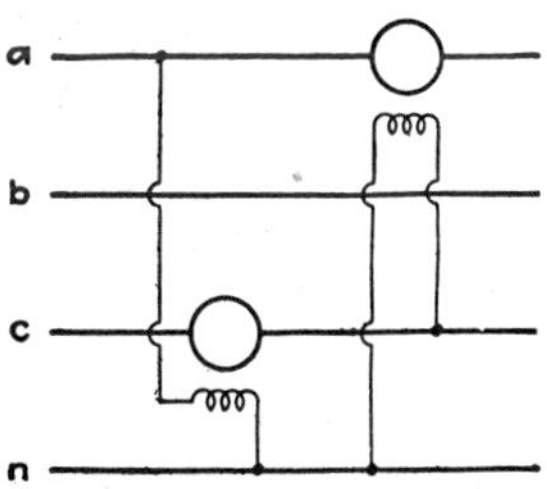

FIG. 179.—Cross-connection scheme for measuring reactive volt-amperes in three-phase systems with balanced voltages and currents.

Another combination which is sometimes considered is that shown in Fig. 179 which will be found to correspond with Fig. 176, with the exception that voltages are taken from other phases which, under balanced conditions, are 90 deg. away from those used for the conventional two-wattmeter scheme for three-phase three-wire service. This scheme may be analyzed with the assistance of Table XVII with the result that

$$\begin{aligned} P_M + jQ_M &= E_c\hat{I}_a - E_a\hat{I}_c \\ &= (a - a^2)E_1\hat{I}_1 + (a - 1)E_1\hat{I}_0 + (a^2 - a)E_2\hat{I}_2 \\ &\quad + (a^2 - 1)E_2\hat{I}_0 + (1 - a^2)E_0\hat{I}_1 + (1 - a)E_0\hat{I}_2 \quad (349) \end{aligned}$$

It will be observed that under *balanced voltage* conditions when $I_0 = 0$, wattmeters used in this connection measure a quantity proportional to the total reactive volt-amperes, irrespective of the value of I_2. Thus the limitations are similar to those of the connection shown in Fig. 178.

147. Meters for the Measurement of Sequence Power Quantities.

Sequence power quantities may be measured in the same manner as phase quantities by impressing a sequence voltage and a sequence current upon the meter element instead of a phase voltage and a phase current. Thus positive-sequence watts, reactive volt-amperes, and power factor may be measured by impressing positive-sequence voltage and positive-sequence current upon the meter elements. In a similar manner negative- and zero-sequence power quantities may also be measured.

Positive-sequence Wattmeter. The method of measuring sequence power quantities is conveniently illustrated by the positive-sequence wattmeter. For obtaining positive-sequence watts any of the general positive-sequence segregating networks for voltage or current, as discussed in Chap. XIV, may be used; the only restriction is that the impedances be so chosen that the same relation between the sequence currents through the meter coils be obtained as in the conventional connections for phase quantities. Thus for measuring positive-sequence watts the current in the potential coil should be in phase with the positive-sequence voltage of the system, and the current in the current coil should be in phase with the positive-sequence current of the corresponding phase of the system. More generally, the meter will measure a quantity proportional to positive-sequence watts provided that the phase displacement of the current through the potential coil with respect to the positive-sequence system voltage is the same as the displacement of the current through the current coil with respect to the positive-sequence current of the system.

In order to illustrate these general relations assume that it is desired to use the positive-sequence voltage network of the common type of Fig. 148 and the bridge-type current network of the type of Fig. 157. A sequence wattmeter with these segregating networks is shown in Fig. 180. For this meter the expression for the current through the potential coil is given in Chap. XIV, equation (306), as

$$I_{M1} = \frac{3E_1}{Z_B + \sqrt{3}M\epsilon^{-j30°}}$$

which may be written as

$$I_{M1} = \frac{3E_1}{\bar{K}\epsilon^{j\theta_K}} \tag{350}$$

The current through the current coil I_{m1} is from (323)

$$I_{m1} = I_1 \tag{351}$$

Let it be further assumed that the impedance M of the potential coil is negligible and that the impedance Z_B is a pure resistance R_B, then

$$I_{M1} = \frac{3E_1}{R_B} \tag{352}$$

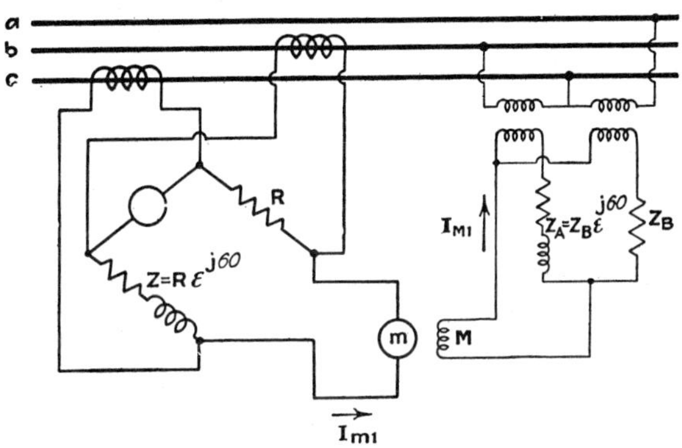

FIG. 180.—Positive-sequence wattmeter using bridge-type current network.

The power quantities measured by the system of Fig. 180 may now be written as

$$P_M + jQ_M = \frac{3E_1}{R_B}\hat{I}_1 \tag{353}$$

Thus the meter of Fig. 180 gives an indication proportional to the total positive-sequence power of the three-phase circuit. Since the resistance term R_B is included in the meter constant, the meter scale will read the total power quantity.

The effect introduced by the impedance of the potential coil having a finite value may readily be corrected by making the power-factor angle of the meter coil branch exactly 30 deg. lagging. Under these conditions it will be seen from equation

(306) that the current through the potential coil is in phase with the positive-sequence voltage when $Z_B = R_B$, a non-inductive resistance. There is also the effect of the potential coil impedance on the calibration which must be taken into account.

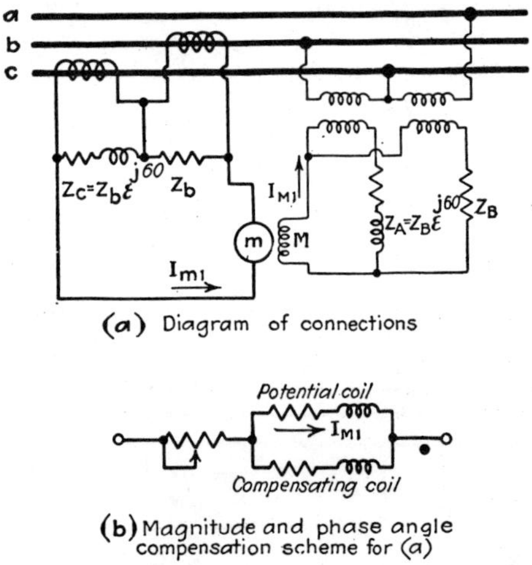

FIG. 181.—Positive-sequence wattmeter using current network of Fig. 155.

Another combination using a different segregating network may be considered in order to further illustrate the method of proportioning the network impedances to measure the desired sequence power quantities. Assume that it is desired to make a positive-sequence wattmeter using the positive-sequence voltage network of Fig. 148 and the positive-sequence current network of Fig. 155, the combination being illustrated in Fig. 181. The expression for the current through the current coil is from equation (318)

$$I_{m1} = \frac{I_1}{1 + \frac{m}{\sqrt{3}Z_b}\epsilon^{-j30°}} \tag{354}$$

in which m is the meter impedance. For this connection the impedance of the current coil cannot be ignored. Consequently, the procedure will be to make a simple current network and adjust the voltage network to give the desired phase relations. Thus, if the impedance Z_b be a non-inductive resistance R_b, equation (354) may be written as

$$I_{m1} = \frac{I_1}{1 + \frac{m}{\sqrt{3}R_b}\epsilon^{-j30°}} = \frac{I_1}{k\epsilon^{j\theta_k}} \tag{355}$$

In order that the meter may measure positive-sequence watts, the phase angles θ_K and θ_k of the right-hand members of equations (350) and (355) must be equal. The determination of the proper impedance values for the network and meter branches of the potential circuits may be carried out conveniently by the method illustrated in Fig. 149 where the angle *DOR* of the voltage network must be equal to the angle θ_k of the current network.

The positive-sequence wattmeter of Fig. 181 is, however, more conveniently adjusted by using voltage and current-segregating networks each having a non-inductive resistance branch. If this is done, it is necessary to adjust the phase relation of the current through the potential coil by shunting the meter element with a suitable impedance, so that the current through the potential coil lags the positive-sequence voltage by the same angle that the current through the current coil lags the positive-sequence current. This may be accomplished by replacing the voltage coil branch of the meter of Fig. 181(*a*) by the connections of (*b*) for which a shunt inductive coil of a suitable value is adjusted to obtain the proper phase-angle correction, and the external resistance is adjusted for final scale calibration.

Under balanced voltage conditions, the sequence voltage segregating network may be eliminated. Thus an ordinary wattmeter may be used in Fig. 180 provided the current coil is made a part of a positive-sequence bridge-type current segregating network.

Positive-sequence Watthour Meters. Watthour meters may be modified by suitable external networks to register a sequence watthour quantity instead of a phase quantity by impressing sequence voltages and currents of the system instead of phase voltages and currents. Watthour meters differ from wattmeters of the dynamometer type in that the current through the potential coil of the former lags approximately 90 deg. behind the impressed voltage, whereas the current is in phase with the voltage in the dynamometer-type meter. The network connections are shown in Fig. 182(*a*) and differ from those of Fig. 181 principally by the fact that the voltage network is connected so that the *c* phase component rather than the *a* phase component of positive-sequence voltage is used in connection with the *a* phase component of positive-sequence current. The reason for this is developed later.

Constructing the impedance diagram of Fig. 156 upon I_{a1} in Fig. 182(*b*), and using a typical phase angle value for the meter impedance m, the phase position of I_{m1} relative to I_{a1} is obtained. Constructing the voltage impedance diagram of Fig. 149 upon $-E_{c1}$, it may be seen that the current through the voltage coil I_M lags behind $-E_{c1}$ by the angle θ_K. For the meter to measure true watts, I_{M1} must lag behind I_{m1} by approximately 90 deg. It will be seen that this condition is fulfilled only when $-E_{c1}$ is used as the reference voltage. The ultimate condition to be fulfilled is that the flux set up by I_{M1} lag I_{m1} by 90 deg. This final adjustment is accomplished by means of a compensating coil in the magnetic circuit of the voltage element. An alternate or supplementary means to

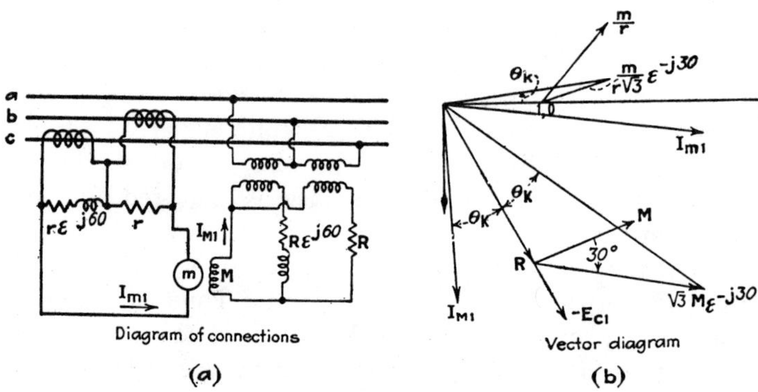

FIG. 182.—Positive-sequence watthour meter.

obtain this adjustment is to shunt the voltage element with a resistance.

148. Flow of Power Quantities Due to Unbalance.

It has been pointed out previously that symmetrical machines generate no negative- or zero-sequence voltages but only positive-sequence voltages; consequently, they can supply to the system only positive-sequence vector power $(P_1 + jQ_1)$. In a symmetrical system with symmetrical loads (or short-circuits), only the positive-sequence power quantities can flow. However, in the case of an unbalanced load (or short-circuit), negative- and zero-sequence power quantities may be produced; thus in a sense positive-sequence power quantities are supplied by the generators and the dissymmetry converts part to negative- and zero-sequence power quantities which are fed into the system at the point of fault. This conception coincides with the equations for the flow of power under unsymmetrical conditions as is readily brought out from the examination of a particular case.

Consider a **single unbalanced load** from line-to-neutral on a system supplied through a grounded generator with positive-, negative-, and zero-sequence impedances of Z_1, Z_2, and Z_0, respectively, and with a generated positive-sequence line-to-neutral voltage of E_{g1}. Assume that this system is supplying a load on phase a whose impedance is Z_L as illustrated in Fig. 183. This problem was solved in Sec. 21 of Chap. III with the result that

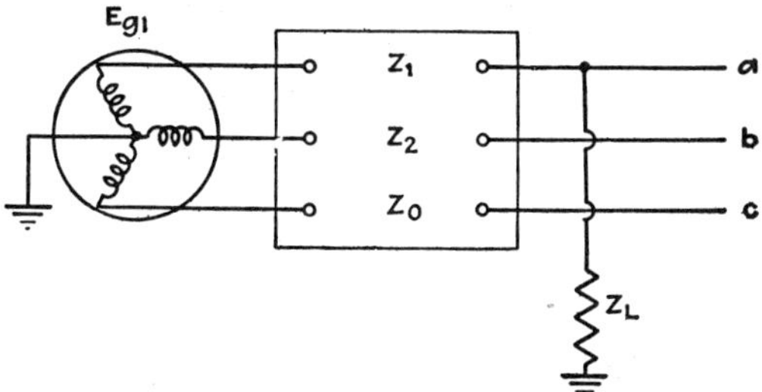

FIG. 183.—Diagram for illustrating the flow of sequence power quantities.

$$E_1 = (Z_2 + Z_0 + 3Z_L)\frac{E_{g1}}{Z_t}$$

$$E_2 = -Z_2\frac{E_{g1}}{Z_t}$$

$$E_0 = -Z_0\frac{E_{g1}}{Z_t}$$

$$I_1 = I_2 = I_0 = \frac{E_{g1}}{Z_t}$$

where

$$Z_t = Z_1 + Z_2 + Z_0 + 3Z_L$$

The total phase quantities at the point of load are readily obtained from the above sequence quantities with the following results

$$E_a = 3Z_L\frac{E_{g1}}{Z_t}$$

$$I_a = \frac{3E_{g1}}{Z_t}$$

The sequence power quantities per phase at the load are

$$P_1 + jQ_1 = E_1\hat{I}_1 = (Z_2 + Z_0 + 3Z_L)\left(\frac{\bar{E}_{g1}}{\bar{Z}_t}\right)^2 \qquad (356)$$

$$P_2 + jQ_2 = -Z_2\left(\frac{\bar{E}_{g1}}{\bar{Z}_t}\right)^2 \tag{357}$$

$$P_0 + jQ_0 = -Z_0\left(\frac{\bar{E}_{g1}}{\bar{Z}_t}\right)^2 \tag{358}$$

The total power quantities may be obtained by combining the sequence quantities as in equation (342), giving

$$P_T + jQ_T = 3[(P_1 + jQ_1) + (P_2 + jQ_2) + (P_0 + jQ_0)] = 9Z_L\left(\frac{\bar{E}_{g1}}{\bar{Z}_t}\right)^2 \tag{359}$$

From the single-phase solution the total power quantities equal

$$P_T + jQ_T = E_a\hat{I}_a + E_b\hat{I}_b + E_c\hat{I}_c = 9Z_L\left(\frac{\bar{E}_{g1}}{\bar{Z}_t}\right)^2 \tag{360}$$

In general, the significance of the negative sign of equations (357) and (358) is that the flow of $P_2 + jQ_2$ and $P_0 + jQ_0$ is in a direction opposite to that of $P_1 + jQ_1$. In other words, an unbalanced load not only draws from a symmetrical system the total power quantities required for the load but also the negative- and zero-sequence power quantities which are fed back into the system at the point of unbalance and which are consumed in producing negative- and zero-sequence losses and in absorbing (in the ordinary system) inductive reactive volt-amperes.

Similar expressions may readily be derived for a line-to-line fault or single-phase load from line-to-line. In this case the negative-sequence current in phase *a* will be the negative of the positive-sequence current of phase *a*. However, the negative-sequence voltage at the fault will have the positive sign so that power quantities are still fed into the network at the point of unbalance.

Combination of Unbalanced Load and Symmetrical Machine Load. In the combination of an unbalanced load, such as due to a single-phase load and a balanced load due to rotating machinery, the unbalanced load produces negative-sequence voltage at the point of unbalance and this negative-sequence voltage impressed on the system causes negative-sequence current to flow not only through the source as described previously but also in all shunt branches including the symmetrical rotating-machinery load. The power due to the flow of the negative-sequence currents produces losses in the rotating-machinery load and increases the total amount of power absorbed by the symmetrical load. Rotating machinery on a system tends to balance the load on a system and to restore the voltages to their normal values. This action takes place with the flow of negative-sequence current in the rotating machinery and with an increased amount of loss and a reduction in the maximum load capacity of the apparatus. Under some conditions negative-sequence current flowing through machines will produce excessive heating in the rotor so that unbalanced operation may require special consideration as discussed for induction motors in Chap. XVII.

In a number of respects, as pointed out by Dr. Fortescue[(17)], positive-sequence power quantities represent more accurately than total power quantities the requirements of power supply systems for handling unbalanced loads; and in addition they give identically the same result for balanced conditions.

Problems

1. The average element of a wattmeter has a resistance of 100 ohms and an inductance of 50 mil-henrys. The external resistance is 15,400 ohms. What capacitance is required across the external resistance to correct the phase angle?

2. The voltage element of a reactive volt-ampere meter has a resistance of 100 ohms and an inductance of 50 mil-henrys. The external reactor has an impedance of $3{,}200 + j14{,}600$ ohms. What capacitance is required across the element to correct the phase angle?

3. A three-phase system, whose voltages across lines are 110 volts, is loaded with the impedances $20 + j3$, $5 - j20$, and $10 + j20$ across lines. Determine $P_T + jQ_T$ and $P_1 + jQ_1$ and $P_2 + jQ_2$.

4. A generator whose synchronous reactance x_d is 1.1 and whose negative-sequence reactance x_2 is 0.15 is loaded across two terminals by an impedance $2 + j0.3$ (in terms of per unit per phase). The voltage across these two terminals is maintained at normal. What is the excitation, neglecting saturation and defining unit excitation as the excitation required to produce normal voltage at no load?

5. What is the positive- and negative-sequence power at the terminals of the machine in Prob. 4?

CHAPTER XVI

MULTIPHASE SYSTEMS

The principal application of the method of symmetrical components is in connection with the solution of the problems of unbalance on three-phase systems. The method, however, is applicable to other multiphase systems used for distribution and conversion. These include the single-phase three-wire system, the two-phase three- and four-wire systems, and the four-phase four- and five-wire systems used for distribution, and also the six- and twelve-phase systems used for synchronous converters and mercury-arc rectifier applications. A study of these multiphase systems will bring out certain points in connection with the fundamental theory of the method of symmetrical components to a better advantage than is possible by consideration of the three-phase systems alone.

149. Resolution of Multiphase Systems into Symmetrical Components.

In order to explain the application of the method of symmetrical components to multiphase systems, it will be necessary to review certain fundamental propositions relating to the theory of numbers and then to illustrate them by a few simple examples. A system of n general numbers may be resolved into n (or more) sets of component numbers, the number of components in each set corresponding to the number of original general numbers. The general numbers may be assumed as the voltage vectors

$$E_a, E_b, E_c, \ldots, E_n$$

The component numbers of each set may be designated by adding a number to the letter subscript identifying the original vectors, as follows:

$$\begin{array}{l} E_{a0}, E_{b0}, E_{c0} \ldots \\ E_{a1}, E_{b1}, E_{c1} \ldots \\ E_{a2}, E_{b2}, E_{c2} \ldots \\ \cdot \quad \cdot \quad \cdot \quad \cdot \quad \cdot \quad \cdot \end{array}$$

The sum of the component numbers associated with each general number is equal to each general number itself. Thus

$$\left.\begin{aligned} E_a &= E_{a0} + E_{a1} + E_{a2} \cdots \\ E_b &= E_{b0} + E_{b1} + E_{b2} \cdots \\ E_c &= E_{c0} + E_{c1} + E_{c2} \cdots \\ &\cdots\cdots\cdots\cdots \end{aligned}\right\} \tag{361}$$

This method of solving a problem involving n general numbers by resolving them into n^2 component numbers or vectors would not, upon casual examination, appear to facilitate the solution. In fact, no advantage is obtained for the most general case though remarkable simplification results when certain kinds of symmetry exist in the relations between these general numbers.

In analyzing the problems involving the general case of the three-phase system, Dr. Fortescue recognized the limitations that existed when considering the two sets of voltages dependent upon phase rotation. Consequently, he went back to a more fundamental consideration based on the theory of numbers which showed that the maximum simplification was obtained when advantage was taken of the symmetry that existed in the relation between these numbers. After recognizing this possibility of simplification he undertook the problem of finding the law governing the formulation of the sets and the components of each set which would reduce the calculations to the maximum degree and yet possess sufficient generality to insure a solution for each particular problem. As a result of this study Dr. Fortescue discovered the principles governing the method of resolving the general numbers into sets of component numbers which constitutes the basis of the method of symmetrical coordinates or symmetrical components.

In the method of symmetrical components, the canonical form of setting up the sets of numbers is to make all the component numbers of equal magnitude in each set. Their phase position, however, is dependent upon the power of a characteristic operator. This characteristic operator is one which produces a phase rotation of $\frac{360}{n}$ deg., where n is the number of phases or the number of the original vectors. The method of setting up the different sets of vectors is to take the vectors in any cyclic order, choosing one vector or phase as a reference. The component vectors of each set may be derived from the vector in the reference phase by rotating it successively by the characteristic angle of each set. The characteristic angles for the different sets are taken as increasing multiples of the characteristic angle of the system. This method of setting up the sequence components for a multiphase system results in

$$\left.\begin{aligned} E_a &= e^{j0}E_{a0} + e^{j0}E_{a1} + e^{j0}E_{a2} + e^{j0}E_{a3} \cdots \\ E_b &= e^{j0}E_{a0} + e^{-j\theta}E_{a1} + e^{-j2\theta}E_{a2} + e^{-j3\theta}E_{a3} \cdots \\ E_c &= e^{j0}E_{a0} + e^{-j2\theta}E_{a1} + e^{-j4\theta}E_{a2} + e^{-j6\theta}E_{a3} \cdots \\ E_d &= e^{j0}E_{a0} + e^{-j3\theta}E_{a1} + e^{-j6\theta}E_{a2} + e^{-j9\theta}E_{a3} \cdots \\ &\cdots\cdots\cdots\cdots \end{aligned}\right\} \quad (362)$$

It will be noted that the characteristic angles have been taken as negative. This choice was made in order that the generated voltages of sequence 1, namely, E_{a1}, E_{b1}, E_{c1}, . . . will have their maximums occurring in time in the cyclic order according to the alphabet. In connection with the above set of equations, it is realized that the vectors E_{a0}, E_{a1}, E_{a2}, . . . may have any magnitude and any phase position; in other words, these vectors have the same number of degrees of freedom as possessed by the original vectors E_a, E_b, E_c, . . .

The set of equations (362) may be written in the form of (363) by using r as a complex number equal to $\epsilon^{j\theta}$.

$$\left.\begin{aligned} E_a &= E_{a0} + E_{a1} + E_{a2} + E_{a3} + \cdots \\ E_b &= E_{a0} + r^{n-1}E_{a1} + r^{n-2}E_{a2} + r^{n-3}E_{a3} + \cdots \\ E_c &= E_{a0} + r^{n-2}E_{a1} + r^{n-4}E_{a2} + r^{n-6}E_{a3} + \cdots \\ E_d &= E_{a0} + r^{n-3}E_{a1} + r^{n-6}E_{a2} + r^{n-9}E_{a3} + \cdots \\ &\cdots\cdots\cdots\cdots \end{aligned}\right\} \quad (363)$$

The analytical expression for the sequence components can be obtained by solving the n equations of (363) resulting in the values

$$\left.\begin{aligned} E_{a0} &= \frac{1}{n}(E_a + E_b + E_c + E_d + \cdots) \\ E_{a1} &= \frac{1}{n}(E_a + rE_b + r^2E_c + r^3E_d + \cdots) \\ E_{a2} &= \frac{1}{n}(E_a + r^2E_b + r^4E_c + r^6E_d + \cdots) \\ &\cdots\cdots\cdots\cdots \end{aligned}\right\} \quad (364)$$

The general method of symmetrical components is therefore applicable to any number of phases. The characteristic features of the different sequence systems for several multiphase systems are summarized in Table XVIII. As a device for visualizing the characteristics of the different sequence systems the component in the reference phase has in every case been taken as of unit magnitude and in phase with the reference. This procedure is used in order to make it easier to remember the relation between the various components in relation to the component in the reference phase. Table XVIII covers multiphase systems with two-, three-, four-, five-, six-, and twelve-phase conductors. The characteristic angles for the various systems are 180, 120, 90, 72, 60, and 30 deg., respectively. That the characteristic angle of the two-phase system is 180 deg. instead of 90 deg. may at first seem surprising, but comparison with the other systems will quickly establish the correctness of this statement. The two-phase three-wire system is in reality an irregular system which will be reviewed in greater detail subsequently. In this analysis the multiphase systems without neutral wire or ground connection are considered as special cases of the system with neutral wire for which the zero-sequence components are zero.

Examination of Table XVIII shows that when the number of phases n is prime as is the case for the three- and five-phase systems, the original vectors may be resolved into one set of zero-sequence vectors and $(n - 1)$ symmetrical sets of vectors which differ only in their subscript designation. Also it may be shown that different sets of components may be obtained from the same set of original vectors, depending upon the original choice of the arbitrary order in which the phases were taken. Consequently, it is possible for certain problems to obtain considerable simplification by choosing the proper arbitrary order, which will normally be taken so that the generated e.m.fs. will always be of sequence 1. Further examination of Table XVIII shows that for the cases in which the number of phases n is not prime, the system may be resolved into multiphase systems of lower order, for example, the twelve-phase system may be resolved into four three-phase systems or into three four-phase systems. The twelve-phase system may also be viewed as made up of four twelve-phase vector systems (sequences 1, 5, 7, and 11), two six-phase vectors (sequences 2 and 10), two four-phase vectors (sequences 3 and 9), two three-phase vectors (sequences 4 and 8), a two-phase vector (sequence 6), and the single-phase vector (sequence 0). In case the twelve-phase system is obtained from three-phase transformers, there may be an appreciable advantage in certain problems in resolving the twelve-phase system into four three-phase systems. Similarly, if the twelve-phase system were derived from a four-phase supply, then there may be a corresponding advantage in

TABLE XVIII.—SEQUENCE COMPONENTS FOR MULTI-PHASE SYSTEMS

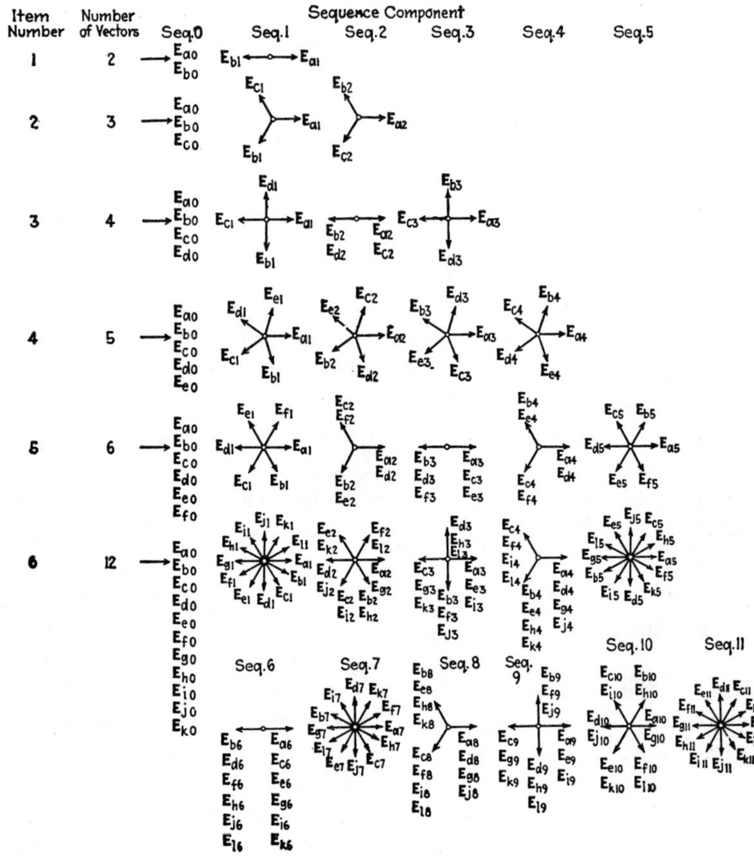

resolving the twelve-phase system into three four-phase systems. Otherwise, it is generally found desirable to consider the twelve-phase system as set up in Table XVIII.

150. Two-phase and Four-phase Systems.

The various forms of the two-phase and four-phase systems will be discussed together since they have the common feature of generated e.m.fs. 90 electrical degrees apart. The four-phase five-wire system is the most general case and may be solved as outlined in item 3 of Table XVIII. The four-phase four-wire system without neutral connection may be viewed as a special case of the five-wire system for which the zero-sequence component is zero. Frequently the four-phase system without neutral connection, or the two-phase system as it is commonly called, is operated with conductors and loads symmetrically disposed

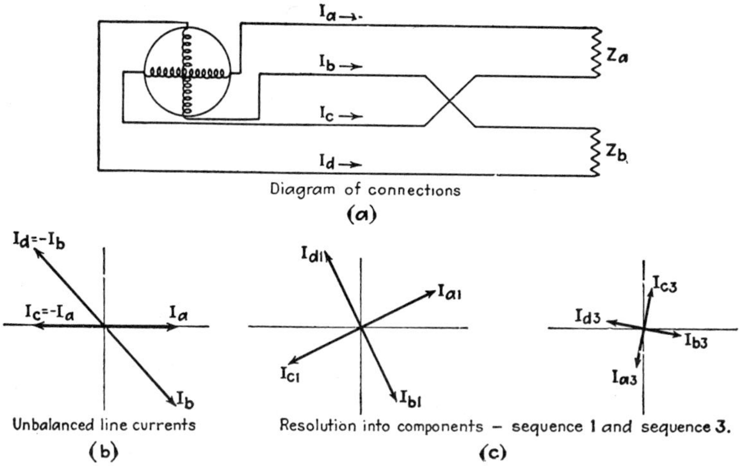

FIG. 184.—Two-phase four-wire system.

with respect to ground and with loads taken only between wires a and c, and b and d, as indicated in Fig. 184. In such a system the currents are

$$\left.\begin{aligned} I_a &= +I_a \qquad & I_c &= -I_a \\ I_b &= +I_b \qquad & I_d &= -I_b \end{aligned}\right\} \tag{365}$$

The above set of equations may be analyzed as a four-phase system with the result that

$$\left.\begin{aligned} I_0 &= 0 \qquad & I_2 &= 0 \\ I_1 &= \frac{I_a + jI_b}{2} \qquad & I_3 &= \frac{I_a - jI_b}{2} \end{aligned}\right\} \tag{366}$$

Thus it will be seen that the four-phase or two-phase system of Fig. 184 may be solved in the manner indicated in item 3 of Table XVIII using only the positive-sequence components (sequence 1) and the negative-sequence component (sequence 3) as illustrated in (b) and (c) of Fig. 184. For balanced conditions the voltages and currents will all be of sequence 1, and only one sequence of generated e.m.fs. will be required for the analysis of the system.

151. Irregular Systems.

The two-phase three-wire system of Fig. 185(a) may be regarded as an irregular system because the characteristic angle

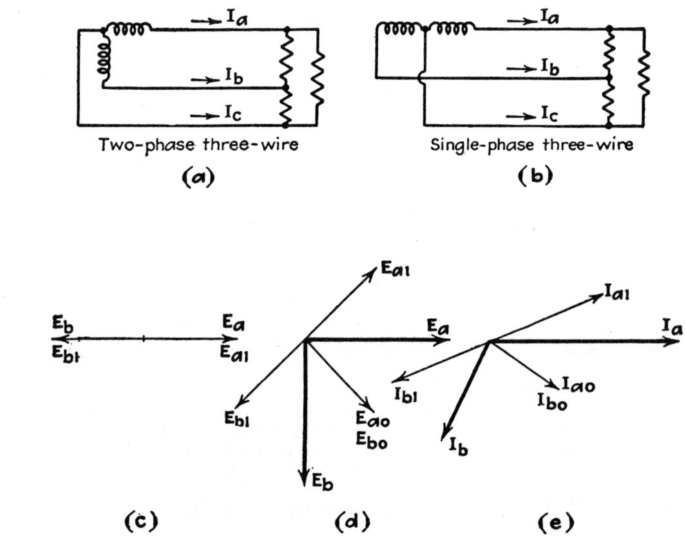

(c) Normal generated e.m.f.'s in a single-phase three-wire system.
(d) Generated e.m.f.'s in a two-phase three-wire system in terms of the symmetrical components of a single-phase three-wire system.
(e) Resolution of phase currents into their symmetrical components.
(f) Analytical expressions

$$
\begin{aligned}
I_{a0} &= \tfrac{1}{2}(I_a + I_b) \\
I_{a1} &= \tfrac{1}{2}(I_a - I_b) \\
I_a &= I_{a0} + I_{a1} \\
I_b &= I_{a0} - I_{a1}
\end{aligned}
$$

FIG. 185.—Representation of a two-phase three-wire system by components of a single-phase three-wire system.

of the system $\frac{360}{n} = 180$ deg. is different from the phase angle of the generated e.m.fs. Because of this fact the generated e.m.fs. cannot be represented by a single sequence of a regular system such as those that have been described. However, by building up certain types of irregular sequence systems it is possible to represent the generated e.m.fs. by a single sequence.

The **two-phase three-wire system** has been selected for analysis to illustrate the general treatment of irregular systems.

The two-phase three-wire system shown in Fig. 185(*a*) can be analyzed in terms of the **single-phase three-wire system** illustrated in (*b*) of the same figure. Each phase of such a system may vary in magnitude and phase position. In the single-phase three-wire system ordinarily the generated e.m.f. is a positive-sequence (sequence 1) only, as shown in Fig. 185(*c*); but for this system to represent the generated voltages of a two-phase three-wire system, two sequences (sequence 1 and sequence 0) are required. This is illustrated in Fig. 185(*d*)

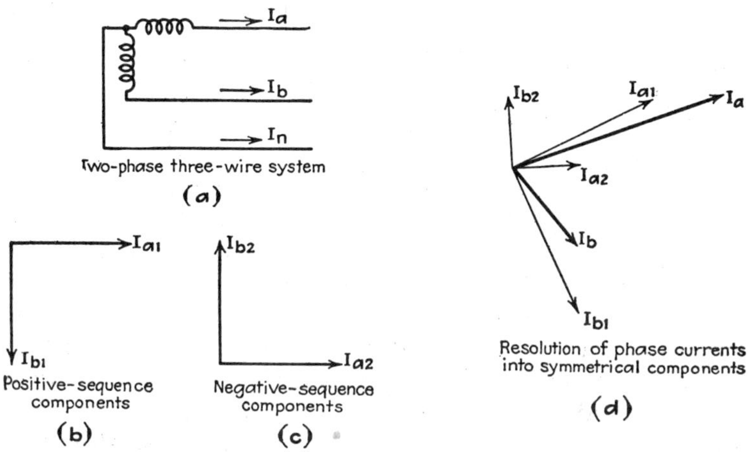

FIG. 186.—Representation of a two-phase three-wire system by components of an irregular sequence system.

in which E_a and E_b represent the generated phase voltages and the other vectors the sequence components. The resolution of two general phase currents into their sequence components is shown in (*e*) of the same figure. The different sequence networks do not have coupling between them if the phase wires are symmetrically transposed with respect to the neutral wire or ground. The generator impedances require some manipulation to convert from one system to the other.

The two-phase three-wire system may also be handled by an **irregular system of sequence components** such as that of Fig. 186. It will be noted that in this irregular system the two sets of sequence components are taken, one set comprising components 90 electrical degrees apart in the positive sense and the other set 90 electrical degrees apart in the negative sense. This irregular system requires only one set of sequence components of the type illustrated in Fig. 185 to take care of the normally generated e.m.fs. of a two-phase three-wire system. The addition of sequence 2 takes care of unbalance. The resolution of the two phase currents into the symmetrical components is shown in Fig. 186(*d*). It is to be realized, however, that coupling will be introduced between positive- and negative-sequence networks in case of an impedance in the neutral connection or in case the impedance of the distribution line requires consideration.

The several methods which have just been outlined for handling the two-phase three-wire system have been described for the purpose of illustrating the types of sequence systems that may be used for solving special problems. The choice of the sequence system to be used with the different types of networks which have been described will be reviewed briefly. For the general case of the four-phase four- or five-wire system, the four-phase system as outlined in item 3 of Table XVIII is preferred. For the ordinary two-phase system of Fig. 184, the same sequence analysis may be used except that sequence 0 and sequence 2 may be omitted. The two-phase three-wire system without neutral impedance or without distribution lines of sufficient impedance to require inclusion in the calculations will be found most convenient by the use of the "irregular system" described in Fig. 185. If the two-phase three-wire system has a neutral impedance or is used for distribution, it will be found more convenient to use the conventional two-phase system as outlined in Fig. 185. This choice for the two-phase three-wire system is suggested because of the fact that it is ordinarily simpler to add generated e.m.fs. in the negative-sequence network than the alternative of requiring coupling between the sequence networks though using but a single source.

152. General Treatment of Polyphase Circuits.

All polyphase systems that are symmetrical, viewed from the standpoint of system constants, may be solved in a manner similar to that for the three-phase systems which have been described previously. The constants can be determined separately for the individual sequence systems and the constants of the individual sequence network set up in a network of its own. For this case there will be no reaction between the different networks. The tie between the different networks will be the relations imposed by the particular kind of fault or unbalanced terminal condition.

Unsymmetrical polyphase systems are more complicated than symmetrical systems in that mutual impedances exist between quantities of the different sequences. This problem will be discussed in Chap. XVIII.

153. Impedances of Symmetrical Polyphase Systems.

The impedances of static equipment, such as transformers and transmission or distribution lines, are so much like those for the three-phase systems as to require no further consideration. Rotating equipment introduces a quite different phenomenon which warrants further discussion.

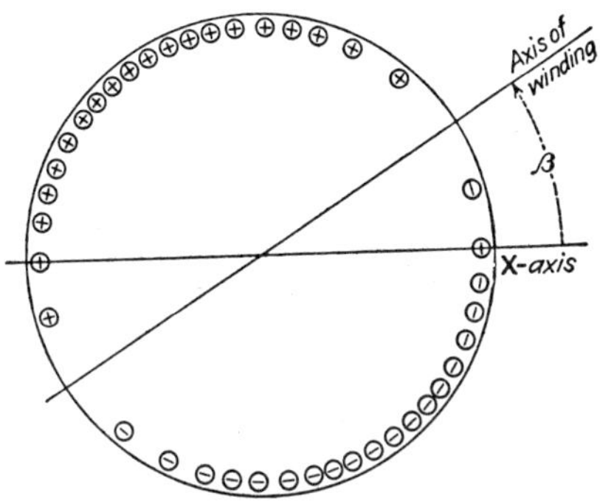

FIG. 187.—Distributed winding whose axis makes an angle β with the X-axis.

For this purpose consider a polyphase machine in which all the windings are distributed, wound so that a sinusoidal space field is produced by a current flowing through each of the phases. In Fig. 187 let β be the phase position of the axis of a winding relative to some arbitrary reference line which will be assumed to lie along the horizontal X-axis. Consider first that this angle is zero and that the current $i = \sqrt{2}\bar{I} \cos(\omega t + \alpha)$ flows through this winding. As shown in Chap. II, equation (16), this current can be expressed as

$$i = \frac{\sqrt{2}}{2}\bar{I}\epsilon^{j(\omega t+\alpha)} + \frac{\sqrt{2}}{2}\bar{I}\epsilon^{-j(\omega t+\alpha)} \tag{367}$$

This current can be represented as the sum of two conjugate vectors rotating in opposite directions as shown in Fig. 188 The imaginary parts thus cancel out, leaving only the real terms which express the instantaneous value of i. For $t = 0$, the value of the current is

$$\frac{\sqrt{2}}{2}[\bar{I}\epsilon^{j\alpha} + \bar{I}\epsilon^{-j\alpha}] \qquad (368)$$

The sinusoidal current flowing through the winding under consideration sets up a pulsating field which at any instant has a sinusoidal space distribution. The axis of this sinusoidal flux wave, since $\beta = 0$, lies along the X-axis. According to the well-known device due to Ferraris, a pulsating sinusoidal flux wave can be resolved into two sinusoidal flux waves of opposite rotation. The magnitude and position of these flux waves coincide with the position of the two vectors of Fig. 188 representing the instantaneous values of current.

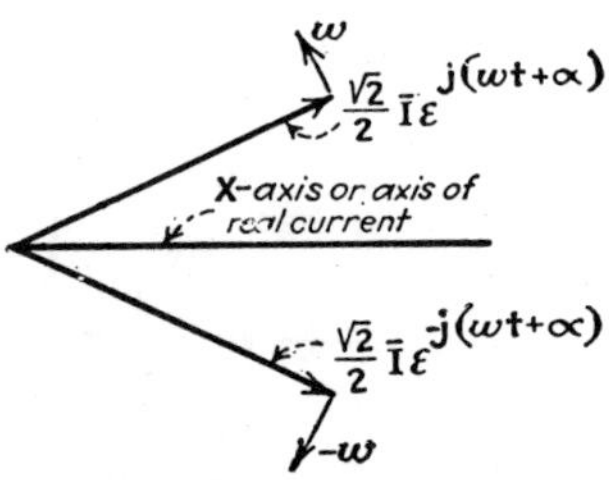

FIG. 188.—Resolution of a pulsating flux in the X-axis into two fluxes rotating in opposite directions.

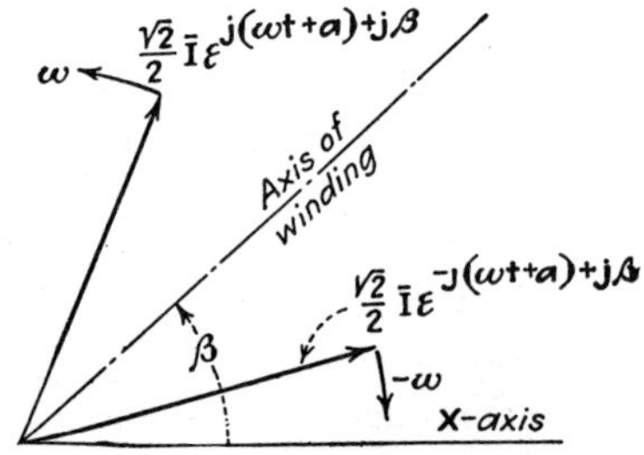

FIG. 189.—Resolution of a pulsating flux in a winding whose axis makes angle β with the X-axis, into two components rotating in opposite directions.

Now consider a similar winding whose axis lies at an angle β from the reference line. If the same current be made to flow through the winding, two rotating fields will again be set up which will have the same positions relative to the axis of the winding. Since, however, this axis is now at an angle β ahead of the reference line, the vector position of the resultant m.m.f. or current field will be given by

$$\frac{\sqrt{2}}{2}\bar{I}\epsilon^{j(\omega t+\alpha)+j\beta} + \frac{\sqrt{2}}{2}\bar{I}\epsilon^{-j(\omega t+\alpha)+j\beta} \qquad (369)$$

which consists of two oppositely rotating fields as shown in Fig. 189.

With this fundamental relation showing the vector position of the two rotating fields as a function of the phase position of the current in the winding, and the action of the winding established, the results may be applied to an n phase machine.

The phase displacement between the windings of such a machine will be $\frac{360}{n}$ deg.; β will then have successively the values 0, $\frac{360}{n}$; $\frac{2}{n}360$; $\cdots$ $\frac{n-1}{n}360$. For the zero-sequence set of currents the phase currents are all equal to the a phase current. The field set up may therefore be written:

For the a phase

$$\frac{\sqrt{2}}{2}\bar{I}\epsilon^{j(\omega t+\alpha)} + \frac{\sqrt{2}}{2}\bar{I}\epsilon^{-j(\omega t+\alpha)} \tag{370}$$

For the b phase

$$\frac{\sqrt{2}}{2}\bar{I}\epsilon^{j\left(\omega t+\alpha+\frac{1}{n}360\right)} + \frac{\sqrt{2}}{2}\bar{I}\epsilon^{-j\left(\omega t+\alpha-\frac{1}{n}360\right)} \tag{371}$$

For the c phase

$$\frac{\sqrt{2}}{2}\bar{I}\epsilon^{j\left(\omega t+\alpha+\frac{2}{n}360\right)} + \frac{\sqrt{2}}{2}\bar{I}\epsilon^{-j\left(\omega t+\alpha-\frac{2}{n}360\right)} \tag{372}$$

For the n phase

$$\frac{\sqrt{2}}{2}\bar{I}\epsilon^{j\left(\omega t+\alpha+\frac{n-1}{n}360\right)} + \frac{\sqrt{2}}{2}\bar{I}\epsilon^{-j\left(\omega t+\alpha-\frac{n-1}{n}360\right)} \tag{373}$$

The resultant field is the sum of the separate fields due to the n phases. Adding up the fields of positive rotation and factoring out $\frac{\sqrt{2}}{2}\bar{I}\epsilon^{j\left(\omega t+\alpha\right)}$, the following coefficient remains:

$$\epsilon^{j0} + \epsilon^{j\frac{1}{n}360} + \epsilon^{j\frac{2}{n}360} + \cdots + \epsilon^{j\frac{n-1}{n}360} \tag{374}$$

Since these terms are all equal except for their arguments which differ by multiples of $\frac{360}{n}$, the sum is zero. Therefore, the net field rotating in a positive sense is zero. The coefficient for the negatively rotating field is the same. It follows that the air-gap flux in such a machine due to a zero-sequence set of currents is zero. The reactance to the zero-sequence is merely the effect of departures from a true sinusoidal wave shape and leakage effects between phases. This reactance will in general be low and be affected greatly by chording.

For the positive-sequence set of currents, calling the a phase the *reference phase,* α will have successively the values α,

$\alpha - \frac{1}{n}360$, $\alpha - \frac{2}{n}360$, $\cdots$, $\alpha - \frac{n-1}{n}360$. The fields set up in the different phases are then the same as for the zero-sequence [equations (370) to (373)] except that the alphas must be given the values indicated. Thus the fields are:

For the a phase

$$\frac{\sqrt{2}}{2}\bar{I}\epsilon^{j(\omega t+\alpha)} + \frac{\sqrt{2}}{2}\bar{I}\epsilon^{-j(\omega t+\alpha)} \tag{375}$$

For the b phase

$$\frac{\sqrt{2}}{2}\bar{I}\epsilon^{j\left(\omega t+\alpha-\frac{1}{n}360+\frac{1}{n}360\right)} + \frac{\sqrt{2}}{2}\bar{I}\epsilon^{-j\left(\omega t+\alpha-\frac{1}{n}360-\frac{1}{n}360\right)}$$

or

$$\frac{\sqrt{2}}{2}\bar{I}\epsilon^{j(\omega t+\alpha)} + \frac{\sqrt{2}}{2}\bar{I}\epsilon^{-j\left(\omega t+\alpha-\frac{2}{n}360\right)} \tag{376}$$

For the c phase

$$\frac{\sqrt{2}}{2}\bar{I}\epsilon^{j\left(\omega t+\alpha-\frac{2}{n}360+\frac{2}{n}360\right)} + \frac{\sqrt{2}}{2}\bar{I}\epsilon^{-j\left(\omega t+\alpha-\frac{2}{n}360-\frac{2}{n}360\right)}$$

or

$$\frac{\sqrt{2}}{2}\bar{I}\epsilon^{j(\omega t+\alpha)} + \frac{\sqrt{2}}{2}\bar{I}\epsilon^{-j\left(\omega t+\alpha-\frac{4}{n}360\right)} \tag{377}$$

For the n phase

$$\frac{\sqrt{2}}{2}\bar{I}\epsilon^{j\left(\omega t+\alpha-\frac{n-1}{n}360+\frac{n-1}{n}360\right)} + \frac{\sqrt{2}}{2}\bar{I}\epsilon^{-j\left(\omega t+\alpha-\frac{n-1}{n}360-\frac{n-1}{n}360\right)}$$

or

$$\frac{\sqrt{2}}{2}\bar{I}\epsilon^{j(\omega t+\alpha)} + \frac{\sqrt{2}}{2}\bar{I}\epsilon^{-j\left(\omega t+\alpha-\frac{2(n-1)}{n}360\right)} \tag{378}$$

It will be observed that for the positively rotating components, the contributions from the individual phases are all equal and add up directly in space phase to produce a net resultant field. This field in the case of a synchronous machine rotates in the same direction and with the same speed as the rotor. The impedances associated with it are of the same nature as the positive-sequence impedances described in Chap. V. With regard to the negatively rotating field, after factoring out the coefficient $\frac{\sqrt{2}}{2}\bar{I}\epsilon^{-j(\omega t+\alpha)}$, the remaining terms, on adding, are

$$\epsilon^{j0} + \epsilon^{+j\frac{2}{n}360} + \epsilon^{+j\frac{4}{n}360} + \cdots \epsilon^{+j\frac{2(n-1)}{n}360} \tag{379}$$

The summation of n terms separated by the angle $\frac{2}{n}360$ is equal to zero. This may be shown by dividing the terms into two groups, the one consisting of the first $\frac{n}{2}$ terms and the other of the remaining terms. The summation of the terms in each group is zero as indicated for the first group in Fig. 190. The negatively rotating field is therefore zero.

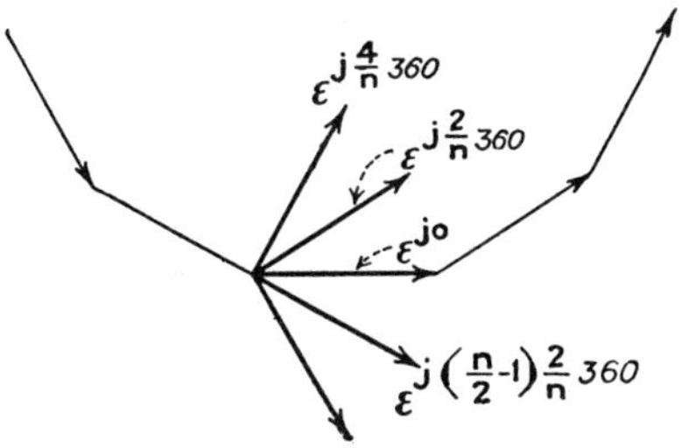

FIG. 190.—Diagram showing that $\frac{n}{2}$ equally spaced vectors separated by the angle, $\frac{2}{n}360$, add to zero.

Corresponding expressions can be set up for the other sequences. For the second-sequence in particular, after factoring out the common terms, the coefficient for the positively rotating field is

$$\epsilon^{j0} + \epsilon^{-j\frac{360}{n}} + \epsilon^{-j\frac{2}{n}360} + \cdots \epsilon^{-j\frac{n-1}{n}360} \tag{380}$$

and for the negatively rotating field is

$$\epsilon^{j0} + \epsilon^{-j\frac{3}{n}360} + \epsilon^{-j\frac{6}{n}360} + \cdots \epsilon^{-j\frac{3(n-1)}{n}360} \tag{381}$$

Both of these summations add to zero, so that the second-sequence air-gap flux in an actual machine is practically zero. The impedance for this sequence partakes of the nature of the zero-sequence.

Similar considerations apply to the other sequences, except the $(n - 1)$th or the negative-sequence. For this sequence the coefficient for the positively rotating field becomes zero, but the components of the negatively rotating field are all in phase and add just as the components of the positively rotating field added together for the positive-sequence. Therefore, since only a negatively rotating field is present for the negative-sequence, the negative-sequence impedance is dependent upon the impedance of the rotor and will depend, therefore, upon the leakage of both the stator and rotor.

Summarizing, in symmetrical machines the positive-sequence current (sequence 1) produces a positively rotating field which reacting with the rotor circuits, gives rise to the various kinds of positive-sequence impedance. The negative-sequence currents (sequence $n - 1$) develop a negatively rotating field which reacts with the rotor circuits and gives rise to the negative-sequence impedance. All other sequences develop no resultant rotating fields. The impedances associated with these sequences are of the nature of armature-leakage impedances and vary with chording.

154. Harmonics.

Distorted waves of a periodic nature can be resolved into a Fourier series consisting of a fundamental and higher harmonics. The result can always be expressed in the form

$$i = \sqrt{2}\bar{I}_1 \cos (\omega t + \alpha_1) + \sqrt{2}\bar{I}_2 \cos 2(\omega t + \alpha_2) + \sqrt{2}\bar{I}_3 \cos 3(\omega t + \alpha_3) + \cdots \quad (382)$$

in which $\bar{I}_1$ $\bar{I}_2$, . . . , represent the r.m.s. values of amplitude of the respective harmonics. This relation can also be expressed as

$$i = \text{real part of } [\sqrt{2}\bar{I}_1\epsilon^{j(\omega t+\alpha_1)} + \sqrt{2}\bar{I}_2\epsilon^{2j(\omega t+\alpha_2)} + \sqrt{2}\bar{I}_3\epsilon^{3j(\omega t+\alpha_3)} + \cdots] \quad (383)$$

Since the impedance to the fundamental component of current in some cases is influenced to a very considerable extent by its sequence, it is highly desirable to determine the sequence of the harmonics. This analysis will be restricted to three-phase systems, but it will be apparent that it may be extended to systems of any number of phases.

Assuming that the system under consideration is symmetrical and of **positive-sequence,** then the three-phase currents will be identical except for the phase shift of 240 and 120 deg. In this case the three phase currents may be expressed analytically by the following:

$$\left.\begin{aligned} i_a &= \text{real part } \sqrt{2}[\bar{I}_1\epsilon^{j(\omega t+\alpha_1)} + \bar{I}_2\epsilon^{j2(\omega t+\alpha_2)} + \bar{I}_3\epsilon^{j3(\omega t+\alpha_3)} + \cdots] \\ i_b &= \text{real part } \sqrt{2}[\bar{I}_1\epsilon^{j(\omega t+\alpha_1+240)} + \bar{I}_2\epsilon^{j2(\omega t+\alpha_2+240)} + \bar{I}_3\epsilon^{j3(\omega t+\alpha_3+240)} + \cdots] \\ i_c &= \text{real part } \sqrt{2}[\bar{I}_1\epsilon^{j(\omega t+\alpha_1+120)} + \bar{I}_2\epsilon^{j2(\omega t+\alpha_2+120)} + \bar{I}_3\epsilon^{j3(\omega t+\alpha_3+120)} + \cdots] \end{aligned}\right\} \quad (384)$$

or

$$\left.\begin{aligned}
i_a &= \text{real part } \sqrt{2}[\bar{I}_1\epsilon^{j(\omega t+\alpha_1)} + \bar{I}_2\epsilon^{j2(\omega t+\alpha_2)} + \\
&\qquad \bar{I}_3\epsilon^{j3(\omega t+\alpha_3)} + \cdots] \\
i_b &= \text{real part } \sqrt{2}[a^2\bar{I}_1\epsilon^{j(\omega t+\alpha_1)} + a\bar{I}_2\epsilon^{j2(\omega t+\alpha_2)} + \\
&\qquad \bar{I}_3\epsilon^{j3(\omega t+\alpha_3)} + \cdots] \\
i_c &= \text{real part } \sqrt{2}[aI_1\epsilon^{j(\omega t+\alpha_1)} + a^2\bar{I}_2\epsilon^{j2(\omega t+\alpha_2)} + \\
&\qquad \bar{I}_3\epsilon^{j3(\omega t+\alpha_3)} + \cdots]
\end{aligned}\right\} \quad (385)$$

Thus it will be seen that the fundamental is of positive-sequence, the second harmonic of negative-sequence, and the third harmonic of zero-sequence. This result is summarized in Table XIX.

For the **negative-sequence,** the three phase currents may be written

$$\left.\begin{aligned}
i_a &= \text{real part } \sqrt{2}[\bar{I}_1\epsilon^{j(\omega t+\alpha_1)} + \bar{I}_2\epsilon^{j2(\omega t+\alpha_2)} \\
&\qquad + \bar{I}_3\epsilon^{j3(\omega t+\alpha_3)} + \cdots] \\
i_b &= \text{real part } \sqrt{2}[\bar{I}_1\epsilon^{j(\omega t+\alpha_1+120)} + I_2\epsilon^{j2(\omega t+\alpha_2+120)} \\
&\qquad + \bar{I}_3\epsilon^{j3(\omega t+\alpha_3+120)} + \cdots] \\
i_c &= \text{real part } \sqrt{2}[\bar{I}_1\epsilon^{j(\omega t+\alpha_1+240)} + \bar{I}_2\epsilon^{j2(\omega t+\alpha_2+240)} \\
&\qquad + I_3\epsilon^{j3(\omega t+\alpha_3+240)} + \cdots]
\end{aligned}\right\} \quad (386)$$

or

$$\left.\begin{aligned}
i_a &= \text{real part } \sqrt{2}[\bar{I}_1\epsilon^{j(\omega t+\alpha_1)} + \bar{I}_2\epsilon^{j2(\omega t+\alpha_2)} \\
&\qquad + \bar{I}_3\epsilon^{j3(\omega t+\alpha_3)} + \cdots] \\
i_b &= \text{real part } \sqrt{2}[a\bar{I}_1\epsilon^{j(\omega t+\alpha_1)} + a^2\bar{I}_2\epsilon^{j2(\omega t+\alpha_2)} \\
&\qquad + \bar{I}_3\epsilon^{j3(\omega t+\alpha_3)} + \cdots] \\
i_c &= \text{real part } \sqrt{2}[a^2\bar{I}_1\epsilon^{j(\omega t+\alpha_1)} + a\bar{I}_2\epsilon^{j2(\omega t+\alpha_2)} \\
&\qquad + \bar{I}_3\epsilon^{j3(\omega t+\alpha_3)} + \cdots]
\end{aligned}\right\} \quad (387)$$

The fundamental is thus of negative-sequence, the second harmonic of positive-sequence, and the third harmonic of zero-sequence. These results are also summarized in Table XIX.

For the **zero-sequence** currents the currents in all three phases are in phase so that the fundamental and all the harmonics are of zero-sequence.

The fundamental frequency positive- and negative-sequence impedance of machines rotating at normal speed differ considerably. This is due to the fact that the rotating field set up by the positive-sequence component is largely magnetizing, no currents are induced in the rotor; whereas for the negative-sequence, currents of double frequency are induced in the rotor and the impedance is determined largely by the leakage fluxes of the stator and rotor. For the harmonics, currents are always

TABLE XIX.—SEQUENCE OF HARMONICS IN THREE-PHASE SYSTEMS

Harmonic	Sequence		
	Positive	Negative	Zero
1	1	2	0
2	2	1	0
3	0	0	0
4	1	2	0
5	2	1	0
6	0	0	0
7	1	2	0
8	2	1	0
9	0	0	0

set up in the rotor, so that the difference between the positive- and negative-sequence impedances decreases as the order of the harmonic increases.

Problems

1. A two-phase system is converted to a three-phase system by means of a "Scott connection." If ρ be the ratio of transformation of the transformers used and m the portion of "teaser" transformer used, prove that

$$E_{a1} = \rho\left\{\frac{E_{x1}}{\sqrt{3}}\left(\frac{m}{\sqrt{3}} + \frac{1}{2}\right) + \frac{E_{x2}}{\sqrt{3}}\left(\frac{m}{\sqrt{3}} - \frac{1}{2}\right)\right\}$$

$$E_{a2} = \rho\left\{\frac{E_{x1}}{\sqrt{3}}\left(\frac{m}{\sqrt{3}} - \frac{1}{2}\right) + \frac{E_{x2}}{\sqrt{3}}\left(\frac{m}{\sqrt{3}} + \frac{1}{2}\right)\right\}$$

in which E_{a1} and E_{a2} are the positive- and negative-sequence voltages to neutral on the three-phase side and E_{x1} and E_{x2} are the corresponding values to the two-phase side.

2. Devise a metering scheme to measure the zero- and positive-sequence voltages of a four-phase system.

3. Devise a metering scheme to measure the zero- and positive-sequence voltages of a six-phase system.

4. A three-phase generator whose impedances Z_1, Z_2, and Z_0 are known is loaded across two terminals by means of an autotransformer, by which process a single-phase three-wire system is produced, the neutral being taken off midway of the transformer. Determine the positive- and zero-sequence impedances of the generator for the single-phase system in terms of Z_1, Z_2, and Z_0.

5. Given $I_a = 20 + j60$ and $I_b = 25 - j10$, determine (*a*) the positive- and zero-sequence components of a single-phase system, and (*b*) the irregular components of a two-phase three-wire system.

CHAPTER XVII

INDUCTION MOTORS

An important application of the theory of symmetrical components is the analysis of induction-motor operation under unbalanced conditions. For this purpose a symmetrically wound machine will be considered. It can readily be understood without discussion that in such a machine there is no reaction between the different sequence quantities. Being symmetrical, it is well known that if balanced voltages of a certain sequence be applied to the stator and rotor, only balanced currents of the same sequence, regardless of slip, will flow. The operation of induction motors under three different conditions will be discussed: (1) application of unbalanced voltages to the stator of a symmetrical machine, (2) one phase of the stator of a symmetrical machine open-circuited, and (3) unbalanced impedances in the rotor of an otherwise symmetrical machine. Before discussing different unbalanced conditions of operation, consider the characteristics of the motor for positive-sequence voltages and currents.

155. Characteristics with Application of Positive-sequence Voltages to the Stator.

Assume a balanced set of positive-sequence voltages applied to the terminals of an induction motor having a certain shaft load. The rotor will rotate in the positive sense (same direction as rotating field) with slip s expressed as a fraction. The characteristics can be best defined in terms of the well-known equivalent diagram. In this development it will be assumed that the rotor has the same number of turns as the stator or has been reduced to the equivalent so that the rotor constants and currents and voltages may be expressed on the same turns basis as the stator. Let

R_s = resistance per phase of the stator
X_s = leakage reactance per phase of the stator at supply frequency.

R_r = resistance per phase of the rotor.
X_r = leakage reactance per phase of the rotor at supply frequency.
I_{s1} = positive-sequence current in the stator referred to phase a.
I_{r1} = positive-sequence current in the rotor referred to phase a.
E_{s1} = positive-sequence line-to-neutral voltage in the stator referred to phase a.

The application of the voltage E_{s1} to the stator sets up a field which rotates with synchronous velocity. This field cuts the rotor windings and sets up therein a current I_{r1} of slip frequency, which produces a field that rotates with slip velocity relative to the rotor. But since the rotor revolves at a speed $(1 - s)$ times synchronous velocity, the field set up by the current I_{r1} rotates also with synchronous velocity relative to the stator.

Designate by E_{g1} the voltage induced in the stator by the resultant synchronously rotating field produced by I_{s1} and I_{r1}. The applied voltage E_{s1} must be equal to the sum of this voltage and the resistance and leakage reactance drops in the stator, or

$$E_{s1} = E_{g1} + (R_s + jX_s)I_{s1} \tag{388}$$

The voltage induced in the rotor must be proportional to the slip, or sE_{g1}. It is this voltage which gives rise to the current I_{r1} of slip frequency. Therefore

$$sE_{g1} = (R_r + jsX_r)I_{r1} \tag{389}$$

where the leakage reactance X_r is measured at supply frequency. Dividing equation (389) by s

$$E_{g1} = \left(\frac{R_r}{s} + jX_r\right)I_{r1} \tag{390}$$

The air-gap voltage E_{g1} is produced as the resultant of I_{s1} and I_{r1}, and if the positive direction of flow of I_{s1} be considered magnetizing and I_{r1} demagnetizing, then E_{g1} is proportional to $(I_{s1} - I_{r1})$. Expressing this proportionality by means of the magnetizing reactance jX_m, there results

$$E_{g1} = jX_m(I_{s1} - I_{r1}) \tag{391}$$

The three conditions expressed analytically by equations (388), (390), and (391), are fulfilled by the **equivalent circuit**

for positive-sequence, shown in Fig. 191(*a*). It follows, therefore, that the solution of this network gives the characteristics of the induction motor. It is convenient to split up the resistance $\frac{R_r}{s}$ into two components R_r and $\frac{1-s}{s}R_r$, as shown in Fig. 191(*b*). By this means the power absorbed by R_r represents the copper loss, and, since this is the only rotor loss considered, the power absorbed by $\frac{1-s}{s}R_r$ represents the shaft power output. Therefore, calling P_1 the total shaft power,

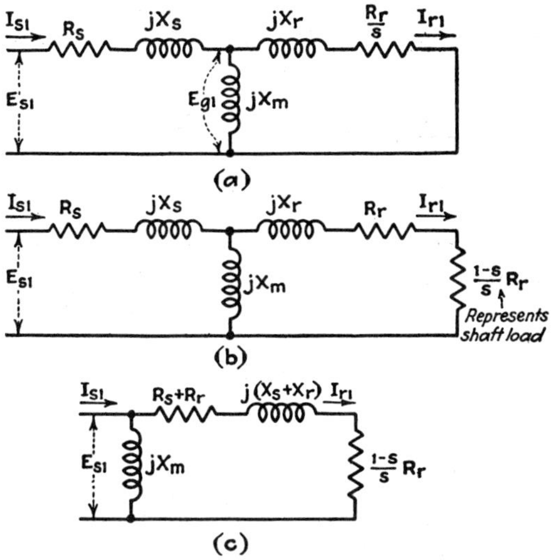

FIG. 191.—Equivalent circuits of induction motors (positive-sequence).

$$P_1 = 3\frac{1-s}{s}R_r\bar{I}^2_{r1} \tag{392}$$

This is equal to the shaft torque T, times the speed, so that

$$T_1 = K\frac{3R_r}{s}\bar{I}^2_{r1} \tag{393}$$

in which K is a constant dependent upon the units. When $3R_r\bar{I}^2_{r1}$ is expressed in watts, and T in pounds at 1 foot,

$$K = \frac{33{,}000}{(746)2\pi(\text{r.p.m.})_{synchronous}} = \frac{33{,}000p}{(746)2\pi f60} = \frac{0.117p}{f} \tag{394}$$

where p is the number of pairs of poles and f the supply frequency.

A usual simplifying assumption is to place the shunt magnetizing reactance jX_m directly across the terminals as shown in

Fig. 191(*c*). In the above development no attempt has been made to include friction, windage, or core losses. For more detailed discussion any standard work may be consulted.

If external impedance is connected in the rotor circuit its effect can be included by inserting the equivalent resistance

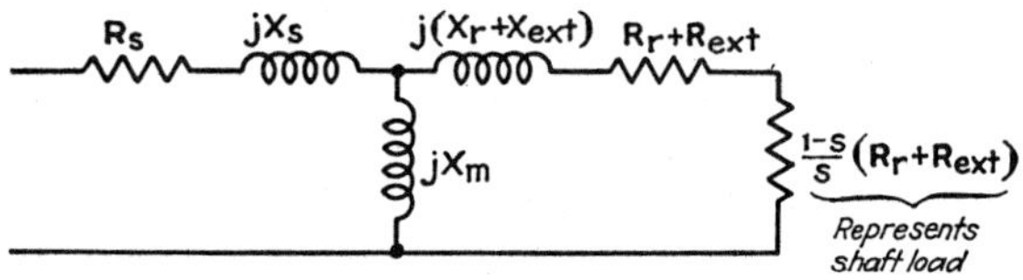

FIG. 192.—Inclusion of external rotor resistance (R_{ext}) and reactance (X_{ext}) in equivalent circuit (positive-sequence).

and reactance (at supply frequency) in the rotor branch of the equivalent diagram. This is illustrated in Fig. 192.

156. Characteristics with Application of Negative-sequence Voltages to the Stator.

The application of negative-sequence voltage is in effect the same as interchanging the leads to two terminals. A field is set up which rotates at synchronous speed in the direction opposite to that set up by the positive-sequence voltages. It follows then that the characteristics for the negative-sequence may be obtained by merely selecting the appropriate value of slip to use in conjunction with the positive-sequence equivalent circuit. Figure 193 shows the relation between the negative-sequence slip s_2, and the positive-sequence slip s. For very small values of s, s_2 is very nearly equal to 2; and, for the rotor blocked, both s and s_2 are equal to unity. The equation relating the slips is

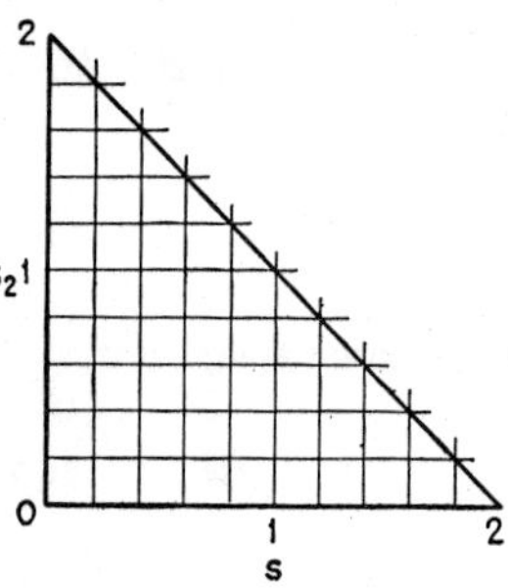

FIG. 193.—Relation between positive- and negative-sequence slips.

$$s_2 = 2 - s \tag{395}$$

The equivalent circuit for the negative-sequence is obtained by replacing s in the positive-sequence network by s_2 from equation (395), as is shown in Fig. 194.

The shaft power of negative-sequence is obtained by calculating the power absorbed in $-\frac{1-s}{2-s}R_r$, namely $-3\frac{1-s}{2-s}R_r\bar{I}^2_{r2}$.

The negative-sequence torque T_2 is

$$T_2 = -K\frac{3R_r\bar{I}^2_{r2}}{2-s} \tag{396}$$

The negative sign indicates that for small slips the torque is opposite to that for positive-sequence voltages; hence, for small slips, the load supplies power to the motor through the shaft.

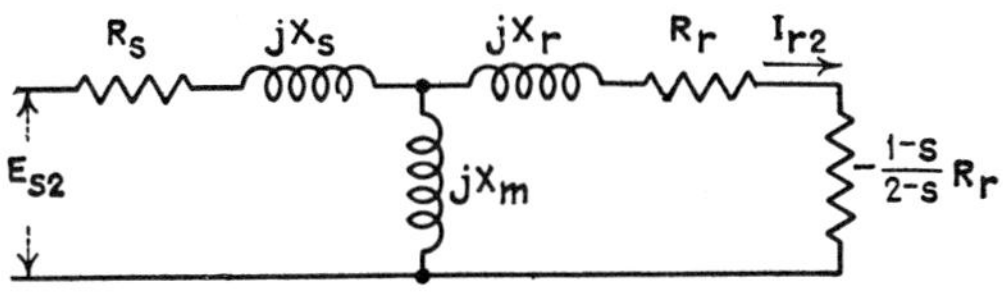

FIG. 194.—Negative-sequence equivalent circuit for an induction motor.

Since $R_r + jX_r$ does not usually exceed 15 per cent and jX_m is of the order of 400 per cent, jX_m may usually be omitted. In addition, $-\frac{1-s}{2-s}R_r$ may be replaced by $-\frac{R_r}{2}$, so that the equivalent circuit may be reduced to that shown in Fig. 195.

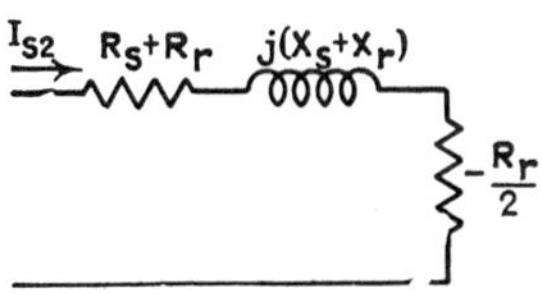

FIG. 195.—Approximate negative-sequence equivalent circuit for an induction motor operating at small values of slip.

157. Characteristics with Application of Both Positive- and Negative-sequence Voltages to the Stator.

When both positive- and negative-sequence voltages are applied simultaneously, the currents of the two sequences do not react upon each other; thus they may be computed separately by the methods discussed. The total shaft power output may therefore be written

$$P_{total\ shaft\ output} = 3\frac{1-s}{s}R_r\bar{I}^2_{r1} - 3\frac{1-s}{2-s}R_r\bar{I}^2_{r2} \tag{397}$$

and the total shaft torque

$$T = 3KR_r\left(\frac{\bar{I}^2_{r1}}{s} - \frac{\bar{I}^2_{r2}}{2-s}\right) \tag{398}$$

The significance of the power, torque, and the loss relations will be made clearer by the consideration of the characteristics of a particular machine.

158. Application of Unbalanced Voltages to the Stator of a Particular Motor.

Consider the case in which both positive- and negative-sequence voltages are applied to the stator of an induction

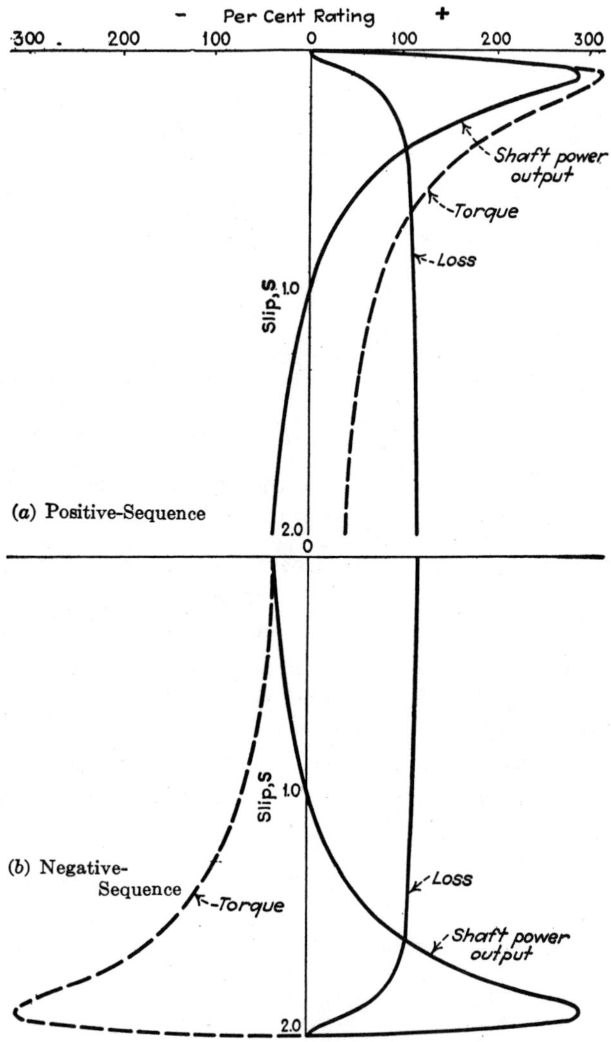

FIG. 196.—Induction-motor characteristics for 100 per cent applied voltage.

motor. In particular, consider a 500-hp. 2,200-volt 25-cycle motor, the constants of which are

$$X_r = X_s = 0.9 \text{ ohm per phase}$$
$$R_r = R_s = 0.2 \text{ ohm per phase}$$
$$X_m = 40 \text{ ohms per phase}$$

The characteristics of this machine for 100 per cent positive- and negative-sequence applied voltages are shown in Fig. 196(a) and (b), respectively, in which shaft output and loss are expressed in per cent of rating, and torque in per cent of the torque at rated speed. It will be observed that the negative-sequence characteristics have the same shape as the positive-sequence characteristics except that they are inverted in various ways.

Assume that 100 per cent positive-sequence and 10 per cent negative-sequence voltages be applied simultaneously, with the load such that the slip is 1.5 per cent. The torque due to the positive-sequence voltage is 93 per cent of normal in a positive sense, a fact which may be verified approximately from Fig. 196(a). The torque due to the negative-sequence voltage (since the torque varies as the square of the applied voltage) is $\frac{1}{100} \times 40$ or 0.40 per cent of the normal positive-sequence torque but in the negative direction. It may be seen, therefore, that the unbalance of voltages that may occur in service will generally have a negligible effect upon the torque characteristics.

The loss due to the positive-sequence set of voltages is 2.8 per cent, and that due to the negative-sequence voltage is $\frac{1}{100} \times 155$ or 1.55 per cent. This loss is quite appreciable when compared with the loss upon which the rating of the machine is based. It follows, therefore, that unbalanced voltages will cause trouble due to heating if the negative-sequence voltage is sufficiently large.

159. Operation with Open-circuited Stator Phase.

An interesting study of induction machines is the phenomena that occur when one phase is accidentally opened, as by the blowing of a fuse in one lead. The treatment is the same whether the machine is star- or delta-connected; in the latter case it is necessary merely to replace the delta connection by its equivalent star. This is a special case of unbalanced applied voltages.

Assuming the a phase to be open, there result from the current relations

$$\left.\begin{aligned} I_a &= 0 \\ I_b &= I \\ I_c &= -I \end{aligned}\right\} \quad (399)$$

or

$$I_{s1} = \frac{1}{3}(0 + aI - a^2I) = \frac{(a - a^2)}{3}I = +j\frac{I}{\sqrt{3}} \tag{400}$$

$$I_{s2} = \frac{1}{3}(0 + a^2I - aI) = -\frac{(a - a^2)}{3}I = -j\frac{I}{\sqrt{3}} = -I_{s1} \tag{401}$$

The only voltage known is the voltage impressed between terminals b and c, namely E_{sA}. This quantity may be constructed from the line-to-neutral components in phase a as follows:

$$E_{sb} = a^2E_{s1} + aE_{s2}$$
$$E_{sc} = aE_{s1} + a^2E_{s2}$$

Therefore

$$E_{sA} = E_{sc} - E_{sb} = (a - a^2)(E_{s1} - E_{s2}) = +j\sqrt{3}(E_{s1} - E_{s2}) \tag{402}$$

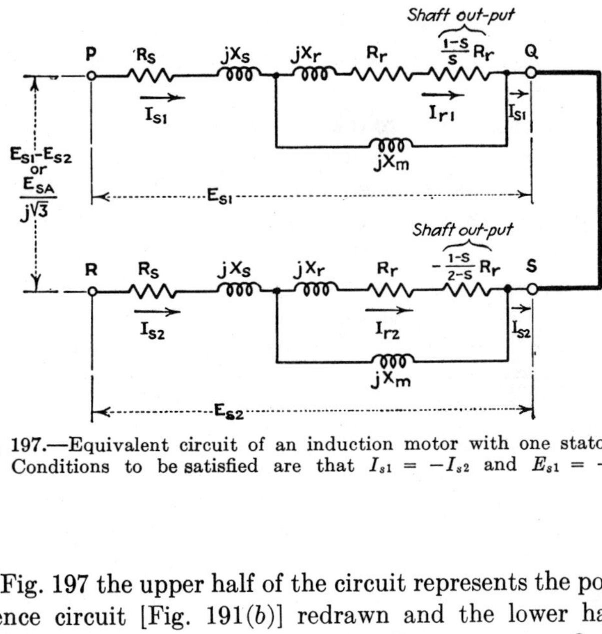

FIG. 197.—Equivalent circuit of an induction motor with one stator phase open. Conditions to be satisfied are that $I_{s1} = -I_{s2}$ and $E_{s1} = -E_{s2} = \dfrac{E_{sA}}{j\sqrt{3}}$.

In Fig. 197 the upper half of the circuit represents the positive-sequence circuit [Fig. 191(b)] redrawn and the lower half the negative-sequence circuit (Fig. 194). By connecting Q and S, the condition expressed by equation (401) that $I_{s2} = -I_{s1}$ is fulfilled. In addition, the voltage across PR is $E_{s1} - E_{s2}$. But from (402) this voltage is also equal to $\dfrac{E_{sA}}{j\sqrt{3}}$. It follows, therefore, that if the voltage $\frac{E_{sA}}{j\sqrt{3}}$ be applied to the circuit shown in Fig. 197 the positive- and negative-sequence currents will be the currents that flow in the respective circuits.

The total shaft output and net torque are then obtained from equations (397) and (398).

A customary approximation in induction-motor theory is to transfer the magnetizing current branch jX_m from across the portion of the circuit representing the rotor to across the source. A similar approximation will be made in this case; the branch jX_m of the positive-sequence network will be placed across the source, that is, across the terminals PR. Since all the impedances in the negative-sequence network but jX_m are of the order of leakage reactances, the magnetizing branch jX_m in this network may be neglected. After making these approximations, and combining, the equivalent circuit becomes that shown in Fig. 198. The shaft power and torque become:

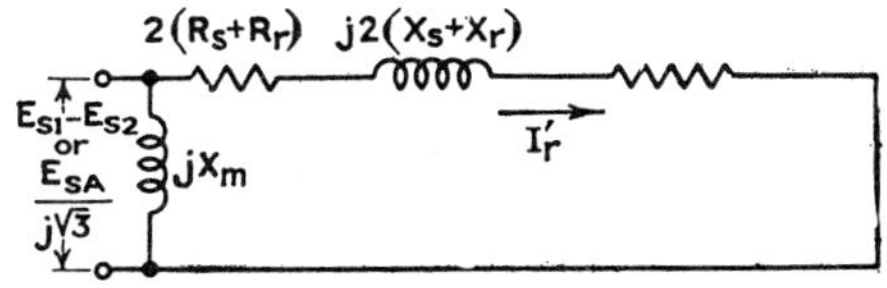

Fictitious resistance of shaft output =

$$\left(\frac{1-s}{s} - \frac{1-s}{2-s}\right)R_r = \frac{2(1-s)^2}{s(2-s)}R_r$$

FIG. 198.—Simplification of equivalent network of Fig. 197 by placing magnetizing branch across the terminals.

$$\text{Total shaft power output} = \frac{6(1-s)^2}{s(2-s)}R_r\bar{I}_r'^2 \qquad (403)$$

$$\text{Net torque} = K\frac{6(1-s)}{s(2-s)}R_r\bar{I}_r'^2 \qquad (404)$$

in which K is defined by equation (394).

From equation (404) it will be observed that for small slips the torque is positive, which interpreted, means that the motor will continue to operate if the load is not too great. For $s = 1$ (standstill), the torque is zero and the machine will not start.

Any of the sequence components of voltage or current, in either the stator or rotor, may be obtained by solving the circuit of Fig. 197 which is clearly marked to indicate the various quantities. Having obtained the sequence components, the individual phase components may readily be computed. Such characteristics as the maximum pull-out torque, loss, or heating can thus be determined satisfactorily.

160. Unbalanced Impedances in the Rotor Circuits.

Another important practical problem involving unbalanced operation of induction motors is that of unbalanced impedances in the secondary rotor circuits. For the analysis of this problem, positive-sequence voltage alone will be assumed applied to the stator terminals of the machine. Currents of fundamental frequency in the stator will give rise to currents of slip frequency in the rotor. The unbalance in the rotor will, however, cause these currents to be unbalanced. These unbalanced currents of slip frequency in the rotor may be resolved into positive- and negative-sequence currents. The latter sets up fields which rotate negatively with respect to the rotor at slip frequency. These currents in turn induce currents in the stator of frequency equal to fundamental frequency minus twice slip frequency. The plan of attack will be to analyze the reactions in the machine into the two groups; first, those associated with fundamental frequency in the stator and slip frequency of positive-sequence in the rotor, and second, those associated with slip frequency of negative-sequence in the rotor and fundamental minus twice slip frequency in the stator. These two groups of reactions will then be tied together through the characteristics of the unbalanced impedances in the rotor.

The diagram in Fig. 199 shows schematically the induction-motor circuits under consideration. With the positive-sequence voltage E_{s1} applied to the stator, the resultant air-gap flux rotating with synchronous velocity induces in the stator the voltage E_{g1} and in the rotor the voltage sE_{g1}, where s is the slip expressed as a fraction. The impedance drop in the stator then provides the relation

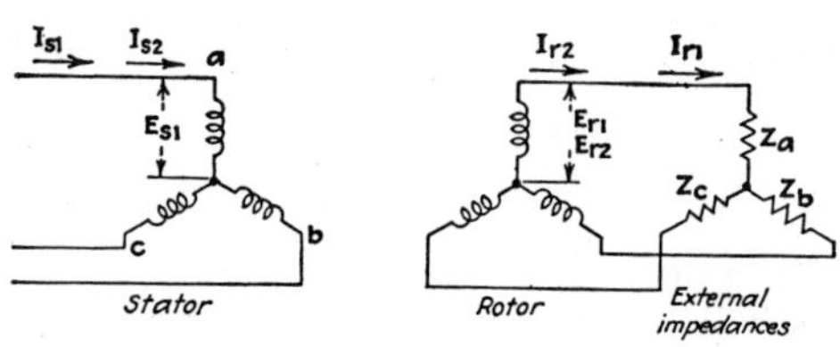

FIG. 199.—Schematic diagram of an induction motor with unbalanced impedances in the rotor circuit.

$$E_{s1} = E_{g1} + (R_s + jX_s)I_{s1} \tag{405}$$

In the rotor, the voltage equation is

$$sE_{g1} = jsX_rI_{r1} + R_rI_{r1} + E_{r1} \tag{406}$$

where E_{r1} equals the rotor terminal voltage of positive-sequence. The voltage E_{r1} is the voltage at slip frequency that would be measured at the terminals of the rotor by means of a positive-sequence measuring device. Dividing equation (406) by s

$$E_{g1} = jX_rI_{r1} + \frac{R_r}{s}I_{r1} + \frac{E_{r1}}{s} \qquad (407)$$

The air-gap flux is produced by the resultant of the currents I_{s1} and I_{r1}. Therefore, considering I_{s1} as positive when magnetizing, and I_{r1} as positive when demagnetizing, the proportionality between these currents and the voltage produced by the air-gap flux may be expressed by

$$E_{g1} = jX_m(I_{s1} - I_{r1}) \qquad (408)$$

From equations (405), (407), and (408), the equivalent diagram of Fig. 200(a) can be obtained just as that of Fig. 191(a) was obtained previously. The approximate diagram is shown in Fig. 200(b). From this circuit it may be seen that

$$\frac{E_{r1}}{s} = E_{s1} - \left[R_s + \frac{R_r}{s} + j(X_s + X_r)\right]I_{r1}$$

or

$$E_{r1} = sE_{s1} - [sR_s + R_r + js(X_s + X_r)]I_{r1} \qquad (409)$$

The development of the equivalent circuit for the negative-sequence may best be approached by assuming a negative-sequence voltage E_{r2} of frequency sf applied to the rotor and assuming the rotor to be stationary and the stator rotating with speed $(1 - s)$ in the negative direction. This gives the same relative condition as though the stator were stationary and the rotor rotating with speed $(1 - s)$. The applied voltage sets up a flux which rotates with velocity $-s$, and, since the assumed velocity of the stator is $-(1 - s)$, these velocities will be equal for $s = 0.5$. For this slip, no voltages are induced in the stator and the characteristics correspond to that of an ordinary induc-

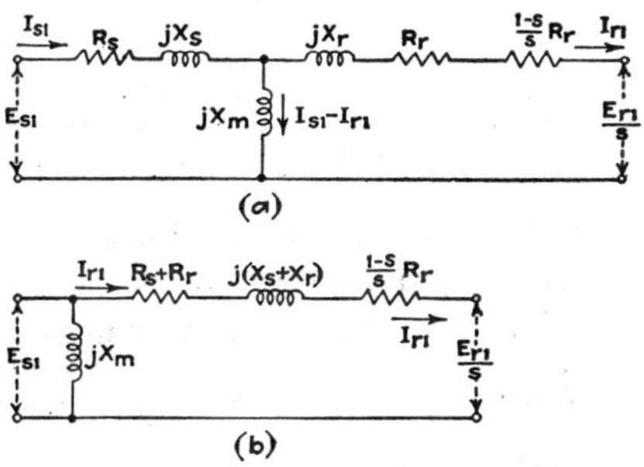

$$\text{Shaft power} = 3\frac{1-s}{s}\left(R_r + \text{real part}\frac{E_{r1}}{I_{r1}}\right)\bar{I}^2_{r1}$$

$$= 3\frac{1-s}{s}(R_r\bar{I}^2_{r1} + \text{positive-sequence power output of rotor})$$

FIG. 200.—Equivalent positive-sequence diagrams for an induction motor with voltage E_{r1} on the rotor. (a) Accurate; (b) Approximate.

tion motor operating at synchronous speed. For $s < 0.5$ the negative velocity of the stator is greater than the negative velocity of the flux set up by the negative-sequence voltage and the machine therefore acts like an induction *generator*, but for $s > 0.5$ the negative velocity of the stator is less than the negative velocity of the flux set up by the negative-sequence voltage and the machine acts like an induction *motor*.

With voltage E_{r2} applied to the rotor and for the present considering current flowing out of the rotor as positive (as indicated in Fig. 199), the voltage equation for the rotor may be written

$$E_{r2} = E_{g2} - (R_r + jsX_r)I_{r2} \tag{410}$$

where I_{r2} = negative-sequence rotor current.

E_{g2} = voltage induced in the rotor due to magnetic flux associated with the negative-sequence reactions within the machine.

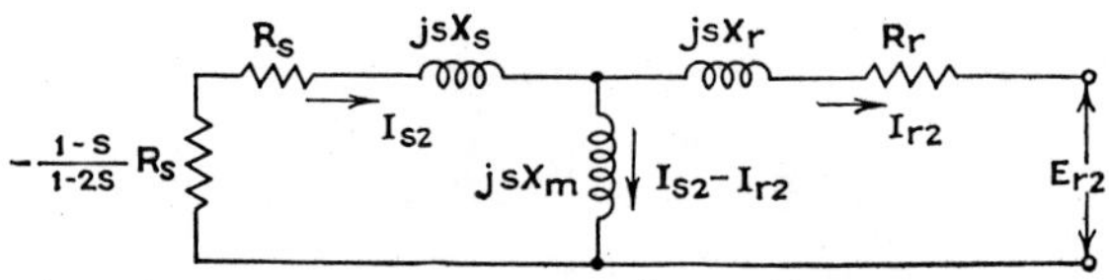

$$\text{Shaft power} = -3\frac{1-s}{1-2s}R_s\bar{I}^2_{s2}$$

FIG. 201.—Equivalent diagram for negative-sequence currents in the rotor of slip frequency voltages of frequency s applied to the rotor of an induction motor operating at speed $(1 - s)$ and stator short-circuited.

The speed of the flux which produces E_{g2} is, relative to the rotor, of course, $-s$ and, with respect to the stator, $(1 - 2s)$. It follows, therefore, that this same flux induces the voltage $-\left(\frac{1-2s}{s}\right)E_{g2}$ in the stator windings. Assuming that the system impedance to the flow of the stator current of frequency $(1 - 2s)f$ is zero, this voltage must be equal to the impedance drop of the stator at the frequency $(1 - 2s)f$, so that

$$-\left(\frac{1-2s}{s}\right)E_{g2} = R_s(-I_{s2}) - j(1 - 2s)X_s(-I_{s2}) \tag{411}$$

where I_{s2} is the stator current associated with the negative-sequence current in the rotor and, to be consistent with previous assumptions, the direction of flow is considered positive as shown in Fig. 201. The sign used in connection with the reactance drop term is negation because at small slips the machine is operating as an induction generator. Dividing through by the coefficient of E_{g2},

$$E_{g2} = \frac{s}{1 - 2s} R_s I_{s2} - jsX_s I_{s2} \tag{412}$$

$$= \frac{1 - s}{1 - 2s} R_s I_{s2} - R_s I_{s2} - jsX_s I_{s2} \tag{413}$$

Assuming that positive values of I_{s2} are magnetizing and positive values of I_{r2} are demagnetizing, E_{g2} is proportional to $(I_{s2} - I_{r2})$. The same factor of proportionality used in the positive-sequence set of currents may be used here also, but since E_{g2} is measured at slip frequency, the equation for E_{g2} is

$$E_{g2} = jsX_m(I_{s2} - I_{r2}) \tag{414}$$

From equations (410), (413), and (414), the equivalent circuit for the negative-sequence may be shown to be represented by that of Fig. 201. The power absorbed by the resistance $-\frac{1 - s}{1 - 2s} R_s$ represents the shaft power output as the total power input from the rotor can represent only copper loss and shaft power. At small slips the quantity is negative which indicates that the shaft power output for the negative-sequence is opposite in sense to that of the positive-sequence. For $s > 0.5$ the contrary is true, and the negative- and positive-sequence powers are additive.

The equivalent circuit for negative-sequence voltages of frequency sf applied to the rotor terminals can also be derived from that of the positive-sequence network, by again assuming that the rotor is stationary and that the stator is rotating with a speed of $-(1 - s)$ in the positive sense. The synchronous velocity of the rotor is $-s$, so that the equivalent slip expressed as a fraction of the synchronous velocity is $\frac{-s - [-(1 - s)]}{-s}$ or $\frac{1 - 2s}{s}$. This expression when substituted for s in the positive-sequence equivalent circuit of Fig. 191 gives the term $-\frac{1 - s}{1 - 2s} R$ for the equivalent resistance which corresponds to the shaft power. Since all of the reactance terms are given for the normal frequency f, it follows that in this equivalent diagram the reactance terms must be multiplied by s since the applied frequency in this case is sf.

The voltage and current relations for the negative-sequence network may be written from Fig. 201 as follows:

$$E_{r2} = I_{r2}\left[R_r + jsX_r + \frac{jsX_m[-R_s + jX_s(1 - 2s)]}{-R_s + j(1 - 2s)(X_s + X_m)}\right] \quad (415)$$

Having developed the two sequence networks, the tie between them and the external impedances may now be considered. For the unsymmetrical impedances Z_a, Z_b, and Z_c, of Fig. 199, the phase voltages and currents at the terminals may be written

$$\left.\begin{aligned} E_a - E_c &= Z_aI_a - Z_cI_c \\ E_b - E_a &= Z_bI_b - Z_aI_a \\ E_c - E_b &= Z_cI_c - Z_bI_b \end{aligned}\right\} \quad (416)$$

The phase voltages and currents may then be expressed in terms of the sequence voltages and currents with the following results:*

$$E_{r1} = \tfrac{1}{3}(Z_a + Z_b + Z_c)I_{r1} + \tfrac{1}{3}(Z_a + a^2Z + aZ_c)I_{r2} \quad (417)$$
$$E_{r2} = \tfrac{1}{3}(Z_a + aZ_b + a^2Z_c)I_{r1} + \tfrac{1}{3}(Z_a + Z_b + Z_c)I_{r2} \quad (418)$$

The sequence voltages at the rotor terminals may be eliminated between equations (417) and (418) and equations (409) and (415). The remaining equations may then be solved for the sequence currents from which the sequence voltages may be determined. The analytical expressions are cumbersome; consequently, it is usually more convenient from this point to carry out the calculations in numerical form. From the sequence currents and voltages the complete performance of the motor as to torque, power loss, phase voltages, and line currents may readily be calculated for various unbalanced external impedances.

161. Operation with Open-circuited Rotor Phase.

The operation of an induction motor with an open-circuited rotor phase may be analyzed as a special case of the problem discussed in the preceding section making use of the equivalent circuits of Figs. 200(b) and 201. Assume that the stator of the motor is connected to an infinite source, and that phase a of the rotor is open-circuited and that phase b and phase c are

* It will be noted that the impedance coefficients of equations (417) and (418) are of symmetrical form similar to that of the sequence voltages and currents. The reason for this is given in the discussion of unsymmetrical impedances in Chap. XVIII.

short-circuited at the rotor terminals in the usual manner. The circuit condition corresponds to that imposed upon the imaginary leads brought out from a transmission line for a line-to-line fault, for which it was found that $I_{r1} = -I_{r2}$, and $E_{r1} = E_{r2}$. It will be observed from the equivalent circuit of the positive-sequence diagram that the rotor voltage is taken into consideration by inserting the voltage $\frac{E_{r1}}{s}$ in the secondary

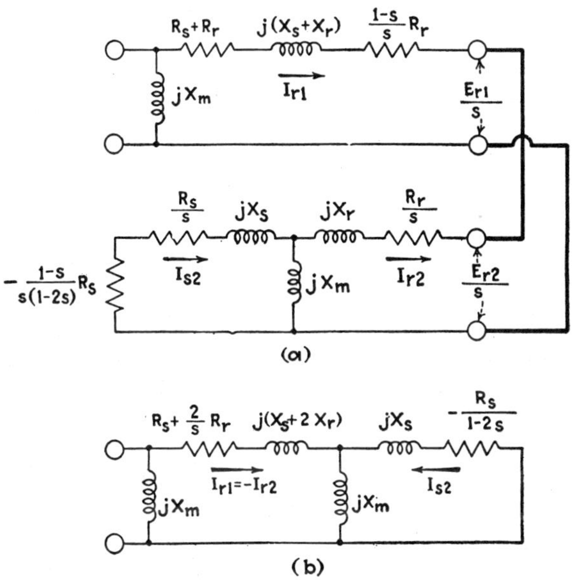

$$\text{Shaft power output} = \left(6\frac{1-s}{s}R_r\bar{I}^2_{r1} - 3\frac{1-s}{1-2s}R_s\bar{I}^2_{s2}\right) + \left(-3\frac{1-s}{1-2s}R_s\bar{I}^2_{s2}\right)$$

$$\text{Shaft torque} = 6K\left[\frac{1}{s}R_r\bar{I}^2_{r1} - \frac{1}{(1-2s)}R_s\bar{I}^2_{s2}\right]$$

$$\text{Loss} = 3(R_s + 2R_r)\bar{I}^2_{r1} + 3R_s\bar{I}^2_{s2}$$

FIG. 202.—Equivalent diagrams for open-circuited rotor phase of an induction motor.

circuit. If the constants of the equivalent diagram for the positive-sequence currents be multiplied by $\frac{1}{s}$, then the same currents will flow when $\frac{E_{r2}}{s}$ is impressed as though E_{r2} is impressed across the unchanged diagram. It follows therefore if this network be connected across $\frac{E_{r1}}{s}$ in the positive-sequence diagram, as shown in Fig. 202(*a*), that the two conditions imposed by the unbalance will be satisfied. The currents indicated are the actual currents flowing in the machine. A word of caution is necessary in connection with this diagram as the power loss in the lower portion of this diagram does not represent the loss in the negative-sequence diagram. No advantage is therefore attached to retaining the identity of the two networks except that it enables one to determine the value of the terminal voltage of the rotor. The diagram may be simplified to that shown in Fig. 202(*b*).

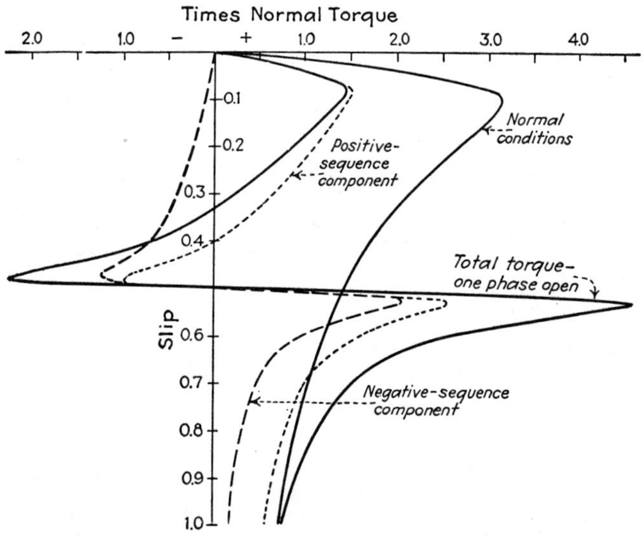

FIG. 203.—Torque of a three-phase induction motor with one rotor phase open-circuited compared with torque for normal conditions.

Since the voltage applied to the stator is by assumption wholly positive-sequence, then the total input is determined by determing the total power input into the two terminals of the network of Fig. 202(*b*). This input is of course merely the sum of the loss associated with the different resistances indicated. Subtracting from this quantity the actual I^2R for the positive- and negative-sequence currents in the stator and rotor gives the total shaft power output shown in Fig. 202. The shaft torque in turn is obtained by multiplying the term by $\frac{K}{1 - s}$ with the result also indicated in Fig. 202.

The operation of a three-phase induction motor with an open-circuited rotor phase has been investigated by B. G. Lamme who showed that this motor has a number of unusual characteristics.* These characteristics are brought out by the consideration of a particular case for which the motor with the characteristics of those given in Sec. 158 will be assumed. The motor torque and shaft output are plotted as functions of speed in Fig. 203 which shows the curves for an open-circuited rotor phase and also for the usual balanced rotor winding. It will be noted that at half speed the motor is capable of developing very little output, in fact, under some conditions, the motor, while capable of starting with a relatively high torque, is not capable of accelerating beyond a speed approximately half normal. Furthermore the pull-out characteristic of the motor in the vicinity of half speed may be considerably more than the pull-out in the vicinity of synchronous speed.

Problems

1. A 50-hp., three-phase, 60-cycle, 1,150-r.p.m., 440-volt, squirrel-cage induction motor was tested at rated voltage with the following results. No-load amperes per terminal = 15.6. Locked rotor current = 313 amp. Locked rotor watts = 112 kw. Stator resistance between two terminals = 0.280 ohm. Determine the constants of the equivalent diagram assuming the magnetizing branch directly across the terminals.

2. A 500-hp., 2,200-volt, 60-cycle wound-rotor induction motor has applied to it the unbalanced voltages of 2,200, 2,000, and 1,800 volts. The magnetizing current per phase is 25 amp., the stator resistance per phase is 0.2 ohm, the rotor resistance in terms of stator 0.15 ohm, the stator and rotor reactance are each 0.6 ohm per phase in terms of the stator. Neglecting friction, windage, and iron loss, plot the curves of torque and loss as a function of slip.

3. The machine of Prob. 2 is operating at rated torque when a fuse blows in one phase of the stator. Determine the current in the sound phases assuming the balanced supply voltage of 2,200 volts is maintained.

4. A 1,000-hp., two-phase, 2,200-volt, 25-cycle induction motor has the following constants:

$$
\begin{aligned}
\text{Stator resistance} &= 0.050 \text{ ohm per phase}\\
\text{Rotor resistance} &= 0.025 \text{ ohm per phase}\\
\text{Stator reactance} &= 0.20 \text{ ohm per phase}\\
\text{Rotor reactance} &= 0.11 \text{ ohm per phase}\\
\text{Turns ratio } \frac{\text{stator}}{\text{rotor}} &= 1.47\\
\text{Magnetizing current} &= 50 \text{ amp. per terminal.}
\end{aligned}
$$

* Polyphase Induction Motor with Single-Phase Secondary, *Elec. Jour.*, September, 1915.

A positive-sequence voltage of 2,000 + $j0$ and a negative-sequence voltage of 0 + $j300$ are applied to the terminals. Determine copper loss and stator currents. Neglect supply system impedance.

5. A 1,000-hp., 25-cycle, 2,200-volt induction motor has the following constants referred to the stator:

Stator resistance	= 0.075 ohm	per phase
Stator reactance	= 0.30 ohm	per phase
Rotor resistance	= 0.10 ohm	per phase
Rotor reactance	= 0.35 ohm	per phase
Magnetizing branch	= 20 ohms	per phase

The rotor is connected Y. With resistors of 0.5, 0.5, and 1.0 ohm in the rotor circuit, what is the resultant speed-torque curve with 2,200 volts applied to the stator? Neglect the supply system impedance.

6. The motor of Prob. 5 has two rotor phases short-circuited and the third open-circuited. Calculate the speed-torque curve.

7. Determine the analytical expressions for the shaft torque due to the positive- and negative-sequence currents for an induction motor with one rotor phase open-circuited. Show from these that the total shaft torque equals that given in Fig. 202.

CHAPTER XVIII

UNSYMMETRICAL SYSTEMS AND PHASE-BALANCERS

In the preceding chapters consideration has been given only to systems in which the constants are symmetrical, *i.e.*, to systems whose constants are the same viewed from any phase. In these systems the voltage drops and current distributions in any one sequence network are not influenced by the currents or voltages in the two other sequence networks. This proposition is not true for unsymmetrical or unbalanced systems; for them the currents of one sequence may give rise to currents or voltages of other sequences. The purpose of the present chapter is twofold: (1) to discuss the treatment of unsymmetrical systems and (2) to consider the problems of unbalance and apparatus for its correction.

162. Voltage Drops in Three Unsymmetrical Self Impedances.

The first case to be considered is the determination of the voltage drops produced by currents flowing through three unsymmetrical impedances. Assume three impedances Z_a, Z_b, and Z_c connected in star to a neutral wire as shown in Fig. 204. Let currents I_a, I_b, and I_c flow in the respective phases. Considering the individual phases, the equations expressing the relations between currents and voltages are

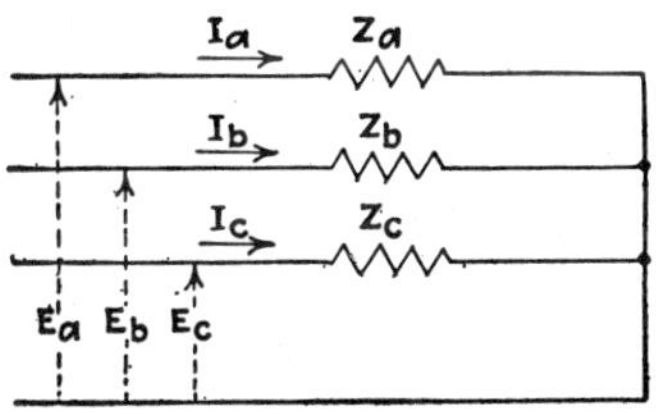

FIG. 204.—Three unsymmetrical impedances.

$$\left.\begin{aligned} E_a &= Z_a I_a \\ E_b &= Z_b I_b \\ E_c &= Z_c I_c \end{aligned}\right\} \quad (419)$$

The currents may be resolved into the three symmetrical components, the equations for which may be written

$$\left.\begin{aligned} I_a &= I_{a0} + I_{a1} + I_{a2} \\ I_b &= I_{a0} + a^2 I_{a1} + a I_{a2} \\ I_c &= I_{a0} + a I_{a1} + a^2 I_{a2} \end{aligned}\right\} \quad (420)$$

Substituting the values of currents from equations (420) in equations (419) gives

$$\left.\begin{aligned} E_a &= (Z_a)I_{a0} + (Z_a)I_{a1} + (Z_a)I_{a2} \\ E_b &= (Z_b)I_{a0} + (a^2Z_b)I_{a1} + (aZ_b)I_{a2} \\ E_c &= (Z_c)I_{a0} + (aZ_c)I_{a1} + (a^2Z_c)I_{a2} \end{aligned}\right\}\quad(421)$$

As shown in Chap. II, the voltages E_a, E_b, and E_c, may be resolved into their symmetrical components [equations (27), (28), and (29)], as follows:

$$\begin{aligned} E_{a0} &= \tfrac{1}{3}(E_a + E_b + E_c) \\ E_{a1} &= \tfrac{1}{3}(E_a + aE_b + a^2E_c) \\ E_{a2} &= \tfrac{1}{3}(E_a + a^2E_b + aE_c) \end{aligned}$$

The various symmetrical components of voltages may now be expressed in terms of impedance drops by substituting the values of E_a, E_b, and E_c, from equations (421) in equations (27) to (29) with the following results:

$$E_{a0} = \tfrac{1}{3}(Z_a + a^2Z_b + aZ_c)I_{a1} + \tfrac{1}{3}(Z_a + aZ_b + a^2Z_c)I_{a2} + \tfrac{1}{3}(Z_a + Z_b + Z_c)I_{a0} \quad (422)$$

$$E_{a1} = \tfrac{1}{3}(Z_a + Z_b + Z_c)I_{a1} + \tfrac{1}{3}(Z_a + a^2Z_b + aZ_c)I_{a2} + \tfrac{1}{3}(Z_a + aZ_b + a^2Z_c)I_{a0} \quad (423)$$

$$E_{a2} = \tfrac{1}{3}(Z_a + aZ_b + a^2Z_c)I_{a1} + \tfrac{1}{3}(Z_a + Z_b + Z_c)I_{a2} + \tfrac{1}{3}(Z_a + a^2Z_b + aZ_c)I_{a0} \quad (424)$$

Examination of these expressions shows that there are only three different coefficients involved, which are in reality the zero-, positive-, and negative-sequence components of the three impedances Z_a, Z_b, and Z_c, formed in a manner analogous to that for currents and voltages. Therefore, let

$$\mathfrak{Z}_0 = \tfrac{1}{3}(Z_a + Z_b + Z_c) \quad (425)$$

$$\mathfrak{Z}_1 = \tfrac{1}{3}(Z_a + aZ_b + a^2Z_c) \quad (426)$$

$$\mathfrak{Z}_2 = \tfrac{1}{3}(Z_a + a^2Z_b + aZ_c) \quad (427)$$

It is to be noted that the impedance terms $\mathfrak{Z}_0$, $\mathfrak{Z}_1$, and $\mathfrak{Z}_2$ have a different significance from the Z_0, Z_1, and Z_2 terms used throughout this volume. The script $\mathfrak{Z}$ has been reserved for this special use. The sequence components of the voltages and impedance drops may be written

$$E_{a0} = \mathfrak{Z}_0I_{a0} + \mathfrak{Z}_2I_{a1} + \mathfrak{Z}_1I_{a2} \quad (428)$$

$$E_{a1} = \mathfrak{Z}_1I_{a0} + \mathfrak{Z}_0I_{a1} + \mathfrak{Z}_2I_{a2} \quad (429)$$

$$E_{a2} = \mathfrak{Z}_2I_{a0} + \mathfrak{Z}_1I_{a1} + \mathfrak{Z}_0I_{a2} \quad (430)$$

It will be noted from these equations that each sequence component of current in the general case gives rise to voltage drops of all three sequences.

These expressions may readily be remembered by observing a relation between subscripts indicating the sequence for voltage, current, and impedance terms. This relation is that the sum of the sequence subscripts for impedance and current in impedance-drop terms gives 0 or 3 for zero-sequence voltage, 1 or 4 for positive-sequence voltage, and 2 for negative-sequence voltage. Thus the impedance drops $\mathfrak{Z}_0 I_{a1}$, $\mathfrak{Z}_2 I_{a2}$, $\mathfrak{Z}_0 I_{a2}$ and $\mathfrak{Z}_2 I_{a1}$ are terms for the sequence voltages E_{a1}, E_{a1}, E_{a2}, and E_{a0}, respectively.

163. Discussion of the Impedances $\mathfrak{Z}_0$, $\mathfrak{Z}_1$, and $\mathfrak{Z}_2$ of Unsymmetrical Self Impedances.

A clearer conception of the meaning of the impedance terms $\mathfrak{Z}_0$, $\mathfrak{Z}_1$, and $\mathfrak{Z}_2$ of unsymmetrical systems may be obtained by considering a few special cases. Consider for a moment the symmetrical system in which

$$Z_a = Z_b = Z_c = Z$$

Then, from equations (425) to (427),

$$\begin{aligned} \mathfrak{Z}_0 &= Z \\ \mathfrak{Z}_1 &= 0 \\ \mathfrak{Z}_2 &= 0 \end{aligned}$$

Substituting in equations (428) to (430)

$$\left.\begin{aligned} E_{a0} &= 0I_{a1} + 0I_{a2} + ZI_{a0} \\ E_{a1} &= ZI_{a1} + 0I_{a2} + 0I_{a0} \\ E_{a2} &= 0I_{a1} + ZI_{a2} + 0I_{a0} \end{aligned}\right\} \quad (431)$$

These equations show that, for this special case, only voltages of the same sequences are set up by the individual sequence currents. This conclusion checks common experience in this regard, and the analysis shows that for the symmetrical system, only the **zero-sequence impedance** terms are present.

Now consider the case in which the three impedances form in themselves a **positive-sequence impedance** set, in which

$$\left.\begin{aligned} Z_a &= Z \\ Z_b &= a^2 Z \\ Z_c &= aZ \end{aligned}\right\} \quad (432)$$

This requires in the impedances both positive and negative reactances and resistances. The physical set up of such impedances offers some difficulty, for while negative reactances may be represented by condensers, some special devices such as vacuum tubes or series commutator machines are necessary to represent negative resistance. These ideas are merely introduced to clear up the conceptions of unbalanced impedances. In this case

$$\left.\begin{aligned} \mathfrak{Z}_0 &= \tfrac{1}{3}[Z + a^2Z + aZ] = 0 \\ \mathfrak{Z}_1 &= \tfrac{1}{3}[Z + a(a^2Z) + a^2(aZ)] = Z \\ \mathfrak{Z}_2 &= \tfrac{1}{3}[Z + a^2(a^2Z) + a(aZ)] = 0 \end{aligned}\right\} \quad (433)$$

Substituting in equations (428) to (430), one obtains

$$\left.\begin{aligned} E_{a0} &= 0I_{a1} + ZI_{a2} + 0I_{a0} \\ E_{a1} &= 0I_{a1} + 0I_{a2} + ZI_{a0} \\ E_{a2} &= ZI_{a1} + 0I_{a2} + 0I_{a0} \end{aligned}\right\} \quad (434)$$

Now if zero-sequence currents alone are present, then, from these equations, it is seen that positive-sequence voltages only are developed. This is to be expected, for since the zero-sequence components in all phases are equal, the voltages in phases a, b, and c, are ZI_{a0}, a^2ZI_{a0}, and aZI_{a0}, respectively. These form a positive-sequence. In this manner positive-sequence voltages are obtained from zero-sequence currents.

Now if the positive-sequence currents alone are made to flow through the three impedances, negative-sequence components alone are set up. Considering the drops in the phases individually, it will be seen that the drop in phase a is ZI; in phase b, $(a^2I_1)(a^2Z) = aZI_1$; and in phase c, $(aI_1)(aZ) = a^2ZI_1$. Thus these voltages form a negative-sequence.

164. General Static Network Containing Unsymmetrical Self and Mutual Impedances.

The most general case of a static network is that in which all the self and mutual impedances are of different values. For the analysis of this case, consider the circuit shown in Fig. 205. It is desired to express the impressed voltages in terms of the symmetrical components of current and the self and mutual impedances. Let the currents I_a, I_b, and I_c flow in the three phases, respectively, and I_n in the neutral (or ground) return. Also, let the self impedances be Z_{aa}, Z_{bb}, Z_{cc}, and Z_{nn}.

In addition, let Z_{ab} be the mutual impedance between phases a and b, Z_{bc} that between phases b and c, and Z_{ca} that between phases c and a; similarly, let the mutual impedances between the conductors and neutral (or ground) return be Z_{an}, Z_{bn}, and Z_{cn}. The equations relating phase-to-neutral voltages and phase currents and the self and mutual impedances may be written from inspection of Fig. 205 as follows:

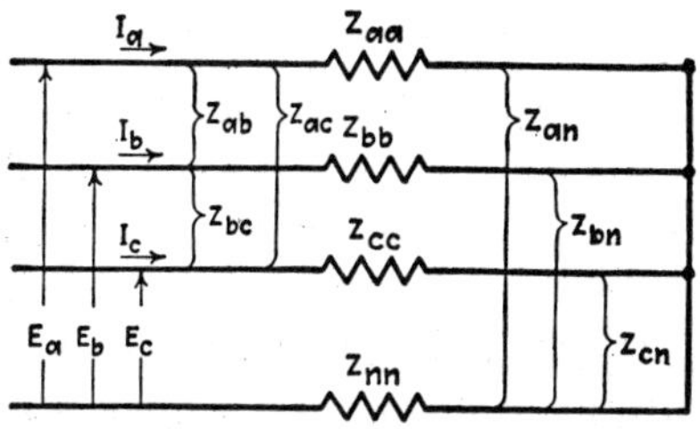

FIG. 205.—General static network.

$$E_a = (Z_{aa} - Z_{an})I_a + (Z_{ab} - Z_{bn})I_b + (Z_{ac} - Z_{cn})I_c + 3(Z_{nn} - Z_{an})I_0 \quad (435)$$

$$E_b = (Z_{ab} - Z_{an})I_a + (Z_{bb} - Z_{bn})I_b + (Z_{bc} - Z_{cn})I_c + 3(Z_{nn} - Z_{bn})I_0 \quad (436)$$

$$E_c = (Z_{ac} - Z_{an})I_a + (Z_{bc} - Z_{bn})I_b + (Z_{cc} - Z_{cn})I_c + 3(Z_{nn} - Z_{cn})I_0 \quad (437)$$

After substituting sequence voltages and currents for phase voltages and currents in equations (435) to (437) these equations may be solved for the sequence voltages. The resulting answer is rather complex and in order to simplify the expressions, let

$$\mathfrak{Z}_{aa0} = \frac{Z_{aa} + Z_{bb} + Z_{cc}}{3} \quad (438)$$

$$\mathfrak{Z}_{aa1} = \frac{Z_{aa} + aZ_{bb} + a^2Z_{cc}}{3} \quad (439)$$

$$\mathfrak{Z}_{aa2} = \frac{Z_{aa} + a^2Z_{bb} + aZ_{cc}}{3} \quad (440)$$

$$\mathfrak{Z}_{bc0} = \frac{Z_{bc} + Z_{ca} + Z_{ab}}{3} \quad (441)$$

$$\mathfrak{Z}_{bc1} = \frac{Z_{bc} + aZ_{ca} + a^2Z_{ab}}{3} \quad (442)$$

$$\mathfrak{Z}_{bc2} = \frac{Z_{bc} + a^2Z_{ca} + aZ_{ab}}{3} \quad (443)$$

$$\mathfrak{Z}_{an0} = \frac{Z_{an} + Z_{bn} + Z_{cn}}{3} \quad (444)$$

$$\mathfrak{Z}_{an1} = \frac{Z_{an} + aZ_{bn} + a^2Z_{cn}}{3} \quad (445)$$

$$\mathfrak{Z}_{an2} = \frac{Z_{an} + a^2Z_{bn} + aZ_{cn}}{3} \quad (446)$$

The solution for the sequence voltages may then be expressed as

$$E_0 = [\mathfrak{Z}_{aa0} + 2(\mathfrak{Z}_{bc0} - 3\mathfrak{Z}_{an0}) + 3\mathfrak{Z}_{nn}]I_0 + [\mathfrak{Z}_{aa2} - \mathfrak{Z}_{bc2} - 3\mathfrak{Z}_{an2}]I_1 + [\mathfrak{Z}_{aa1} - \mathfrak{Z}_{bc1} - 3\mathfrak{Z}_{an1}]I_2 \quad (447)$$

$$E_1 = [\mathfrak{Z}_{aa1} - \mathfrak{Z}_{bc1} - 3\mathfrak{Z}_{an1}]I_0 + [\mathfrak{Z}_{aa0} - \mathfrak{Z}_{bc0}]I_1 + [\mathfrak{Z}_{aa2} + 2\mathfrak{Z}_{bc2}]I_2 \quad (448)$$

$$E_2 = [\mathfrak{Z}_{aa2} - \mathfrak{Z}_{bc2} - 3\mathfrak{Z}_{an2}]I_0 + [\mathfrak{Z}_{aa1} + 2\mathfrak{Z}_{bc1}]I_1 + [\mathfrak{Z}_{aa0} - \mathfrak{Z}_{bc0}]I_2 \quad (449)$$

It will be noted that only six separate impedance constants are involved for this general case of the unsymmetrical static three-phase network. The various constants may be grouped as follows:

Zero-sequence impedance terms, type $\mathfrak{Z}_0$

$$\mathfrak{Z}_{aa0} + 2(\mathfrak{Z}_{bc0} - 3\mathfrak{Z}_{an0}) + 3Z_{nn} = \frac{E_0}{I_0} \text{ (component)} \quad (450)$$

$$\mathfrak{Z}_{aa0} - \mathfrak{Z}_{bc0} = \frac{E_1}{I_1} \text{ (component)} \quad (451)$$

$$\mathfrak{Z}_{aa0} - \mathfrak{Z}_{bc0} = \frac{E_2}{I_2} \text{ (component)} \quad (452)$$

Positive-sequence impedance terms, type $\mathfrak{Z}_1$

$$\mathfrak{Z}_{aa1} - \mathfrak{Z}_{bc1} - 3\mathfrak{Z}_{an1} = \frac{E_0}{I_2} \text{ (component)} \quad (453)$$

$$\mathfrak{Z}_{aa1} - \mathfrak{Z}_{bc1} - 3\mathfrak{Z}_{an1} = \frac{E_1}{I_0} \text{ (component)} \quad (454)$$

$$\mathfrak{Z}_{aa1} + 2\mathfrak{Z}_{bc1} = \frac{E_2}{I_1} \text{ (component)} \quad (455)$$

Negative-sequence impedance terms, type $\mathfrak{Z}_2$

$$\mathfrak{Z}_{aa2} - \mathfrak{Z}_{bc2} - 3\mathfrak{Z}_{an2} = \frac{E_0}{I_1} \text{ (component)} \quad (456)$$

$$\mathfrak{Z}_{aa2} + 2\mathfrak{Z}_{bc2} = \frac{E_1}{I_2} \text{ (component)} \quad (457)$$

$$\mathfrak{Z}_{aa2} - \mathfrak{Z}_{bc2} - 3\mathfrak{Z}_{an2} = \frac{E_2}{I_0} \text{ (component)} \quad (458)$$

In symmetrical static networks the positive- and negative-sequence impedance terms of the type $\mathfrak{Z}_1$ and $\mathfrak{Z}_2$ corresponding to equations (453) to (458) are all zero. Thus the only terms left are the zero-sequence terms of the type $\mathfrak{Z}_0$ corresponding to equations (450) to (452). These zero-sequence impedance terms correspond to Z_0, Z_1, and Z_2, in the notation used throughout this volume. It will be observed that the positive- and negative-sequence impedances Z_1 and Z_2 of symmetrical static networks are always equal to each other as shown by the identity of equations (451) and (452).

165. Symmetrical Transformations.

In the preceding section it was shown that for symmetrical static networks only the zero-sequence impedance terms of the type $\mathfrak{Z}_0$ are present. Also, it was shown that the sequences do not react upon each other; in other words, positive-sequence

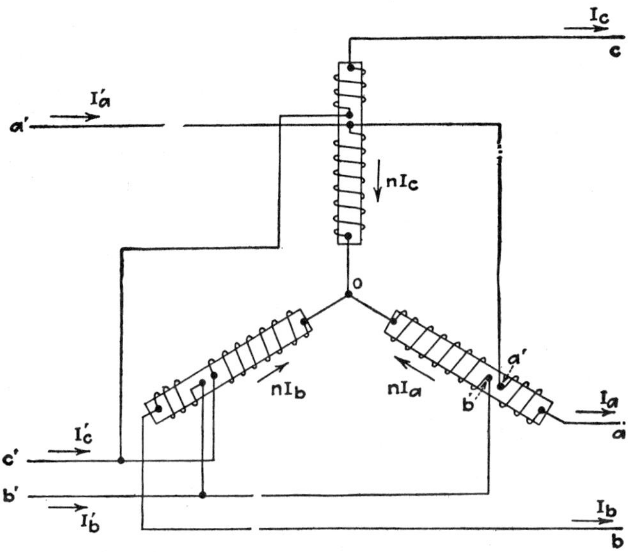

FIG. 206.—Interconnec ed symmetrical booster transformer.

currents produce only positive-sequence voltage drops, and similarly for the other sequences. It is of interest, however, to consider the case of static networks involving symmetrical transformations which is discussed in this section.

A type of transformation that is occasionally encountered is that shown in Fig. 206 in which both turns ratio and phase angle are altered by interconnecting the phases in the same sequence as the voltages. By this means, both the power and reactive volt-ampere loading for balanced conditions may be controlled. Questions, however, arise as to (1) the performance of such a combination under unbalanced conditions, and (2) the introduction of the characteristics of such transformations into single-line diagrams of the equivalent network.

The transformation may be assumed to involve three single-phase transformers, each of which possesses a series and a shunt winding. For the present assume the transformers to be ideal to the extent that exciting currents and series impedances are both zero. In short-circuit studies it is common practice to ignore the exciting current; the transformer impedances will be considered subsequently. Calling n the turns ratio between the series and shunt windings, it will be observed from the lower right-hand leg of Fig. 206, that the voltage from a' to a is n times the voltage from O to b' or is equal to nE_b'. The voltage of a may then be written as the sum of the voltages between O and a', and a' and a; or,

$$E_a = E_a' + nE_b'$$

Similarly,

$$E_b = E_b' + nE_c' \tag{459}$$

and

$$E_c = E_c' + nE_a'$$

Or

$$E_a' + nE_b' = E_a \tag{460}$$
$$E_b' + nE_c' = E_b \tag{461}$$
$$E_c' + nE_a' = E_c \tag{462}$$

By adding equations (460), (461) and (462) it may be seen that

$$E_{a0}' = \frac{1}{1+n}E_{a0} \tag{463}$$

Multiplying equation (460) by 1, (461) by a, and (462) by a^2, and adding:

$$E_{a1}' = \frac{1}{1+na^2}E_{a1} \tag{464}$$

Similarly, multiplying (460) by 1, (461) by a^2, and (462) by a, and adding:

$$E_{a2}' = \frac{1}{1+na}E_{a2} \tag{465}$$

For the upper leg, the current through the series winding, I_c, produces a demagnetizing effect which to annul requires the current nI_c in the shunt winding of the same leg. Equating the currents about the junction point of this shunt winding gives the relation

$$I_a' = I_a + nI_c$$

Similarly,

$$I_b' = I_b + nI_a$$

and

$$I_c' = I_c + nI_b$$

Operating on these equations in the same manner as for equations (460) to (462) there result

$$I_{a0}' = (1 + n)I_{a0} \tag{466}$$
$$I_{a1}' = (1 + na)I_{a1} \tag{467}$$
$$I_{a2}' = (1 + na^2)I_{a2} \tag{468}$$

Thus it is seen that the different sequence networks do not react upon each other; the equations relating current and voltage on the two sides involve only terms of the same sequence.

Either side of the transformer may be chosen as a reference base. If the right-hand side be chosen as the reference voltage, the currents and voltages on that side will be the actual quantities but the actual voltages and currents on the left-hand side are related to the base voltages and currents by the equations

$$\left.\begin{aligned} E_{a0}'{}_{(actual)} &= \frac{1}{1+n}E_{a0}'{}_{(base)} \\ E_{a1}'{}_{(actual)} &= \frac{1}{1+na^2}E_{a1}'{}_{(base)} \\ E_{a2}'{}_{(actual)} &= \frac{1}{1+na}E_{a2}'{}_{(base)} \end{aligned}\right\} \tag{469}$$

and

$$\left.\begin{aligned} I_{a0}'{}_{(actual)} &= (1+n)I_{a0}'{}_{(base)} \\ I_{a1}'{}_{(actual)} &= (1+na)I_{a1}'{}_{(base)} \\ I_{a2}'{}_{(actual)} &= (1+na^2)I_{a2}'{}_{(base)} \end{aligned}\right\} \tag{470}$$

Equivalent Circuit for Symmetrical Transformations. To complete the analysis of the symmetrical transformation, it is necessary to determine the equivalent circuit taking into account the presence of an impedance on either side of the series winding. If the positive-sequence impedance $Z_1'{}_{(actual\ impedance)}$ be inserted in the left-hand side, and the positive-sequence current $I_{a1}'{}_{(actual)}$ be made to flow through it, the actual drop across the impedance is $Z_1'{}_{(actual)} I_{a1}'{}_{(actual)}$. In terms of the base voltage from equation

(469) this is equal to $(1 + na^2)Z_1'_{(actual)}\ I_{a1}'_{(actual)}$, and in terms of the base current from equation (470) this equals

$$(1 + na)(1 + na^2)Z_1'_{(actual)}I_{a1}'_{(base)}.$$

This voltage must be equal to $Z_1'_{(equiv.)}I_{a1}'_{(base)}$, so that the equivalent impedance for positive-sequence is

$$Z_1'_{(equiv.)} = (1 + na)(1 + na^2)Z_1'_{(actual)} \qquad (471)$$

The two terms in the parentheses are conjugates of each other, so that their product is a real number whose magnitude is the square of the magnitude of either one. The same expression is obtained for negative-sequence and $(1 + n)^2$ is obtained for zero-sequence. Thus

$$Z_2'_{(equiv.)} = (1 + na)(1 + na^2)Z_2'_{(actual)} \qquad (472)$$

and

$$Z_0'_{(equiv.)} = (1 + n)^2Z_0'_{(actual)} \qquad (473)$$

To include the effect of transformer impedance, the series impedance Z_t between the series or booster windings and the shunt winding on any leg should be determined in terms of the turns of the series winding as the base. Thus the impedance will be expressed directly in terms of the right-hand side. The impedances may then be combined in the same manner as any other external impedance of the same value.

Thus the transformer itself may be replaced by a series impedance Z_t in each of the sequence networks; impedances on right-hand side by their actual values, and the impedances on the left-hand side by values defined by equations (471), (472), and (473). To obtain actual currents and voltages in the left-hand side from the equivalent voltages and currents, use the relations defined by equations (469) and (470). Voltages and currents on the right-hand side are already on their own base. It is interesting to note the similarity of these results with those of the special case of the star-delta transformation defined in Chap. II, Figs. 6 and 7, in which the positive- and negative-sequence quantities in passing through the transformer were rotated in opposite directions.

166. General Case of Unsymmetrical Systems.

Having indicated the general voltage, current, and impedance relations that exist in general static networks, it is now practicable to consider the general case of the unsymmetrical system involving rotating machines. The basic equations are similar in form to equations (428) to (430) except that the nine impedance coefficients may be different in value. To take care of this, the notation for the impedance coefficients $\mathfrak{Z}_0$, $\mathfrak{Z}_1$, and $\mathfrak{Z}_2$ will be modified by the addition of a subscript to indicate the current term with which they are to be used. Thus $\mathfrak{Z}_{01}$ indicates the zero-sequence impedance term to be used with positive-sequence current. The general equations for the unsymmetrical system may be written from comparison with equations (428) to (430) as follows:

$$E_{a0} = \mathfrak{Z}_{00}I_{a0} + \mathfrak{Z}_{21}I_{a1} + \mathfrak{Z}_{12}I_{a2} \tag{474}$$

$$E_{a1} = \mathfrak{Z}_{10}I_{a0} + \mathfrak{Z}_{01}I_{a1} + \mathfrak{Z}_{22}I_{a2} \tag{475}$$

$$E_{a2} = \mathfrak{Z}_{20}I_{a0} + \mathfrak{Z}_{11}I_{a1} + \mathfrak{Z}_{02}I_{a2} \tag{476}$$

The determination of the impedance coefficients of equations (474) to (476) may readily be accomplished by finding the sequence components of voltage produced by the circulation of unit current of the proper sequence. Thus $\mathfrak{Z}_{00}$, $\mathfrak{Z}_{10}$, and $\mathfrak{Z}_{20}$ may be obtained by the circulation of unit zero-sequence current; the zero-, positive-, and negative-sequence voltages produced are equal to the desired impedances.

The most general network represented by the equations (474) to (476) requires nine separate impedances, *i.e.*, the nine impedances have independent values only when no symmetry exists in the system. If the impedances have the same value for each of the three phases, then all the positive- and negative-sequence impedance terms ($\mathfrak{Z}_{10}$, $\mathfrak{Z}_{11}$, $\mathfrak{Z}_{12}$, $\mathfrak{Z}_{20}$, $\mathfrak{Z}_{21}$, $\mathfrak{Z}_{22}$) will be zero. Consequently, the number of the separate impedances varies with the system under consideration. The number of impedances may vary from nine for the most general case down to three equal impedances for the simple case of three identical series or shunt impedances.

In ordinary commercial rotating machines the positive- and negative-sequence impedances are zero. This is due to the fact that these machines are symmetrical from phase to phase. Consequently, only three impedances are involved, namely, $\mathfrak{Z}_{01}$, $\mathfrak{Z}_{02}$, and $\mathfrak{Z}_{00}$ which correspond to Z_1, Z_2, and Z_0, in terms of the notation used in Chaps. I to XVII.

Static systems not infrequently involve unsymmetrical impedances, principally unsymmetrical coupling between phases.

In many cases transpositions are introduced to avoid the effects of such asymmetry.

167. Independence of the Sequence Components in a Symmetrical General Network.

It has been pointed out on several occasions that a great simplification in the analysis by symmetrical components takes place if the system is symmetrical except for terminal conditions such as an unbalanced load or fault at a single location. This simplification resides in the fact that in the symmetrical parts of the network the current of one sequence produces impedance drops of that sequence only. Stated differently, in a symmetrical network all the positive- and negative-sequence impedance terms ($\mathfrak{Z}_{10}$, $\mathfrak{Z}_{11}$, $\mathfrak{Z}_{12}$, $\mathfrak{Z}_{20}$, $\mathfrak{Z}_{21}$, $\mathfrak{Z}_{22}$) are zero. This leaves only the zero-sequence terms ($\mathfrak{Z}_{00}$, $\mathfrak{Z}_{01}$, $\mathfrak{Z}_{02}$). When these relations are satisfied, the different sequence networks may be set up independently as discussed previously, principally in Chaps. III, IV, XI, and XII.

The relation which must be satisfied to obtain symmetry is that for every set of self or mutual impedances, the value for each phase, considering the phases in cyclic order, must be equal. Thus on a four-phase system the zero-sequence impedance terms only are present when

$$\left.\begin{aligned}
Z_{aa} &= Z_{bb} = Z_{cc} = Z_{dd}\\
Z_{ab} &= Z_{bc} = Z_{cd} = Z_{da}\\
Z_{ac} &= Z_{bd} = Z_{ca} = Z_{db}\\
Z_{ad} &= Z_{ba} = Z_{cb} = Z_{dc}\\
Z_{an} &= Z_{bn} = Z_{cn} = Z_{dn}
\end{aligned}\right\} \qquad (477)$$

where the notation corresponds to that used in connection with Fig. 205, extended to include a fourth phase. It is not necessary, in general, that the coupling between all phases be equal though this is true for the particular case of the three-phase system.

The condition for the independence of the sequences for the general case of a symmetrical three-phase network will now be examined. For the three-phase system the general relations for symmetry as defined in equations (477) reduce to

$$\left.\begin{aligned}
Z_{aa} &= Z_{bb} = Z_{cc}\\
Z_{ab} &= Z_{bc} = Z_{ca}\\
Z_{an} &= Z_{bn} = Z_{cn}
\end{aligned}\right\} \qquad (478)$$

It can be seen from equations (438) to (446) that the positive- and negative-sequence impedance terms given by equations (453) to (458) are all equal to zero. There remain only the zero-sequence impedance terms which relate voltages and currents of the same sequence. Consequently, the sequences are independent.

Independence of the sequences is a basic premise for the simplified method of symmetrical components which has been used throughout this volume for the calculation of unsymmetrical loads or faults on systems which otherwise are symmetrical. Since independence of the sequences arises because of symmetry, only the zero-sequence impedance terms require consideration, namely terms of the type $\mathfrak{Z}_{01}$, $\mathfrak{Z}_{02}$, and $\mathfrak{Z}_{00}$. Hence for these cases the first subscript is unnecessary and has been omitted in the notation used in all the preceding chapters. To prevent confusion, the script $\mathfrak{Z}$ is used only for impedance coefficients determined according to equations (425) to (427). In all analyses of symmetrical systems the general equations for the unsymmetrical system, equations (474) to (476), reduce to the following:

$$\left.\begin{aligned} E_{a0} &= Z_0 I_0 \\ E_{a1} &= Z_1 I_1 \\ E_{a2} &= Z_2 I_2 \end{aligned}\right\} \quad (479)$$

168. Systems Combining Symmetrical and Unsymmetrical Phase Impedances.

Special problems distinct from the ordinary short-circuit or stability problem arise in which a portion of the system is unsymmetrical. The solution of such problems usually hinges upon the point in the system where the dissymmetry begins. To illustrate the handling of such problems, the particular case of a generator feeding an induction motor load and an unbalanced impedance load will be considered.

Figure 207 shows diagrammatically the system and the nomenclature to be used. The generator on the left with impedances Z_{g0}, Z_{g1}, and Z_{g2} supplies the motor with impedances Z_{m1} and Z_{m2} and the shunt load with phase impedances Z_a, Z_b, and Z_c. At the junction point the sequence components of current must add to zero, giving

$$\left.\begin{aligned} I_{g0} &= I_{m0} + I_{L0} \\ I_{g1} &= I_{m1} + I_{L1} \\ I_{g2} &= I_{m2} + I_{L2} \end{aligned}\right\} \quad (480)$$

In general it will be found simpler to solve as much of the symmetrical portion of the network as possible and leave the evalua-

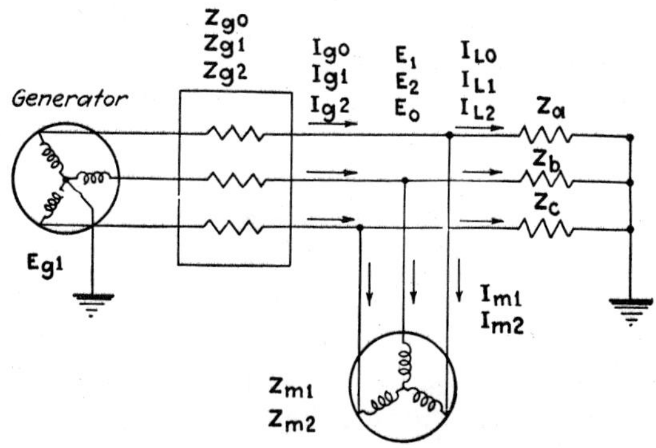

FIG. 207.—Diagram illustrating method of combining symmetrical and unsymmetrical network elements.

tion of the unsymmetrical portion as the final step. Following this suggestion,

$$\left.\begin{aligned} I_{m0} &= 0 \\ I_{m1} &= \frac{E_1}{Z_{m1}} \\ I_{m2} &= \frac{E_2}{Z_{m2}} \end{aligned}\right\} \quad (481)$$

$$\left.\begin{aligned} I_{g0} &= -\frac{E_0}{Z_{g0}} \\ I_{g1} &= \frac{E_{g1} - E_1}{Z_{g1}} \\ I_{g2} &= -\frac{E_2}{Z_{g2}} \end{aligned}\right\} \quad (482)$$

in which E_0, E_1, and E_2 are the sequence components of voltage at the junction point. Substituting in (480), and solving for the junction point voltages, gives the results

$$\left.\begin{aligned} E_0 &= -Z_{g0}I_{L0} \\ E_1 &= \frac{Z_{m1}}{Z_{g1} + Z_{m1}}E_{g1} - \frac{Z_{g1}Z_{m1}}{Z_{g1} + Z_{m1}}I_{L1} \\ E_2 &= -\frac{Z_{g2}Z_{m2}}{Z_{g2} + Z_{m2}}I_{L2} \end{aligned}\right\} \quad (483)$$

he junction point voltage can also be expressed in terms ae shunt-impedance characteristics, giving

$$\left.\begin{aligned} E_0 &= \mathfrak{Z}_0 I_{L0} + \mathfrak{Z}_2 I_{L1} + \mathfrak{Z}_1 I_{L2} \\ E_1 &= \mathfrak{Z}_1 I_{L0} + \mathfrak{Z}_0 I_{L1} + \mathfrak{Z}_2 I_{L2} \\ E_2 &= \mathfrak{Z}_2 I_{L0} + \mathfrak{Z}_1 I_{L1} + \mathfrak{Z}_0 I_{L2} \end{aligned}\right\} \quad (484)$$

hich $\mathfrak{Z}_0$, $\mathfrak{Z}_1$, and $\mathfrak{Z}_2$, are defined by equations (425) to (427). ating (483) and (484) and combining terms

$$\left.\begin{aligned} &+ Z_{g0})I_{L0} + \mathfrak{Z}_2 I_{L1} + \mathfrak{Z}_1 I_{L2} = 0 \\ &{}_{L0} + \left(\mathfrak{Z}_0 + \frac{Z_{g1}Z_{m1}}{Z_{g1} + Z_{m1}}\right)I_{L1} + \mathfrak{Z}_2 I_{L2} = \frac{Z_{m1}}{Z_{g1} + Z_{m1}}E_{g1} \\ &{}_{L0} + \mathfrak{Z}_1 I_{L1} + \left(\mathfrak{Z}_0 + \frac{Z_{g2}Z_{m2}}{Z_{g2} + Z_{m2}}\right)I_{L2} = 0 \end{aligned}\right\} \quad (485)$$

hese expressions provide a sufficient number of equations ı which it is possible to solve for the three unknown quanti- I_{L0}, I_{L1}, and I_{L2}.

. Unsymmetrical Series Impedances.

his same method of attack may be applied to other problems ılving conditions that may be reduced to unbalanced ter- al restraints. E. L. Harder[(80)] has analyzed the problem aserting three impedances of different values in series relation he three phases of an otherwise symmetrical system. For present, consideration will be given to only the problem in ch the impedance Z_a is inserted in series with phase a of an erwise symmetrical system. A special case of this problem vhich Z_a is infinite represents the case of an open-circuited se, such as may arise from the mechanical failure of a con- tor without grounding or the opening of a single pole of a ker. The latter example is important because of the atten- that has been given recently to the use of single-pole breakers ncrease the stability limit of systems.

igure 208(a) represents the schematic diagram of the problem. two points on either side of the impedance Z_a will be desig- ed x and y, respectively. From this diagram it may be seen

$$\left.\begin{aligned} E_{xa} &= E_{ya} + Z_a I_a \\ E_{xb} &= E_{yb} \\ E_{xc} &= E_{yc} \end{aligned}\right\} \quad (486)$$

From these relations

$$E_{x0} = E_{y0} + \tfrac{1}{3}Z_aI_a \quad (\cdot$$
$$E_{x1} = E_{y1} + \tfrac{1}{3}Z_aI_a \quad (\cdot$$
$$E_{x2} = E_{y2} + \tfrac{1}{3}Z_aI_a \quad (4$$

The line currents and consequently the sequence compone also, are equal at both x and y when the positive sense is ta as indicated. In the tation for current the s scripts indicating position have been o ted. Equations (4 (488), and (489) s that the drops in the t sequences must e $\frac{1}{3}Z_a(I_0 + I_1 + I_2)$. necessary conditi therefore, for an equ lent network are that currents flowing *into* respective networks a must equal the curr flowing *out* at x and t the voltage drops in t networks between x a must all equal

$$\tfrac{1}{3}Z_a(I_0 + I_1 + I_2).$$

These conditions are filled by the connecti shown in Fig. 208(*b*).

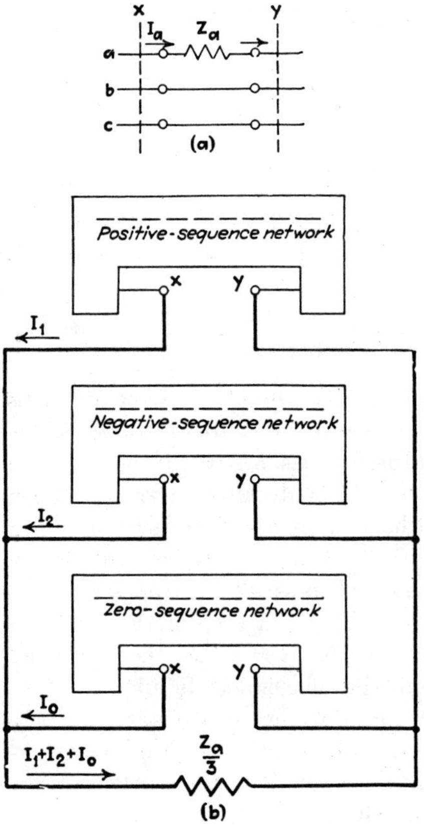

FIG. 208.—Equivalent circuit for insertion of impedance Z_a in phase a.

The special case for open-circuited phase represented by letting be infinite, which equivalent to omitting Z_a branch. Since x y are any two points, by letting the point y be the neutral p or ground, the case for $Z_a = \infty$ represents a double line-to-gro fault. It will be seen that the equivalent network then beco the same as that for the double line-to-ground fault develope Chap. XII.

:. L. Harder has also shown that the equivalent diagram two open-circuited phases is that shown in Fig. 209. The y difference between this diagram and that for only one se open-circuited is that for the latter the negative- and)-sequence networks are in parallel, and for the former the ative- and zero-sequence networks are in series. If the three ses on the y side of the open-circuit were grounded, then nt y in each of the networks would be connected to their pective zero potential buses and the network reduces to that a single line-to-ground fault discussed previously. How- r, if the phase conductors on the y side are merely connected

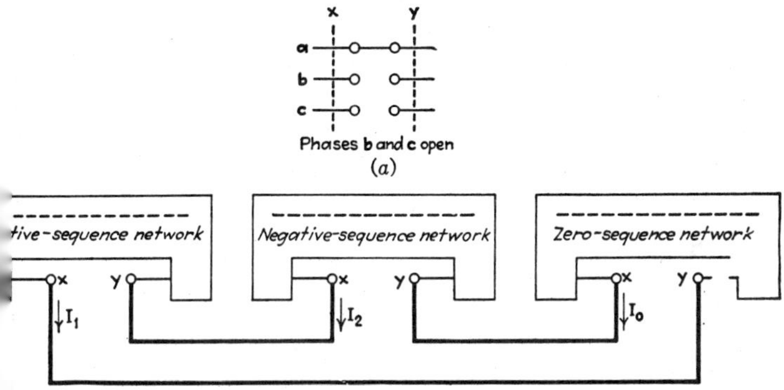

FIG. 209.—Equivalent circuit for phases b and c open.

;ether and not grounded, then the y points in the positive- d negative-sequence networks alone are connected to their o potential buses.

). Power Supply to Single-phase Loads.

In commercial power systems the generating stations, the ostations, and the transmission systems are, with but few eptions, designed for balanced load. Individual single-phase ds on distribution systems are allocated to the different ases in such a manner that substantial balance is achieved. ere are, however, certain applications such as power supply alternating-current railways and electric furnaces where the gle-phase load may be so large in comparison with the total d that it is necessary to give special consideration to the control the unbalanced effects of the load. Two general methods of wer supply are applicable.

1. Separate generation for the single-phase loads.
2. Common generation for both balanced and unbala loads with phase-balancing equipment for the latter.

171. Single-phase Generation.

Single-phase power supply is usually obtained from a sepa generator driven by its own prime mover or by a three-p 60-cycle motor. The single-phase generator differs from ordinary three-phase generator for balanced loads princip in the use of a heavy damper winding to minimize rotor loss to double frequency currents and to prevent local heat particularly at the end rings in the case of turbogenerat Motor-generator sets of the larger sizes supplying single-p loads are provided with spring mounting to absorb the pulsat in power due to the single-phase loads. Single-phase genera are commonly designed on a three-phase basis with the t phase available for emergency operation or replaced by "dumr coils.

172. Phase-balancing of Single-phase Loads.

The flow of power in a single-phase circuit pulsates at a quency equal to twice that of the alternating-current sup Consequently, it is readily appar that some means of energy storag necessary in order to convert a sin phase load with pulsating power t balanced load of constant power. order to reduce the periodic variat in power, it is necessary, in general utilize load from some other phase the system. Failure to recognize significance of the pulsating charac of single-phase loads has led to frequ proposals to draw equal currents fr the different phases. Typical of th schemes is the one shown in Fig. which has one winding reversed a which draws equal currents in the th phases. A little consideration will show that additional ap ratus capacity is required for handling single-phase load but spite of this the power remains single-phase in character. C

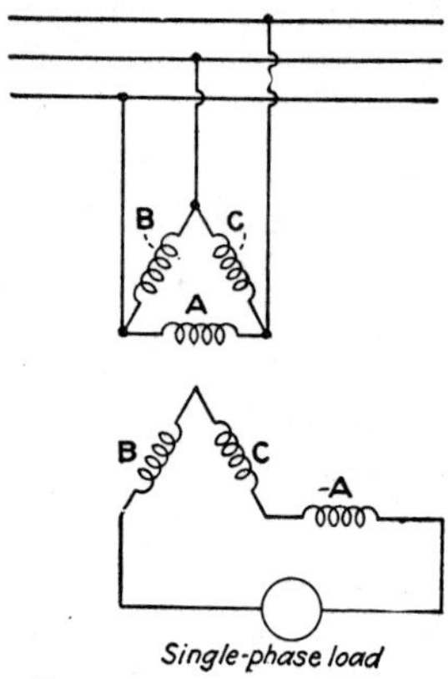

FIG. 210.—Unsound attempt to balance a single-phase load.

sequently, nothing is gained by the use of the three single-phase transformers and a single transformer is preferable.

In case the load can be subdivided and distributed among the different phases, the balance of the system is of course greatly improved. It is not necessary, however, to balance the loads by using identical impedances in the different phases. All that is necessary is that the total pulsating power be balanced. The fact that loads of different power factor on different phases can produce balanced power is illustrated in the connections of

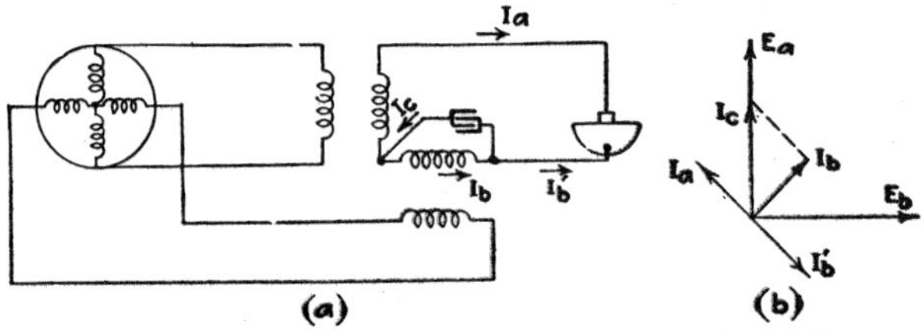

FIG. 211.—Diagram illustrating method of balancing single-phase resistance (furnace) load by adding capacitor load on one phase.

Fig. 211 which illustrates a two-phase generator supplying a transformer whose secondary is connected to form a two-phase three-wire system. With the *a* and *b* phases connected to the single-phase furnace load as illustrated, the vector diagrams will be as shown in Fig. 211(*b*). By adding a capacitor connected across phase *b*, the total current on the supply lines will be indicated by the terms I_a and I_b which form a balanced system.

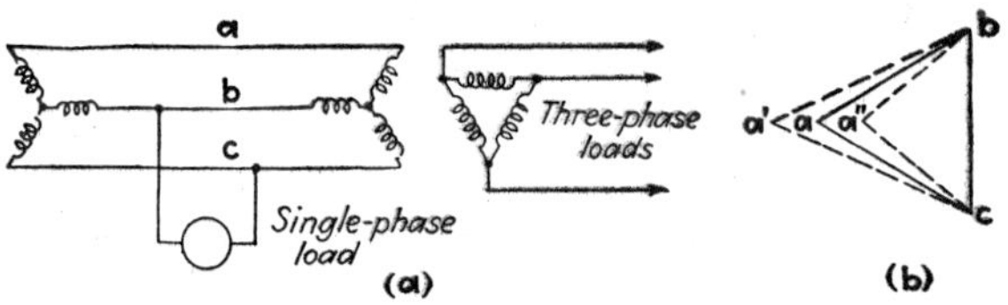

FIG. 212.—Method of using unsymmetrical transformer taps to improve balance on a three-phase circuit.

Another method of improving the balance of systems by static means is to alter the transformer ratios from their normal values. For example, consider a three-phase generator supplying the principal load between phases *b* and *c* as illustrated in Fig. 212. Assume that voltage is maintained constant across phases *b* and *c* under all load conditions. Then under light load conditions the voltage triangle *abc* will be substantially balanced, as shown in Fig. 212(*b*). However, under heavy load conditions the unloaded phases will be of higher value and the voltage triangle is $a'bc$. Consequently, it is possible to choose the transformer ratios so that the system is approximately balanced under an average load condition, being unbalanced in one direction under no load and unbalanced in the other direction at full load. This may be accomplished by adjusting the taps on transformer windings to give $a''bc$ under light load.

Another method of balancing the power drawn by unbalanced loads is illustrated in Fig. 213 which shows an electric furnace with two electrodes and one hearth connection. The load may be represented by the star-connected resistance shown in the diagram. It will be noted, however, that the resistance in series with the upper electrode will be of higher value than the one in series with the hearth since the upper electrodes involve arc drop. By distorting the voltage triangle applied to the furnace, it is possible to obtain balanced power. This has been demonstrated by means of a theoretical investigation described in the literature.* Either the delta-star or the star-delta connections may be used to obtain balanced currents and voltages on the high-voltage side of the transformer assuming the furnace load consists of arc and hearth resistances of different values.

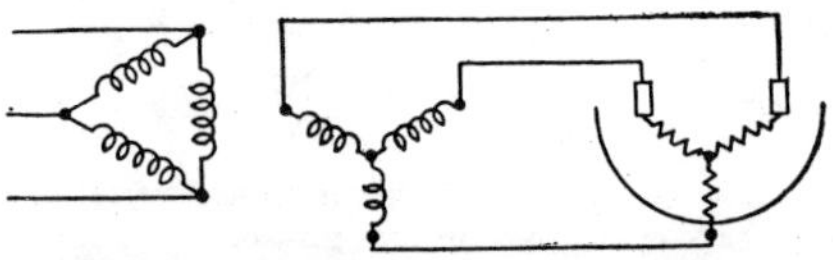

FIG. 213.—Method of using unsymmetrical transformer voltages to balance the load drawn by an electric furnace with two electrodes and one hearth connection.

In connection with static methods for phase-balancing, it is to be realized that if one load is variable, the other loads must be correspondingly adjusted if balance is to be maintained.

173. Rotating Balancers.

Rotating balancers tend to balance the voltages and currents on a power system by periodically absorbing and restoring energy to the system using in this process the energy stored by the inertia of rotating parts. Thus rotating machines tend to provide balancing by inherent action and do not require the

* EVANS, R. D., Electric Furnace as a Central Station Load with Particular Reference to Phase-balancing Systems, *Elec. Jour.*, vol. 17, September, 1920.

adjustable feature characteristic of static balancing systems as discussed in the previous paragraph.

Rotating balancers are of two general types:

1. Negative-sequence e.m.f. generator.
2. Impedance-type balancer.

Balancers may also be classified as to their connection to the system which may involve either series or shunt connections or their combinations. The principal types of balancers will be taken up and discussed separately.

174. Negative-sequence E.M.F.-type Phase-balancer.

Probably the earliest proposal for obtaining accurate phase-balancing is that due to E. F. W. Alexanderson[(1)] and illustrated in Fig. 214. The method is based on the idea of generating a

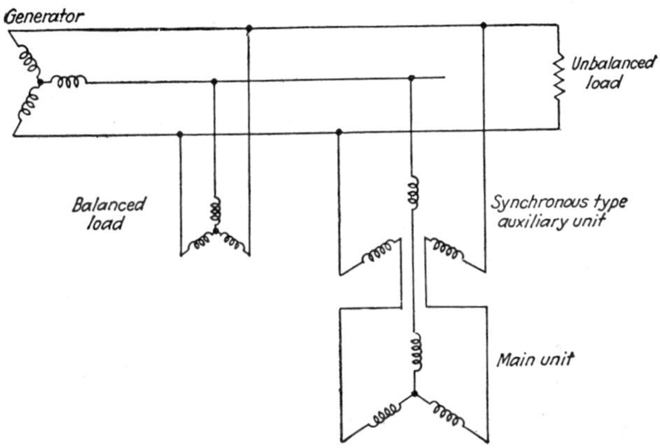

FIG. 214.—Negative-sequence e.m.f. type of phase balancer. (*Alexanderson.*)

negative-sequence e.m.f. of the proper magnitude and phase position to cancel the negative-sequence currents due to the single-phase loads which flow through the generator and other symmetrical portions of the system. This negative-sequence e.m.f. is generated in the auxiliary machine shown in the diagram which machine is mounted on the same shaft as the main unit and is provided with excitation in two axes so that the desired magnitude and phase relation can be controlled. The auxiliary machine is in series with the main unit which is of the ordinary synchronous condenser construction except for the heavy damper windings provided to take care of negative-sequence current.

The main machine draws balanced positive-sequence power from the system which the auxiliary generator converts to negative-sequence power and supplies to the system, thus canceling the pulsating component of load in the generator and other symmetrical parts of the system.

175. Impedance-type Phase-balancers.

The first proposal to use impedance-type balancers was due to C. L. Fortescue who suggested the **series impedance balancer** illustrated in Fig. 215. The auxiliary machine in this case is of the induction-motor type and is connected so that its phase

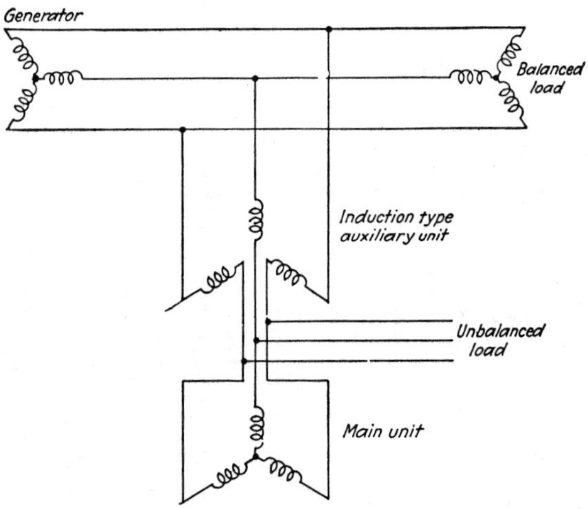

FIG. 215.—Series impedance type of phase balancer. (*Fortescue.*)

rotation is opposite to that in which it would normally run as an induction motor. Consequently, the auxiliary machine offers very low impedance to positive-sequence currents and very high impedance to negative-sequence currents, and advantage is taken of this fact. The negative-sequence currents required by the unbalanced load must, therefore, be supplied by the main unit, which may be of either the induction or synchronous types. The rating of the auxiliary machine is determined by the impedance drop due to the negative-sequence current flowing through the main unit and the positive-sequence current flowing into the load.

Power-factor correction may be secured by the main units of either the Fortescue or Alexanderson type of phase-balancer.

Due to the fact that single-phase loads are frequently of low power factor, the phase-balancing unit would normally be designed to give power-factor correction as well.

It might be pointed out that the shunt-type balancer requires automatic adjustments in the voltage regulator to correct for the change in the unbalanced condition. Consequently, it will not be so rapid in its action generally as the inherent type of phase-balancer making use of the impedance principle, such as is illustrated in Fig. 215. If the single-phase load can be segregated from the remainder, then the series machine will have its current rating determined by the positive-sequence component of that load. If the single-phase load cannot be segregated from a considerable amount of balanced load, the shunt-type balancer may be more attractive.

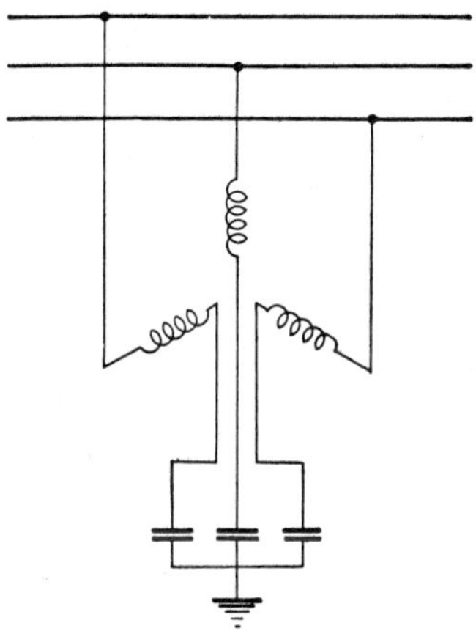

FIG. 216.—Shunt impedance balancer with series capacitor. (*Slepian.*)

The shunt impedance balancer of Fig. 216 is the simplest scheme which has been proposed for phase-balancing. This scheme, due to J. Slepian, uses a synchronous machine similar to the main unit of either of the previously described balancers and in addition, in series with each phase, a set of capacitors of such value that the impedance to negative-sequence is made negligible. This arrangement will, therefore, prevent negative-sequence current from flowing past this shunt machine to the generator or other balanced machines on the system. The scheme is also inherent in action. The scheme has not been used commercially but looks promising. The principal problem involved is in the protection of the series capacitor units at times of short-circuit. It has been proposed to take care of this by connecting the capacitors in the circuit through transformers which would saturate for loads in excess of the normal rating of the balancer and thus prevent full balancing action, which greatly relieves the stresses due to short-circuit currents that would otherwise flow.

176. Phase-converter.

It is not necessary to balance the entire single-phase load in all cases. For example, it may be desirable to use a generator without a damper winding in parallel with several machines equipped with damper windings. The amount of the single-phase load that can be carried by the combination may be increased greatly by connecting a negative-sequence auxiliary machine in series with the generator without a damper winding as illustrated in Fig. 217. A negative-sequence machine used with this connection is known as a phase-converter. The negative-sequence current will be compelled by the auxiliary machine to flow wholly (or partially) through the machine equipped with a damper winding. The positive-sequence current, on the other hand, may be divided between the machines in the ordinary manner according to the excitation and the governor setting of the prime mover.

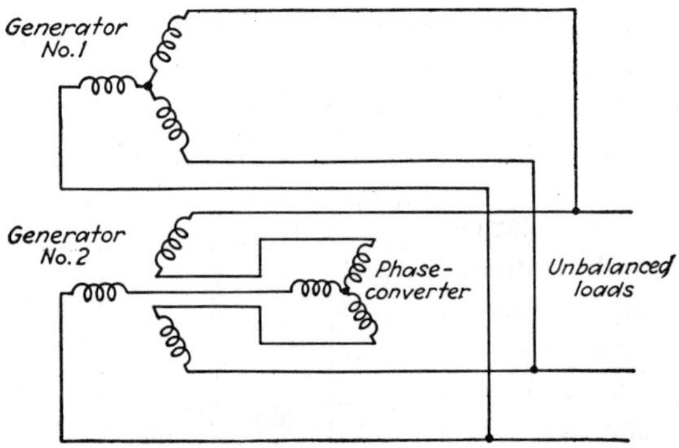

FIG. 217.—Use of phase-converter to minimize negative-sequence current flowing through generator No. 2.

177. Balanced Polyphase Voltages from a Single-phase Source.

For certain applications it is desirable to obtain balanced polyphase voltages from single-phase sources; thus, for control purposes a single-phase voltage may be applied to the polyphase sequence segregating network of Fig. 166, Chap. XIV, and a balanced set of positive-sequence voltages will be obtained from the terminals which are connected to the polyphase relay. However, rotating machines will in general be required for practical schemes for power purposes. The principal application at the present time has been in connection with alternating-current railway electrification where a single-phase contact system is desirable, but polyphase voltages simplify the motor equipment. This has led to the phase-converter locomotive, typical connections of which are shown in Fig. 218. The phase-converter may be of either the induction or synchronous type; with the latter, power-factor correction also may be provided. The phase-converter is not self-starting, but when operating at normal speed it will provide polyphase voltages for starting and running the induction motors. The phase-converter will not provide balanced voltages under all conditions because of its regulation. To overcome this difficulty the motor side of the transformer winding is provided with several taps so that tap changing under load can be used to secure better voltage balance. Taps on either side of one line terminal and the mid-point of the transformer winding are required to insure balance during both motoring and regenerating operations.

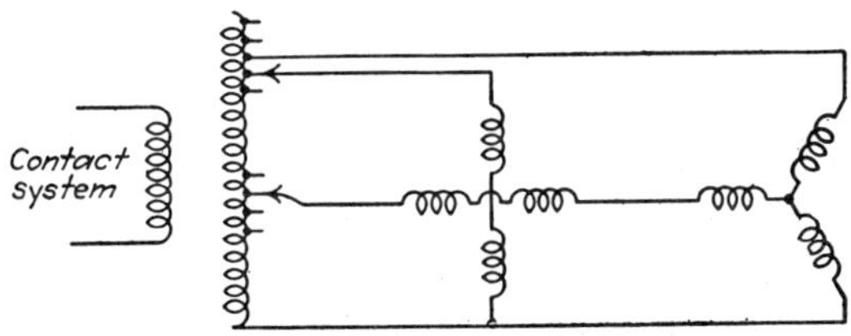

FIG. 218.—Schematic diagram for phase-converter locomotives such as are used on the Norfolk and Western and Virginian Railway electrifications.

Problems

1. One form of phasing device consists of two lamps and a reactor connected in star across the three leads of a three-phase system. The lamps are connected to the *a* and *c* phases and the reactor to the *b* phase. Assuming the line-to-line voltage to be 110 volts and the impedance of each lamp and reactor to be $10 + j0$ and $0 + j10$ ohms, respectively, determine the current through the lamps when positive-sequence voltage alone and negative-sequence voltage alone are applied to the phasing device.

2. Three resistors of 20, 10, and 5 ohms are connected in star to the *a*, *b*, and *c* phases, respectively. The junction point of the three resistors is free. Determine the voltage between this junction point and the neutral of the system voltage when a positive-sequence voltage of 100 volts to neutral is applied to the resistors.

3. A three-electrode furnace is supplied by a 3,500-kva., 2,200–200-volt, three-phase, 60-cycle transformer. The primary is star-connected and the secondary delta-connected. At a given instant the characteristics of the furnace may be replaced by three star-connected resistances of 0.010, 0.010 and 0.002 ohm in the *a*, *b*, and *c* phases, respectively. If the transformer reactance is 30 per cent, what are the magnitudes of current in the three electrodes, neglecting transformer resistance and the impedance of the leads?

4. An interconnected booster transformer of the type shown in Fig. 206 is rated at 20,000 kva., three phase, 110 kv., 60 cycle. The rating is determined by the voltage of the shunt windings and the current-carrying capacity of the series windings, and on this basis the leakage reactance is 70 per cent. Assume the ratio of transformation to be 15 per cent and two three-phase reactors of 12 per cent and 15 per cent (on 20,000-kva., 110-kv. base) are inserted in a series, one on each side of the transformer. With the voltage maintained at 110 kv. outside of the reactors on the left-hand side, compute the fault current for (*a*) a three-phase fault on the right-hand side; (*b*) a line-to-line fault on the right-hand side.

5. A symmetrical three-phase generator having the constants $Z_1 = j100$ per cent, $Z_2 = j30$ per cent, and $Z_0 = j5$ per cent is loaded between terminals and neutral by the following impedances, $100 + j0$ per cent for phase *a*, $0 + j100$ per cent for phase *b*, and $100 + j100$ per cent for phase *c*. The excitation is 120 per cent of the no-load voltage. Determine the line-to-neutral and line-to-line voltages.

6. Two similar ungrounded generators are synchronized with phase *a* open and the positive-sequence internal voltages 30 deg. out of phase. Determine the *b* and *c* phase currents and the synchronizing power in terms of the ratings of the machines and compare with the case for phase *a* closed. Use $Z_1 = j0.40$ and $Z_2 = j0.25$ in per unit and the internal voltage of unit magnitude.

7. Two similar grounded-neutral generators are paralleled through the *a* phase only. The constants of the machines are $Z_1 = j1.0$, $Z_2 = j0.30$, and $Z_0 = j0.05$. If the positive-sequence voltages are 30 deg. out of phase, what are the voltages between the *b* phase terminals and between the *c* phase terminals? Assume the internal voltage to be of unit magnitude.

8. A three-phase electric boiler is supplied from a three-phase source, but, due to certain dissymmetry which arises from time to time, a negative-sequence voltage is developed which reflects in its effect throughout the entire system. It is proposed to place in series with the load an induction machine capable of limiting the negative-sequence current so that the voltage on the supply side is substantially balanced. It may be assumed that the amount of the negative-sequence voltage of 3 per cent across the furnace will not be appreciably changed by a reduction in the negative-sequence current. Thus a machine of essentially induction-motor construction of 3 per cent of the rating of the boiler is required. This machine will be driven at a speed of 200 per cent slip which corresponds to synchronous speed for the negative-sequence. Such an arrangement will limit the negative-sequence current flowing back into the system to the magnitude of the magnetizing current. This can be further reduced by shunting the series machine by static condensers. If the constants of the series machine in per unit on its own rating are

$$R_{rotor} = 0.01 \qquad R_{stator} = 0.01$$
$$X_{rotor} = 0.15 \qquad X_{stator} = 0.15$$
$$X_m = 4.00$$

what is the rating of the capacitors required to resonate with the magnetizing current and thus making the impedance to negative-sequence current a maximum? What is the rating of the driving motor in terms of the rating of the series machine neglecting windage and friction, and iron losses?

9. Two single-phase transformers of impedance Z ohms on the primary turns basis are connected in open delta between the a and b and the c and a phases. With positive-sequence voltage E_{A1} applied to the primary, determine the expression for the secondary voltage on the primary turns basis when positive- and negative-sequence currents I_{A1} and I_{A2} on the primary turns basis are drawn from the secondary.

10. When positive-sequence current alone is drawn from the open-delta connection of Prob. 9, dissymmetry is introduced by the unbalanced circuit. This condition may be corrected by the introduction of a series reactance in one of the primary leads. Determine the lead in which this reactance should be placed and also its magnitude in terms of the transformer impedance Z.

BIBLIOGRAPHY

This bibliography is divided into two parts, the first listing references in the English language and the second those in foreign languages.

While the Bibliography is presented without pretensions as to completeness, the authors believe that it is fairly so with respect to American publications. Many of the foreign references were obtained from the bibliographic study of symmetrical components made by Dr. Hague.[190]

The following is a list of the abbreviations used in the Bibliography:

Arch. f. Elekt.	*Archiv für Elektrotechnik* (Berlin).
Bull. assoc. ing. élec. Liège	*Bulletin de l'Association des Ingénieurs électriciens sortis de l'Institut électrotechnique Montefiore* (Liège).
Bell System Tech. Jour.	*Bell System Technical Journal* (New York).
Bull. soc. franç. élect.	*Bulletin de la Société française des Électriciens* (Paris).
Compt. rend.	*Comptes rendus hebdomadaires des Séances de l'Académie des Sciences* (Paris).
Elec. Rev.	*The Electrical Review* (London).
Elec. Times	*The Electrical Times* (London).
Elec. World	*Electrical World* (New York).
Elec. Eng.	*Electrical Engineering* (New York).
Elec. Jour.	*The Electric Journal* (Pittsburgh).
Elec. Lab. Min. Comms. Researches, Tokyo	*Researches of the Electrotechnical Laboratory of the Ministry of Communications* (Tokyo).
E T Z	*Elektrotechnische Zeitschrift* (Berlin).
E.u.M.	*Elektrotechnik und Maschinenbau* (Wien).
Gen. Elec. Rev.	*General Electric Review* (Schenectady).
Ind. élec.	*Industrie électrique* (Paris).
Inst. Civil Eng. Selected Eng. Papers	*Selected Engineering Papers of the Institution of Civil Engineers* (London).
Jour. A.I.E.E.	*Journal of the American Institute of Electrical Engineers* (New York).
Jour. I.E.E.	*Journal of the Institution of Electrical Engineers* (London).
Jour. I.E.E. Japan	*Journal of the Institution of Electrical Engineers of Japan* (Tokyo).
Jour. Inst. Eng. Australia	*Journal of the Institution of Engineers of Australia.*

Elettrotec.	*L'Elettrotecnica. Giornale ed Atti dell' Associazione Elettrotecnica Italiana* (Milan).
R.G.E.	*Revue générale de l'Électricité* (Paris).
Tijd. v. Elec.	*Tijdschrift voor Electrotechniek.*
Trans. A.I.E.E.	*Transactions of the American Institute of Electrical Engineers* (New York).
Wiss. Veröfftl. Siemens Konz.	*Wissenschaftliche Veröffentlichungen aus dem Siemens Konzern* (Berlin).
World Power	*World Power* (London).
Z. f. angew. Math. u. Mech.	*Zeitschrift für angewande Mathematik und Mechanik* (Berlin).

Part I. English Language References

1913

1. ALEXANDERSON, E. F. W.: Phase-balancer for Single-phase Load on Polyphase Systems, *Gen. Elec. Rev.*, vol. 16, No. 12, pp. 962–964, December.

1915

2. STOKVIS, L. G.: Analysis of Unbalanced Three-phase Systems. Reactions in a Generator Carrying an Unbalanced Load Treated as Equivalent to Two Balanced Loads, *Elec. World*, vol. 65, pp. 1111–1115, May 1.

1916

3. ALEXANDERSON, E. F. W., and G. H. HILL: Single-phase Power Production, *Trans. A.I.E.E.*, vol. 35, pp. 1315–1327.
4. GILMAN, R. E., and C. L. FORTESCUE: Single-phase Power Service from Central Stations, *Trans. A.I.E.E.*, vol. 35, Pt. II, pp. 1329–1347.

1918

5. FORTESCUE, C. L.: Method of Symmetrical Coordinates Applied to the Solution of Polyphase Networks, *Trans. A.I.E.E.*, vol. 37, Pt. II, pp. 1027–1140.

1919

6. WALKER, MILES: The Supply of Single-phase Power from Three-phase Systems, *Jour. I.E.E.*, vol. 57, pp. 109–139.

1920

7. SLEPIAN, J.: Induction Motors on Unbalanced Voltages, *Elec. World*, vol. 75, pp. 313–315, Feb. 7.
8. LYON, W. V.: Unbalanced Three-phase Circuits, *Elec. World*, vol. 75, No. 23, pp. 1304–1308, June 5.
9. SLEPIAN, J.: Reactive Power and Magnetic Energy, *Trans. A.I.E.E.*, vol. 39, pp. 1115–1133.
10. FECHHEI ER, C. J.: Power Factor and Unbalance on a Polyphase System, *Trans. A.I.E.E.*, vol. 39, Pt. II, pp. 1469–1473.
11. FORTESCUE, C. L.: Polyphase Power Representation by Means of Symmetrical Coordinates, *Trans. A.I.E.E.*, vol. 39, Pt. II, pp. 1481–1484.

12. EVANS, R. D.: Measurement of Power Factor on Unbalanced Polyphase Circuits, *Trans. A.I.E.E.*, vol. 39, Pt. II, pp. 1485–1487.
13. LYON, W. V.: Reactive Power and Unbalanced Circuits, *Elec. World*, vol. 75, pp. 1417–1420.

1921

14. TAKATSU, KYOSHI: Measurement of Power Factor in Polyphase Unbalanced Circuits, *Elec. Lab., Min. Comms., Tokyo, Researches*, No. 90, pp. 1–23, February.

1922

15. FORTESCUE, C. L.: Transformers and Rectifiers, in F. F. Fowle, "Standard Handbook for Electrical Engineers," 5th ed., Sec. 6, p. 427, McGraw-Hill Book Company, Inc., New York.
16. LAWRENCE, R. R.: "Principles of Alternating Currents," Chap. 12, p. 337, McGraw-Hill Book Company, Inc., New York.

1923

17. FORTESCUE, C. L.: The Measurement of Power in Polyphase Circuits, *Trans. A.I.E.E.*, vol. 42, pp. 358–371.
18. EVANS, R. D.: New Sequence System of Polyphase Meters, *Elec. World*, vol. 81, No. 6, pp. 333–334, Feb. 10.
19. BEKKU, SADATOSHI: Theoretical Researches on Transmission Line Problems with Special Reference to the Arc Suppressing Reactor, *Elec. Lab., Min. Comms., Tokyo, Researches*, No. 129, April.
20. HAY, ALFRED: "Alternating Currents," p. 375.

1924

21. BREISKY, J. V.: A Phase-balance Current Relay of the Negative Phase Sequence Type, *Elec. Jour.*, vol. 21, No. 2, pp. 77–81, February.
22. BEKKU, SADATOSHI: Synchronous Operation of Two Alternators through Unsymmetrical Impedances, *Electrotechnical Laboratory, Min. Comms., Tokyo, Researches*, No. 138, pp. 1–11, February.
23. DUDLEY, A. M.: Induction Motors on Unbalanced Circuits. Vector Methods of Analysis of Unsymmetrical Systems, *Elec. Jour.*, vol. 21, pp. 339–343, July.

1925

24. SCHOENFELD, O. C.: Effect of Unbalanced Voltages on Operation of Induction Motors, *Elec. Jour.*, vol. 22, pp. 30–33, January.
25. EVANS, R. D.: Finding Single-phase Short-circuit Currents on Calculating Boards, *Elec. World*, vol. 85, pp. 761–764, April 11.
26. BEKKU, SADATOSHI: Calculation of Short-circuit Ground Currents on Three-phase Power Networks, Using the Method of Symmetrical Coordinates, *Gen. Elec. Rev.*, vol. 28, No. 7, pp. 472–478, July.
27. LUNDHOLM, R.: "A Generalized Vector Theory for Electrical Alternating-current Circuits with Its Application to Certain Practical Problems."

1926

28. EVANS, R. D., and C. F. WAGNER: Further Studies of Transmission Stability, *Trans. A.I.E.E.*, vol. 45, pp. 51–94.
29. MACKERRAS, A. P.: Calculation of Single-phase Short-circuits by the Method of Symmetrical Components, *Gen. Elec. Rev.*, vol. 29, No. 4, pp. 218–231, April; No. 7, pp. 468–481, July.
30. BEKKU, S., M. DOTE, and M. URUSHIBATA: On the Method of Measurement of Zero and Negative Phase Sequence Impedance of the Three-phase Alternator, *Elec. Lab.*, *Min. of Comms.*, *Tokyo*, *Researches*, No. 170, pp. 1–20, April.
31. BEKKU, S., and K. MAYEKAWA: Calculation of Ground Current Caused by One Line Grounded in a Three-phase Transmission Network, *Electrotechnical Laboratory*, *Ministry of Communications*, *Tokyo*, *Researches*, vol. 181, pp. 1–114, September.
32. CARSON, J. R.: Wave Propagation in Overhead Wires with Ground Return, *Bell System Tech. Jour.*, vol. 5, pp. 539–555, October.
33. HAGUE, BERNARD: The Method of Symmetrical Coordinates in the Theory of Polyphase Circuits, *Inst. Civil Eng.* (*London*), *Selected Eng. Papers*, No. 40, pp. 1–38.

1927

34. WAGNER, C. F., and A. DOVJIKOV: Impedance of a Rotating Synchronous Machine to Negative Sequence Voltage, *Elec. Jour.*, vol. 24, No. 3, pp. 117–121, March.
35. GRISCOM, S. B.: Characteristics of Ground Faults on Three-phase Systems, *Elec. Jour.*, vol. 24, No. 4, pp. 151–156, April.
36. KARAPETOFF, V.: "Experimental Electrical Engineering," 3d ed., vol. 2, Chap. 39, p. 118, John Wiley & Sons, Inc., New York.

1928

37. KENNELLY, A. E.: Computation of the Unbalance Factor of a Three-phase Triangle When the Lengths of the Three Sides Are Given, *Trans. A.I.E.E.*, vol. 47, No. 1, pp. 341–342, January.
38. SAH, A. P.: A Note on the Unbalancing Factor of Three-phase Systems, *Trans. A.I.E.E.*, vol. 47, No. 1, p. 343, January.
39. WAGNER, C. F., and R. D. EVANS: Symmetrical Components, *Elec. Jour.*, vol. 25, No. 3, pp. 151–157, March; No. 4, pp. 194–197, April; No. 6, pp. 307–311, June; No. 7, pp. 359–362, July. (Continued in references 50 and 61.)
40. NORMAN, H. M.: Mathematical Solution of an Unbalanced Three-phase Voltage System, *Jour. A.I.E.E.*, vol. 47, No. 6, pp. 449–451, June.
41. PARK, R. H., and B. L. ROBERTSON: Reactances of Synchronous Machines, *Trans. A.I.E.E.*, vol. 47, No. 2, pp. 514–536, April.
42. STUBBINGS, G. W.: Symmetrical Components of Unbalanced Three-phase Systems, *World Power*, vol. 10, pp. 512–517.
43. DAHL, O. G. C.: "Electric Circuits," vol. I, Theory and Application, Chap. 3, p. 68, McGraw-Hill Book Company, Inc., New York.
44. LEWIS, W. W.: "Transmission Line Engineering," 1st ed., Chap. 4, p. 163, McGraw-Hill Book Company, Inc., New York.

1929

45. Park, R. H., and E. H. Bancker: System Stability as a Design Problem, *Trans. A.I.E.E.*, vol. 48, No. 1, pp. 170–194, January.
46. N.E.L.A.: Calculation of Ground Currents, *Serial Rep. Elec. Apparatus Comm., Eng. Sec. Southeastern Division, N.E.L.A.*, March.
47. Dovjikov, A.: Application of Automatic Oscillographs, *Elec. Jour.*, vol. 26, No. 4, pp. 175–179, April; No. 5, pp. 202–206, May.
48. Stewart, H. R.: How the Positive Phase Sequence Network Works, *Elec. Jour.*, vol. 26, No. 6, pp. 260–263, June.
49. Ku, Y. H.: Transient Analysis of A.C. Machinery, *Trans. A.I.E.E.*, vol. 48, No. 3, pp. 707–715, July.
50. Wagner, C. F., and R. D. Evans: Symmetrical Components, *Elec. Jour.*, vol. 26, No. 9, pp. 426–431, September; No. 12, pp. 571–581, December. (Continuation of reference 39 completed in reference 61.)
51. Goto, M.: Various Expressions for Symmetrical Three-phase Circuits in Terms of Symmetrical Coordinates and a Symbolic Method for Circular Functions, *Elec. Lab., Min. Comms., Tokyo, Researches*, No. 255, pp. 1–24.

1930

52. Summers, I. H., and J. B. McClure: Progress in the Study of System Stability, *Trans. A.I.E.E.*, vol. 49, No. 1, pp. 132–161, January.
53. Joint Devel. and Res. Sub-committee: An Investigation of Ground Faults on 33-kv. Transmission System and the Resulting Voltages in a Paralleling Telephone System, *Eng. Rep.* 4, pp. 7–47, *N.E.L.A. and Bell System*, April.
54. Monseth, I. E., and P. H. Robinson: Theory and Application of Relay Systems, *Elec. Jour.*, vol. 27, pp. 411–415, 473–480.
55. Stigant, S. A.: Phase Sequence on Three-phase Systems. Notes on the Phase Sequence Idea in the Study of Unbalanced Three-phase A.C. Systems, *Elec. Times*, vol. 78, pp. 1021–1023.
56. Goto, M.: Undamped Electric Oscillation of an Alternator Connected to a Transmission Line, *Elec. Lab., Min. Comms., Tokyo, Researches*, No. 281, pp. 1–47.
57. Bowen, A. E., and C. L. Gilkeson: Mutual Impedances of Ground Return Circuits, Some Experimental Studies, *Trans A.I.E.E.*, vol. 49, No. 4, pp. 1370–1383, October.

1931

58. Wagner, C. F.: Damper Windings for Waterwheel Generators, *Trans. A.I.E.E.*, vol. 50, pp. 140–152, March.
59. Evans, R. D., and S. H. Wright: Power System Voltages and Currents under Fault Conditions—Effect of Sequence Impedances, *A.I.E.E. Paper*, No. 31–59; Abstract, *Elec. Eng.*, pp. 415–420, June.
60. Lewis, W. A., and L. S. Tippett: Fundamental Basis for Distance Relaying on Three-phase Systems, *A.I.E.E. Paper*. No. 31–61; Abstract, *Elec. Eng.*, p. 420, June.

61. Wagner, C. F., and R. D. Evans: Symmetrical Components, *Elec. Jour.*, vol. 28, No. 4, pp. 239–244, April; No. 5, pp. 308–312, May; No. 10, pp. 586–590, October; No. 11, pp. 624–630, November. (Completes references 39 and 50.)

62. Monseth, I. T., and P. H. Robinson: Theory and Application of Relay Systems, *Elec. Jour.*, vol. 28, pp. 375–379, 431–435.

63. Stubbings, G. W.: Unbalanced Three-phase Circuits, *Elec. Rev. (London)*, vol. 108, pp. 783–784.

64. Wilson, W.: Vector Algebra, *World Power*, vol. 15, pp. 192–195.

65. Stigant, S. A.: Vector Operators in Three-phase Systems, *Elec. Times*, vol. 80, pp. 735–736.

66. Hayward, A. P.: Measurement of Symmetrical Components, *Elec. Jour.*, vol. 28, No. 6, pp. 351–357, June.

67. Bohner, C. W., and G. W. Vaughan: Phase Faults Analyzed on Direct-current Calculating Board, *Elec. World*, vol. 97, No. 23, pp. 1068–1071, June 6.

68. Tippett, L. S. C.: Application of Impedance Relays to Earth Fault Protection, *Jour. Inst. Eng. Australia*, vol. 3, pp. 233–243.

69. Clem, J. E.: Reactance of Transmission Lines with Ground Return, *Trans. A.I.E.E.*, vol. 50, No. 3, pp. 901–918, September.

70. Clarke, Edith: Simultaneous Faults on Three-phase Systems, *Trans. A.I.E.E.*, vol. 50, No. 3, pp. 919–941, September.

71. Garza, E. de la: A Note on Symmetrical Components, *Elec. Eng.*, vol. 50, p. 910.

72. Kilgore, L. A.: Calculation of Synchronous Machine Constants, Reactances and Time Constants Affecting Transient Characteristics, *Trans. A.I.E.E.*, vol. 50, pp. 1201–1214.

73. Wright, S. H.: Determination of Synchronous Machine Constants by Test, *Trans. A.I.E.E.*, vol. 50, pp. 1331–1351.

74. Dannatt, C., and J. W. Dalgleish: "Electrical Power Transmission and Interconnection," Chap. 9, Sir Isaac Pitman & Sons, New York.

1932

75. Hahn, W. C., and C. F. Wagner: Standard Decrement Curves, *Trans. A.I.E.E.*, vol. 51, pp. 353–362.

76. Wagner, C. F., and S. H. Wright: Calculations of Short-circuits on Power Systems, *A.I.E.E. Paper* No. 32M–4; Abstract, *Elec. Eng.*, vol. 51, No. 2, p. 131, February.

77. Hahn, W. C.: Decrement Curves for Specific Systems, *A.I.E.E. Paper*, 32M–5; Abstract, *Elec. Eng.*, vol. 51, p. 131, February.

78. Sels, H. K.: Proposed Definitions of Terms Used in Power System Studies, *Rept. Subject Comm. on Definitions, A.I.E.E. Paper*, 32M–2; Abstract *Elec. Eng.*, vol. 51, No. 2, p. 106.

79. Joint Devel. and Res. Sub-committee: Coupling Factors for Ground-return Circuits—General Considerations and Methods of Calculations, vol. 2, *Eng. Rept.* No. 14, pp. 121–185, *N.E.L.A. and Bell System*, April.

80. Harder, E. L.: Peak Voltages on Saturating Reactances in Three-Phase Circuits, *Elec. Jour.*, vol. 29, pp. 124–129, March.

Part II. Foreign Language References

1912

101. STOKVIS, L. G.: "Der Spannungsabfall des synchronen Drehstromgenerators bei unsymmetrischer Belastung," R. Oldenbourg, Munich.

1914

102. STOKVIS, L. G.: Sur la création des harmoniques 3 dans les alternateurs par suite des deséquilibrages des phases, *Compt. rend.*, vol. 159, pp. 46–49.

1915

103. STOKVIS, L. G.: En cirkeldiagram voor cheeve belasting von draastroommachines, *De Ingénieur*, Jan. 9.

1917

104. SALDA, C. DELLA: The Three-phase Induction Motor, Fed by Asymmetrical Voltage, *Elettrotec.*, vol. 5, p. 294, June 5.

1918

105. MULLER, P.: Unsymmetrische Mehrphasensysteme, *E T Z*, vol. 39, pp. 343–356, Aug. 29; pp. 353–356, Sept. 5.

1921

106. EMDE, F.: Zur Definition der Scheinleistung und der Blindleistung bei ungleichmässig belasteten Mehrphasensystemen, *E.u.M.*, vol. 39, pp. 545–547.
107. BRUDERLIN, R.: Die Induktionsmaschine als Phasenumformer, *E.u.M.*, vol. 39, pp. 361–366.

1922

108. STOKVIS, L. G.: Sur la production des harmoniques 3 dans les machines à induction a charge deséquilibrée, *R.G.E.*, vol. 12, pp. 619–629, Oct. 28; pp. 661–670, Nov. 4.
109. STOKVIS, L. G.: Sur les diagrammes circulaires des systèmes triphases deséquilibrés et la définition de leur degré de deséquilibrage, *Compt. rend.*, vol. 174, pp. 1418–1420; *R.G.E.*, vol. 11, pp. 957–959.
110. STOKVIS, L. G.: Sur la définition de la tension, de l'intensité et du facteur de puissance d'un système triphase, *R.G.E.*, vol. 12, pp. 307–312, Sept. 2.

1923

111. ZACHRISSON, E.: Zur Definition des Begriffes Unsymmetrie eines Drehstromsystems, *E T Z*, vol. 44, pp. 897–899, Sept. 27.
112. STOKVIS, L. G.: Essai d'une théorie générale des machines synchronés à courant alternatif basée sur les champs tournants sinusidaux, *R.G.E.*, vol. 13, pp. 953–967, June 9; pp. 995–999, June 16.
113. STOKVIS, L. G.: Sur le diagramme de la connection Scott utilisée pour l'alimentation de deux réseaux monophasés indépendants, *Bull. soc. franç. élect.*, vol. 3, pp. 729–750; Over de Scott'sche schakeling

als transformatie van draastroom in twee eenphasenstroomen, *Tijd. v. Elect.*, May, June, and July, 1925.

114. Stokvis, L. G., and C. Lavanchy: Diagramme de la chute de tension dans un conducteur de ligne triphasée de tension moyenne, cas général d'un système deséquilibré geómetriquement et électriquement, *R.G.E.*, vol. 13, pp. 3–5.

1924

115. Thomalen, A.: Die Zerlegung unsymmetrischer Drehstromsysteme in symmetrische, *E T Z*, vol. 44, pp. 500–501.

116. Stokvis, L. G.: Sur la décomposition des systèmes triphases ne satisfaisant pas à la condition que la somme des vecteurs représentatifs est zéro, *Compt. rend.*, vol. 178, pp. 1074–1076, March 24.

117. Stokvis, L. G.: Sur la décomposition des systèmes triphases dont la somme des vecteurs n'est pas nulle, *R.G.E.*, vol. 16, pp. 611–614, Oct. 18.

118. Stokvis, L. G.: Décomposition de l'énergie d'un système triphase ou une énergie constante à une énergie pulsatoire, *Compt. rend.*, vol. 179, pp. 595–597.

119. Stokvis, L. G.: Sur la transmission de l'énergie par un système triphase deséquilibré pour lequel ni la somme des tensions de phase, ni la somme des courants de phase n'est égale à zéro, *R.G.E.*, vol. 16, pp. 979–982, Dec. 20.

120. Dubusc, R.: Relais de deséquilibrage pour circuits triphases, *R.G.E.*, vol. 16, pp. 659–670, Oct. 25.

121. Genkin, V.: Théorie du deséquilibre et son application aux systèmes de protection, *R.G.E.*, vol. 16, pp. 907–916, Dec. 6; pp. 951–959, Dec. 13.

122. Bottani, E.: La moderna matematica dei circuiti trifasi, *Elettrotec.* vol. 11, pp. 738–745, 768–774, 928–937, 966–973, 1013–1023.

1925

123. Bekku, Sadatoshi: Synchronisierung zweier Drehstrom Generatoren durch unsymmetrische Impedanzen, *E T Z*, vol. 46, p. 656, April 30.

124. Rasch, G.: Unsymmetrische Mehrphasensysteme, *E T Z*, vol. 46, pp. 1446–1447, Sept. 17.

125. Darrieus, G.: Protection des grand réseaux, *Bull. soc. franç. élect.*, October.

126. Bekku, Sadatoshi: Method of Symmetrical Coordinates and Its Application to the Transmission Line Problem, *Jour. Inst. Elec. Eng. Japan*, November (in Nipponese).

127. Genkin, V.: Deséquilibre électrique dans les installations polyphases. Phasemètre pour circuits triphases deséquilibrés, *R.G.E.*, vol. 18, pp. 890–892, Nov. 28.

128. Genkin, V.: Deséquilibre électrique dans les installations polyphases, Supplément au *bull. soc. franç., élect.*, vol. 5, p. 45.

129. Bekku, Sadatoshi: Methode der symmetrischen Koordinaten und allgemeine Theorie der Erdschlusslöscheinrichtungen, *Arch. f. Elekt.*, vol. 14, pp. 543–555.

130. GERSTMANN, B.: Ueber die Dimensionierung von Phasenspaltern, *E.u.M.*, vol. 43, pp. 653–658.

1926

131. GENKIN, V.: Réseau électrique filtrant et autorégulateur pour circuits triphases, *R.G E.*, vol. 19, pp. 123–129, Jan. 23.
132. GENKIN, V.: Étude des dispositifs électriques applicables aux appareils de mesure et aux relais et destinés à modifier un courant dans un rapport complexe constant, *R.G.E.*, vol. 20, May 1.
133. GENKIN, V.: Récepteurs dyssymétriques et procédés d'équilibrage des circuits triphases à phases inégalement chargées, *R.G.E.*, vol. 20, pp. 123–136, July 24.
134. GENKIN, V.: Deséquilibre électrique dans les réseaux, *Ind. élec.*, p. 389, Sept. 10; p. 419, Sept. 25.
135. FALLOU, J.: Calcul des courants de court-circuit dans les réseaux triphases, *Bull. soc. franç. élect.*, vol. 6, pp. 969–1002, September.
136. NATALIS, F.: Unsymmetrische Drehstromsysteme, *Wiss. Veröfftl. Siemens Konz.*, vol. 5, Pt. I, pp. 1–6.
137. RUDENBERG, REINHOLD: Die Ausbrietung der Erdströme in der Umgebung von Wechselstromleitungen, *Z. f. angew. Math. u. Mech.*, vol. 5, p. 361, October.
138. GENKIN, V.: Mesure du facteur de puissance d'un circuit triphase non équilibre. Phasemètre de la Compagnie des Compteurs, *Bull. soc. franç. élect.*, vol. 6, pp. 1158–1166.
139. GENKIN, V.: La puissance électrique dans un circuit triphase non équilibre, *R.G.E.*, vol. 19, pp. 165–170.
140. GENKIN, V.: La mesure indirecte de l'énergie réactive, *Bull. soc. franç. élect.*, vol. 6, pp. 1148–1157.
141. STOKVIS, L. G.: Over de energie van draaistroomsystemen, *Tijd. v. Elect.*, July-August.
142. GENKIN, V.: Sur les procédés statiques d'équilibrage d'une charge monophasée dans un réseau triphase, *Bull. soc. franç., élect.*, vol. 6, pp. 841–853.

1927

143. DUBUSC, R.: Les idées modernes en matière de protection contre les surintensités, *Conférence internationale des grand réseaux*, vol. 2, p. 469.
144. OBERDORFER, G.: Das Rechnen nach der Methode der symmetrischen Koordinaten, *E.u.M.*, vol. 45, pp. 296–301.
145. GENKIN, V.: Contribution à la théorie du fonctionnement des appareils de mesure ou des relais sensibles au deséquilibre électrique, *R.G.E.*, vol. 21, pp. 525–527.
146. HAUFFE, G.: Unsymmetrische Drehstromnetze, *E T Z*, vol. 48, pp. 1734.
147. DUBUSC, R.: Les idées modernes en matière de protection contre les surintensités, *R.G.E.*, vol. 22, pp. 293–300.
148. FALLOU, J.: À propos de la mesure de la réactance de dispersion des alternateurs triphases, *R.G.E.*, vol. 22, pp. 543–550.

1928

149. FALLOU, J., L. DUPUIS, and G. TINNEL: Note sur le système employé par l'Union d'Électricité pour la protection sélective des lignes en parallèle, *R.G.E.*, vol. 24, pp. 29–34, July 7.
150. ILIOVICI, A.: Dispositifs permettant d'obténir les composantes symétriques des courants des tensions ou des puissances dans un système polyphase sans l'employé des filtres, *Bull. soc. franç. élect.*, vol. 8, pp. 839–868, August.
151. FORTESCUE, C. L., and G. CALABRESE: L'applicazione delle coordinate simmetriche alla risoluzione delle equazioni algebriche, *Atti del congresso internazionale dei matematici*, Bologna, p. 159, September.
152. FOCACCIA, B., and G. CERILLO: Determinazione delle constanti delle linee aeree triphase per l'applicazione del metodo delle coordinate simmetriche, *Elettrotec.*, Dec. 15 and 25.
153. OBERDORFER, G.: Einige Erdschlussgrundprobleme in symmetrischer Darstellung, *E.u.M.*, pp. 969–976, Oct. 7.
154. GORBUNOFF, M. M.: Calculation of Fault Currents and Voltages in High Voltage Systems under Unsymmetrical Short-circuit Conditions, *Theoretical and Experimental Electricity*, vol. 1 (in Russian).
155. BARBAGELATA, A.: Extensione del metodo della quattro letture alla determinazione delle, componenti simmetriche, di un sistema trifase, *Elettrotec.*, vol. 15, pp. 482–490.
156. OBERDORFER, G.: Die Spannungsverlagerung in Netzen mit Löschtransformatoren nach System Bauch, *Arch. f. Elekt.*, vol. 19, pp. 405–423.
157. ILIOVICI, A.: Protection sélective des réseaux, *Bull. soc. franç. élect.*, vol. 8, pp. 1063–1102.

1929

158. DOVJIKOV, A.: Application of Oscillographs in Electric Power Systems, *American Commerce and Industry*, January (in Russian).
159. OBERDORFER, G.: Die Leistung in unsymmetrischen Dreiphasensystemen, *E T Z*, vol. 50, pp. 265–267, Feb. 21.
160. GENKIN, V.: Sur la protection sélective des réseaux de distribution d'énergie contre les surintensités de courants, *R.G.E.*, pp. 285–289, Feb. 23; pp. 323–329, Mar. 2.
161. HAUFFE, G.: Unsymmetrische Drehstromsysteme, *E T Z*, vol. 50, p. 1446, Oct. 3.
162. ORTIZ, JOAQUIN GARCIA: Aplicación de la teoría de las coordenadas simetricas al studio general de las sistemas trifasicos equilibrados o no equilibrados, *Vease Anales, Tomoll*, vol. 3, p. 580.
163. TOLVENSKY, V. A.: Voltage Diagram of Three-phase System under Unsymmetrical Loads." *Theoretical and Experimental Electricity*, vol. 5 (in Russian).
164. OBERDORFER, G.: "Das Rechnen mit symmetrischen Komponenten," Verlag B.G. Teubner, Leipzig (72 pages).
165. BARBAGELATA, A.: La mesure et la tarification de l'énergie dans les réseaux triphases, *R.G.E.*, vol. 26, pp. 470–472.

166. NUTZELBERGER, H.: Drei neue Verfahren der Zerlegung eines unsymmetrischen Mehrphasensystems in zwei symmetrische, *Arch. f. Elekt.*, vol. 23, pp. 119–123.

167. ILIOVICI, A.: Les protections sélectives des réseaux basées sur le deséquilibrage des courants, des tensions et des puissances, *R.G.E.*, vol. 26, pp. 517–518.

1930

168. FALLOU, J.: Mise à la terre du point neutre par l'intermédiaire d'une inductance, étude des régimes transitoires asymétriques succédant a l'établissement et à la rupture du court-circuit, *R.G.E.*, vol. 27, pp. 739–751, May 10.

169. YERNAUX, J.: À propos de la décomposition des systèmes de trois vecteurs deséquilibrés, *R.G.E.*, vol. 28, pp. 259–260.

170. ZORN, M.: Bestimmung der Unsymmetrie von Drehstromnetzen, *E T Z*, vol. 51, pp. 1233–1238, Aug. 28.

171. TOLWINSKI, W.: Das Problem der unsymmetrischen Belastung des Synchrondrehstromgenerators und des Drehstromtransformators, *Arch. f. Elekt.*, vol. 23, pp. 497–521.

172. RICCIA, A. DELLA: Décomposition et effets des courants polyphases non équilibrés, *Bull. assoc. ing. élec.*, *Liège*, vol. 8, pp. 141–148.

173. RIMINI, C.: Sistemi trifasi simmetrici es quilibrati, *Elettrotec.*, vol. 17, pp. 273–278.

174. OBERDORFER, G.: Der Doppelerdschluss in einer zweifach gespeisten Einfachleitung im Lichte der Rechnung mit symmetrischen Komponenten, *Wiss. Veröfftl.*, *Siemens Konz.*, vol. 9, Pt. II, pp. 77–87.

175. FROIDEVAUX, J.: Étude fonctionnement d'une machine asynchrone à charge sur un réseau à charge deséquilibré, *R.G.E.*, vol. 28, pp. 389–402.

1931

176. BLONDEL, A.: Sur l'étude directe des systèmes triphases deséquilibrés au moyen d'impédances et admittances mutuelles de phases dans les problèmes de chutes de tension et de mise en court-circuit, *R.G.E.*, vol. 29, pp. 3–11, 53–63, 90–100.

177. FALLOU, J.: Succession de phases des harmoniques dans les branches d'un réseau équilibré à nombre de phases quelconque, *Bull. soc. franç. élect.*, vol. 1, pp. 123–127, February.

178. HESSENBERG, K.: Die Berechnung von Symmetriestörungen in Drehstromnetzen mit Hilfe von symmetrischen Komponenten und Ersatzschaltungen, *E.u.M.*, vol. 49, pp. 273–276, Apr. 2; pp. 299–304, Apr. 19.

179. STOKVIS, L. G.: Over het rekenen met symmtrische componenten, *De Ingénieur*, No. 36.

180. FRIEDLÄNDER, E., and O. SCHMUTZ: Ueber Drehfeldscheider zur Aufspaltung unsymmetrischer Drehstromsysteme in die symmetrischen Komponenten, *Wiss. Veröfftl.*, *Siemens Konz.*, vol. 10, Pt. I, pp. 24–41.

181. ILIOVICI, A.: Phasemètre a fer tournant pour circuits triphases non équilibrés, échelle 360°, *Bull. soc. franç. élect.*, vol. 1, pp. 978–995.

182. Wild, W.: Der Doppelerdschlusstrom in Drehstromkabeln und seine Einwirkung auf benachbarte Fernmeldkabel, *Wiss. Veröfftl., Siemens Konz.*, vol. 10, Pt. I, pp. 51–77.

183. Genkin, V.: Courants de court-circuit dans les canalisations en cable armé, *Bull. soc. franç. élect.*, vol. 1, pp. 886–897.

184. Hauffe, G.: Ueber Drehfelder bei unsymmetrischen Drehstromsystemen, *E.u.M.*, vol. 49, pp. 223–224.

185. Genkin, V.: Sur la détermination graphique approchée du fonctionnement des systèmes triphases deséquilibrés en étoile avec point neutre isolé, *Ind. élec.* pp. 281–283, June 25.

1932

186. Hauffe, G.: Die Symmetrierung des Drehstromsystems, *E.u.M.*, vol. 50, pp. 85–87.

187. Margand, F.: Relations entre les coefficients fondamentaux des machines synchrones dans l'état variable et les diverses impédances apparaissant dans l'étude des régimes synchronés deséquilibrés, *R.G.E.*, vol. 31, pp. 241–248.

188. Fourmarier, P.: Calcul des caractéristiques des moteurs asynchronés à rotor monophase, *R.G.E.*, vol. 31, pp. 376–380.

189. Stokvis, L. G.: "La théorie des composantes Symétriques," 4e Section, Communication No. 4-C-1, Congrès international d'électricité, Paris.

190. Hague, Bernard: "Méthode des coordonnées symétriques," 4e section, Rapport 4, Congrès international d'électricité, Paris.

APPENDIX

This appendix includes material that has been collected to provide a convenient reference for use in calculations by the methods that have been outlined in this volume. The material is arranged as follows:

I. NOTATION

The principal features of the notation used for voltage, current, impedance, and admittance symbols are given below.

Voltage Symbols: E, vector expression for voltage (r.m.s. value); $\hat{E}$, conjugate vector expression for voltage (r.m.s. value); $\bar{E}$, absolute value of vector E; e, instantaneous voltage.

Current Symbols: I, vector expression for current (r.m.s. value); $\hat{I}$, conjugate vector expression for current (r.m.s. value); $\bar{I}$, absolute value of vector I; i, instantaneous current.

Impedance Symbols: $Z = R + jX$, total impedance usually; $z = r + jx$ for impedance of network element or unit length of transmission line or cable; $\bar{Z}$, absolute value of vector Z.

Admittance Symbols: $Y = G + jB$, used generally to indicate system constants; $y = g + jb$, used generally to indicate constants per unit length of transmission line or cable; $\bar{Y}$, absolute value of vector Y.

Phase Symbols: Subscripts a, b, and c indicate phase a to neutral, phase b to neutral, and phase c to neutral, respectively, and subscripts A, B, and C indicate phase between conductors bc, phase between conductors ca, phase between conductors ab, respectively.

Sequence Symbols: Subscripts 0, 1, and 2 indicate zero-, positive-, and negative-sequence, respectively, in a three-phase system.

Letter Subscripts: These refer to either phase or circuit elements; the exact significance is to be determined by the context.

II. GENERAL FORMULAS

1. *Fundamental Equations:*

$$E_a = E_{a0} + E_{a1} + E_{a2}$$
$$E_b = E_{a0} + a^2E_{a1} + aE_{a2}$$
$$E_c = E_{a0} + aE_{a1} + a^2E_{a2}$$

$$I_{a0} = \tfrac{1}{3}(I_a + I_b + I_c)$$
$$I_{a1} = \tfrac{1}{3}(I_a + aI_b + a^2I_c)$$
$$I_{a2} = \tfrac{1}{3}(I_a + a^2I_b + aI_c)$$

2. *Star-delta Current Transformations:*

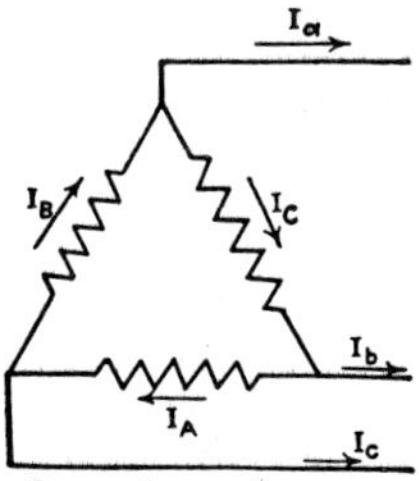

$$I_a = I_B - I_C$$
$$I_b = I_C - I_A$$
$$I_c = I_A - I_B$$

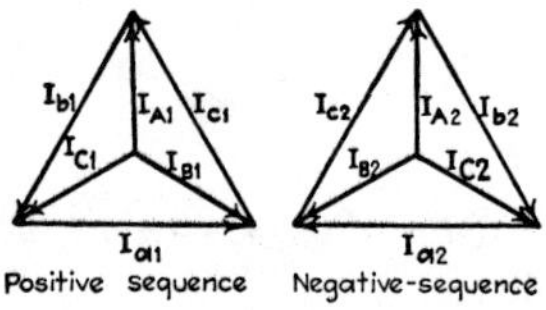

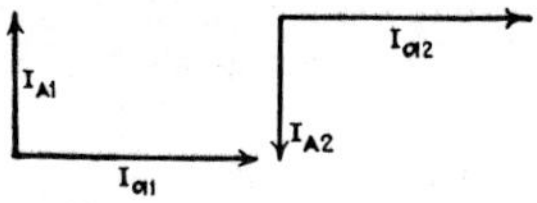

$$I_{A0} = \text{indeterminate}$$
$$I_{A1} = \frac{j}{\sqrt{3}}I_{a1}$$
$$I_{A2} = -\frac{j}{\sqrt{3}}I_{a2}$$
$$I_{a0} = 0$$
$$I_{a1} = -j\sqrt{3}I_{A1}$$
$$I_{a2} = +j\sqrt{3}I_{A2}$$

3. *Star-delta Voltage Transformations:*

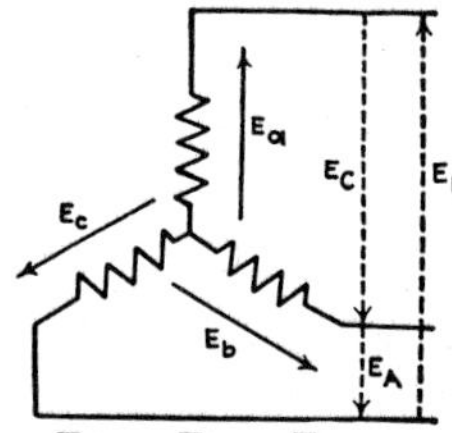

$$E_A = E_c - E_b$$
$$E_B = E_a - E_c$$
$$E_C = E_b - E_a$$

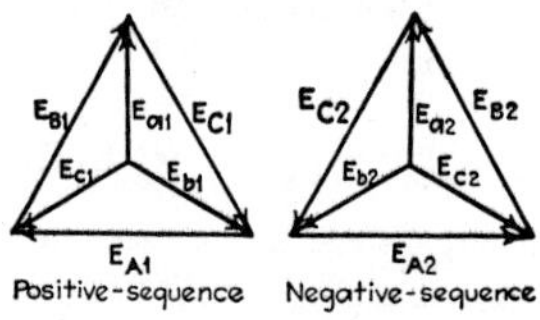

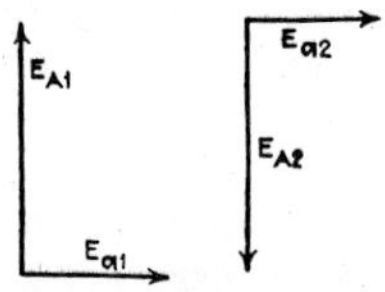

$$E_{A0} = 0$$
$$E_{A1} = j\sqrt{3}E_{a1}$$
$$E_{A2} = -j\sqrt{3}E_{a2}$$
$$E_{a0} = \text{indeterminate}$$
$$E_{a1} = -\frac{j}{\sqrt{3}}E_{A1}$$
$$E_{a2} = +\frac{j}{\sqrt{3}}E_{A2}$$

4. *Connections of Sequence Networks for Various Types of Faults:*

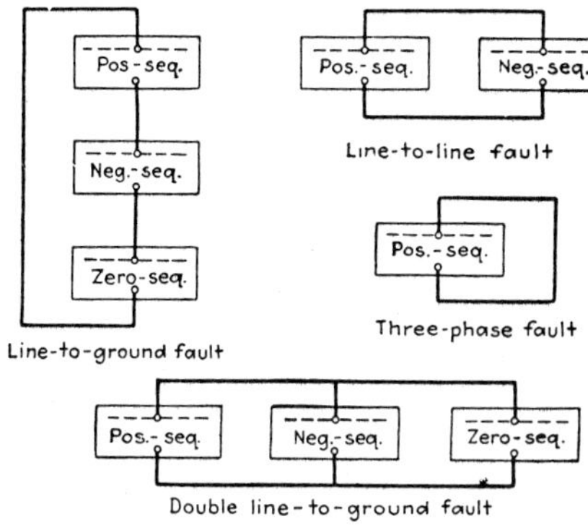

5. *Properties of the Vector a:*

$$1 = \epsilon^{j0} = 1 + j0.0$$
$$a = \epsilon^{j120} = -0.5 + j0.866$$
$$a^2 = \epsilon^{j240} = -0.5 - j0.866$$
$$a^3 = \epsilon^{j360} = \epsilon^{j0} = 1 + j0.0$$
$$1 + a + a^2 = 0$$
$$a - a^2 = \sqrt{3}\epsilon^{j90} = j\sqrt{3}$$
$$a^2 - 1 = \sqrt{3}\epsilon^{j210} = ja\sqrt{3}$$
$$1 - a = \sqrt{3}\epsilon^{j330} = ja^2\sqrt{3}$$
$$a^4 = a$$
$$a^5 = a^2$$
$$-a = \epsilon^{j300} = 0.5 - j0.866$$
$$-a^2 = \epsilon^{j60} = 0.5 + j0.866$$
$$a^2 - a = \sqrt{3}\epsilon^{j270} = -j\sqrt{3}$$
$$1 - a^2 = \sqrt{3}\epsilon^{j30} = -ja\sqrt{3}$$
$$a - 1 = \sqrt{3}\epsilon^{j150} = -ja^2\sqrt{3}$$

III. APPARATUS CONSTANTS

1. *Conversion Formulas:*

$$X_{ohms\ per\ phase} = \frac{X_{per\ unit}E^2_{kv.}}{\text{Mva.}} \text{ or } \frac{X_{per\ cent}E^2_{kv.}(10)}{\text{Kva.}}$$

$$X_{per\ unit} = \frac{X_{ohms\ per\ phase}\text{Mva.}}{E^2_{kv.}}$$

NOTE. These formulas are for three-phase circuits and when the line-to-line voltage is used for E, Mva. and Kva. should express total values, but when line-to-neutral values are used for E, Mva. and Kva. should express phase values.

2. *Constants of Typical Three-phase Synchronous Machines:*

Type of machine	X_d	X_d'	X_d''	X_2	X_0	Time constants
Turbine generators	110	19	12	12	3	See Fig. 55
Salient-pole motors and generators (with damper windings)	110	33	22	22	6	See Fig. 56
Waterwheel generators (without damper windings)	110	35	30	50	7	See Fig. 56
Condensers	180	37	25	24	8	See Fig. 56

3. *Constants of Typical Transformers:*

	Reactance, Per Cent
Distribution	3
Network	5
Power	
Up to 66 kv	5– 7
88 and 110 kv	6– 9
132 and 154 kv	8–10
187 and 220 kv	10–14

The direct-current resistance varies from 0.35 to 0.50 per cent.

4. *Three-winding Transformers:*

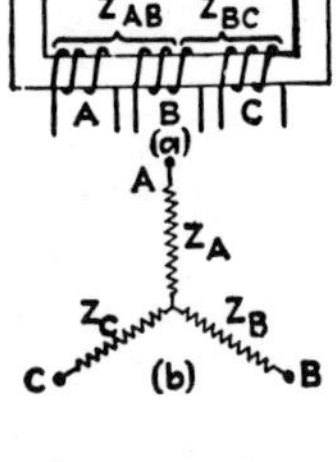

$$Z_{AB} = Z_A + Z_B$$
$$Z_{BC} = Z_B + Z_C$$
$$Z_{CA} = Z_C + Z_A$$
$$Z_A = \tfrac{1}{2}(Z_{AB} + Z_{AC} - Z_{BC})$$
$$Z_B = \tfrac{1}{2}(Z_{AB} + Z_{BC} - Z_{AC})$$
$$Z_C = \tfrac{1}{2}(Z_{CA} + Z_{BC} - Z_{AB})$$

IV. TRANSMISSION-LINE CONSTANTS

1. *Positive- and Negative-sequence Impedance:*

$$r_1 + jx_1 = r_{one\ conductor} + j0.279 \log_{10} \frac{D}{G.M.R._{conductor}} \text{ ohms per mile per phase at 60 cycles.}$$

in which D = geometric mean of three separations in feet.
$G.M.R._{conductor}$ = geometric mean radius of one conductor in feet (see Table VI).

Ground wires have negligible effect.

2. *Zero-sequence Impedance:*

a. Self impedance:

$$r_0 + jx_0 = 3 \text{ (resistance of all conductors in parallel considered as a group)} + 0.286 + j0.838 \log_{10} \frac{280\sqrt{\rho}}{G.M.R.} \text{ ohms per mile per phase at 60 cycles.}$$

in which $G.M.R.$ = geometric mean radius of conductors in feet.
ρ = earth resistivity in meter-ohms.
$\rho_{average}$ = 100.

b. Mutual impedance:

$$r_{m0} + jx_{m0} = 0.286 + j0.838 \log_{10} \frac{280\sqrt{\rho}}{d_{ab}} \text{ ohms per mile per phase at 60 cycles}$$

in which d_{ab} = geometric mean separation between the two groups of conductors in feet.

NOTE. For very wide separations it may be necessary to use the more general relations developed in Chap. VII

c. Ground wire:

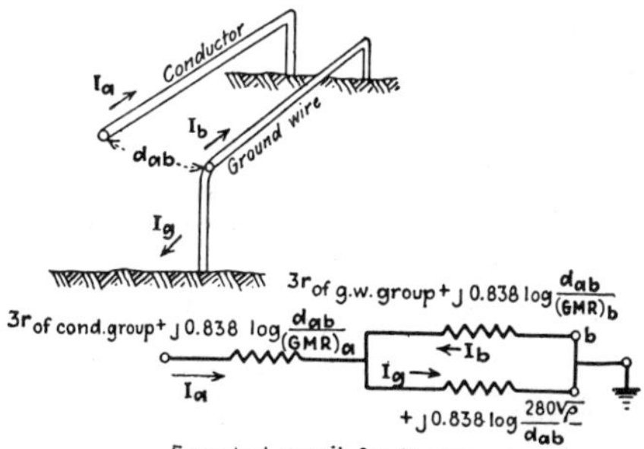

Equivalent circuit for 60 cycle

3. *Equivalent Circuit for Three Parallel Unsymmetrical Lines in Space:*

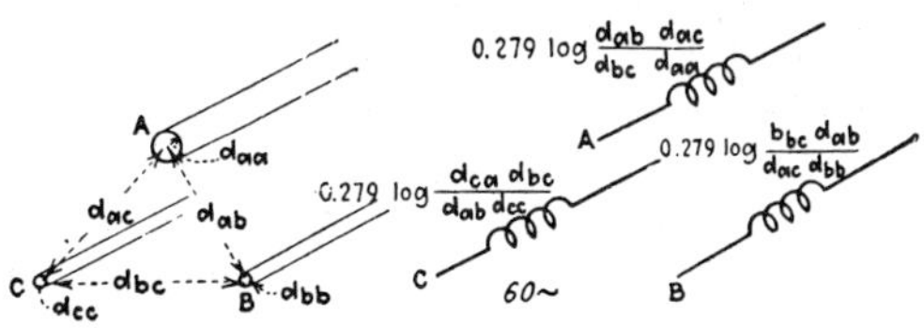

The above constants represent the reactances for 60 cycles. The resistance of each line may be added to each branch.

4. *Positive- and Negative-sequence Capacitive Susceptance:*

$$b_1 = \frac{14.64 \times 10^{-6}}{\log_{10} \dfrac{D}{a}} \text{ mhos per mile per phase at 60 cycles}$$

in which D = equivalent spacing in feet.
a = radius of conductor in feet.

5. *Zero-sequence Capacitive Susceptance:*

$$b_0 = \frac{3.883 \times 10^{-6}}{\log_{10} \dfrac{G.M.D.}{G.M.R.}} \text{ mhos per mile per phase at 60 cycles}$$

in which $G.M.D.$ = 2 (average height of conductors) in feet.
$G.M.R.$ = geometric mean radius of conductors in feet. (In susceptance calculations the $G.M.R.$ of individual conductors is equal to the physical radius.)

6. *Equivalent Circuit of Long Lines:*

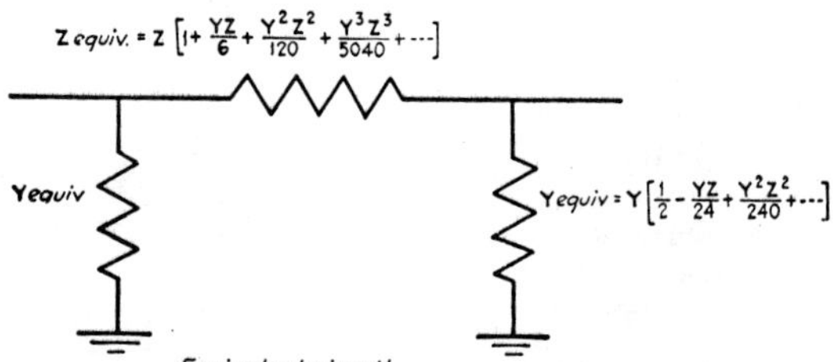

Equivalent circuit

$$Z = [(r_1 + jx_1) \text{ or } (r_0 + jx_0)] \times \text{length in miles}$$
$$Y = [(jb_1) \text{ or } (jb_0)] \times \text{length in miles}$$

TABLE XX.—CONSTANTS OF TYPICAL TRANSMISSION LINES

Voltage	Equivalent spacing, feet	Conductor	Positive- and negative-sequence			
			Reactance per mile		Susceptance, micromhos per mile	Charging, kva. per mile
			Ohms	Per cent based on 100,000 kva.		
33,000	7	00 A.C.S.R.	0.80	7.35	5.69	6.20
66,000	11	00 copper	.823	1.89	5.61	24.5
110,000	15	0000 copper	.833	0.69	5.16	62.4
154,000	19	477,000 A.C.S.R.	.764	0.32	5.40	128
220,000	24	795,000 A.C.S.R.	.767	0.16	5.37	260

Without ground wire or with magnetic ground wire:
X_0 of single circuit = $(2.7 \text{ to } 3.5)X_1$ of single circuit
X_0 of double circuit = $(3.6 \text{ to } 6.0)X_1$ of double circuit

With non-magnetic ground wire:
X_0 of single circuit = $(1.7 \text{ to } 2.7)X_1$ of single circuit
X_0 of double circuit = $(2.0 \text{ to } 4.0)X_1$ of double circuit

V. CABLE CONSTANTS

1. *Single Conductor:*

a. Positive- and negative-sequence resistance.

(1) Conductor (see wire tables such as X).

(2) Sheath currents (solidly bonded).

ΔR = increase in resistance due to sheath currents

$$= \frac{X_m^2 R_s}{X_m^2 + R_s^2} \text{ ohms per mile per phase}$$

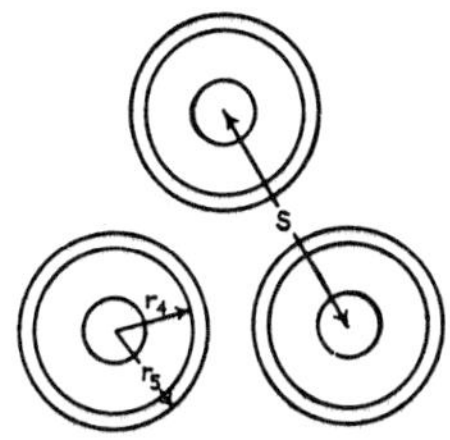

Dimensions in inches

$$X_m = 0.279 \log_{10} \frac{2S}{r_4 + r_5} \text{ ohms per mile per phase at 60 cycles}$$

$$R_s = \frac{0.200}{(r_5 + r_4)(r_5 - r_4)} \text{ ohms per mile per phase for lead at 50°C.}$$

b. Positive- and negative-sequence reactance.

(1) Without sheath currents.

$$X_1 = 0.279 \log_{10} \frac{S}{(G.M.R)_{conductor}} \text{ ohms per mile at 60 cycles}$$

(2) With sheath currents (solidly bonded).

$$X_1 = 0.279 \log_{10} \frac{S}{(G.M.R)_{conductor}} - \frac{X_m^3}{X_m^2 + R_s^2} \text{ ohms per mile at 60 cycles}$$

c. Positive- and negative- and zero-sequence shunt capacitive reactance.

(See Fig. 108, Chap. X.)

d. Zero-sequence impedance

(1) Solidly bonded sheaths.

Equivalent circuit similar to that shown in Fig. 105; however, ground circuit may usually be neglected, in which case $Z_0 = Z_1$

(2) Sheaths insulated.

(See Fig. 82 or Appendix IV-2.)

2. *Three Conductor:*

Positive-, negative- and zero-sequence resistance, reactance, and shunt capacitive reactance.

(See Table X, Chap. X.)

3. *Type H:*

Positive-, negative-, and zero-sequence resistance and reactance, same as three-conductor cable.

(See Table X, Chap. X.)

Positive-, negative-, and zero-sequence shunt capacitive reactance, same as single-conductor cable with sheath diameter equal to shield diameter.

(See Fig. 108, Chap. X.)

TABLE XXI.—RECOMMENDED THICKNESS OF INSULATION[1] FOR SINGLE-CONDUCTOR CABLE AND THREE-CONDUCTOR TYPE H CABLE

A partial list from D. M. Simmons, "Calculation of the Electrical Problems of Underground Cables," *Elec. Jour.*, p. 237, May, 1932.

Rated voltage[2]	Size of conductor A.W.G. number or 1,000 cir. mils	Paper				Varnished cambric[6]				Rubber[7]			
		Grounded neutral		Ungrounded neutral		Grounded neutral		Ungrounded neutral		Grounded neutral		Ungrounded neutral	
		64ths	Mils	64ths	Mils	64ths	Mils	64ths	Mils	64ths	Mils	64ths	Mils
600	1–4/0	5	78	5	78	5	78	5	78	5	78	5	78
	225–500	6	94	6	94	6	94	6	94	6	94	6	94
	525–1,000	7	109	7	109	7	109	7	109	7	109	7	109
	Over 1,000	8	125	8	125	8	125	8	125	8	125	8	125
1,000	1–4/0	5	78	5	78	5	78	5	78	6	94	6	94
	225–500	6	94	6	94	6	94	6	94	7	109	7	109
	525–1,000	7	109	7	109	7	109	7	109	8	125	8	125
	Over 1,000	8	125	8	125	8	125	8	125	9	141	9	141
2,000	1–4/0	6	94	6	94	6	94	6	94	7	109	7	109
	225–500	6	94	6	94	6	94	6	94	8	125	8	125
	525–1,000	7	109	7	109	7	109	7	109	8	125	8	125
	Over 1,000	8	125	8	125	8	125	8	125	9	141	9	141
3,000	7–4/0	6	94	6	94	7	109	7	109	8	125	8	125
	225–500	6	94	6	94	8	125	8	125	9	141	9	141
	525–1,000	7	109	7	109	8	125	8	125	9	141	9	141
	Over 1,000	8	125	8	125	9	141	9	141	10	156	10	156

5,000	14–4/0	7	109	7	109	9	141	10	156	10	156	10	156
	225–1,000	7	109	7	109	10	156	11	172	11	172	11	172
	Over 1,000	8	125	8	125	10	156	11	172	12	188	12	188
7,000	8–1,000	9	141	10	156	11	172	13	203	11	172	14	219
	Over 1,000	9	141	10	156	11	172	13	203	12	188	15	234
10,000[3]	7–1,000	10	156	13	203	14	219	18	281	14	219	18	281
	Over 1,000	10	156	13	203	14	219	18	281	15	234	19	297
15,000	4–1,000	14[1]	219	17	266	19	297	26	406	19[1]	297	27	422
	Over 1,000	14	219	17	266	19	297	26	406	20	313	28	438
23,000	2 and larger	19	297	24	375	27	422						
34,000	2/0 and larger	26	406	33	516								
46,000[4]	4/0 and larger	33	516										
69,000	See Note 5	48	750										

[1] The insulation thicknesses for paper and varnished cambric are I.P.C.E.A. Standards. The rubber walls, though not I.P.C.E.A. Standards, are believed to be representative of modern practice under average conditions.

[2] All cables are rated on conductor to conductor basis for three-phase circuits. All cables have an operating tolerance of 5 per cent above the rated voltage except those rated at 15,000 volts and below, which have no operating tolerance. For intermediate voltages take the wall for the next higher listed voltage.

Paper

[3] For voltages 10,000 to 35,000 the thicknesses given apply to both single-conductor and three-conductor type H cable.

[4] For voltages 40,000 and higher, type H single-conductor cable is recommended.

[5] Above 46,000 volts the ratio of the core diameter to the conductor diameter shall not exceed 3 to 1.

Varnished Cambric

[6] For braided or special designs consult I.P.C.E.A. specifications or manufacturer.

Rubber

[7] These thicknesses also apply to multi-conductor cables. For multi-conductor lead-covered cables for voltages 10,000 and higher, shielding is advisable. For braided or special designs consult the manufacturer.

TABLE XXII.—THICKNESS OF INSULATION FOR THREE-CONDUCTOR BELTED CABLES
Recommended by the Insulated Power Cable Engineers Association Standards (January, 1932)
A partial list from D. M. Simmons, "Calculation of the Electrical Problems of Underground Cables," *Elec. Jour.*, p. 237, May, 1932.

Rated voltage*	Size of conductor, A.W.G. number or 1,000 cir. mils	Paper						Varnished cambric‡					
		Conductor		Belt				Conductor		Belt			
				Grounded neutral		Ungrounded neutral				Grounded neutral		Ungrounded neutral	
		64ths	Mils	64ths	Mils	64ths	Mils	64ths	Mils	64ths	Mils	64ths	Mils
1,000	1–4/0	4	63	2	31	2	31	5	78	0	0	0	0
	225–500	5	78	3	47	3	47	6	94	0	0	0	0
	525–1,000	5	78	3	47	3	47	6	94	2	31	2	31
	Over 1,000	5	78	3	47	3	47	7	109	2	31	2	31
2,000	8–2	6	94	3	47	3	47	5	78	0	0	0	0
	1–500	6	94	3	47	3	47	6	94	0	0	0	0
	525–1,000	6	94	3	47	3	47	6	94	2	31	2	31
	Over 1,000	6	94	3	47	3	47	7	109	2	31	2	31
3,000	8–2	6	94	3	47	3	47	5	78	2	31	2	31
	1–500	6	94	3	47	3	47	6	94	2	31	2	31
	525–1,000	6	94	3	47	3	47	6	94	3	47	3	47
	Over 1,000	6	94	3	47	3	47	7	109	3	47	3	47
5,000	8–4/0	7	109	4	63	4	63	6	94	4	63	4	63
	225–1,000	7	109	4	63	4	63	7	109	4	63	4	63
	Over 1,000	7	109	4	63	4	63	7	109	5	78	5	78
7,000	8 and larger	8	125	4	63	8	125	7	109	5	78	6	94
10,000†	6 and larger	9	141	4	63	9	141	9	141	6	94	9	141
15,000†	4 and larger	11	172	5	78	11	172	13	203	7	109	13	203
17,000	4 and larger	..	...	..	..	..	...	14	219	7	109	14	219

* *General.* All cables for three-phase circuits are rated on conductor to conductor basis. All cables have an operating tolerance of 5 per cent except those rated at 15,000 volts and below which have no operating tolerance.
For intermediate voltage take wall for next higher listed voltage.
† *Paper.* For voltages 10,000 to 15,000 type H cable is recommended where applicable; for voltages 16,000 and higher type H cable only is recommended.
‡ *Varnished cambric.* For braided or special designs consult I.P.C.E.A. specifications or manufacturer.
Rubber. For rubber multi-conductor lead-covered cables use thicknesses given in Table XXI.

TABLE XXIII.—THICKNESS OF LEAD SHEATH

Recommended by The Insulated Power Cable Engineers Association

Paper lead power cables

Core diameter under lead, mils	Lead thickness	
	64ths	Mils
0– 400	5	78
401–1,000	6	94
1,001–1,800	7	109
1,801–2,800	8	125
2,801 and over	9	141

Varnished cambric* lead power cables

Core diameter under lead, mils	Lead thickness	
	64ths	Mils
0– 425	3	47
426– 700	4	63
701–1,050	5	78
1,051–1,500	6	94
1,501–2,000	7	109
2,001–3,000	8	125
3,001 and over	9	141

* Usually considered representative for rubber lead power cables also.

VI. EQUIVALENT NETWORKS

1. *Star-delta Transformations:*

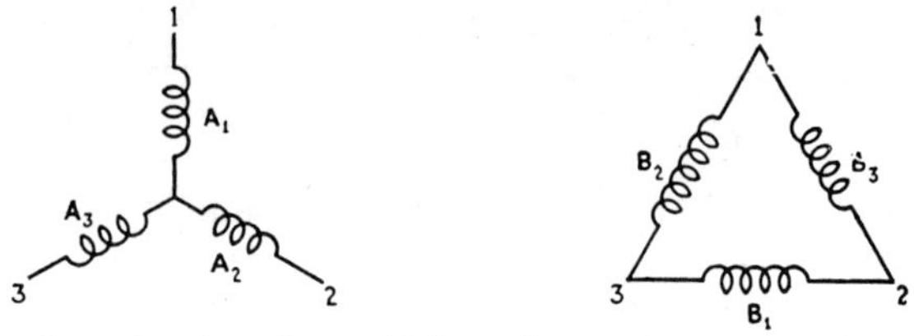

Notation for using *impedances* in formulas

$$A_1 = \frac{B_2B_3}{B_1 + B_2 + B_3}$$
$$A_2 = \frac{B_3B_1}{B_1 + B_2 + B_3}$$
$$A_3 = \frac{B_1B_2}{B_1 + B_2 + B_3}$$
$$B_1 = \frac{A_1A_2 + A_2A_3 + A_3A_1}{A_1} = A_2 + A_3 + \frac{A_2A_3}{A_1}$$
$$B_2 = \frac{A_1A_2 + A_2A_3 + A_3A_1}{A_2} = A_3 + A_1 + \frac{A_3A_1}{A_2}$$
$$B_3 = \frac{A_1A_2 + A_2A_3 + A_3A_1}{A_3} = A_1 + A_2 + \frac{A_1A_2}{A_3}$$

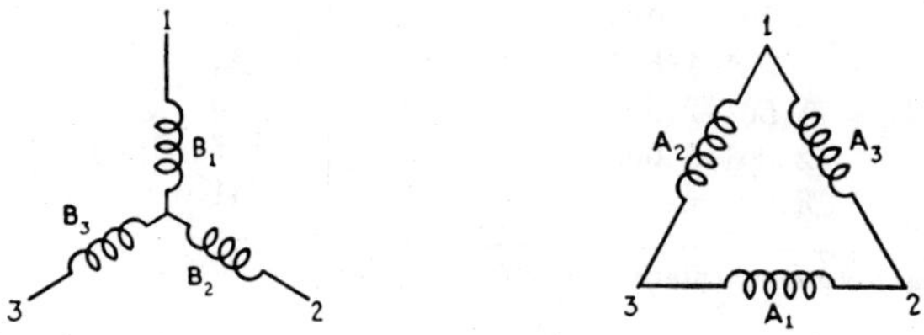

Notation for using *admittances* in formulas

2. *Elimination of Star Point:*

Express network in terms of admittances, as

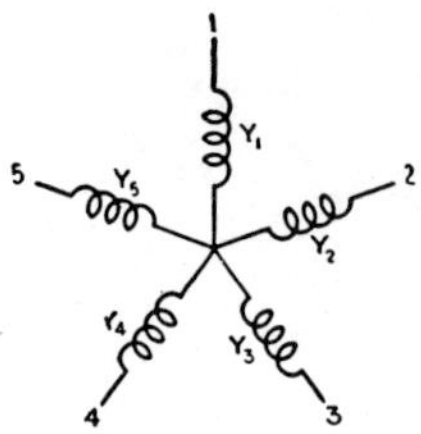

The equivalent delta without the junction point is

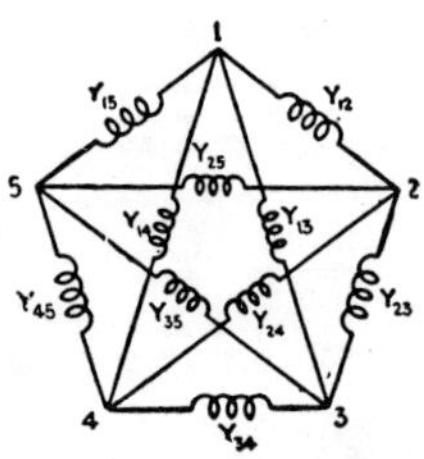

$$\text{in which } Y_{12} = \frac{Y_1 Y_2}{Y_1 + Y_2 + Y_3 + Y_4 + Y_5}$$

$$Y_{23} = \frac{Y_2 Y_3}{Y_1 + Y_2 + Y_3 + Y_4 + Y_5}, \text{ etc.,}$$

These relations may be generalized to include any number of terminals.

3. *Equivalent Network for Two Mutually Coupled Circuits:*

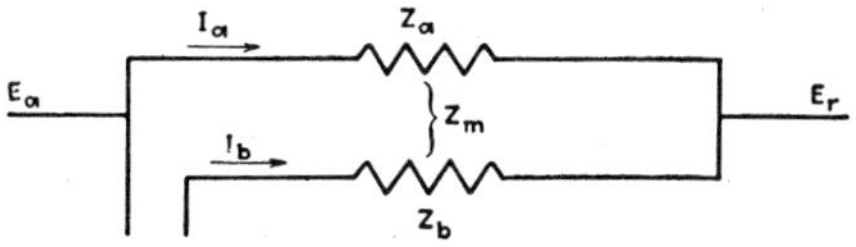

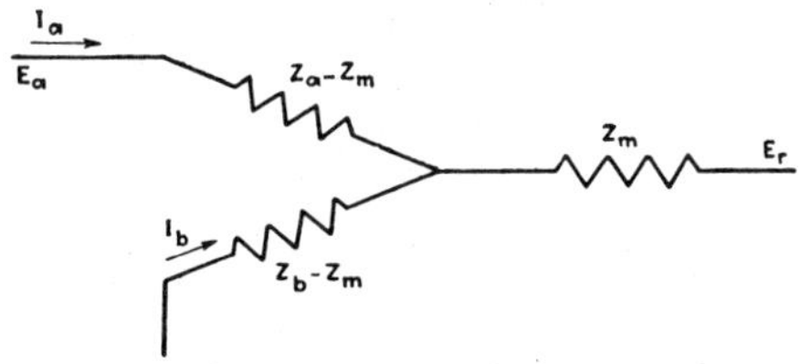

4. *Equivalent Network for Three Branches Connected in Star and Mutually Coupled:*

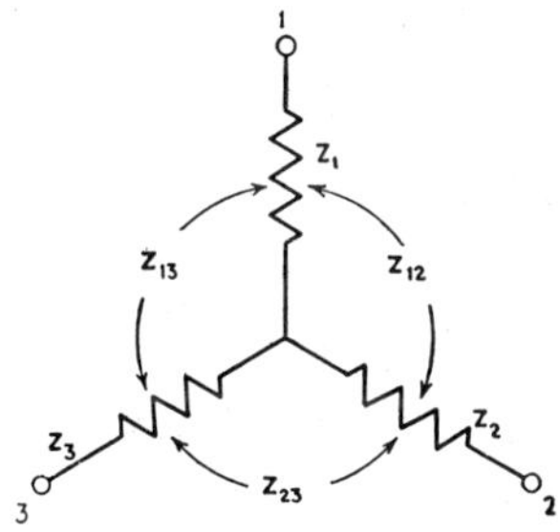

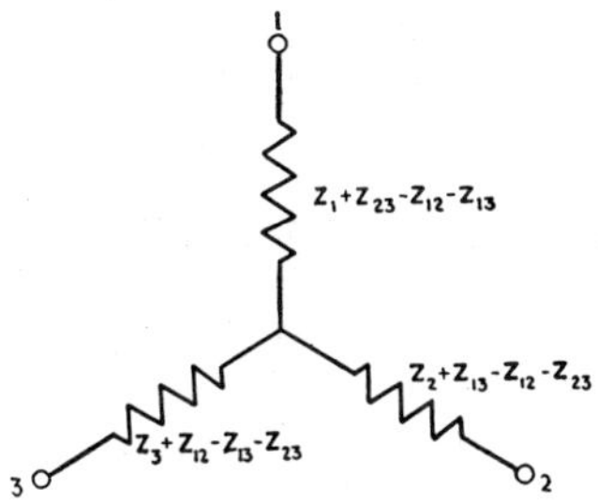

NOTE. The signs of the mutual impedances may be ± depending upon the physical arrangement.

TABLE XXIV.—GENERAL CIRCUIT CONSTANTS FOR DIFFERENT TYPES OF NETWORKS

Network number	Type of network		Equations for general circuit constants in terms of constants of component networks			
			$A =$	$B =$	$C =$	$D =$
1	Series impedance	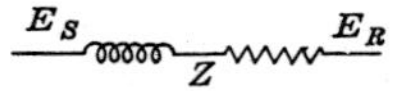	1	Z	0	1
2	Shunt admittance	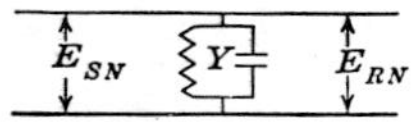	1	0	Y	1
3	Transformer	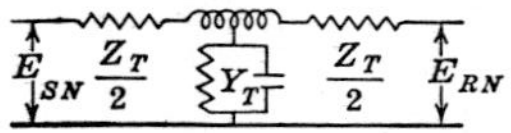	$1 + \frac{Z_T Y_T}{2}$	$Z_T\left(1 + \frac{Z_T Y_T}{4}\right)$	Y_T	$1 + \frac{Z_T Y_T}{2}$
4	Transmission line	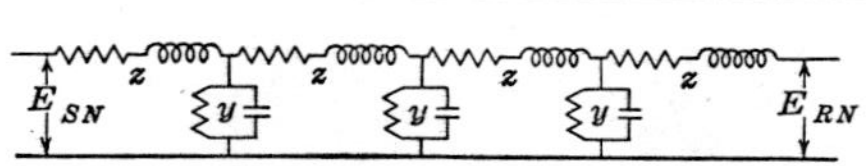	$\text{Cosh}\ \sqrt{ZY} = \left(1 + \frac{ZY}{2} + \frac{Z^2Y^2}{24} + \cdots\right)$	$\sqrt{Z/Y}\ \text{Sinh}\ \sqrt{ZY} = Z\left(1 + \frac{ZY}{6} + \frac{Z^2Y^2}{120} + \cdots\right)$	$\sqrt{Y/Z}\ \text{Sinh}\ \sqrt{ZY} = Y\left(1 + \frac{ZY}{6} + \frac{Z^2Y^2}{120} + \cdots\right)$	Same as A
5	General network	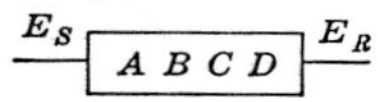	A	B	C	D
6	General network and transformer impedance at receiving end	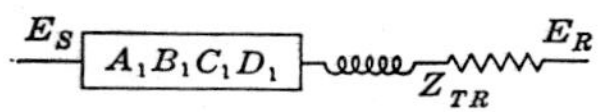	A_1	$B_1 + A_1 Z_{TR}$	C_1	$D_1 + C_1 Z_{TR}$
7	General network and transformer impedance at sending end	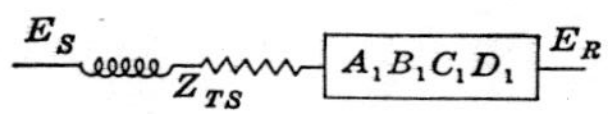	$A_1 + C_1 Z_{TS}$	$B_1 + D_1 Z_{TS}$	C_1	D_1
8	General network and transformer impedance at both ends—referred to high voltage	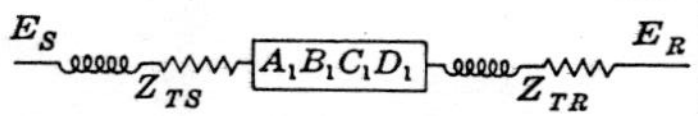	$A_1 + C_1 Z_{TS}$	$B_1 + A_1 Z_{TR} + D_1 Z_{TS} + C_1 Z_{TR} Z_{TS}$	C_1	$D_1 + C_1 Z_{TR}$

9	General network and transformer impedance at both ends—transformers having different ratios T_R and T_S referred to low voltage	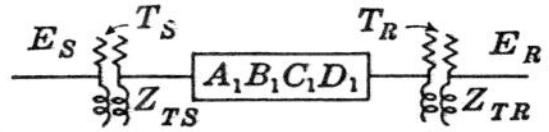	$\frac{T_R}{T_S}(A_1 + C_1Z_{TS})$	$\frac{1}{T_RT_S}(B_1 + A_1Z_{TR} + D_1Z_{TS} + C_1Z_{TR}Z_{TS})$	$C_1T_RT_S$	$\frac{T_S}{T_R}(D_1 + C_1Z_{TR})$
10	General network and shunt admittance at receiving end	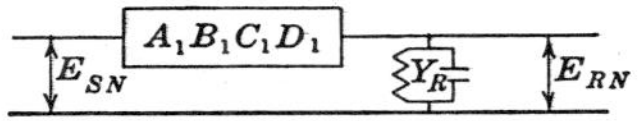	$A_1 + B_1Y_R$	B_1	$C_1 + D_1Y_R$	D_1
11	General network and shunt admittance at sending end	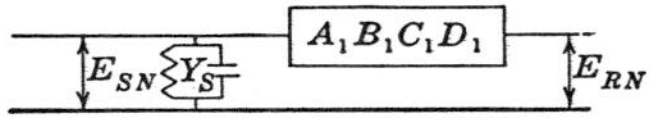	A_1	B_1	$C_1 + A_1Y_S$	$D_1 + B_1Y_S$
12	General network and shunt admittance at both ends	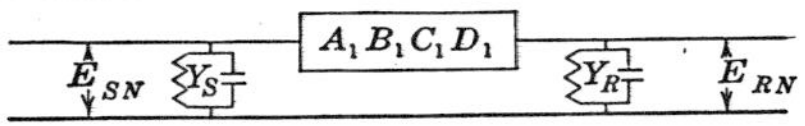	$A_1 + B_1Y_R$	B_1	$C_1 + A_1Y_S + D_1Y_R + B_1Y_RY_S$	$D_1 + B_1Y_S$
13	Two general networks in series	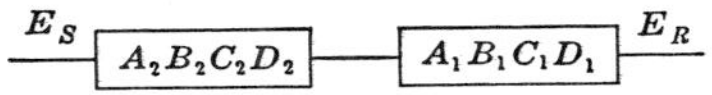	$A_1A_2 + C_1B_2$	$B_1A_2 + D_1B_2$	$A_1C_2 + C_1D_2$	$B_1C_2 + D_1D_2$
14	Two general networks in series with intermediate impedance	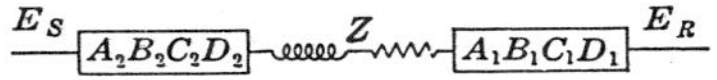	$A_1A_2 + C_1B_2 + C_1A_2Z$	$B_1A_2 + D_1B_2 + D_1A_2Z$	$A_1C_2 + C_1D_2 + C_1C_2Z$	$B_1C_2 + D_1D_2 + D_1C_2Z$
15	Two general networks in series with intermediate shunt admittance	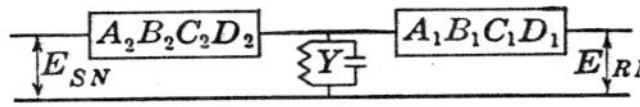	$A_1A_2 + C_1B_2 + A_1B_2Y$	$B_1A_2 + D_1B_2 + B_1B_2Y$	$A_1C_2 + C_1D_2 + A_1D_2Y$	$B_1C_2 + D_1D_2 + B_1D_2Y$
16	Three general networks in series	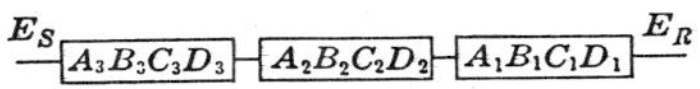	$A_3(A_1A_2 + C_1B_2) + B_3(A_1C_2 + C_1D_2)$	$A_3(B_1A_2 + D_1B_2) + B_3(B_1C_2 + D_1D_2)$	$C_3(A_1A_2 + C_1B_2) + D_3(A_1C_2 + C_1D_2)$	$C_3(B_1A_2 + D_1B_2) + D_3(B_1C_2 + D_1D_2)$
17	Two general networks in parallel	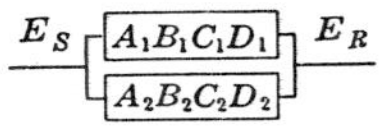	$\frac{A_1B_2 + B_1A_2}{B_1 + B_2}$	$\frac{B_1B_2}{B_1 + B_2}$	$C_1 + C_2 + \frac{(A_1 - A_2)(D_2 - D_1)}{B_1 + B_2}$	$\frac{B_1D_2 + D_1B_2}{B_1 + B_2}$

NOTE. The exciting current of the receiving end transformers should be added vectorially to the load current, and the exciting current of the sending end transformers should be added vectorially to the sending end current.

General equations: $E_S = E_RA + I_RB$; $E_R = E_SD - I_SB$; $I_S = I_RD + E_RC$; $I_R = I_SA - E_SC$. As a check in the numerical calculation of the A, B, C, and D constants note that in all cases $AD - BC = 1$.

$P_1 + jQ_1 = \frac{\hat{A}}{\hat{B}}\bar{E}_1^2 - \frac{1}{\hat{B}}E_1\hat{E}_2$	$= \hat{Y}_{11}\bar{E}_1^2 - \hat{Y}_{12}E_1\hat{E}_2$	$= Z_{11}\bar{I}_1^2 - Z_{12}\hat{I}_1 I_2$
$P_2 + jQ_2 = \frac{\hat{C}}{\hat{B}}\bar{E}_2^2 - \frac{1}{\hat{B}}\hat{E}_1E_2$	$= \hat{Y}_{22}\bar{E}_2^2 - \hat{Y}_{12}\hat{E}_1E_2$	$= Z_{22}\bar{I}_2^2 - Z_{12}I_1\hat{I}_2$

NOTE 1. P_1 and P_2 are positive in all cases for power flowing *into* the network from the point considered.
NOTE 2. P and Q of same sign indicates lagging power factor; that is $P + jQ = E\hat{I}$.

TABLE XXV.—NETWORK CONVERSION FORMULAS

To		To convert from				
		ABCD	Admittance	Impedance	Equivalent π	Equivalent T
ABCD	$A =$	*ABCD* constants 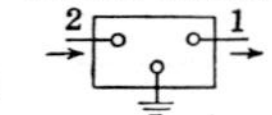 $E_2 = AE_1 + BI_1$ $I_2 = CE_1 + DI_1$ $E_1 = DE_2 - BI_2$ $I_1 = -CE_2 + AI_2$	$\frac{Y_{11}}{Y_{12}}$	$-\frac{Z_{22}}{Z_{12}}$	$1 + ZY_R$	$1 + Z_SY$
	$B =$		$\frac{1}{Y_{12}}$	$-\frac{Z_{11}Z_{22} - Z_{12}^2}{Z_{12}}$	Z	$Z_R + Z_S + YZ_RZ_S$
	$C =$		$\frac{Y_{11}Y_{22} - Y_{12}^2}{Y_{12}}$	$-\frac{1}{Z_{12}}$	$Y_R + Y_S + ZY_RY_S$	Y
	$D =$		$\frac{Y_{22}}{Y_{12}}$	$-\frac{Z_{11}}{Z_{12}}$	$1 + ZY_S$	$1 + Z_RY$
Admittance	$Y_{11} =$	$\frac{A}{B}$	Admittance constants 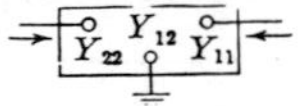	$\frac{Z_{22}}{Z_{11}Z_{22} - Z_{12}^2}$	$Y_R + \frac{1}{Z}$	$\frac{1 + Z_SY}{Z_R + Z_S + YZ_RZ_S}$
	$Y_{12} =$	$\frac{1}{B}$		$-\frac{Z_{12}}{Z_{11}Z_{22} - Z_{12}^2}$	$\frac{1}{Z}$	$\frac{1}{Z_R + Z_S + YZ_RZ_S}$
	$Y_{22} =$	$\frac{D}{B}$	$I_1 = Y_{11}E_1 - Y_{12}E_2$ $I_2 = Y_{22}E_2 - Y_{12}E_1$	$\frac{Z_{11}}{Z_{11}Z_{22} - Z_{12}^2}$	$Y_S + \frac{1}{Z}$	$\frac{1 + YZ_R}{Z_R + Z_S + YZ_RZ_S}$
Impedance	$Z_{11} =$	$\frac{D}{C}$	$\frac{Y_{22}}{Y_{11}Y_{22} - Y_{12}^2}$	Impedance constants 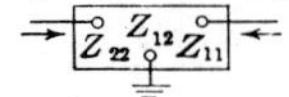	$\frac{1 + ZY_S}{Y_R + Y_S + ZY_RY_S}$	$Z_R + \frac{1}{Y}$
	$Z_{12} =$	$-\frac{1}{C}$	$-\frac{Y_{12}}{Y_{11}Y_{22} - Y_{12}^2}$		$-\frac{1}{Y_R + Y_S + ZY_RY_S}$	$-\frac{1}{Y}$
	$Z_{22} =$	$\frac{A}{C}$	$\frac{Y_{11}}{Y_{11}Y_{22} - Y_{12}^2}$	$E_1 = Z_{11}I_1 - Z_{12}I_2$ $E_2 = Z_{22}I_2 - Z_{12}I_1$	$\frac{1 + ZY_R}{Y_R + Y_S + ZY_RY_S}$	$Z_S + \frac{1}{Y}$
Equivalent π	$Y_R =$	$\frac{A - 1}{B}$	$Y_{11} - Y_{12}$	$\frac{Z_{22} + Z_{12}}{Z_{11}Z_{22} - Z_{12}^2}$	Equivalent π 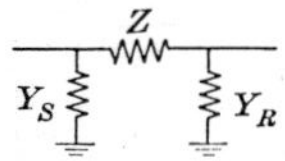	$\frac{YZ_S}{Z_R + Z_S + YZ_RZ_S}$
	$Z =$	B	$\frac{1}{Y_{12}}$	$-\frac{Z_{11}Z_{22} - Z_{12}^2}{Z_{12}}$		$Z_R + Z_S + YZ_RZ_S$
	$Y_S =$	$\frac{D - 1}{B}$	$Y_{22} - Y_{12}$	$\frac{Z_{11} + Z_{12}}{Z_{11}Z_{22} - Z_{12}^2}$		$\frac{YZ_R}{Z_R + Z_S + YZ_RZ_S}$
Equivalent T	$Z_R =$	$\frac{D - 1}{C}$	$\frac{Y_{22} - Y_{12}}{Y_{11}Y_{22} - Y_{12}^2}$	$Z_{11} + Z_{12}$	$\frac{ZY_S}{Y_R + Y_S + ZY_RY_S}$	Equivalent T 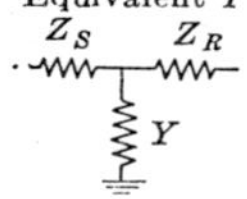
	$Y =$	C	$\frac{Y_{11}Y_{22} - Y_{12}^2}{Y_{12}}$	$-\frac{1}{Z_{12}}$	$Y_R + Y_S + ZY_RY_S$	
	$Z_S =$	$\frac{A - 1}{C}$	$\frac{Y_{11} - Y_{12}}{Y_{11}Y_{22} - Y_{12}^2}$	$Z_{22} + Z_{12}$	$\frac{ZY_R}{Y_R + Y_S + ZY_RY_S}$	

VII. CHARACTERISTICS OF CONDUCTORS

With the development of the method of symmetrical components, it became advantageous to prepare a new type of table of the electrical characteristics of transmission and distribution circuit conductors in order to give the *zero-sequence* characteristics as well as the ordinary or positive-sequence characteristics.

Another innovation is the introduction of the characteristics of the recently developed hollow types of conductors. These tables give the characteristics of four types as follows: (1) stranded copper conductors, (2) aluminum (A.C.S.R.) conductors, (3) Anaconda hollow conductors, and (4) General Cable Type HH hollow conductors.

In preparing the new tables the authors have attempted to simplify previous tables. They are indebted to W. A. Lewis for the suggestion of splitting the reactances into components associated with the conductor, the conductor spacing, and the equivalent depth of earth return. After consideration of loading and ambient temperatures and a discussion with the engineers of the wire companies, the single value of 50°C. for conductor temperature was agreed upon as representing average operation and as being the most satisfactory value for calculations required in power-system studies. In addition, a column has been added to give the approximate maximum permissible current-carrying capacity. The calculations are based on tarnished conductor surface and air velocity of 2 ft. per second and a conductor temperature of 75°C. with 25°C. ambient, using the method described by Schurig and Frick (*Gen. Elec. Rev.*, March, 1930).

The conductors included in the tables were selected after discussion with wire-company engineers. Advantage has been taken of a number of refinements in the calculation of inductance of stranded conductors which were recently developed by W. A. Lewis and which will be described by him in the near future. These tables, within their scope, are intended to supersede previous tables of electrical characteristics of conductors.

The tables were prepared in cooperation with W. A. Lewis of the Westinghouse Electric and Manufacturing Company with the assistance of the engineers of the wire companies, particularly F. R. Dallye of the Aluminum Company of America, R. B. Steinmetz of the Anaconda Wire and Cable Company, and D. M. Simmons of the General Cable Corporation.

Illustration. Determine the positive- and zero-sequence impedances of the following line at 60 cycles, no ground wires:

Conductor: 795,000 cir. mils A.C.S.R. 54 aluminum strands.

Effective separation: 26 ft. Earth resistivity: 100 meter-ohms.

From Table XXVI, p. 420: $r_a = 0.138$ ohms per phase per mile;
$x_a = 0.401$ ohms per phase per mile.

From p. 421 (or p. 423): $x_d = 0.395$ ohms per phase per mile.

From p. 420 (or p. 422): $r_e = 0.286$ ohms per phase per mile;
$x_e = 2.89$ ohms per phase per mile.

Positive-sequence impedance: $z_1 = r_a + j(x_a + x_d) = 0.138 + j0.796$ ohms per phase per mile.

Zero-sequence impedance: $z_0 = r_a + r_e + j(x_a + x_e - 2x_d) = 0.424 + j2.50$ ohms per phase per mile.

At a given frequency, changing conductor changes only r_a and x_a; changing spacing changes only x_d; changing earth resistivity changes only x_e.

TABLE XXVI.—CHARACTERISTICS OF ALUMINUM CABLE STEEL REINFORCED

Size of conductor, circular mils or A.W.G.	Number of wires		Copper equivalent,* circular mils or A.W.G.	Outside diameter, inches	Geometric mean radius, feet (60 cycles)	Approximate maximum† current carrying capacity at 60 cycles	r_a = resistance, ohms per phase per mile at 50°C. 25°C. rise above 25°C. ambient				x_a = reactance at 1 ft., ohms per phase per mile		
	Aluminum	Steel					D.c.	25 cycles	50 cycles	60 cycles	25 cycles	50 cycles	60 cycles
1,590,000	54	19	1,000,000	1.545	0.0520	1,380	0.0646	0.0656	0.0675	0.0684	0.150	0.299	0.359
1,510,500	54	19	950,000	1.506	.0507	1,340	.0680	.0690	.0710	.0720	.151	.302	.362
1,431,000	54	19	900,000	1.465	.0493	1,300	.0718	.0729	.0749	.0760	.152	.304	.365
1,351,500	54	19	850,000	1.424	.0479	1,250	.0761	.0771	.0792	.0803	.154	.307	.369
1,272,000	54	19	800,000	1.382	.0465	1,200	.0808	.0819	.0840	.0851	.155	.310	.372
1,192,500	54	19	750,000	1.338	.0450	1,160	.0862	.0872	.0894	.0906	.157	.314	.376
1,113,000	54	19	700,000	1.293	.0435	1,110	.0924	.0935	.0957	.0969	.159	.317	.380
1,033,500	54	7	650,000	1.246	.0420	1,060	.0994	.101	.103	.104	.160	.321	.385
954,000	54	7	600,000	1.196	.0403	1,010	.108	.109	.112	.113	.162	.325	.390
900,000	54	7	566,000	1.162	.0391	966	.115	.116	.118	.119	.164	.328	.393
874,500	54	7	550,000	1.146	.0386	949	.118	.119	.122	.123	.165	.329	.395
795,000	54	7	500,000	1.093	.0368	897	.129	.131	.136	.138	.167	.334	.401
795,000	26	7	500,000	1.108	.0375	901	.129	.129	.129	.129	.166	.332	.399
795,000	30	19	500,000	1.140	.0393	909	.129	.129	.129	.129	.164	.327	.393
715,500	54	7	450,000	1.036	.0349	834	.144	.145	.147	.148	.170	.339	.407
715,500	26	7	450,000	1.051	.0355	838	.144	.144	.144	.144	.169	.337	.405
715,500	30	19	450,000	1.081	.0372	845	.144	.144	.144	.144	.166	.333	.399
666,600	54	7	419,000	1.000	.0337	800	.154	.157	.159	.160	.172	.343	.412
636,000	54	7	400,000	0.977	.0329	774	.162	.164	.168	.169	.173	.345	.414
636,000	26	7	400,000	0.990	.0335	777	.162	.162	.162	.162	.172	.344	.412
636,000	30	19	400,000	1.019	.0351	781	.162	.162	.162	.162	.169	.339	.406
605,000	54	7	380,500	0.953	.0321	748	.170	.172	.176	.178	.174	.348	.417
556,500	26	7	350,000	.927	.0313	734	.185	.186	.186	.186	.175	.350	.420
556,500	30	7	350,000	.953	.0328	726	.185	.186	.186	.186	.173	.346	.415
500,000	30	7	314,500	.904	.0311	692	.206	.206	.206	.206	.175	.351	.421
477,000	26	7	300,000	.858	.0290	666	.216	.216	.216	.216	.179	.358	.430
477,000	30	7	300,000	.883	.0304	671	.216	.216	.216	.216	.177	.353	.424
397,500	26	7	250,000	.783	.0265	591	.259	.259	.259	.259	.184	.367	.441
397,500	30	7	250,000	.806	.0278	596	.259	.259	.259	.259	.181	.362	.435
336,400	26	7	0000	.721	.0244	530	.306	.306	.306	.306	.188	.376	.451
336,400	30	7	0000	.741	.0255	535	.306	.306	.306	.306	.186	.371	.445
300,000	26	7	188,700	.680	.0230	493	.342	.342	.342	.342	.191	.382	.458
300,000	30	7	188,700	.700	.0241	497	.342	.342	.342	.342	.188	.377	.452
266,800	26	7	000	.642	.0217	457	.385	.385	.385	.385	.194	.387	.465
0000	6	1	00	.563	.00814	340	.485	.514	.567	.592	.242	.484	.581
000	6	1	0	.502	.00600	303	.612	.642	.697	.723	.259	.517	.621
00	6	1	1	.447	.00510	266	.773	.806	.866	.895	.267	.534	.641
0	6	1	2	.398	.00446	233	.974	1.01	1.08	1.12	.273	.547	.656
1	6	1	3	.355	.00418	199	1.23	1.27	1.34	1.38	.277	.554	.665
2	6	1	4	.316	.00418	179	1.55	1.59	1.66	1.69	.277	.554	.665
3	6	1	5	.281	.00430	157	1.95	1.98	2.04	2.07	.275	.551	.661
4	6	1	6	.250	.00437	137	2.47	2.50	2.54	2.57	.274	.549	.659
5	6	1	7	.223	.00416	118	3.10	3.12	3.16	3.18	.279	.557	.665
6	6	1	8	.198	.00394	102	3.92	3.94	3.97	3.98	.281	.561	.673

* Based upon copper, 97 per cent; aluminum, 61 per cent.
† Based on a conductor temperature of 75°C. and an ambient of 25°C.

	r_e*	x_e,* ohms per phase per mile								
ρ, meter-ohms	All	1	5	10	50	**100†**	500	1,000	5,000	10,000
25 cycles	0.119	0.923	1.04	1.10	1.22	**1.27**	1.39	1.44	1.57	1.62
50 cycles	.239	1.74	1.98	2.09	2.33	**2.44**	2.68	2.78	3.03	3.13
60 cycles	286	2.05	2.35	2.47	2.77	**2.89**	3.19	3.31	3.61	3.73

* Based on Eq. (163).
† This represents an average value which can be used in the absence of more definite information.

TABLE XXVI (*Continued*).—CHARACTERISTICS OF COPPER CABLES

Hard drawn 97.3 per cent conductivity

Size of conductor: Circular mils	Size of conductor: A.W.G. or B. & S.	Number of wires	Outside diameter, inches	Geometric mean radius, feet	Approximate maximum* current-carrying capacity at 60 cycles	r_a = resistance, ohms per phase per mile at 50°C. 25°C. rise above 25°C. ambient: D.c.	r_a: 25 cycles	r_a: 50 cycles	r_a: 60 cycles	x_a = reactance at 1 ft., ohms per phase per mile: 25 cycles	x_a: 50 cycles	x_a: 60 cycles
1,000,000	....	61	1.152	0.0370	1,300	0.0640	0.0648	0.0672	0.0685	0.167	0.333	0.400
900,000	....	61	1.093	.0352	1,220	.0711	.0718	.0740	.0752	.169	.339	.406
800,000	....	61	1.031	.0332	1,130	.0800	.0806	.0826	.0837	.172	.344	.413
750,000	....	61	0.998	.0321	1,090	.0853	.0860	.0878	.0888	.174	.348	.417
700,000	....	61	.964	.0311	1,040	.0914	.0920	.0937	.0947	.176	.351	.421
600,000	....	37	.891	.0285	945	.107	.107	.109	.109	.180	.360	.432
500,000	....	37	.814	.0260	842	.128	.128	.130	.130	.184	.369	.443
400,000	....	19	.725	.0229	730	.160	.160	.161	.162	.191	.382	.458
300,000	....	19	.628	.0198	607	.213	.214	.214	.215	.198	.397	.476
250,000	....	19	.574	.0181	540	.256	.256	.257	.257	.203	.406	.487
211,600	0000	19	.528	.0167	485	.302	.303	.303	.303	.207	.414	.497
211,600	0000	7	.522	.0158	484	.302	.303	.303	.303	.210	.420	.504
167,806	000	7	.464	.0140	416	.381	.381	.382	.382	.216	.432	.518
133,077	00	7	.414	.0124	359	.481	.481	.481	.481	.222	.443	.532
105,535	0	7	.368	.0111	309	.606	.606	.607	.607	.227	.455	.546
83,693	1	7	.328	.0099	266	.765	.765	.765	.765	.233	.467	.560
66,371	2	7	.292	.0088	230	.964	.964	.964	.964	.239	.478	.574
52,635	3	7	.260	.0079	198	1.22	1.22	1.22	1.22	.245	.490	.588
41,741	4	7	.232	.0070	171	1.53	1.53	1.53	1.53	.251	.502	.602
33,102	5	7	.206	.0062	147	1.93	1.93	1.93	1.93	.257	.514	.616
26,251	6	7	.184	.0056	127	2.44	2.44	2.44	2.44	.263	.526	.630

* Based on a conductor temperature of 75°C. and an ambient of 25°C.

x_d, ohms per phase per mile

Separation, feet		0	1	2	3	4	5	6	7	8	9
25 cycles	0	.....	0	0.035	0.056	0.070	0.081	0.091	0.098	0.105	0.111
	10	0.116	0.121	.126	.130	.133	.137	.140	.143	.146	.149
	20	.151	.154	.156	.159	.161	.163	.165	.167	.168	.170
	30	.172	.174	.175	.177	.178	.180	.181	.183	.184	.185
50 cycles	0	.....	0	.070	.111	.140	.163	.181	.197	.210	.222
	10	.233	.242	.251	.259	.267	.274	.280	.286	.292	.298
	20	.303	.308	.313	.317	.321	.325	.329	.333	.337	.340
	30	.344	.347	.350	.354	.357	.359	.362	.365	.368	.370
60 cycles	0	.....	0	.084	.133	.168	.195	.217	.236	.252	.267
	10	.279	.291	.302	.311	.320	.329	.336	.344	.351	.357
	20	.364	.369	.375	.380	.386	.391	.395	.400	.404	.409
	30	.413	.417	.421	.424	.428	.431	.435	.438	.441	.445

Positive- and negative-sequence impedance:

$$z_1 = z_2 = r_a + j(x_a + x_d)$$

Zero-sequence impedance:

$$z_0 = (r_a + r_e) + j(x_a + x_e - 2x_d)$$

TABLE XXVII.—CHARACTERISTICS OF ANACONDA HOLLOW COPPER CONDUCTORS

Design No.	Size of conductor, circular mils or A.W.G.	Number of wires	Outside diameter, inches	Geometric mean radius, feet	Approximate maximum* current-carrying capacity at 60 cycles	r_a = resistance, ohms per phase per mile at 50°C. 25°C. rise above 25°C. ambient				x_a = reactance at 1 ft., ohms per phase per mile		
						D.c.	25 cycles	50 cycles	60 cycles	25 cycles	50 cycles	60 cycles
163	800,000	22 + 28	1.269	0.0464	1,230	0.0798	0.0800	0.0804	0.0806	0.155	0.311	0.373
167	800,000	16 + 22	1.185	.0420	1,200	.0798	.0801	.0806	.0810	.160	.321	.385
96	750,000	18 + 24	1.176	.0422	1,160	.0852	.0854	.0858	.0861	.160	.320	.384
362	750,000	12 + 18	1.092	.0376	1,130	.0852	.0855	.0862	.0867	.166	.332	.398
559	700,000	21 + 27	1.171	.0426	1,120	.0912	.0914	.0918	.0920	.160	.319	.383
361	700,000	16 + 20	1.100	.0388	1,100	.0912	.0915	.0920	.0923	.164	.329	.394
560	650,000	19 + 25	1.106	.0399	1,060	.0983	.0984	.0988	.0990	.163	.326	.391
561	650,000	12 + 18	1.015	.0349	1,030	.0983	.0986	.0992	.0996	.170	.339	.407
360	600,000	22 + 28	1.100	.0402	1,020	.106	.107	.107	.107	.163	.325	.390
30	600,000	16 + 22	1.027	.0365	996	.106	.107	.107	.107	.168	.335	.402
563	550,000	24 + 30	1.075	.0396	968	.116	.116	.116	.117	.163	.327	.392
379	550,000	18 + 24	1.008	.0361	949	.116	.116	.117	.117	.168	.336	.403
4-B	500,000	22 + 28	1.002	.0366	903	.128	.128	.128	.128	.167	.334	.401
401	500,000	12	0.995	.0356	901	.128	.128	.128	.128	.169	.337	.405
405	450,000	26	1.180	.0455	902	.142	.142	.142	.142	.156	.312	.375
426	450,000	15	1.000	.0367	857	.142	.142	.142	.142	.167	.334	.401
51	400,000	19	1.005	.0378	810	.160	.160	.160	.160	.166	.331	.397
565	350,000	21	0.968	.0367	749	.182	.183	.183	.183	.167	.334	.401
378	350,000	12 + 18	.742	.0255	690	.182	.183	.183	.183	.185	.371	.445
178	300,000	12	.763	.0273	645	.213	.213	.213	.213	.182	.364	.437
567	300,000	9	.710	.0244	631	.213	.213	.213	.213	.188	.376	.451
37	250,000	16	.754	.0279	587	.255	.256	.256	.256	.181	.362	.434
415	250,000	14	.725	.0264	580	.255	.256	.256	.256	.184	.367	.441
446	0000	21	.755	.0287	540	.302	.302	.302	.302	.180	.359	.431
569	0000	9	.598	.0205	504	.302	.302	.302	.302	.196	.393	.471
158	000	16	.611	.0226	452	.381	.381	.381	.381	.192	.383	.460
570	000	12	.570	.0204	443	.381	.381	.381	.381	.197	.394	.472
10	00	17	.550	.0205	390	.480	.480	.480	.480	.197	.393	.472

* Based on a conductor temperature of 75°C. and an ambient of 25°C.

	r_e*	x_e,* ohms per phase per mile								
ρ, meter-ohms	All	1	5	10	50	**100†**	500	1,000	5,000	10,000
25 cycles	0.119	0.923	1.04	1.10	1.22	**1.27**	1.39	1.44	1.57	1.62
50 cycles	.239	1.74	1.98	2.09	2.33	**2.44**	2.68	2.78	3.03	3.13
60 cycles	.286	2.05	2.35	2.47	2.77	**2.89**	3.19	3.31	3.61	3.73

* Based on Eq. (163).
† This represents an average value which can be used in the absence of more definite information.

TABLE XXVII (*Continued*).—CHARACTERISTICS OF GENERAL CABLE TYPE HH HOLLOW COPPER CONDUCTORS

Size of conductor, circular mils or A.W.G.	Number of segments	Outside diameter, inches*	Wall thickness, inches	Geometric mean radius, feet	Approximate maximum† current-carrying capacity at 60 cycles	r_a = resistance, ohms per phase per mile at 50°C. 25°C. rise above 25°C. ambient				x_a = reactance at 1 ft., ohms per phase per mile		
						D.c.	25 cycles	50 cycles	60 cycles	25 cycles	50 cycles	60 cycles
1,000,000	12	1.923	0.142	0.0763	1,580	0.0633	Same as for d.c.	Same as for d.c.	Same as for d.c.	0.130	0.260	0.312
950,000	12	1.865	.139	.0739	1,520	.0666				.132	.263	.316
900,000	12	1.809	.136	.0717	1,470	.0703				.133	.266	.320
850,000	12	1.745	.133	.0691	1,410	.0745				.135	.270	.324
800,000	12	1.687	.130	.0668	1,350	.0791				.137	.273	.328
750,000	12	1.622	.126	.0641	1,290	.0844				.139	.277	.333
700,000	10	1.554	.123	.0614	1,230	.0904				.141	.282	.338
650,000	10	1.490	.120	.0588	1,170	.0974				.143	.286	.343
600,000	10	1.418	.116	.0560	1,110	.106				.146	.291	.350
550,000	10	1.343	.113	.0529	1,040	.115				.149	.297	.357
500,000	9	1.267	.109	.0498	977	.127				.152	.303	.364
450,000	9	1.188	.105	.0466	909	.141				.155	.310	.372
400,000	8	1.105	.101	.0433	838	.158				.159	.317	.381
350,000	7	1.016	.096	.0397	764	.181				.163	.326	.391
300,000	7	0.920	.091	.0359	686	.211				.168	.336	.404
250,000	7	.818	.086	.0318	604	.253				.174	.349	.418
0000	7	.733	.082	.0283	538	.299				.180	.360	.432
000	7	.629	.077	.0241	458	.377				.188	.376	.452

* The conductor diameters for these areas can be increased by as much as 10 per cent and decreased by as much as 25 per cent if desired.

† Based on a conductor temperature of 75°C. and an ambient of 25°C.

x_d, ohms per phase per mile

Separation, feet		0	1	2	3	4	5	6	7	8	9
25 cycles	0	.....	0	0.035	0.056	0.070	0.081	0.091	0.098	0.105	0.111
	10	0.116	0.121	.126	.130	.133	.137	.140	.143	.146	.149
	20	.151	.154	.156	.159	.161	.163	.165	.167	.168	.170
	30	.172	.174	.175	.177	.178	.180	.181	.183	.184	.185
50 cycles	0	.....	0	.070	.111	.140	.163	.181	.197	.210	.222
	10	.233	.242	.251	.259	.267	.274	.280	.286	.292	.298
	20	.303	.308	.313	.317	.321	.325	.329	.333	.337	.340
	30	.344	.347	.350	.354	.357	.359	.362	.365	.368	.370
60 cycles	0	.....	0	.084	.133	.168	.195	.217	.236	.252	.267
	10	.279	.291	.302	.311	.320	.329	.336	.344	.351	.357
	20	.364	.369	.375	.380	.386	.391	.395	.400	.404	.409
	30	.413	.417	.421	.424	.428	.431	.435	.438	.441	.445

Positive- and negative-sequence impedance:

$$z_1 = z_2 = r_a + j(x_a + x_d)$$

Zero-sequence impedance:

$$z_0 = (r_a + r_e) + j(x_a + x_e - 2x_d)$$

TABLE XXVIII.—SHUNT CAPACITIVE REACTANCE

Aluminum cable steel reinforced

Circular mils or A.W.G. (B. & S.)	Number of aluminum strands	x_a' Shunt capacitive reactance at 1 ft., megohms per phase per mile — 25 cycles	50 cycles	60 cycles
1,590,000	54	0.195	0.0977	0.0814
1,510,000	54	.197	.0986	.0821
1,431,000	54	.199	.0996	.0830
1,351,500	54	.201	.101	.0838
1,272,000	54	.203	.102	.0847
1,192,500	54	.206	.103	.0857
1,113,000	54	.208	.104	.0867
1,033,500	54	.211	.105	.0878
954,000	54	.214	.107	.0890
900,000	54	.216	.108	.0898
874,500	54	.217	.108	.0903
795,000	54	.220	.110	.0917
795,000	26	.219	.110	.0912
795,000	30	.217	.109	.0904
715,500	54	.224	.112	.0932
715,500	26	.223	.111	.0928
715,500	30	.221	.110	.0920
666,600	54	.226	.113	.0943
636,000	54	.228	.114	.0950
636,000	26	.227	.114	.0946
636,000	30	.225	.113	.0937
605,000	54	.230	.115	.0957
556,500	26	.232	.116	.0965
556,500	30	.230	.115	.0957
500,000	30	.234	.117	.0973
477,000	26	.237	.119	.0988
477,000	30	.235	.118	.0980
397,500	26	.244	.122	.102
397,500	30	.242	.121	.101
336,400	26	.250	.125	.104
336,400	30	.248	.124	.103
300,000	26	.254	.127	.106
300,000	30	.252	.126	.105
266,800	26	.258	.129	.107
0000	6	.267	.133	.111
000	6	.275	.138	.115
00	6	.284	.142	.118
0	6	.292	.146	.122
1	6	.300	.150	.125
2	6	.308	.154	.128
3	6	.317	.158	.132
4	6	.325	.162	.135
5	6	.333	.167	.139
6	6	.342	.171	.142

Stranded copper conductors

Circular mils or A.W.G. (B. & S.)	Number of copper strands	x_a' Shunt capacitive reactance at 1 ft., megohms per phase per mile — 25 cycles	50 cycles	60 cycles
1,000,000	61	0.216	0.108	0.0901
900,000	61	.220	.110	.0916
800,000	61	.224	.112	.0934
750,000	61	.226	.113	.0943
700,000	61	.229	.115	.0954
600,000	37	.235	.117	.0977
500,000	37	.241	.121	.100
400,000	19	.249	.125	.104
300,000	19	.259	.130	.108
250,000	19	.266	.133	.111
0000	19	.272	.136	.113
0000	7	.273	.136	.113
000	7	.282	.141	.117
00	7	.289	.145	.120
0	7	.298	.149	.124
1	7	.306	.153	.127
2	7	.314	.157	.131
3	7	.322	.161	.134
4	7	.330	.165	.138
5	7	.339	.169	.141
6	7	.347	.173	.145

Illustration. Determine the positive- and zero-sequence capacitive reactances of the line in the illustration, page 419.

Height above ground: 60 ft.

$x_a' = 0.0917$ megohms per phase per mile.

$x_d' = 0.0967$ megohms per phase per mile.

$x_e' = 0.426$ megohms per phase per mile.

$x_1' = x_a' + x_d' = 0.1884$ megohms per phase per mile.

$x_0' = x_a' + x_e' - 2x_d' = 0.324$ megohms per phase per mile.

For other lengths than one mile *divide* by the length.

x_e', megohms per phase per mile

Height above ground, feet	10	15	20	25	30	40	50	60	70	80	90	100
25 cycles.....	0.640	0.727	0.788	0.836	0.875	0.936	0.984	1.02	1.06	1.08	1.11	1.13
50 cycles.....	.320	.363	.394	.418	.437	.468	.492	0.511	0.528	0.542	0.555	0.566
60 cycles.....	.267	.303	.328	.348	.364	.390	.410	.426	.440	.452	.462	.472

TABLE XXVIII.—SHUNT CAPACITIVE REACTANCE.—(*Continued*)

Anaconda hollow copper conductors					General Cable Type HH hollow copper conductors			
Design No.	Circular mils or A.W.G. (B. & S.)	x_a' Shunt capacitive reactance at 1 ft., megohms per phase per mile			Circular mils or A.W.G. (B. & S.)	x_a' Shunt capacitive reactance at 1 ft., megohms per phase per mile		
		25 cycles	50 cycles	60 cycles		25 cycles	50 cycles	60 cycles
163	800,000	0.209	0.105	0.0872	1,000,000	0.180	0.0899	0.0749
167	800,000	.214	.107	.0893	950,000	.182	.0910	.0758
96	750,000	.215	.107	.0895	900,000	.184	.0920	.0767
362	750,000	.220	.110	.0917	850,000	.187	.0933	.0778
559	700,000	.215	.108	.0896	800,000	.189	.0945	.0789
361	700,000	.220	.110	.0915	750,000	.192	.0959	.0799
560	650,000	.219	.110	.0913	700,000	.195	.0975	.0812
561	650,000	.225	.113	.0938	650,000	.198	.0990	.0825
360	600,000	.220	.110	.0915	600,000	.201	.101	.0839
30	600,000	.224	.112	.0935	550,000	.205	.103	.0855
563	550,000	.221	.111	.0921	500,000	.209	.105	.0873
379	550,000	.226	.113	.0941	450,000	.214	.107	.0892
4-B	500,000	.226	.113	.0942	400,000	.219	.110	.0913
401	500,000	.227	.113	.0944	350,000	.225	.113	.0938
405	450,000	.215	.107	.0894	300,000	.232	.116	.0968
426	450,000	.226	.113	.0943	250,000	.241	.120	.100
51	400,000	.226	.113	.0941	0000	.248	.124	.104
565	350,000	.229	.114	.0953	000	.259	.130	.108
378	350,000	.248	.124	.103				
178	300,000	.246	.123	.102				
567	300,000	.251	.125	.104				
37	250,000	.245	.122	.102				
415	250,000	.249	.125	.104				
446	0000	.246	.123	.103				
569	0000	.263	.132	.110				
158	000	.261	.131	.109				
570	000	.266	.133	.111				
10	00	.269	.134	.112				

x_d', megohms per phase per mile

Separation, feet		0	1	2	3	4	5	6	7	8	9
25 cycles....	0	......	0	0.0494	0.0782	0.0987	0.115	0.128	0.139	0.148	0.157
	10	0.164	0.171	.177	.183	.188	.193	.197	.202	.206	.210
	20	.213	.217	.220	.223	.226	.229	.232	.235	.237	.240
	30	.242	.245	.247	.249	.251	.253	.255	.257	.259	.261
50 cycles....	0	......	0	.0247	.0391	.0494	.0573	.0638	.0693	.0740	.0782
	10	.0820	.0854	.0885	.0913	.0940	.0964	.0987	.101	.103	.105
	20	.107	.108	.110	.112	.113	.115	.116	.117	.119	.120
	30	.121	.122	.123	.125	.126	.127	.128	.129	.130	.130
60 cycles....	0	......	0	.0206	.0326	.0411	.0478	.0532	.0577	.0617	.0652
	10	.0683	.0711	.0737	.0761	.0783	.0804	.0823	.0841	.0858	.0874
	20	.0889	.0903	.0917	.0930	.0943	.0955	.0967	.0978	.0989	.0999
	30	.101	.102	.103	.104	.105	.106	.106	.107	.108	109

Positive- and negative-sequence shunt capacitive reactance:

$$x_1' = x_2' = x_a' + x_d'$$

Zero-sequence shunt capacitive reactance:

$$x_0' = x_a' + x_e' - 2x_d'$$

VIII. EXPONENTIAL FUNCTIONS

TABLE XXIX.—EXPONENTIAL FUNCTIONS ϵ^{-x}

X	0	0.01	0.02	0.03	0.04	0.05	0.06	0.07	0.08	0.09
0	1.00000	0.99005	0.98020	0.97045	0.96079	0.95123	0.94177	0.93239	0.92312	0.91393
0.1	.90484	.89583	.88692	.87810	.86936	.86071	.85214	.84367	.83527	.82696
.2	.81873	.81058	.80252	.79453	.78663	.77880	.77105	.76338	.75578	.74826
.3	.74082	.73345	.72615	.71892	.71177	.70469	.69768	.69073	.68386	.67706
.4	.67032	.66365	.65705	.65051	.64404	.63763	.63128	.62500	.61878	.61263
.5	0.60653	.60050	.59452	.58861	.58275	.57695	.57121	.56553	.55990	.55433
.6	.54881	.54335	.53794	.53259	.52729	.52205	.51685	.51171	.50662	.50158
.7	.49659	.49164	.48675	.48191	.47711	.47237	.46767	.46301	.45841	.45384
.8	.44933	.44486	.44043	.43605	.43171	.42741	.42316	.41895	.41478	.41066
.9	.40657	.40252	.39852	.39455	.39063	.38674	.38289	.37908	.37531	.37158
1.0	.36788	.36422	.36059	.35701	.35345	.34994	.34646	.34301	.33960	.33622
1.1	.33287	.32956	.32628	.32303	.31982	.31664	.31349	.31037	.30728	.30422
1.2	.30119	.29820	.29523	.29229	.28938	.28650	.28365	.28083	.27804	.27527
1.3	.27253	.26982	.26714	.26448	.26185	.25924	.25666	.25411	.25158	.24908
1.4	.24660	.24414	.24171	.23931	.23693	.23457	.23224	.22993	.22764	.22537
1.5	.22313	.22091	.21871	.21654	.21438	.21225	.21014	.20805	.20598	.20393
1.6	.20190	.19989	.19790	.19593	.19398	.19205	.19014	.18825	.18637	.18452
1.7	.18268	.18087	.17907	.17728	.17552	.17377	.17204	.17033	.16864	.16696
1.8	.16530	.16365	.16203	.16041	.15882	.15724	.15567	.15412	.15259	.15107
1.9	.14957	.14808	.14661	.14515	.14370	.14227	.14086	.13946	.13807	.13670
2.0	.13534	.13399	.13266	.13134	.13003	.12873	.12745	.12619	.12493	.12369
2.1	.12246	.12124	.12003	.11884	.11765	.11648	.11533	.11418	.11304	.11192
2.2	.11080	.10970	.10861	.10753	.10646	.10540	.10435	.10331	.10228	.10127
2.3	.10026	.09926	.09827	.09730	.09633	.09537	.09442	.09348	.09255	.09163
2.4	.09072	.08982	.08892	.08804	.08716	.08629	.08544	.08458	.08374	.08291
2.5	.08209	.08127	.08046	.07966	.07887	.07808	.07730	.07654	.07577	.07502
2.6	.07427	.07353	.07280	.07208	.07136	.07065	.06995	.06925	.06856	.06788
2.7	.06721	.06654	.06588	.06522	.06457	.06393	.06329	.06266	.06204	.06142
2.8	.06081	.06021	.05961	.05901	.05843	.05784	.05727	.05670	.05613	.05558
2.9	.05502	.05448	.05393	.05340	.05287	.05234	.05182	.05130	.05079	.05029
3.0	.04979	.04929	.04880	.04832	.04783	.04736	.04689	.04642	.04596	.04550
3.1	.04505	.04460	.04416	.04372	.04328	.04285	.04243	.04200	.04159	.04117
3.2	.04076	.04036	.03996	.03956	.03916	.03877	.03839	.03801	.03763	.03725
3.3	.03688	.03652	.03615	.03579	.03544	.03508	.03474	.03439	.03405	.03371
3.4	.03337	.03304	.03271	.03239	.03206	.03175	.03143	.03112	.03081	.03050
3.5	.03020	.02990	.02960	.02930	.02901	.02872	.02844	.02816	.02788	.02760
3.6	.02732	.02705	.02678	.02652	.02625	.02599	.02573	.02548	.02522	.02497
3.7	.02472	.02448	.02423	.02399	.02375	.02352	.02328	.02305	.02282	.02260
3.8	.02237	.02215	.02193	.02171	.02149	.02128	.02107	.02086	.02065	.02045
3.9	.02024	.02004	.01984	.01964	.01945	.01925	.01906	.01887	.01869	.01850
4.0	.01832	.01813	.01795	.01777	.01760	.01742	.01725	.01708	.01691	.01674
4.1	.01657	.01641	.01624	.01608	.01592	.01576	.01561	.01545	.01530	.01515
4.2	.01500	.01485	.01470	.01455	.01441	.01426	.01412	.01398	.01384	.01370
4.3	.01357	.01343	.01330	.01317	.01304	.01291	.01278	.01265	.01253	.01240
4.4	.01228	.01216	.01203	.01191	.01180	.01168	.01156	.01145	.01133	.01122
4.5	.01111	.01100	.01089	.01078	.01067	.01057	.01046	.01036	.01025	.01015
4.6	.01005	.00995	.00985	.00975	.00966	.00956	.00947	.00937	.00928	.00919
4.7	.00910	.00900	.00892	.00883	.00874	.00865	.00857	.00848	.00840	.00831
4.8	.00823	.00815	.00807	.00799	.00791	.00783	.00775	.00767	.00760	.00752
4.9	.00745	.00737	.00730	.00723	.00715	.00708	.00701	.00694	.00687	.00681
5.0	.00674									

INDEX

W

Z